THEORETICAL AND COMPUTATIONAL PLASMA PHYSICS

INTERNATIONAL CENTRE FOR THEORETICAL PHYSICS, TRIESTE

THEORETICAL AND COMPUTATIONAL PLASMA PHYSICS

SELECTED LECTURES PRESENTED AT
TWO MEETINGS, BOTH HELD AT THE
INTERNATIONAL CENTRE FOR THEORETICAL PHYSICS, TRIESTE

THE ICTP COLLEGE ON THEORETICAL AND
COMPUTATIONAL PLASMA PHYSICS
22 MARCH TO 9 APRIL 1977
ORGANIZED BY THE ICTP

AND

THE THIRD INTERNATIONAL ('KIEV') CONFERENCE
ON PLASMA THEORY
5 TO 9 APRIL 1977
ORGANIZED BY THE ACADEMY OF SCIENCES
OF THE UKRAINIAN SSR AND THE
ACADEMY OF SCIENCES OF THE USSR

INTERNATIONAL ATOMIC ENERGY AGENCY
VIENNA, 1978

THE INTERNATIONAL CENTRE FOR THEORETICAL PHYSICS (ICTP) in Trieste was established by the International Atomic Energy Agency (IAEA) in 1964 under an agreement with the Italian Government, and with the assistance of the City and University of Trieste.

The IAEA and the United Nations Educational, Scientific and Cultural Organization (UNESCO) subsequently agreed to operate the Centre jointly from 1 January 1970.

Member States of both organizations participate in the work of the Centre, the main purpose of which is to foster, through training and research, the advancement of theoretical physics, with special regard to the needs of developing countries.

THEORETICAL AND COMPUTATIONAL PLASMA PHYSICS,
IAEA, VIENNA, 1978
STI/PUB/474
ISBN 92–0–130078–6

Printed by the IAEA in Austria
July 1978

FOREWORD

The first major activity of the International Centre for Theoretical Physics in Trieste was an International Seminar on Plasma Physics in October 1964, and the Proceedings were published by the International Atomic Energy Agency. With that Seminar began a tradition of extended contacts between plasma physicists from the United States, West European and Soviet schools. The present Proceedings contain selected lectures from the ICTP College on Theoretical and Computational Plasma Physics held in Trieste from 22 March to 9 April 1977 and from the Third International ('Kiev') Conference on Plasma Theory held in Trieste from 5 to 9 April 1977, concurrently with the last week of the College.

The decision to hold in Trieste the Third International ('Kiev') Conference on Plasma Theory, directed by B.B. Kadomtsev and V.N. Tsytovich (Moscow, USSR), was made for two reasons: first, to take advantage of the fact that many regular participants would already be lecturing at the ICTP College, and, second, to include a number of plasma physicists from developing countries who would be attending and lecturing at the College.

The College provided a comprehensive survey of the state of current fusion research, with the aim of training participants in the use of the principal tools of plasma theory and computation. About half the papers are based on extensive computer calculations, following the present trend in plasma theory. The Directors of the College were B.B. Kadomtsev (Moscow, USSR), B. McNamara (Livermore, USA) and M.N. Rosenbluth (Princeton, USA).

Financial support from the International Union of Pure and Applied Physics (IUPAP) for the College and a special grant from the IAEA for the 'Kiev' Conference are gratefully acknowledged by the organizers of these activities.

Abdus Salam

EDITORIAL NOTE

The papers and discussions have been edited by the editorial staff of the International Atomic Energy Agency to the extent considered necessary for the reader's assistance. The views expressed and the general style adopted remain, however, the responsibility of the named authors or participants. In addition, the views are not necessarily those of the governments of the nominating Member States or of the nominating organizations.

Where papers have been incorporated into these Proceedings without resetting by the Agency, this has been done with the knowledge of the authors and their government authorities, and their cooperation is gratefully acknowledged. The Proceedings have been printed by composition typing and photo-offset lithography. Within the limitations imposed by this method, every effort has been made to maintain a high editorial standard, in particular to achieve, wherever practicable, consistency of units and symbols and conformity to the standards recommended by competent international bodies.

The use in these Proceedings of particular designations of countries or territories does not imply any judgement by the publisher, the IAEA, as to the legal status of such countries or territories, of their authorities and institutions or of the delimitation of their boundaries.

The mention of specific companies or of their products or brand names does not imply any endorsement or recommendation on the part of the IAEA.

Authors are themselves responsible for obtaining the necessary permission to reproduce copyright material from other sources.

CONTENTS

ICTP COLLEGE ON THEORETICAL AND COMPUTATIONAL PLASMA PHYSICS

BASIC PLASMA THEORY

MHD THEORY

KINETIC THEORY

THIRD INTERNATIONAL ('KIEV') CONFERENCE ON PLASMA THEORY

STATISTICS AND KINETICS

WAVES AND INSTABILITIES

NON-LINEARITIES

ICTP COLLEGE ON THEORETICAL AND COMPUTATIONAL PLASMA PHYSICS

Directors

B.B. KADOMTSEV
Union of Soviet Socialist Republics

B. McNAMARA
M.N. ROSENBLUTH
United States of America

ICTP COLLEGE ON THEORETICAL AND COMPUTATIONAL PLASMA PHYSICS 1977

Research on power from nuclear fusion is moving firmly into the development stage, and world expenditure is now about one thousand million US dollars per year. In a three-week College at the International Centre for Theoretical Physics, Trieste, from 22 March to 9 April 1977, the major theoretical and computational advances of the last five years were reviewed in some depth. The level of the lectures was rather advanced, with only a few basic talks to lay the groundwork. A few lectures were given on the younger but exciting developments in electron beams and laser fusion. The complete programme of the College will be found in Appendix A.

The instabilities which plagued the fusion experiments of the 1960s are now fairly well understood and catalogued, and non-linear theories have been developed to explore their consequences. In mirror machines the cyclotron frequency instabilities which feed on the non-Maxwellian ion distributions have been stabilized experimentally and the stabilization mechanisms and time development of the experiments explained theoretically. In tokamaks, in which the major fusion experiments have taken place, the upper limits to the machines' ability to contain hot plasma have been explained in great detail, including the non-linear plasma motions and topological changes in the plasma-magnetic field system at the stability limits.

A major contribution to the understanding and design of these experiments has come from the computational physics studies of plasma transport, equilibrium, stability, and the non-linear development of instabilities. The US National Magnetic Fusion Energy Computer Center has been operating since 1975 and consists of a central CDC 7600, limited by 50 kilobit lines linked to a DEC PDP-10 computer at each of the major fusion research sites. European laboratories have a variety of computing facilities, and Japan is bringing a national centre into operation very soon. The computational methods and results in fusion research were reviewed extensively at the College. Techniques have advanced rapidly in the last few years, to the point where three-dimensional, time-dependent magnetohydrodynamics of high-pressure pinches and five-phase space-dimensional electromagnetic particle models of laser-plasma interactions are routine calculations. In a sense, these huge calculations are like experiments, and the results are widely available in the literature to guide the theorist in non-linear calculations and to help the experimentalist in understanding scaling laws with more realistic physics and geometry than the theorist can provide.

The ICTP College was a welcome opportunity for many people from developing countries to review and discuss current progress in plasma physics and fusion research. Although computing is becoming dramatically cheaper each year, only the largest institutions in developing countries have computers, and fusion research has become so expensive that only the commitment of the advanced countries to solving their long-term energy needs will allow it to come to fruition. Nevertheless, there is a rich diversity of plasma problems which any well-informed physicist can solve, and developing countries have made many significant contributions in the field.

It is a pleasure to acknowledge the excellence of the facilities at ICTP and the generous efforts of the staff there which made the meeting so effective. Professor Salam's guidance in our interactions with the participants and lecturers from developing countries was especially valuable. We look forward to many fruitful events in the future.

B. McNAMARA

Basic plasma theory

IAEA-SMR-31/1

SINGLE-PARTICLE BEHAVIOUR IN PLASMAS*

Brendan McNAMARA
Lawrence Livermore Laboratory,
University of California,
Livermore, California,
United States of America

Abstract

SINGLE-PARTICLE BEHAVIOUR IN PLASMAS.

1. Motion of charged particles in electromagnetic fields; 2. Adiabatic invariants; 3. Analogy with magnetic field structures; 4. Resonant effects on adiabatic invariants; 5. Jumps in adiabatic invariants; 6. Superadiabaticity.

INTRODUCTION

Since one cannot expect to memorize the content of such an intense course of lectures as this, considerable emphasis has been placed on collecting and cataloguing key results and on principal references. The behaviour of single particles in a plasma is basic to the rest of the lectures from the ICTP College in these Proceedings, but is not elementary. The subject is well covered in many textbooks and the purpose of this paper is merely to collect in a brief form the essential formulae and mathematical methods.

The paper essentially discusses the motion of charged particles in electromagnetic fields. The various useful forms of the method of averaging are displayed and applied to calculation of constants of motion. The breakdown of these constants is discussed along with some of the implications for fusion systems.

1. MOTION OF CHARGED PARTICLES IN ELECTROMAGNETIC FIELDS

Charged-particle motions are generally complicated, and in designing fusion devices one tries to simplify the motions by use of symmetries or constants of the

* Work performed under the auspices of the US Department of Energy by the Lawrence Livermore Laboratory under Contract No. W-7405-ENG-48.

particle motions to provide confinement within the device. The equations of motion in an electric field $\vec{E}(\vec{X},t)$ and in a magnetic field $\vec{B}(\vec{X},t)$ are, in Gaussian units,

$$\frac{d\vec{X}}{dt} = \vec{V} \ , \quad \frac{d\vec{V}}{dt} = \frac{e}{m}\left(\vec{E} + \frac{\vec{V} \times \vec{B}}{c}\right) \tag{1.1}$$

where $\vec{X}$ is the position of the particle and $\vec{V}$ its velocity. The equations obviously separate into motion parallel and perpendicular to $\vec{B}$. In constant $\vec{E}$ and $\vec{B}$ fields the equations are trivially solved to give:

$$X_{\parallel} = X_{\parallel 0} + V_{\parallel 0}\, t + \frac{q}{2m} E_{\parallel} t^2$$

$$V_{\parallel} = V_{\parallel 0} + \frac{e}{m} E_{\parallel} t \tag{1.2}$$

$$\vec{X}_{\perp} = \vec{V}_D\, t + \vec{\rho}$$

where the electric drift velocity is $V_D = (\vec{E} \times \vec{B}/B^2)\, c$; $\vec{\rho}$ is a circular motion in the drift frame with frequency $\Omega = eB/mc$; and Larmor radius $\rho = v_{\perp}/\Omega$. It is important to notice that the drift velocity is the same for ions and electrons, being independent of mass and charge, but that the cyclotron frequencies and gyroradii are not. The electric field only *accelerates* the particle parallel to $\vec{B}$, and because electrons respond so quickly it is difficult to maintain a constant $E_{\parallel}$, except in a potential well generated by a collection of (magnetically trapped) ions. We assume $E_{\parallel} = 0$ for the moment.

In a real device we need to understand the motion of particles in electromagnetic fields which vary on time scale t and space scales $L_{\perp}$, $L_{\parallel}$. In the case where $\Omega\tau$, $\rho/L_{\perp}$, $\rho/L_{\parallel}$ are all 0(1), only a high degree of symmetry will save you from needing a computer, but otherwise there are various forms of perturbation theory which give approximate solutions. The most useful cases will be discussed.

1.1. Small Larmor radius, slow time scales

The case with $\Omega\tau \gg 1$, $\rho/L_{\perp}, \rho/L_{\parallel} \ll 1$ is of the most interest. Since the gyrofrequency is large, it seems appropriate to average out this rapid motion and develop equations for the mean drift of the particle. As the particle moves, its local gyrofrequency will change; consider the Taylor series expansion of the function

$$\begin{aligned} x &= x_0 \cos(\Omega_0 + \epsilon\Omega)\, t \\ &= x_0 \cos \Omega_0 t + x_0 t\, \epsilon\Omega \sin \Omega_0 t - \frac{x_0}{2}(t\epsilon\Omega)^2 \cos \Omega_0 t + \dots \end{aligned} \tag{1.3}$$

The successive terms are not purely oscillatory, but are 'secular' with coefficients $0(t^n)$. The radius of convergence of the series is $0(\Omega_0^{-1})$, so simple Taylor expansion of the equation of motion is of little value and we prefer the method of averaging which, although it is only asymptotic, has a range $0(1/\epsilon\Omega_0)$ or better. The method is required in many applications and so is worth giving here in detail.

The original equations of motion must be normalized and transformed $(\vec{x}, \vec{v} \to \vec{y}, \nu)$ to display the phase angle ν of the gyromotion as follows [1]:

$$\vec{Y}_t = \epsilon\vec{g}(\vec{Y}, \nu) \quad , \quad \nu_t = 1 + \epsilon f(\vec{Y}, \nu) \tag{1.4}$$

where $\vec{g}$ and f are periodic in ν, period τ_0, and ϵ is the small expansion parameter, $0(\rho/L_\perp, \ \rho/L_\parallel, \ \Omega\tau^{-1})$. We construct a transformation to new variables $(\vec{Z}, \phi)$ with $\vec{Z}$ periodic in ν, and ϕ being an angle variable:

$$\begin{aligned} &\vec{Z}(\vec{Y}, \nu) = Z(\vec{Y}, \nu + \tau_0) \\ &\phi(\vec{Y}, \nu + \tau_0) = \tau_0 + \phi(\vec{Y}, \nu) \end{aligned} \tag{1.5}$$

The equations for the drift variables $(\vec{Z}, \phi)$ should not contain the angle variable which is to be averaged out:

$$\vec{Z}_t = \epsilon\vec{h}(\vec{Z}) \quad , \quad \phi_t = 1 + \epsilon\omega(\vec{Z}) \tag{1.6}$$

The original Eqs (1.4) and the transformation give

$$\begin{aligned} &\vec{Z}_t = \epsilon\vec{g} \cdot \vec{Z}_{\vec{Y}} + (1 + \epsilon f)\vec{Z}_\nu \equiv \epsilon\vec{h} \\ &\phi_t = \epsilon\vec{g} \cdot \phi_{\vec{Y}} + (1 + \epsilon f)\,\phi_\nu \equiv 1 + \epsilon\omega \end{aligned} \tag{1.7}$$

which can be integrated over ν. We combine the equations into a single vector equation by setting $\vec{Z} = (\phi, \vec{Z})$, $\vec{H} = (\omega, \vec{h})$, $\vec{Y} = (\nu, \vec{Y})$ and $G(\vec{Y}) = (f, \vec{g})$, and the integration, with boundary condition $\vec{Z}(0) = \vec{Y}$, gives

$$\vec{Z} = \vec{Y} + \epsilon \int_0^\nu d\nu(\vec{H}(\vec{Z}) - \vec{G} \cdot \vec{Z}_{\vec{Y}}) \tag{1.8}$$

The condition that $\vec{Z}$ should have no secular terms in ν is that the average of the integrand should vanish:

$$\int_0^{\tau_0} d\nu(\vec{H} - \vec{G} \cdot \vec{Z}_{\vec{Y}}) = 0 \tag{1.9}$$

Equations (1.8) and (1.9) are sufficient to generate a power series expansion for the transformation $\vec{Z}$ and the averaged driving terms $\vec{H}$. The result is conveniently written down in terms of the integrating and averaging operators (~,–) defined as follows: for any function f, periodic in ν,

$$
\begin{aligned}
\tilde{f} &= \int_0^{\nu} (f - \bar{f})\, d\nu' \\
\bar{f} &= \frac{1}{\tau_0} \int_0^{\tau_0} f d\nu'
\end{aligned}
\qquad (1.10)
$$

Notice that $\tilde{f}$ contains a constant of integration or initial condition and so $\bar{\tilde{f}} = 0$ but $\tilde{\tilde{f}} \neq 0$. Without further ado, we write the transformation to $0(\epsilon^3)$ as

$$
\begin{aligned}
\vec{Z} &= \vec{Y} - \epsilon\tilde{\vec{G}} + \epsilon^2 \left(\widetilde{\vec{G}\cdot\tilde{\vec{G}}_Y - \tilde{\tilde{\vec{G}}}\cdot\bar{\vec{G}}_Y} \right) + 0(\epsilon^3) \\
\vec{Y} &= \vec{Z} + \epsilon\tilde{\vec{G}} + \epsilon^2 \left(\widetilde{\tilde{\vec{G}}\cdot\vec{G}_Z - \bar{\vec{G}}\cdot\tilde{\tilde{\vec{G}}}_Z} \right) + 0(\epsilon^3)
\end{aligned}
\qquad (1.11)
$$

The average co-ordinates $\vec{Z}$ have the equation of motion

$$
\vec{Z}_t = \vec{a} + \epsilon\bar{\vec{G}}(\vec{Z}) - \epsilon^2 \left(\overline{\vec{G}\cdot\tilde{\vec{G}}_Z} - \bar{\tilde{\vec{G}}} \cdot \bar{\vec{G}}_Z \right) + 0(\epsilon^3) \qquad (1.12)
$$

where $\vec{a} = (1,0,0,\ldots,0)$. Notice that the phase ϕ does not appear on the right of this equation, as desired. There are many descriptions of the method of averaging in the textbooks but Eqs (1.11) and (1.12) are the answer for the plasma physicist. In celestial mechanics one is usually interested in a high order of accuracy and so requires many orders of the expansion. The best method is due to Deprit [2] and is well described in Nayfeh's book [3]. The method uses a generating function or Lie transform, $\vec{W}(\vec{Y})$, which allows the manipulations to be computerized on an algebraic manipulator. This generating function approach also allows any function of the old variables to be expanded directly in the new variables.

We observe that the original problem has merely been transformed to a simpler one which still must be solved (Eq.(1.12)). As a final answer, which will appear many times in these Proceedings, we can write down the result of applying this method to eliminating the gyrorotation from the equations of motion of a charged particle (1.1). These are the well-known drift equations [4]:

$$\left.\begin{aligned}
\frac{d\vec{X}}{dt} &= v_{\parallel}\frac{\vec{B}}{B} + \frac{c}{B^2}(\vec{E}\times\vec{B}) + \frac{mcv_{\parallel}^2}{eB^4}\vec{B}\times(\vec{B}\cdot\nabla)\cdot\vec{B} + \frac{mcv_{\perp}^2}{2eB^3}\vec{B}\times\nabla B \\
\frac{d\epsilon}{dt} &= e\vec{E}\cdot\frac{d\vec{X}}{dt} + \frac{mv_{\perp}^2}{2B}\frac{\partial B}{\partial t} \\
\frac{d}{dt}\left(\frac{cm^2v_{\perp}^2}{m_0^2Be}\right) &= 0 \equiv \frac{d\mu_0}{dt} \\
\frac{d\phi}{dt} &= \Omega
\end{aligned}\right\} \quad (1.13)$$

where the energy of the relativistic particle is

$$\epsilon = \frac{m_0c^3}{(c^2 - v_{\parallel}^2 + v_{\perp}^2)^{\frac{1}{2}}}$$

or, for a non-relativistic particle,

$$\epsilon = \frac{m}{2}(v_{\parallel}^2 + v_{\perp}^2)$$

We observe that, to this order, the perpendicular velocity is determined by the constant of the motion, the adiabatic invariant μ_0. When $\nabla\times\vec{B} = 0$ the magnetic drifts are of the same form and we get

$$\frac{d\vec{X}}{dt} = v_{\parallel}\frac{\vec{B}}{B} + \frac{c}{B^2}(\vec{E}\times\vec{B}) + \frac{mc^2}{2eB^3}(\vec{B}\times\nabla B)(2v_{\parallel}^2 + v_{\perp}^2) \quad (1.14)$$

One essential assumption in the derivation was that $E \ll vB/c$. If we allow for a large drift velocity, $\vec{v}_E = (\vec{E}\times\vec{B}/B^2)c$, the equations are modified to

$$\frac{d\vec{X}}{dt} = \vec{U} - \frac{mc}{eB^2}\left[\frac{\partial\vec{U}}{\partial t} + (\vec{U}\cdot\nabla)\cdot\vec{U}\right]\times\vec{B} + \frac{mcv_{\perp}^2}{2eB^3}\vec{B}\times\nabla B$$

$$\epsilon = \left(\frac{m}{2}v_{\parallel}^2 + v_{\perp}^2 + v_E^2\right), \qquad \vec{U} = v_{\parallel}\frac{\vec{B}}{B} + \vec{v}_E \quad (1.15)$$

The second term in (1.15) is the drift due to the inertial effect of the large electric drift. These drift equations are very useful in determining the dynamics of a plasma on time scales long compared with the cyclotron period. In some cases the drift equations themselves will describe a still slower period oscillation. The same technique can be used to average over this oscillation and reduce the system still further.

1.2. High-frequency fields

In the case when $\Omega_\tau \ll 1$, $\tau v/L \ll 1$, the equations of motion can be averaged over the high-frequency field variation. In the non-relativistic case we get

$$m\frac{d\vec{V}}{dt} = e\vec{\bar{E}} + \frac{e}{c}(\vec{V} \times \vec{\bar{B}}) - \frac{e^2}{2m\omega^2}\,\overline{\nabla\tilde{E}^2} \tag{1.16}$$

The high-frequency part of the field appears as a potential $u = (e^2/2m\omega^2)\,\overline{\tilde{E}^2}$, independent of the sign of the charge. This additional force is of prime importance in laser fusion. When $\bar{E}$, $\bar{B}$ vary slowly on the scale of the Larmor period, the method of averaging can be applied, as before.

2. ADIABATIC INVARIANTS

The six equations of motion have six constants of the motion, namely the initial conditions on the motion. These are in general useless for making further deductions and we seek a better choice of constants in systems with sufficient symmetry. A typical example for a charged particle in a time-independent field is the total energy or the Hamiltonian

$$H = \frac{1}{2m}\left(\vec{P} - \frac{e}{c}\vec{A}\right)^2 + e\phi \tag{2.1}$$

If the fields $\vec{B} = \nabla \times \vec{A}$, $\vec{E} = -\nabla\phi$, are independent of a co-ordinate θ, then the corresponding canonical momentum P_θ is a constant of the motion. Such constants confine a particle to a surface in phase space which, if we are lucky or chose the configuration carefully, will confine the particle in configuration space. One method for finding constants in less symmetric situations is to transform the Hamiltonian to momentum co-ordinates which display the Larmor angle. We could then seek a canonical transformation, as a power series in r_L/L, which would make the Hamiltonian independent of the new phase angle and hence the corresponding momentum would be a constant. Unfortunately, it would be expressed in terms of the averaged variables and we would then have to find its expansion as a series in the original phase-space co-ordinates.

We shall demonstrate a different formulation which is generally useful. Consider systems in which the particles execute closed orbits in the unperturbed system, so the Hamiltonian can be reduced to

$$H = P_1 + \epsilon\Omega(q_i, P_i) \tag{2.2}$$

where Ω is (almost) periodic in the angle co-ordinate q_1. Then we look for a constant of the motion J by solving the *linear, partial* differential equation

$$\frac{dJ}{dt} \equiv [J,H] = 0 \tag{2.3}$$

where [J,H] is the Poisson bracket,

$$[J,H] \equiv \sum_i \left(\frac{\partial J}{\partial P_i} \frac{\partial H}{\partial q_i} - \frac{\partial J}{\partial q_i} \frac{\partial H}{\partial P_i} \right) \tag{2.4}$$

J is expanded as a power series

$$\sum_0^\infty \epsilon^n J_n$$

to give the recursion

$$\frac{\partial J_0}{\partial q_1} = 0 \quad , \quad \frac{\partial J_n}{\partial q_1} = \left[J_{n-1}, \Omega \right] \tag{2.5}$$

The nth equation is easily integrated:

$$J_n = \int \left[J_{n-1}, \Omega \right] dq_1 + G_n \tag{2.6}$$

where G_n is independent of q_1. We finally require that J_n be periodic in q_1 and so the average of the Poisson bracket must be made to vanish by choice of J_0 and the integration constants G_n:

$$\overline{\left[J_{n-1}, \Omega \right]} = 0 \tag{2.7}$$

The first equation is

$$\overline{\left[J_0, \Omega\right]} = 0 = \left[J_0, \overline{\Omega}\right] \qquad (2.8)$$

since J_0 is independent of q_1. The obvious non-trivial solution is

$$J_0 = J_0(\overline{\Omega}) \qquad (2.9)$$

The series can be developed in terms of the Poisson bracket operator, the averaging operator and the indefinite integrator

$$\hat{J} = \int (J - \overline{J})\, dq_1 \qquad (2.10)$$

and the general answer, correct to $0(\epsilon^2)$, is

$$J = \overline{\Omega} + \epsilon[\overline{\Omega}, \hat{\Omega}] + \frac{\epsilon}{2}\left[\overline{\hat{\Omega}, \Omega}\right] + \epsilon^2\left[\widehat{[\overline{\Omega}, \hat{\Omega}] + \frac{1}{2}\,\overline{[\hat{\Omega}, \Omega]}}, \Omega\right]$$
$$+ \frac{\epsilon^2}{3}\left[\overline{\hat{\Omega}\,[\hat{\Omega}, \Omega]}\right] + \frac{2\epsilon^2}{3}\left[\overline{\hat{\Omega}, [\hat{\Omega}, \overline{\Omega}]}\right] + 0(\epsilon^3) \qquad (2.11)$$

The most general Hamiltonian for which we have developed such an adiabatic invariant is of the form [5]:

$$H = \psi(\epsilon q_2, \ldots, \epsilon q_N, P_1, \epsilon P_2, \ldots, \epsilon P_N, \epsilon t) + \epsilon\Omega(q_i, P_i, \epsilon t) \qquad (2.12)$$

In terms of the rotation frequency $\lambda = \partial\psi/\partial P_1$ and the 'slow' bracket $\{\ \}$ defined as

$$[\psi, f] = \lambda\,\frac{\partial f}{\partial q_1} + \epsilon\{\psi, f\} \qquad (2.13)$$

the result for a general oscillatory system is

$$J = P_1 + \epsilon\left(\frac{\Omega - \overline{\Omega}}{\lambda}\right) + \epsilon^2\left(\frac{1}{\lambda}\left[\frac{\hat{\Omega}}{\lambda}, \psi\right] + \left[\frac{1}{2\lambda^2}, \frac{\hat{\Omega}^2}{2}\right] - \frac{1}{\lambda}\left[\frac{\overline{\Omega}}{\lambda}, \hat{\Omega}\right]\right.$$
$$\left. - \frac{1}{\lambda}\frac{\partial}{\partial \epsilon t}\left(\frac{\hat{\Omega}}{\lambda}\right) + \frac{1}{2}\left[\overline{\frac{\Omega}{\lambda}, \frac{\hat{\Omega}}{\lambda}}\right]\right) + 0(\epsilon^3) \qquad (2.14)$$

As far as plasma theory is concerned, we can regard the calculation of adiabatic invariants as solved. However, the Hamiltonian formulation is most inconvenient since H is a function of the potentials (A, ϕ) and not the fields (B,E).

The above method really only required the equation $[J,H]=0$ to be expressed in co-ordinates which display the phase angle over which we average. Hastie, Taylor and Haas [6] have applied the method to the Vlasov equation to generate the magnetic moment μ, the longitudinal invariant J, and the flux invariant Φ as given below. The algebra involved was formidable and the results are worth some comments.

A charged particle in a time-independent electromagnetic field will have as constants of the motion the energy ϵ and the canonical moments corresponding to any symmetries of the configuration. If there are no symmetries, then, in a strong magnetic field such that $r_L/L \ll 1$, the magnetic moment will be an adiabatic constant:

$$\mu = \mu_0 + \frac{m}{e}\mu_1 + 0\left(\frac{r_L}{L}\right)^2 \qquad (2.15)$$

where

$$\mu_0 = \frac{v_\perp^2}{2B}$$

$$\mu_1 = -\frac{1}{B}\left[\vec{V}_\perp \cdot \vec{W}_D + \frac{\vec{V}\cdot b}{4}\left\{\vec{V}_\perp \cdot (\vec{a}\cdot\nabla)\vec{b} + \vec{a}\cdot(\vec{V}_\perp\cdot\nabla)\vec{b} + 4\mu_0\vec{b}\cdot\nabla\times\vec{b}\right\}\right]$$

and

$$W_D = \frac{\vec{B}}{B^2} \times (V_\parallel^2\vec{\rho} + \mu\nabla B) \quad , \quad \vec{a} = \frac{\vec{V}\times\vec{B}}{B^2}$$

$\vec{\rho} = -\vec{b}\cdot\nabla\vec{b}$, the field line curvature

Notice that μ contains $V_\perp$ and oscillates on the cyclotron period. When the particles are trapped in a magnetic mirror field to a particular region of a field line, we have to introduce the sign $\sigma = \pm 1$ of the parallel velocity. The bounce motion, at frequency ω_b, can be averaged to yield a second invariant provided $\Omega \gg \omega_b \gg |W_D/L|$:

$$J = \oint\left(2\left(\epsilon - \left(\mu_0 + \frac{m}{e}\mu_1\right)B\right)\right)^{\frac{1}{2}} ds - \sigma\int_{S_0}^{S}\frac{dS}{q} V_D\cdot\nabla J_0 + 0\left(\frac{r_L}{L}\right)^2 \qquad (2.16)$$

where

$$J_0 = \oint 2(\epsilon - \mu_0 B)^{\frac{1}{2}} dS$$

Finally, if the fields are allowed to change very slowly in time over a period much longer than the drift of a particle around a J surface, the energy ϵ is no longer a constant of motion on this time scale. The total flux Φ through a drift surface, J = const, is another adiabatic invariant:

$$\Phi = \oint_{J=\mathrm{const}} \alpha d\beta \tag{2.17}$$

where (α, β) are the field line co-ordinates, $\vec{B} = \nabla\alpha \times \nabla\beta$.

For a plasma to be in equilibrium in a given static magnetic field, the plasma distribution function must be a function of the constants of motion, $f = f(\epsilon, \mu, J)$.

3. ANALOGY WITH MAGNETIC FIELD STRUCTURES

As an aside to the main business of particle motions, the structure of magnetic fields can be analysed in the same fashion. Compare the general invariants for a divergence-free magnetic field and a Hamiltonian system, namely the flux Φ through an arbitrary curve C, which always passes through the same set of field lines, and the action integral J round an arbitrary loop in phase space which always passes through the same trajectories:

$$\Phi = \iint_C B_z \, dxdy \quad , \quad J = \oint pdq \tag{3.1}$$

This suggests that a magnetic field might be described in canonical co-ordinates:

$$p \equiv \int^y B_z dy = B_z^{\dagger} \quad , \quad q \equiv x \quad , \quad t \equiv z \tag{3.2}$$

The Hamiltonian may be found from the equations of motion:

$$-\frac{\partial H}{\partial q} = \frac{dp}{dt} \quad , \quad \frac{\partial H}{\partial p} = \frac{dq}{dt} \equiv \frac{dx}{dz} = \frac{B_x}{B_z} \tag{3.3}$$

and the constraint $\nabla \cdot \vec{B} = 0$. It turns out that we need to separate the field component $B_y = B_{y1}(x,y,z) + B_{yz}(x,z)$ to choose the constants of integration correctly when solving for H:

$$H = \int^y B_x dy - \int^x B_{yz} dx \equiv B_x^{\dagger} - B_y^{*} \tag{3.4}$$

As a simple example, we can now apply the Hamiltonian formalism to a stellarator field which we write as

$$\vec{B} = B_0\hat{z} + \epsilon \vec{b}_1(x,y,z,\epsilon) \tag{3.5}$$

where the field is principally in the z direction ($\hat{z}$ is the unit vector) and $\vec{b}_1$ is periodic in z and may be further expandable in ϵ. The momentum and Hamiltonian are

$$p = B_0 y + \epsilon b_z^\dagger \quad , \qquad H = \epsilon b_x^\dagger - \epsilon b_y^* \equiv \epsilon h \tag{3.6}$$

We seek an adiabatic invariant ψ to describe the magnetic surfaces by solving the Hamiltonian form of $\vec{B} \cdot \nabla\psi = 0$:

$$\frac{d\psi}{dt} = 0 = \frac{\partial\psi}{\partial t} + \epsilon[h,\psi] \tag{3.7}$$

The solution can be written down easily from Eq.(2.11):

$$\psi = \overline{b_x^\dagger} - \overline{b_y^*} + \epsilon\, \overline{b_x \hat{b}_{y1}} + \epsilon \left[\overline{b_x^\dagger} - \overline{b_y^*},\ \hat{b}_x^\dagger - \hat{b}_y^*\right] + 0(\epsilon^2) \tag{3.8}$$

In the stellarator configuration it is usually assumed that $\overline{b_x^\dagger}$, $\overline{b_y^*}$ are $0(\epsilon)$ and so the form of the surfaces is determined by

$$\psi = \epsilon\, \overline{b_x \hat{b}_{y1}} \tag{3.9}$$

This work is one illustration of how to take a conservative $(\nabla \cdot \vec{B} = 0)$ system and express it in Hamiltonian form and it shows how the discussion of particle orbits relates to magnetic surfaces. Typical magnetic surfaces in an L = 3 stellarator are shown in Fig.1 (from Ref.[7]).

4. RESONANT EFFECTS ON ADIABATIC INVARIANTS

The theory of invariants so far described shows how to average over a single frequency. In systems where there is more than one fundamental frequency or where the fundamental varies in phase space, it is possible for beats between the various frequencies to produce a slow variation. Terms like $\cos(n\omega_1 - m\omega_2)q$ arise in the series expansions and, when integrated, have a denominator $(n\omega_1 - m\omega_2)$ which could be very small for large values of n,m. The series can only be shown to be asymptotic and one simply has to stop the expansion when a small denominator arises. If this happens in the second or third term, the whole procedure must be modified.

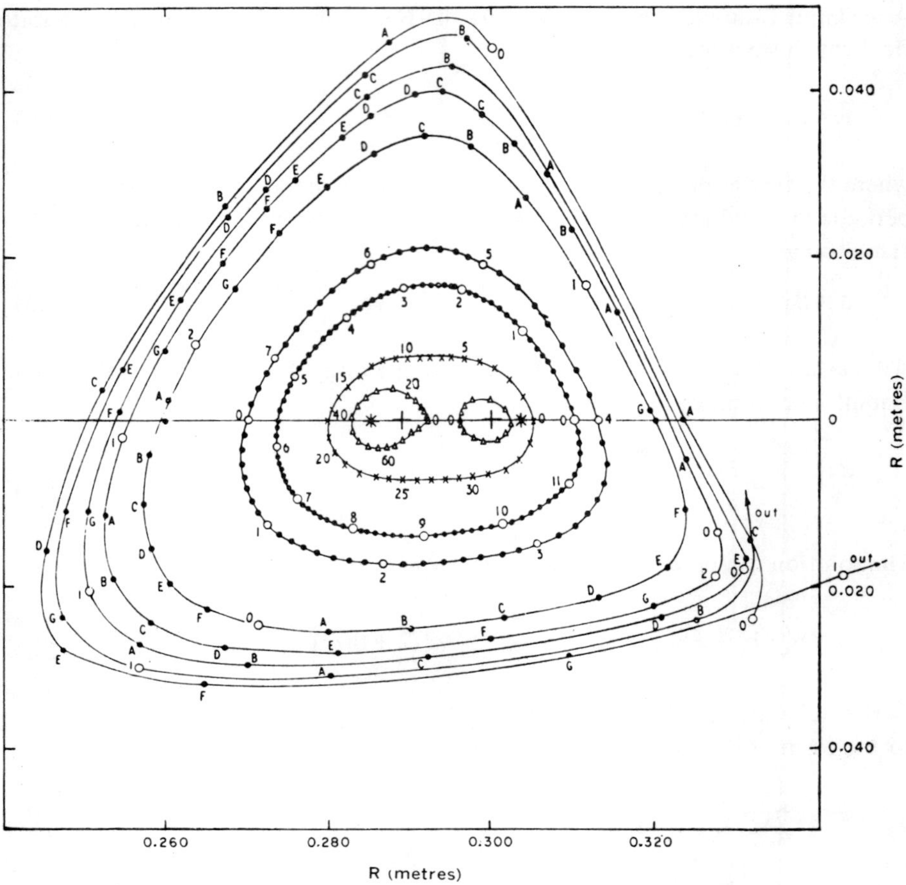

FIG.1. Intersection of magnetic field lines with radial planes, at field period intervals, in a toroidal L = 3 stellarator. There are 8 field periods round the torus and the numbers indicate revolutions about the major axis. The splitting of the axis arises from a small L = 1 field component due to the helical windings. (From Ref.[7].)

A nice example [8,9] is of a charged particle in a uniform magnetic field interacting with an electrostatic plasma wave. The Hamiltonian is

$$H(\vec{r},\vec{p}) = (\vec{p} - m\Omega x\hat{y})^2/2m + e\Phi_0 \sin(kz + k_\perp x) \tag{4.1}$$

where $\Omega = eB/mc$ and Φ_0 is the wave amplitude. This is first transformed to the action-angle co-ordinates of the gyromotion, $P_\phi = mv_\perp^2/2\Omega$, with gyroradius $\rho = (2P_\phi/m\Omega)^{\frac{1}{2}}$:

$$H = P_z^2/2m + \Omega P_\phi + e\Phi_0 \sin(kz - k_\perp \rho \sin\phi) \tag{4.2}$$

In the case of propagation at 45° ($k = k_\perp$), the Hamiltonian may be non-dimensionalized and, using a Bessel function identity, becomes

$$H = P_\phi + \frac{P^2}{2} + \epsilon \sum J_L(\rho)\sin(z - \ell\phi) \equiv H_0 + \epsilon H_1 \tag{4.3}$$

The recurrence relations in (2.5) for an invariant I become

$$\frac{\partial I_0}{\partial \phi} + P\frac{\partial I_0}{\partial z} = 0 \ , \qquad \frac{\partial I_n}{\partial \phi} + P\frac{\partial I_n}{\partial z} = \left[H_1, I_{n-1}\right] \tag{4.4}$$

Observe that the zeroth order orbit depends on P!

The solution to $0(\epsilon)$ of Eq.(4.4) is

$$I_0 + \epsilon I_1 = I_0(p) + \epsilon \frac{dI_0}{dp} \sum J_L(\rho) \frac{\sin(z - \ell\phi)}{P - L} \tag{4.5}$$

The expansion clearly fails at every integral resonance $P = L$ unless I_0 is chosen to vanish in the same way at each resonance. An appropriate choice is $I_0 = \cos(\pi P)/\pi$:

$$I_0 + \epsilon I_1 = \pi^{-1}\cos(\pi p) - \epsilon\sin(\pi p) \sum J_L \frac{\sin(z - L\phi)}{(P - L)} \tag{4.6}$$

This invariant is shown in Figs 2 and 3, in the plane $\phi = \pi$ for two values of ϵ and $\rho = (1.48^2 - P^2)^{\frac{1}{2}}$. We observe that resonances at $p = 0,1$ overlap strongly in the second case.

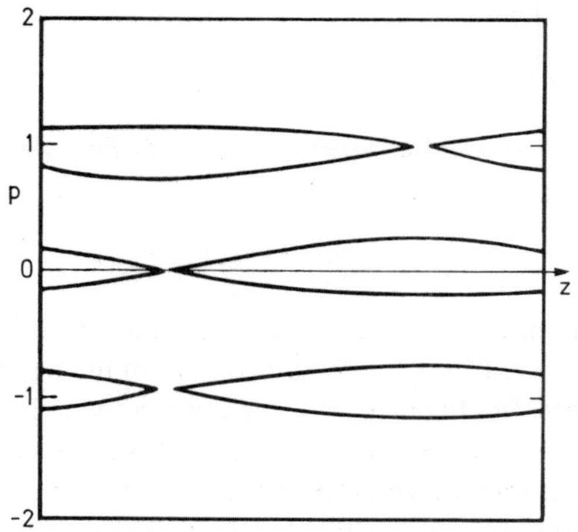

FIG.2. Surface of section plot of $I_0 + \epsilon I_1$, $\epsilon = 0.025$ (from Ref.[9]).

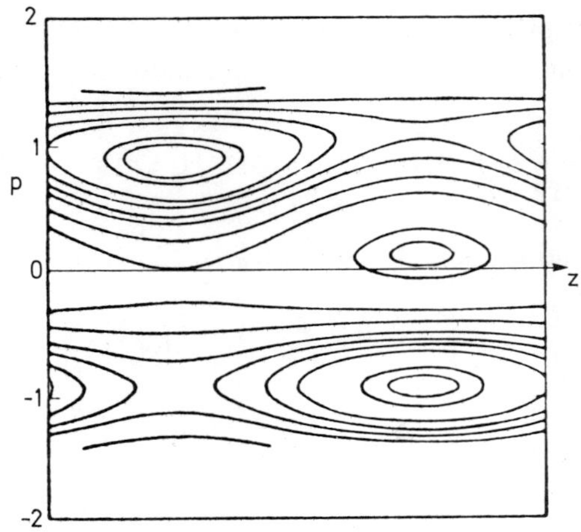

FIG.3. Surface of section plot at ε = 0.1. Primary resonances interact strongly. (From Ref.[9].)

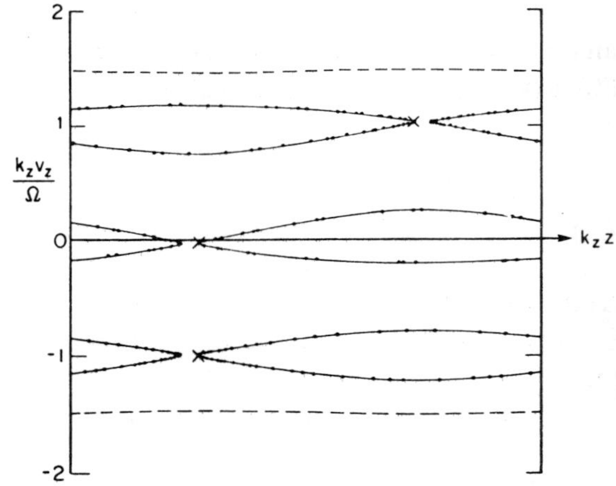

FIG.4. Numerical orbit computations at ε = 0.025, by Smith and Kaufman [10].

These curves can now be compared with the numerical orbit calculations of Smith and Kaufman [10], Figs 4 and 5, at the same parameter values. The orbits are plotted as they intersect the plane $\phi = \pi$, and when the points lie on a smooth curve it is clear that the invariant is a good one. In Fig.5 the surfaces have broken up, leaving only islands round certain fixed points of the phase plane. Jaeger and Lichtenberg [11] have examined a number of simpler examples and distinguish two possible ways in which the resonant surfaces can break up (see also Ref.[12]).

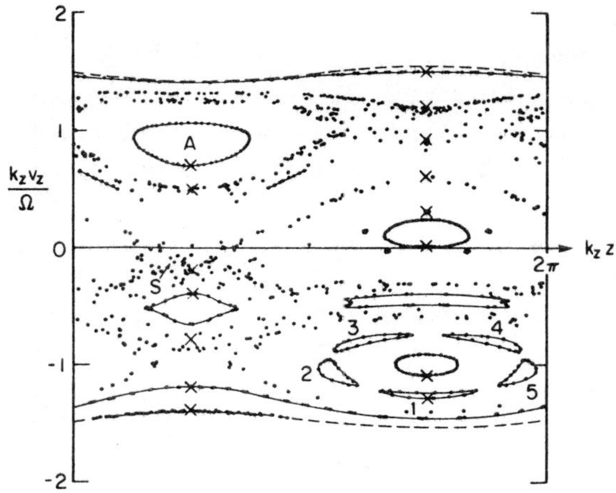

FIG.5. Surface of section of orbits showing break-up of primary resonances and formation of n = 5 secondary resonances at ε = 0.1 (from Ref.[10]).

In the case of an exact resonance, a 2D oscillator problem can be reduced to the averaged Hamiltonian:

$$\bar{H} = K + \epsilon\bar{\Omega}(Q_2, K, P_2) + 0(\epsilon^2) \tag{4.7}$$

where $K = \tau H - \epsilon\tau J$ is the canonical invariant conjugate to Q_1 in Section 2 of this paper. The remaining motion, in Q_2, P_2, may have an elliptic fixed point $(\bar{Q}_2, \bar{P}_2)$, where

$$\frac{\partial\bar{H}}{\partial Q_2} = 0 \quad , \quad \frac{\partial\bar{H}}{\partial P_2} = 0 \tag{4.8}$$

Expanding about this point we get the Hamiltonian for the local motion $\delta Q_2 = Q_2 - \bar{Q}_2$, $\delta P_2 = P_2 - \bar{\bar{P}}_2$:

$$\bar{H} \cong K + \epsilon\bar{\Omega}(\bar{Q}_2, K, \bar{P}_2) + \epsilon\frac{\delta Q_2^2}{2}\left(\frac{\partial^2\bar{\Omega}}{\partial Q_2^2}\right) + \epsilon\frac{\delta\bar{P}_2^2}{2}\left(\frac{\partial^2\bar{\Omega}}{\partial P_2^2}\right) + 0(\delta^3) \tag{4.9}$$

The frequency of these oscillations is clearly $0(\epsilon)$ and the ratio of the semi-axes of the orbits in the $(\delta Q_2, \delta P_2)$ phase plane is

$$R = \frac{(\delta P_2)_{MAX}}{(\delta Q_2)_{MAX}} = 0\left(\left(\frac{\partial^2\bar{\Omega}}{\partial Q_2^2}\right)\Bigg/\left(\frac{\partial^2\bar{\Omega}}{\partial P_2^2}\right)\right)^{\frac{1}{2}} \tag{4.10}$$

If a high harmonic of these oscillations resonates with the primary oscillations in the (K,Q_1) phase space, the invariant is altered just as described above for the magnetized particle in a wave. This example was more complicated in that the resonance between the ϕ and z oscillations depended on p_z. The best we can do with the invariant is to write the Hamiltonian as

$$\bar{H} = \psi(I, P_2) + \epsilon\bar{\Omega}(I, P_2, Q_2) + 0(\epsilon^2) \tag{4.11}$$

The expansion about an elliptic point in this case gives

$$\bar{H} \cong \psi + \epsilon\bar{\Omega} + \frac{\delta P_2^2}{2}\left(\frac{\partial^2\psi}{\partial P_2^2} + \epsilon\frac{\partial^2\bar{\Omega}}{\partial P_2^2}\right) + \epsilon\frac{\delta Q_2^2}{2}\frac{\partial^2\bar{\Omega}}{\partial Q_2^2} + 0(\delta^3) \tag{4.12}$$

The frequency of the drift in the P_2, Q_2 plane is now $0(\sqrt{\epsilon})$, and R is also $0(\sqrt{\epsilon})$. This resonates more readily with the fundamental and it is the overlap of these *secondary islands,* in either case, which leads to the stochastic behaviour. The criterion used by Smith and Kaufman [10], based on overlap of the primary resonances, it not accurate and their computations clearly show a secondary chain of five islands round one of the fixed points. The start of the required transformations must be done with the usual generating function approach:

$$S = S(P_{old}, Q) = -P_\phi Q_\theta - \frac{Q_X}{\pi}\cos\pi p \tag{4.13}$$

so that

$$H = P_\theta + \tfrac{1}{2}\left(\cos^{-1}\pi P_X\right)^2 + \epsilon\sum J_L(\rho)\sin\left(Q_X\sqrt{1-\pi^2P_X^2} - \rho\phi\right)$$

which makes $P_X = (1/\pi)\cos\pi P$ the leading order invariant in the new co-ordinates. I have not carried out the rest of the analysis of this case, but it shows how to bring together the elements of the modern theory. Jaeger et al. [13] have applied the theory to electron cyclotron resonance heating in mirror machines. They show that, as the electron energy increases, the high-order (5th) resonance of the bounce motion with the cyclotron heating breaks up the invariant surfaces and places an upper limit on the attainable electron energy.

5. JUMPS IN ADIABATIC INVARIANTS

The question arises whether the adiabatic invariant series are approximations to some true constant or whether they are merely approximate constants. In fusion plasmas we certainly want to contain the particles much longer than a few hundred cyclotron periods, and the question of the convergence of the series is important. There are cases of simple dynamical systems where an exact constant can be found which, on expansion in the appropriate small parameter, gives the adiabatic series. In general, the best that can be done is to show that the series is asymptotically convergent. The general invariant J, of Eq.(2.11), summed to n terms, can be shown to vary like

$$\frac{dJ^{[n]}}{dt} = 0\,(\epsilon^{n+1}) \tag{5.1}$$

A key assumption in the whole development was that J could be expanded in a power series in ϵ, which, a priori, eliminates small non-expandable terms like $ae^{-b/\epsilon}$. The magnetic moment μ of a particle displays just such jumps at each bounce of the particle in a mirror machine [14–19]. This is easily deduced by examining the change in μ_0 over one bounce. The exact equations of motion give

$$\frac{d\mu_0}{dt} = -\frac{v_\perp}{2B}\,(v_\perp^2 + 2v_\parallel^2)\,\rho_\perp \cos\psi + \frac{v_\parallel^2}{B}\,v_\perp \rho_J \cos\psi_J$$

$$-\frac{v_\parallel}{B}\left[\mu_0 \frac{\partial B}{\partial s} + \vec{v}_\perp\cdot(\vec{v}_\perp\cdot\nabla)\cdot\hat{b}\right] \tag{5.2}$$

where

$$\cos\psi = (v_\perp\cdot\nabla B)/(v_\perp|\nabla B|) \quad , \qquad \cos\psi_J = \vec{v}\cdot\vec{\rho}_J/(v_\perp\rho_J)$$

$$\vec{\rho}_J = \frac{\vec{B}\times(\nabla\times\vec{B})}{B^2} \quad , \qquad \rho_\perp = |\nabla_\perp B|/B$$

$$\vec{v} = v_\perp \cos\overline{\psi}\,\hat{e}_1 - v_\perp \sin\overline{\psi}\,\hat{e}_2 + v_\parallel\hat{b}$$

This equation is integrated along a field line, the zeroth-order motion of the particle, to give the change in μ_0:

$$\Delta\mu_0 = \sqrt{2\mu_0}\;v^2\,\mathrm{Re}\oint \frac{ds}{v_\parallel}\,\frac{\rho_\perp e^{i\psi} + \rho_J}{\sqrt{B}}\;e^{i\psi_J} \tag{5.3}$$

The phases ψ, ψ_J are rotating rapidly at the gyrofrequency and the integral is close to zero. A more careful analysis is done by deforming the path of integration S into the complex plane to pick up the residues round the zeros of B. The details vary for each plasma configuration, and so we shall display the results for a finite β, P(B) equilibrium in a mirror machine when it can be shown that

$$\nabla \times y(B)\,\vec{B} = 0 \quad , \qquad \rho_J = \rho_\perp \frac{B}{y}\frac{\partial y}{\partial B} \tag{5.4}$$

The field can then be expanded about the jth zero in the complex S-plane in the form

$$B = B_j\,\epsilon^\nu(\psi - \psi_j)^\nu \quad , \qquad \epsilon = \frac{V}{\Omega L} \quad , \qquad L = \left(2B/\frac{\partial^2 B}{\partial S^2}\right)^{\frac{1}{2}} \tag{5.5}$$

and the general result is

$$\frac{\Delta\mu_0}{\mu_0} = \frac{4\pi}{y}\sum_j \frac{\nu\epsilon^{-\nu/2}}{\Gamma\left(\frac{\nu}{2}+1\right)}\,\mathrm{Re}\left[\left(\frac{B_0}{B_j}\right)^{\frac{1}{2}}\left(1+\frac{By'}{y}\right)_j\right.$$

$$\left.\times\ \exp\left[i\left(\psi_0 + \nu\frac{\pi}{4}\right)\right]\exp\left[-K_j/\epsilon\right]\right] \tag{5.6}$$

where

$$K_j = -i\int_{s=0}^{S_j}\frac{B}{LB_0}\frac{ds}{v_\parallel} + i\epsilon\int_0^{S_j} ds\,\hat{e}_i\cdot\frac{\partial\hat{e}_2}{\partial s} \tag{5.7}$$

These small jumps in μ_0 can lead to a diffusion in velocity space and rapid loss of containment for the most energetic particles. The maximum energy particle which can be contained in a mirror machine has

$$W_{max} = \frac{10^{-3}K^2B_0^2L^2Z^2}{M(1-0.036\,\ln A_0)^2}\ \mathrm{keV} \tag{5.8}$$

where

$$A_0 = \frac{M}{2}\left(\frac{50\ \mathrm{keV}}{W}\right)\left[\frac{L}{170\ \mathrm{cm}}\ \frac{0.5\ v}{\langle v_\parallel \rangle_{\mathrm{bounce}}}\ \frac{10^{-2}\mathrm{sec}}{\tau}\right]^2$$

$$\times\left[\frac{15(1-R)}{A}\left(\frac{v_\perp}{v}\right)_0 \frac{1}{0.83\,\langle \cos\psi\rangle_{\mathrm{rms}}}\right]^4 \qquad (5.9)$$

(A is atomic mass, τ is bounce time, M is mass, z is charge, and W is energy.)

Notice that these results arise from a non-resonant coupling between the bounce motion and the cyclotron motion and account for the stochastic motions in phase space when the adiabatic invariant has broken down. One remaining question is whether or not these exponentially small jumps destroy the invariance of μ over a long time scale even in the adiabatic region.

6. SUPERADIABATICITY

This has been investigated by Aamodt [20] and by Rosenbluth [21] for the case of mirror-trapped particles in the presence of electrostatic fluctuations near a harmonic of the cyclotron frequency which produces similar jumps in μ. The key point is that jumps in μ_0 are periodic in ψ_0. Let ψ_n be the phase on the nth bounce and μ_n the magnetic moment on the prior-to-the-n^{th} scattering, then (5.6) can be rewritten as

$$\mu_{n+1} = \mu_n + \alpha \sin \psi_n \qquad (6.1)$$

The particle makes many gyrations between bounces and we need a simple model to describe ψ_{n+1} in terms of ψ_n and μ. Following Rosenbluth [21], let us consider a simple quadratic variation in field strength so that the cyclotron frequency is

$$\Omega = \Omega_0 (1 + s^2/L^2) \qquad (6.2)$$

Constancy of the total energy gives

$$\frac{1}{2}\, v_\parallel^2 = \frac{1}{2}\left(\frac{ds}{dt}\right)^2 + \mu B_0 \frac{s^2}{L^2} \qquad (6.3)$$

where $v_\parallel$ is the parallel velocity at $s = 0$. The change in $\delta\psi$ between bounces is then

$$\Delta\psi = \int (\Omega - \Omega_0)\, dt = 2 \int_0^{x_{max}} \frac{\Omega_0 (s^2/L^2)\, ds}{(v_\parallel^2 - 2\mu B_0 s^2/L^2)^{\frac{1}{2}}} = \left(\frac{L}{\rho_i}\right) \frac{\pi v_\parallel^3}{(2\mu B_0)^{\frac{3}{2}}} \tag{6.4}$$

Expressing μB_0 in units of $v_\parallel^2/2$, the phase change between bounces is

$$\psi_{n+1} = \psi_n + \left(\frac{L}{\rho}\right) \frac{\pi}{\mu_{n+1}^{3/2}} \tag{6.5}$$

There are clearly many fixed points of the mapping (6.1), (6.5) whenever $\mu_{n+1} = \mu_n = (L/\rho m)^{\frac{2}{3}}$, $\psi_{n+1} = \psi_n + m\pi$. Let us linearize the motion about one such point, $\psi_n = \psi_F + \delta\psi_n$, $\mu_n = \mu_F + \delta\mu_n$, to get

$$\begin{aligned} \delta\mu_{n+1} &= \delta\mu_n + \alpha\delta\psi_n \\ \delta\psi_{n+1} &= \delta\psi_n - \frac{3}{2}\left(\frac{L}{\rho\mu_F^{5/2}}\right)\delta\mu_{n+1} \end{aligned} \tag{6.6}$$

Eliminating $\delta\psi_n$ gives

$$\delta\mu_{n+1} - \left(2 - \frac{\alpha 3}{2}\left(\frac{L}{\rho\mu_F^{5/2}}\right)\right)\delta\mu_n + \delta\mu_{n-1} = 0 \tag{6.7}$$

and we look for solutions of the form $\delta\mu_n \sim \lambda^n$. For stability, $|\lambda| \leqslant 1$, which gives

$$\alpha < \frac{2}{3}\left(\frac{L}{\rho}\right)^{\frac{2}{3}} m^{-5/3} \tag{6.8}$$

When this condition is met, the particle orbits do not diffuse in velocity space, owing to the non-adiabatic jumps in μ, and the orbits are called superadiabatic. Numerical calculations by Cohen [14] show that particle orbits in typical mirror fields are indeed superadiabatic up to about twice the energy at which the jumps could compete with Coulomb scattering in a fusion plasma (Eq.(5.8)). The adiabatic invariants of a charged particle are indeed approximations to good constants of the motion.

Note added in proof:

The Lie transform methods mentioned in Section 1 have been applied to the resonant problems described here and in generalizing the invariants of Section 2 [22].

REFERENCES

[1] McNAMARA, B., WHITEMAN, K.J., J. Math. Phys. **8** (1967) 2029.

[2] DEPRIT, A., J. Celestial Mech. **1** (1969) 12.

[3] NAYFEH, A., Perturbation Methods, Wiley, New York (1973).

[4] MOROZOV, A.I., SOLOV'EV, L.S., Plasma Phys. **8** (1966) 593.

[5] WHITEMAN, K.J., McNAMARA, B., J. Math. Phys. **9** (1968) 1385.

[6] HASTIE, R.J., TAYLOR, J.B., HAAS, F.A., Ann. Phys. **41** (1967) 302.

[7] GIBSON, A., Phys. Fluids **10** (1967) 1553.

[8] ROSENBLUTH, M.N., SAGDEEV, R.Z., TAYLOR, J.B., ZASLAVSKI, G.M., Destruction of magnetic surfaces by magnetic field irregularities, Nucl. Fusion **6** (1966) 297.

[9] TAYLOR, J.B., LAING, E.W., Phys. Rev. Lett. **35** (1976) 1306.

[10] SMITH, G., KAUFMAN, A.N., Phys. Rev. Lett. **34** (1976) 1613.

[11] JAEGER, E.F., LICHTENBERG, A.J., Ann. Phys. **71** (1972) 319.

[12] DUNNETT, D.A., LAING, E.W., Stochastic motion of particles in a non-adiabatic magnetic trap, Plasma Phys. **8** (1966) 399.

[13] JAEGER, F., LICHTENBERG, A.J., LIEBERMAN, M.A., Theory of electron cyclotron resonance heating, Plasma Phys. **14** (1972) 1073.

[14] COHEN, R., Univ. of California Rep. UCRL–78889 (1976) (submitted to Phys. Fluids).

[15] HASTIE, R.J., HOBBS, G.D., TAYLOR, J.B., "Non-adiabatic behaviour of particles in inhomogeneous magnetic fields", Plasma Physics and Controlled Nuclear Fusion Research 1968 (Proc. 3rd Int. Conf. Novosibirsk, 1968) 1, IAEA, Vienna (1968) 389.

[16] FOOTE, J.H., Nonadiabatic energy limit versus mirror ratio in magnetic-well geometry, Plasma Phys. **14** (1972) 543.

[17] LIEBERMAN, M.A., LICHTENBERG, A.J., Stochastic and adiabatic behavior of particles accelerated by periodic forces, Phys. Rev. **A5** (1971) 1852.

[18] HOWARD, J.E., Phys. Fluids **14** (1971) 2378.

[19] LICHTENBERG, A.J., BERK, H.L., Nucl. Fusion **15** (1975) 999.

[20] AAMODT, R.E., Phys. Rev. Lett. **27** (1971) 135.

[21] ROSENBLUTH, M.N., Phys. Rev. Lett. **29** (1972) 408.

[22] McNAMARA, B., Super-convergent Adiabatic Invariants with Resonant Denominators by Lie Transforms, Lawrence Livermore Lab. Rep. UCRL 79843 (1978), submitted to J. Math. Phys.

NUMERICAL SOLUTION OF THE FOKKER-PLANCK EQUATIONS FOR A MULTI-SPECIES PLASMA*

J. KILLEEN, A.A. MIRIN
Lawrence Livermore Laboratory,
Livermore, California
and
Department of Applied Science,
University of California,
Davis/Livermore, California,
United States of America

Abstract

NUMERICAL SOLUTION OF THE FOKKER-PLANCK EQUATIONS FOR A MULTI-SPECIES PLASMA.

Two numerical models used for studying collisional multi-species plasmas are described. The mathematical model is the Boltzmann kinetic equation with Fokker-Planck collision terms. A one-dimensional code and a two-dimensional code, used for the solution of the time-dependent Fokker-Planck equations for ion and electron distribution functions in velocity space, are described. The required equations and boundary conditions are derived and numerical techniques for their solution are given.

1. INTRODUCTION

In the simulation of magnetically confined plasmas where the ions are not Maxwellian and where a knowledge of the distribution functions is important, kinetic equations must be solved. The proposition that a stable mirror plasma will yield net thermonuclear power depends on the rate at which particles are lost out of the ends of the device. At number densities and energies typical of mirror machines, the end losses are due ideally to the scattering of charged particles into the loss cones in velocity space by classical Coulomb collisions. The

* Work performed under the auspices of USERDA, Contract No. W-7405-Eng-48.

kinetic equation describing this process is the Boltzmann equation with Fokker-Planck collision terms [1, 2].

The use of this equation is not restricted to mirror systems. The heating of plasmas by energetic neutral beams, the thermalization of α-particles in DT plasmas, the study of runaway electrons and ions in tokamaks, and the performance of two-energy component fusion reactors are other examples where the solution of the Fokker-Planck equation is required [3].

The problem is to solve a nonlinear partial differential equation for the distribution function of each charged species in the plasma, as functions of seven independent variables (three spatial coordinates, three velocity coordinates, and time). Such an equation, even for a single species, exceeds the capability of any present computer, so several simplifying assumptions are therefore required to treat the problem. Typical approximations that are made in present-day codes are to neglect spatial dependence and to assume that the distribution functions are azimuthally invariant in velocity space (about the direction of the magnetic field). These assumptions reduce the number of independent variables to three — two velocity space coordinates, v and θ; the speed and pitch angle; and the time, t. Even with these basic assumptions there has been an evolution of numerical Fokker-Planck calculations for the past fifteen years [3].

A multi-species code [2] was developed in order to study D-T and and D-^{3}He mirror reactors, including the effects of reaction products. The principal assumptions of this code are that the "Rosenbluth potentials" (see Section 2) are isotropic and that the distribution functions can be represented by their lowest angular eigenfunction. The resulting set of coupled equations for $f_a(v,t)$ are then solved numerically, using an implicit finite-difference scheme described in Section 3. An extensive parameter study [2] was conducted yielding values of the confinement parameter $n\tau$ and the figure of merit Q (the ratio of thermonuclear power to injected power) as a function of mirror ratio and injection energy.

In Section 4 we describe a highly versatile, two-dimensional multi-species code in which the ion kinetic equations are solved by finite-difference methods on a two-dimensional (v,θ) domain. The electron distribution function is still represented using the lowest angular eigenfunction (isotropic in a full velocity space); hence, a one-dimensional electron kinetic equation is solved. In Ref. [3] we give a variety of applications of this code and describe diagnostics that have been added for the special problems. For mirror systems with beam injection, we find values of the containment parameter $n\tau$ as much as 40% higher than those given by the one-dimensional Fokker-Planck code. We also consider two-component toroidal systems and calculate the energy multiplication factor for a number of different scenarios.

2. MULTI-SPECIES FOKKER-PLANCK EQUATIONS

The appropriate kinetic equations are Boltzmann equations with Fokker-Planck collision terms, often referred to simply as Fokker-Planck equations:

$$\frac{\partial f_a}{\partial t} + \vec{v} \cdot \frac{\partial f_a}{\partial \vec{r}} + \frac{\underline{F}}{m_a} \cdot \frac{\partial f_a}{\partial \vec{v}} = \left(\frac{\partial f_a}{\partial t}\right)_c + S_a + L_a \tag{1}$$

Here f_a is the distribution function in 6-dimensional phase space for particles of species a; S_a is the source term; $(\partial f_a/\partial t)_c$ is the collision term; and L_a contains loss terms.

The Fokker-Planck collision term for an inverse-square force was derived by Rosenbluth et al. [4] in the form

$$\frac{1}{\Gamma_a}\left(\frac{\partial f_a}{\partial t}\right)_c = - \frac{\partial}{\partial v_i}\left(f_a \frac{\partial h_a}{\partial v_i}\right) + \frac{1}{2}\frac{\partial^2}{\partial v_i \partial v_j}\left(f_a \frac{\partial^2 g_a}{\partial v_i \partial v_j}\right) \tag{2}$$

where $\Gamma_a = 4\pi Z_a^4 e^4/m_a^2$. In the present work we write the "Rosenbluth potentials" [2]

$$g_a = \sum_b \left(\frac{Z_b}{Z_a}\right)^2 \ln \Lambda_{ab} \int f_b(\vec{v}')|\vec{v} - \vec{v}'|d\vec{v}' \tag{3}$$

$$h_a = \sum_b \frac{m_a + m_b}{m_b}\left(\frac{Z_b}{Z_a}\right)^2 \ln \Lambda_{ab} \int f_b(\vec{v}')|\vec{v} - \vec{v}'|^{-1}d\vec{v}' \tag{4}$$

The transformation of Eq. (2) to spherical polar coordinates (v, θ, ϕ) in velocity space has been given by Rosenbluth et al. [4]. With our assumption of azimuthal symmetry, the resulting distribution functions are of the form $f_a(v, \mu, t)$, where $\mu = \cos\theta$ and $v = |v|$. The equation for each species is

$$\begin{aligned}
\frac{1}{\Gamma_a}\left(\frac{\partial f_a}{\partial t}\right) = &- \frac{1}{v^2}\frac{\partial}{\partial v}\left(f_a v^2 \frac{\partial h_a}{\partial v}\right) - \frac{1}{v^2}\frac{\partial}{\partial \mu}\left[f_a(1-\mu^2)\frac{\partial h_a}{\partial \mu}\right] + \frac{1}{2v^2}\frac{\partial^2}{\partial v^2}\left(f_a v^2 \frac{\partial^2 g_a}{\partial v^2}\right) \\
&+ \frac{1}{2v^2}\frac{\partial^2}{\partial \mu^2}\left\{f_a\left[\frac{1}{v^2}(1-\mu^2)^2\frac{\partial^2 g_a}{\partial \mu^2} + \frac{1}{v}(1-\mu^2)\frac{\partial g_a}{\partial v}\right.\right. \\
&\left.\left. - \frac{1}{v^2}\mu(1-\mu^2)\frac{\partial g_a}{\partial \mu}\right]\right\} + \frac{1}{v^2}\frac{\partial^2}{\partial \mu \partial v}\left\{f_a(1-\mu^2)\left[\frac{\partial^2 g_a}{\partial \mu \partial v} - \frac{1}{v}\frac{\partial g_a}{\partial \mu}\right]\right\} \\
&+ \frac{1}{2v^2}\frac{\partial}{\partial v}\left\{f_a\left[-\frac{1}{v}(1-\mu^2)\frac{\partial^2 g_a}{\partial \mu^2} - 2\frac{\partial g_a}{\partial v} + \frac{2\mu}{v}\frac{\partial g_a}{\partial \mu}\right]\right\} \\
&+ \frac{1}{2v^2}\frac{\partial}{\partial \mu}\left\{f_a\left[\frac{1}{v^2}\mu(1-\mu^2)\frac{\partial^2 g_a}{\partial \mu^2} + \frac{2\mu}{v}\frac{\partial g_a}{\partial v}\right.\right. \\
&\left.\left. + \frac{2}{v}(1-\mu^2)\frac{\partial^2 g_a}{\partial \mu \partial v} - \frac{2}{v^2}\frac{\partial g_a}{\partial \mu}\right]\right\}
\end{aligned} \tag{5}$$

The functions g_a and h_a, defined by Eqs (3) and (4), can be represented by expansions in Legendre polynomials (Rosenbluth et al. [4]). For this purpose we let

$$f_a(v,\mu,t) = \sum_{j=0}^{\infty} V_j^a(v,t)P_j(\mu) \tag{6}$$

where

$$V_j^a(v,t) = \frac{2j+1}{2}\int_{-1}^{+1} f_a(v,\mu,t)P_j(\mu)d\mu \tag{7}$$

The expansions for g_a and h_a are

$$g_a(v,\mu,t) = \sum_{j=0}^{\infty}\sum_b \left(\frac{Z_b}{Z_a}\right)^2 \ln \Lambda_{ab}B_j^b(v,t)P_j(\mu) \tag{8}$$

$$h_a(v,\mu,t) = \sum_{j=0}^{\infty}\sum_b \frac{m_a+m_b}{m_b}\left(\frac{Z_b}{Z_a}\right)^2 \ln \Lambda_{ab}A_j^b(v,t)P_j(\mu) \tag{9}$$

where

$$A_j^a = \frac{4\pi}{2j+1}\left[\int_0^v \frac{(v')^{j+2}}{v^{j+1}} V_j^a(v',t)dv' + \int_v^{\infty} \frac{v^j}{(v')^{j-1}} V_j^a(v',t)dv'\right] \tag{10}$$

$$B_j^a = -\frac{4\pi}{4j^2-1}\left[\int_0^v \frac{(v')^{j+2}}{v^{j-1}}\left(1-\frac{j-1/2}{j+3/2}\frac{(v')^2}{v^2}\right)V_j^a(v')dv'\right.$$
$$\left. + \int_v^{\infty} \frac{v^j}{(v')^{j-3}}\left(1+\frac{j-1/2}{j+3/2}\frac{v^2}{(v')^2}\right)V_j^a(v')dv'\right] \tag{11}$$

In the computations one takes a finite number of terms in the Legendre expansions of g_a and h_a.

Loss Cone Domain, Ambipolar Potential

Since examples studied are devoted to the problem of plasma confinement within systems of magnetic mirrors, we consider the mathematical description of such systems.

The loss cone angle is (Spitzer [5])

$$\sin^2 \theta_{LC} = 1/R_m \tag{12}$$

where $R_m = B_m/B(z)$; B_m is the magnetic field at the mirror; and $B(z)$ is the magnetic field at the interior point being considered. A particle whose angle in velocity space is less than θ_{LC} will be immediately lost from the mirror system. θ_{LC} is independent of velocity as well as particle mass and charge. Eq. (12) is derived under the assumption that no electrostatic potential exists, and θ_{LC} is the actual loss angle only under that condition.

However, because of their greater mobility, the scattering rate of electrons will be greater than that of ions, and more electrons than ions will tend to leak out of the ends of the device. Hence, an ambipolar potential will build up, being greatest at the center and decreasing towards the ends. The fact that this potential is established leads to a fundamental change in the loss characteristics for the two types of particles. The loss regions are then defined by a loss angle which is a function of speed and charge. If $Z_a e$ is charge and ϕ is electrostatic potential, the loss angle is given by

$$\sin^2 \theta_{LC} = 1/R_a \tag{13}$$

where

$$R_a = \left(\frac{1 + Z_a e\phi / \frac{1}{2} m_a v^2}{R_m} \right)^{-1} \tag{14}$$

is the "effective" mirror ratio.

Equation (13) approaches Eq. (12) asymptotically as $v \to \infty$. For ions, the right hand side of Eq. (13) can exceed unity; no ion in such a velocity regime can be contained. Conversely, for electrons, this term can be less than zero; all electrons at such velocities will be electrostatically trapped. The loss region for ions is transformed from a cone into a hyperboloid of one sheet. Its minimum radius occurs at $\theta = \pi/2$, and is equal to the minimum ion velocity possible for confinement. The electron loss region is transformed into a hyperboloid of two sheets.

3. SOLUTION OF THE ONE-DIMENSIONAL MULTI-SPECIES EQUATIONS

In this section we describe a numerical model which has proved very useful [2], in which we assume that the Rosenbluth potentials given by Eqs (8) and (9) are isotropic, i.e.

$$\frac{\partial g_a}{\partial \mu} = \frac{\partial h_a}{\partial \mu} = 0 \tag{15}$$

With this assumption Eq. (5) becomes

$$\frac{1}{\Gamma_a}\frac{\partial f_a}{\partial t} = -\frac{1}{v^2}\frac{\partial}{\partial v}\left(f_a v^2 \frac{\partial h_a}{\partial v}\right) + \frac{1}{2v^2}\frac{\partial^2}{\partial v^2}\left(f_a v^2 \frac{\partial^2 g_a}{\partial v^2}\right)$$

$$+ \frac{1}{2v^3}\frac{\partial g_a}{\partial v}\left[(1-\mu^2)\frac{\partial^2 f_a}{\partial \mu^2} - 4\mu\frac{\partial f_a}{\partial \mu} - 2 f_a\right]$$

$$-\frac{1}{v^2}\frac{\partial}{\partial v}\left(f_a \frac{\partial g_a}{\partial v}\right) + \frac{1}{v^3}\frac{\partial g_a}{\partial v}\left(\mu\frac{\partial f_a}{\partial \mu} + f_a\right)$$

The above equation is separable, and if we let $f_a(v,\mu,t) = U_a(v,t)M_a(\mu)$, then

$$\frac{M_a}{\Gamma_a}\frac{\partial U_a}{\partial t} = -\frac{M_a}{v^2}\frac{\partial}{\partial v}\left(U_a v^2 \frac{\partial h_a}{\partial v}\right) + \frac{M_a}{2v^2}\frac{\partial^2}{\partial v^2}\left(U_a v^2 \frac{\partial^2 g_a}{\partial v^2}\right) - \frac{M_a}{v^2}\frac{\partial}{\partial v}\left(U_a \frac{\partial g_a}{\partial v}\right)$$

$$+ \frac{U_a}{2v^3}\frac{\partial g_a}{\partial v}\left[(1-\mu^2)\frac{\partial^2 M_a}{\partial \mu^2} - 2\mu\frac{\partial M_a}{\partial \mu}\right]$$

We obtain an eigenvalue problem for $M_a(\mu)$

$$(1 - \mu^2) \frac{d^2 M_a}{d\mu^2} - 2\mu \frac{dM_a}{d\mu} + \Lambda_a M_a = 0 \tag{16}$$

which is Legendre's equation on the domain $-\cos\theta_{LC}^a \le \mu \le \cos\theta_{LC}^a$, where θ_{LC}^a is the loss cone angle for species a. For each eigenvalue Λ_a we have an equation of the form [in the following we have replaced $U_a(v,t)$ by $f_a(v,t)$]:

$$\frac{1}{\Gamma_a} \frac{\partial f_a}{\partial t} = -\frac{1}{v^2} \frac{\partial}{\partial v} \left[f_a \left(v^2 \frac{\partial h_a}{\partial v} + \frac{\partial g_a}{\partial v} \right) \right] + \frac{1}{2v^2} \frac{\partial^2}{\partial v^2} \left(v^2 f_a \frac{\partial^2 g_a}{\partial v^2} \right) - \frac{\Lambda_a}{2v^3} \frac{\partial g_a}{\partial v} f_a \tag{17}$$

The functions $h_a(v,t)$ and $g_a(v,t)$ are given by the equations

$$h_a(v,t) = 4\pi \sum_b \left(\frac{Z_b}{Z_a} \right)^2 \frac{m_a + m_b}{m_b} \ln \Lambda_{ab} \left[\int_0^v f_b(v',t) \frac{v'^2}{v} dv' + \int_v^\infty f_b(v',t) v' dv' \right] \tag{18}$$

$$g_a(v,t) = 4\pi \sum_b \left(\frac{Z_b}{Z_a} \right)^2 \ln \Lambda_{ab} \left[\int_0^v f_b(v',t) v \left(1 + \frac{1}{3} \frac{v'^2}{v^2} \right) v'^2 dv' + \int_v^\infty f_b(v',t) \left(1 + \frac{1}{3} \frac{v^2}{v'^2} \right) v'^3 dv' \right] \tag{19}$$

The summations are taken over all the species of particles being considered, including type a. If we use Eqs (18) and (19) to evaluate the coefficients of Eq. (17), then the equation for $f_a(v,t)$ becomes

$$\frac{\partial f_a}{\partial t} = \frac{1}{v^2}\frac{\partial}{\partial v}\left[\alpha_a f_a + \beta_a \frac{\partial f_a}{\partial v}\right] - \frac{\gamma_a}{v^2} f_a \qquad (20)$$

where

$$\alpha_a = 4\pi\Gamma_a \sum_b \left[\left(\frac{Z_b}{Z_a}\right)^2 \frac{m_a}{m_b} \ln \Lambda_{ab} \int_0^v f_b(v',t)v'^2 dv'\right]$$

$$\beta_a = 4\pi\Gamma_a \sum_b \left(\frac{Z_b}{Z_a}\right)^2 \ln \Lambda_{ab}\left[\frac{1}{3v}\int_0^v f_b(v',t)v'^4 dv' + \frac{v^2}{3}\int_v^\infty f_b(v',t)v' dv'\right]$$

$$\gamma_a = \frac{\Lambda_a}{2v} 4\pi\Gamma_a \sum_b \left(\frac{Z_b}{Z_a}\right)^2 \ln \Lambda_{ab}\left[\int_0^v f_b(v',t)v'^2\left(1 - \frac{1}{3}\frac{v'^2}{v^2}\right)dv' + \frac{2v}{3}\int_v^\infty f_b(v',t)v' dv'\right]$$

The number density of particles of type a is

$$n_a(t) = 4\pi \int_0^\infty f_a(v,t)v^2 dv \qquad (21)$$

The last term on the right of Eq. (20) is the particle loss term due to scattering into the loss cone. If $\gamma_a = 0$ then $n_a(t)$ should remain constant in time, which gives the appropriate boundary conditions for the solution of Eq. (20), i.e.

$$\frac{dn_a}{dt} = 4\pi \int_0^\infty \frac{1}{v^2}\frac{\partial}{\partial v}\left[\alpha_a f_a + \beta_a \frac{\partial f_a}{\partial v}\right] v^2 dv = 0$$

is satisfied by the conditions

$$\left[\alpha_a f_a + \beta_a \frac{\partial f_a}{\partial v}\right]\Bigg|_{v=0}^{v=v_{max}} = 0 \qquad (22)$$

for $t > 0$, where $0 \leq v \leq v_{max}$ is the domain of Eq. (20). At $v = 0$, Eq. (22) is consistent with $\partial f_a/\partial v = 0$, since $\lim_{v \to 0} \alpha_a(v)/\beta_a(v) = 0$. In multispecies problems with an ambipolar potential defined, the effective velocity domain will be the interval $[v_{min}, v_{max}]$. In this case v_{min} will depend on the ion species and will equal zero for the electrons.

It is convenient to introduce dimensionless variables. Let $x = v/v_0$, where v_0 is a characteristic velocity, $\tau = t/t_N$ where t_N is a characteristic time, and let $F_a = (4\pi v_0^3/K_a)f_a$, where

$$n_a(0) = K_a \int_0^\infty F_a(x,0)x^2 dx$$

(The normalizing constants K_a are determined by the initial density and distribution for each species.) We generally take t_N to be defined by the electron equation, i.e. $t_N = (2v_0^3/K_e)(m_e^2/4\pi e^4)$.

We define the functionals

$$M(F) = \int_x^\infty F(y,\tau)y\ dy \tag{23}$$

$$N(F) = \int_0^x F(y,\tau)y^2\ dy \tag{24}$$

$$E(F) = \int_0^x F(y,\tau)y^4\ dy \tag{25}$$

In terms of the dimensionless variables, Eq. (20) becomes

$$\frac{\partial F_a}{\partial \tau} = \frac{1}{x^2}\frac{\partial}{\partial x}\left[A_a F_a + B_a \frac{\partial F_a}{\partial x}\right] - \frac{C_a}{x^2} F_a + D_a \tag{26}$$

where

$$A_a = \frac{t_N \Gamma_a}{v_0^3} \sum_b \left[\left(\frac{Z_b}{Z_a}\right)^2 \left(\frac{m_a}{m_b}\right) K_b \ln \Lambda_{ab} N(F_b) \right]$$

$$B_a = \frac{t_N \Gamma_a}{v_0^3} \sum_b \left(\frac{Z_b}{Z_a}\right)^2 K_b \ln \Lambda_{ab} \left[\frac{1}{3x} E(F_b) + \frac{x^2}{3} M(F_b) \right]$$

$$C_a = \frac{t_N \Gamma_a}{v_0^3} \frac{\Lambda_a}{2x} \sum_b \left(\frac{Z_b}{Z_a}\right)^2 K_b \ln \Lambda_{ab} \left[N(F_b) - \frac{1}{3x^2} E(F_b) + \frac{2x}{3} M(F_b) \right]$$

To the right side of Eq. (26) we have added a source term $D_a(x,\tau)$ for each species.

Equation (26) is a system of coupled partial differential equations. For each particle species, a, we have an equation corresponding to Λ_a, the eigenvalue of Eq. (16). In principle we could consider several eigenvalues for each species, but in this paper we have used only the lowest eigenvalue corresponding to the first normal mode. Hence we have one equation for each species. In solving the system given by Eq. (26) we consider it as a vector equation of the form

$$\frac{\partial F}{\partial \tau} = \frac{1}{x^2} \frac{\partial G}{\partial x} - \frac{C}{x^2} F + D \qquad (27)$$

where

$$F = \begin{bmatrix} F_1 \\ \cdot \\ \cdot \\ \cdot \\ F_p \end{bmatrix} \quad ; \quad G = AF + B \frac{\partial F}{\partial x}$$

and A, B, C are diagonal p x p matrices and D is the source vector. Without source and loss terms, Eq. (27) becomes

$$\frac{\partial F}{\partial \tau} = \frac{1}{x^2} \frac{\partial G}{\partial x} \qquad (28)$$

We see that this equation is in conservation form (divergence of a flux), which is consistent with the correct boundary conditions.

We solve Eq. (27) by finite-difference methods. On the domain $0 \leq x \leq x_J$, $t \geq 0$, we have a finite-difference mesh denoted by x_j, $j = 0, \cdots, J$ and t_n, $n = 0, 1, 2, \cdots$. The x spacing is variable and we define $\Delta x_{j+1/2} = x_{j+1} - x_j$, $\Delta x_{j-1/2} = x_j - x_{j-1}$, $\Delta x_j = \frac{1}{2}(x_{j+1} - x_{j-1})$. We approximate Eq. (27) by the following implicit difference equation:

$$\frac{F_j^{n+1} - F_j^n}{\Delta\tau} = \rho\left[\frac{1}{x_j^2}\,\frac{G_{j+1/2}^{n+1} - G_{j-1/2}^{n+1}}{\Delta x_j} - \frac{C_j^{n+1}}{x_j^2} F_j^{n+1} + D_j^{n+1}\right]$$

$$+ (1 - \rho)\left[\frac{1}{x_j^2}\,\frac{G_{j+1/2}^{n} - G_{j-1/2}^{n}}{\Delta x_j} - \frac{C_j^{n}}{x_j^2} F_j^{n} + D_j^{n}\right] \qquad (29)$$

where

$$G_{j+1/2}^n = \frac{1}{2} A_{j+1/2}^n (F_{j+1}^n + F_j^n) + B_{j+1/2}^n \frac{F_{j+1}^n - F_j^n}{\Delta x_{j+1/2}}$$

$$G_{j-1/2}^n = \frac{1}{2} A_{j-1/2}^n (F_j^n + F_{j-1}^n) + B_{j-1/2}^n \frac{F_j^n - F_{j-1}^n}{\Delta x_{j-1/2}}$$

For numerical stability we must have $1/2 \leq \rho \leq 1$; we usually take $\rho = 1$. Without source and loss terms, i.e. $C_j^n = D_j^n = 0$ for all j and n, we have

$$\sum_{j=1}^{J-1} \left(\frac{F_j^{n+1} - F_j^n}{\Delta\tau}\right) x_j^2 \Delta x_j = 0$$

for all n, independent of the mesh spacing, as long as $G_{+1/2}^n = G_{J-1/2}^n = 0$, for all n. This condition is the boundary condition given by Eq. (22). Thus we see that our difference scheme rigorously conserves particle density in the absence of source and loss terms.

In order to solve the difference equations given by Eq. (29), we write it as the linear algebraic system:

$$-\alpha_j^{n+1} F_{j+1}^{n+1} + \beta_j^{n+1} F_j^{n+1} - \gamma_j^{n+1} F_{j-1}^{n+1} = \delta_j^n \tag{30}$$

$j = 1, \cdots, J - 1$, where

$$\alpha_j^{n+1} = \frac{\rho}{x_j^2} \cdot \frac{1}{\Delta x_j}\left(\frac{A_{j+1/2}^{n+1}}{2} + \frac{B_{j+1/2}^{n+1}}{\Delta x_{j+1/2}}\right)$$

$$\beta_j^{n+1} = \frac{1}{\Delta t} - \frac{\rho}{x_j^2}\left[\frac{1}{\Delta x_j}\left(\frac{A_{j+1/2}^{n+1} - A_{j-1/2}^{n+1}}{2} - \frac{B_{j+1/2}^{n+1}}{\Delta x_{j+1/2}} - \frac{B_{j-1/2}^{n+1}}{\Delta x_{j-1/2}}\right) - C_j^{n+1}\right]$$

$$\gamma_j^{n+1} = \frac{\rho}{x_j^2} \cdot \frac{1}{\Delta x_j}\left(-\frac{A_{j-1/2}^{n+1}}{2} + \frac{B_{j-1/2}^{n+1}}{\Delta x_{j-1/2}}\right)$$

$$\delta_j^n = F_{j+1}^n\left[\frac{(1-\rho)}{x_j^2} \cdot \frac{1}{\Delta x_j}\left(\frac{A_{j+1/2}^n}{2} + \frac{B_{j+1/2}^n}{\Delta x_{j+1/2}}\right)\right] + F_j^n\left\{\frac{1}{\Delta t}\right.$$

$$\left. + \frac{(1-\rho)}{x_j^2}\left[\frac{1}{\Delta x_j}\left(\frac{A_{j+1/2}^n - A_{j-1/2}^n}{2} - \frac{B_{j+1/2}^n}{\Delta x_{j+1/2}} - \frac{B_{j-1/2}^n}{\Delta x_{j-1/2}}\right) - C_j^n\right]\right\}$$

$$+ F_{j-1}^n\left[\frac{(1-\rho)}{x_j^2} \cdot \frac{1}{\Delta x_j}\left(-\frac{A_{j-1/2}^n}{2} + \frac{B_{j-1/2}^n}{\Delta x_{j-1/2}}\right)\right]$$

$$+ \rho D_j^{n+1} + (1-\rho) D_j^n$$

In order to linearize the system (30) we extrapolate the α, β, γ defined above to the new time step, $n + 1$, using their values at n and $n - 1$. The method used to solve Eq. (30) is the standard tridiagonal method given by RICHTMYER and MORTON [6].

At every time step we calculate the number density and average energy of each type of particle from the equations:

$$n_a(\tau) = K_a I_2^{\,a}(\tau)$$

$$E_a(\tau) = \frac{3}{2}\, kT_a = \frac{1}{2}\, m_a v_0^{\,2} I_4^{\,a}(\tau)/I_2^{\,a}(\tau)$$

where

$$I_2^{\,a}(\tau) = \int_0^{\infty} F_a(x,\tau)x^2 dx$$

$$I_4^{\,a}(\tau) = \int_0^{\infty} F_a(x,\tau)x^4 dx$$

The coefficient of the scattering loss term, Λ_a, can be obtained by solving the eigenvalue problem, Eq. (16), for a given θ_{LC}. For the eigenvalue of the Legendre equation corresponding to the first normal mode we shall use the approximate value

$$\Lambda_a = (\log_{10} R_a)^{-1} \tag{31}$$

where R_a is the effective mirror ratio for particles of type a and depends on the ambipolar potential $e\Phi$.

For electrons we have

$$R_e = R\left(1 - \frac{e\Phi}{\frac{1}{2}\, m_e v^2}\right)^{-1} \tag{32}$$

where $R = B_{max}/B_0$, $e\Phi = \frac{1}{2}\, m_e v_{cr}^{\,2}$. For ions we have

$$R_a = R\left(1 + \frac{Z_a e\Phi}{\frac{1}{2}\, m_a v^2}\right)^{-1} \tag{33}$$

The procedure for determining $e\Phi$ follows. Let

$$Q^-(\tau) = n_e(\tau)$$

$$Q^+(\tau) = \sum_b Z_b n_b(\tau) \qquad (b \neq e)$$

(the sum taken over the ion species). At every time step, Q^- and Q^+ are computed and the difference $(Q^+ - Q^-)/Q^-$ is compared to a specified small number. If the difference exceeds this number, the v_{cr} is increased by an amount Δv_{cr} and the time step is repeated. This process is repeated until the condition is satisfied. The term Λ_e is then

$$\Lambda_e = \begin{cases} \left[\log_{10}\left\{R\left(1 - \dfrac{x_{cr}^2}{x^2}\right)^{-1}\right\}\right]^{-1} & x > x_{cr} \\ 0 & x \leq x_{cr} \end{cases} \tag{34}$$

where $v_0 x_{cr} = v_{cr}$. In the ion equations for those values of x such that $x \leq (Z_a m_e/m_a)^{1/2}\left[1/(R - 1)^{1/2}\right]x_{cr}$, we set the corresponding values of the distribution function $F_a(x, \tau)$ equal to zero. For $x > (Z_a m_e/m_a)^{1/2}\left[1/(R - 1)^{1/2}\right]x_{cr}$

$$\Lambda_a = \left[\log_{10}\left\{R\left(1 + \frac{Z_a e\Phi}{\frac{1}{2} m_a v_0^2 x^2}\right)^{-1}\right\}\right]^{-1} \tag{35}$$

This separated approach can be generalized in the following sense. We represent the solutions of Eq. (5) by the orthogonal series

$$f_a(v,\mu,t) = \sum_{\ell=1}^{N} U_\ell^a(v,t)\, M_\ell^a(\mu) \tag{36}$$

where $M_\ell^a(\mu)$ are Legendre functions with appropriate boundary conditions. We relax condition (15) and expand the Rosenbluth potentials in a Legendre polynomial series. We then obtain a system of equations in the form

$$\frac{\partial U_k^a}{\partial t} = \sum_{\ell=1}^{N} \left[\frac{1}{v^2} \frac{\partial}{\partial v} \left(\alpha_{k\ell}^{\ a} U_\ell^a + \beta_{k\ell}^{\ a} \frac{\partial U_\ell^a}{\partial v} \right) - \frac{\gamma_{k\ell}^{\ a}}{v^2} U_\ell^a \right] \quad \text{for } k = 1, \cdots, N \tag{37}$$

A detailed description of this formulation appears in Killeen et al. [3].

4. SOLUTION OF THE MULTI-SPECIES EQUATIONS IN A TWO-DIMENSIONAL VELOCITY SPACE

Equation (5) in (v,θ) coordinates, written in conservative form, is

$$\frac{1}{\Gamma_a} \left(\frac{\partial f_a}{\partial t} \right)_c = \frac{1}{v^2} \frac{\partial G_a}{\partial v} + \frac{1}{v^2 \sin\theta} \frac{\partial H_a}{\partial \theta} \tag{38}$$

where

$$G_a = A_a f_a + B_a \frac{\partial f_a}{\partial v} + C_a \frac{\partial f_a}{\partial \theta} \tag{39}$$

$$H_a = D_a f_a + E_a \frac{\partial f_a}{\partial v} + F_a \frac{\partial f_a}{\partial \theta} \tag{40}$$

and

$$A_a = \frac{v^2}{2} \frac{\partial^3 g_a}{\partial v^3} + v \frac{\partial^2 g_a}{\partial v^2} - \frac{\partial g_a}{\partial v} - v^2 \frac{\partial h_a}{\partial v} - \frac{1}{v} \frac{\partial^2 g_a}{\partial \theta^2} + \frac{1}{2} \frac{\partial^3 g_a}{\partial v \partial \theta^2}$$
$$- \frac{\cot\theta}{v} \frac{\partial g_a}{\partial \theta} + \frac{\cot\theta}{2} \frac{\partial^2 g_a}{\partial v \partial \theta} \tag{41}$$

$$B_a = \frac{v^2}{2} \frac{\partial^2 g_a}{\partial v^2} \tag{42}$$

$$C_a = - \frac{1}{2v} \frac{\partial g_a}{\partial \theta} + \frac{1}{2} \frac{\partial^2 g_a}{\partial v \partial \theta} \tag{43}$$

$$D_a = \frac{\sin\theta}{2v^2}\frac{\partial^3 g_a}{\partial\theta^3} + \frac{\sin\theta}{2}\frac{\partial^3 g_a}{\partial v^2 \partial\theta} + \frac{\sin\theta}{v}\frac{\partial^2 g_a}{\partial v \partial\theta} - \frac{1}{2v^2 \sin\theta}\frac{\partial g_a}{\partial\theta}$$

$$+ \frac{\cos\theta}{2v^2}\frac{\partial^2 g_a}{\partial\theta^2} - \sin\theta \frac{\partial h_a}{\partial\theta} \tag{44}$$

$$E_a = \sin\theta \left[-\frac{1}{2v}\frac{\partial g_a}{\partial\theta} + \frac{1}{2}\frac{\partial^2 g_a}{\partial v \partial\theta} \right] \tag{45}$$

$$F_a = \frac{\sin\theta}{2v^2}\frac{\partial^2 g_a}{\partial\theta^2} + \frac{\sin\theta}{2v}\frac{\partial g_a}{\partial v} \tag{46}$$

The functions g_a and h_a are given by Eqs (8) and (9).

Equation (38) is integrated using the method of splitting, or fractional timesteps. We first advance

$$\frac{1}{\Gamma_a}\frac{\partial f_a}{\partial t} = \frac{1}{v^2}\frac{\partial G_a}{\partial v} \tag{47}$$

using an implicit difference algorithm [6], and then advance

$$\frac{1}{\Gamma_a}\frac{\partial f_a}{\partial t} = \frac{1}{v^2 \sin\theta}\frac{\partial H_a}{\partial\theta} \tag{48}$$

in an analogous manner.

Equation (47) is differenced as follows:

$$\frac{f^{n+1}_{i,j} - f^n_{i,j}}{\Gamma_a \Delta t} = \frac{A^n_{i,j+1} f^{n+1}_{i,j+1} - A^n_{i,j+1} f^{n+1}_{i,j-1}}{2v_j^2 \Delta v_j}$$

$$+ \frac{1}{v_j^2 \Delta v_j}\left[\frac{B^n_{i,j+1/2}(f^{n+1}_{i,j+1} - f^{n+1}_{i,j})}{\Delta v_{j+1/2}} - \frac{B^n_{i,j-1/2}(f^{n+1}_{i,j} - f^{n+1}_{i,j-1})}{\Delta v_{j-1/2}}\right]$$

$$+ \frac{1}{2v_j^2 \Delta v_j}\left[\frac{C^n_{i,j+1}(f^n_{i+1,j+1} - f^n_{i-1,j+1})}{2\Delta\theta_i}\right.$$

$$\left. - \frac{C^n_{i,j-1}(f^n_{i+1,j-1} - f^n_{i-1,j-1})}{2\Delta\theta_i}\right] \tag{49}$$

Here, $f^n_{i,j} = f(v_j, \theta_i, n\Delta t)$.

By rearranging terms, Eq. (49) may be put into the tri-diagonal form:

$$-\alpha_{i,j}^{n} f_{i,j+1}^{n+1} + \beta_{i,j}^{n} f_{i,j}^{n+1} - \gamma_{i,j}^{n} f_{i,j-1}^{n+1} = \delta_{i,j}^{n} \qquad (50)$$

We see that the terms of mixed second derivative type may not be written fully implicitly if we wish to maintain a tri-diagonal form. Equation (48) is integrated in a similar manner, with the roles of v and θ reversed.

The direct two-dimensional approach for obtaining the ion velocity-space distributions in mirror-confined plasmas can be used to analyse many different physical situations which cannot be studied with the one-dimensional lowest-normal-mode approach described in Section 3. In particular, transient plasmas [3] typically encountered in mirror experiments may undergo rapid changes which would not allow the establishment of a lowest-normal-mode ion distribution. Furthermore, the injection of well collimated neutral beams gives rise to ion sources that are sharply peaked in velocity space so that many normal modes would be required for adequate representation.

The two-dimensional multi-species Fokker-Planck code is a useful tool for studying the physics involved in the thermalization of directed monoenergetic neutral beams injected into a dense tokamak plasma. The essential physics of a two-component toroidal reactor (TCT) was first described by Dawson et al. [7]. Variations and refinements of this concept were given by Furth and Jassby [8]. The authorized construction of a "breakeven" TCT experiment has given added impetus to the search for a detailed understanding of the plasma physics involved in the design of such a system.

An idealized model in which the plasma is assumed to be spatially uniform over some finite toroidal volume allows one to analyse the system in terms of velocity-space variables only. In particular, electron and ion distribution functions are obtained from solutions of the time-dependent velocity-space Fokker-Planck equations. These solutions can be used to compute the energy multiplication factor

$$Q = \frac{\text{(Energy from Fusion Reactions)}}{\text{(Energy in the Injected Deuterons)}} \tag{51}$$

which serves as the figure of merit for a pulsed TCT system [3].

Several features of the Fokker-Planck model are especially appropriate for representing physically significant effects in a TCT. The nonlinear nature of the kinetic equations ensures that the collisional interaction of all species, including self-interactions, is properly accounted for regardless of the form of the distribution functions. Alpha particles and impurity ions are treated on an equal footing with the deuterium and tritium ions since there is no inherent restriction on the number of species which can be handled in the code. Major radius compression is a useful technique for supplementing neutral beam injection in toroidal plasmas, and the two-dimensional nature of our velocity space allows us to accurately account for distortions of the distribution functions due to this anisotropic driving force.

In axisymmetric toroidal systems the constants of the particle motion are

$$\frac{mv_\perp^2}{2B} = \text{magnetic moment} \tag{52}$$

$$mv_\parallel R = \text{toroidal angular momentum} \tag{53}$$

where R is the major radius of the torus. For a time-varying major radius we derive $(\dot{v}_\parallel, \dot{v}_\perp)$ by taking the time derivative of these equations, obtaining

$$\dot{v}_\perp = + \frac{1}{2} v_\perp \frac{\dot{B}}{B} \tag{54}$$

$$\dot{v}_\parallel = - v_\parallel \frac{\dot{R}}{R} \tag{55}$$

Since B is essentially just the toroidal field strength, it varies inversely with R, yielding

$$\frac{\dot{B}}{B} = - \frac{\dot{R}}{R} \tag{56}$$

With these results the Fokker-Planck equation becomes

$$\frac{\partial f}{\partial t} + \frac{\dot{R}}{R} \left[- \left(1 - \frac{1}{2} \sin^2\theta \right) v \frac{\partial f}{\partial v} + \frac{1}{2} \sin\theta \cos\theta \frac{\partial f}{\partial \theta} \right] = \left(\frac{\partial f}{\partial t} \right)_c + S + L \tag{57}$$

where $(\partial f/\partial t)_c$ is given by Eq. (38). For isotropic electrons we have

$$\frac{\partial f_e}{\partial t} + \left(- \frac{2}{3} v \frac{\partial f_e}{\partial v} \right) \frac{\dot{R}}{R} = \left(\frac{\partial f_e}{\partial t} \right)_c + S_e + L_e \tag{58}$$

The source term S_a in Eq. (1) is of the form

$$S_a(v,\theta,t) = \sum_{\ell} J_a^{\ell}(t)\, s_a(v,\theta)\, \delta_s^{a,\ell}(t) \tag{59}$$

where the shape function $s_a(v,\theta)$ is a Gaussian in v and cosθ of density 1, $\delta_s^{a,\ell}(t)$ is either 0 or 1, and $J_a^{\ell}(t)$ is a current.

The loss term L_a may combine several loss processes. Losses due to charge exchange with the beam are expressed as

$$L_a^c = - \left[\sum_{b,\ell} D_{ab}^{\ell}\, \delta_s^{b,\ell}(t) \right] f_a(v,\theta,t) \tag{60}$$

where the quantities D_{ab}^{ℓ} are constant parameters. The presence of the vectors $\delta_s^{b,\ell}$ allows the whole charge-exchange process to be implemented as a unit; that is, a given charge-exchange term along with its corresponding source term depends on the same temporal function $\delta_s^{b,\ell}(t)$.

The effects of finite particle and energy confinement times in a tokamak are incorporated by adding a contribution to L_a of the form

$$L_a^\tau = -\frac{f_a(v,\theta,t)}{\tau_p} + \frac{1}{v^2}\frac{\partial}{\partial v}\left[\left(\frac{1}{\tau_e} - \frac{1}{\tau_p}\right)\frac{v^3 f_a(v,\theta,t)}{2}\right] \qquad (61)$$

where τ_p and τ_e are particle and energy confinement times, respectively. If we ignore all other terms in Eq. (1) and compute moments, we find

$$\frac{\partial n_a}{\partial t} = -\frac{n_a}{\tau_p} \qquad (62)$$

$$\frac{\partial (n_a \bar{E}_a)}{\partial t} = -\frac{n_a \bar{E}_a}{\tau_e} \qquad (63)$$

In problems where one wants to observe the relaxation of a distribution in the absence of a loss cone, or where one assumes that ions in a loss cone domain are not lost instantly, full velocity space boundary conditions are applied, namely:

(a) $f(v=0,\theta)$ is independent of θ (64)

(b) $\frac{\partial f}{\partial v}(v=0,\theta=\pi/2) = 0$ (65)

(c) $\frac{\partial f}{\partial \theta}(v,\theta=0) = \frac{\partial f}{\partial \theta}(v,\theta=\pi) = 0$ (66)

The first condition is a result of continuity at v=0. The second and third conditions are a result of the requirement that the distribution be azimuthally symmetric.

REFERENCES

[1] KILLEEN, J., MARX, K.D., "The solution of the Fokker-Planck equation for a mirror-confined plasma", Methods in Computational Physics 9, Academic Press, New York, (1970) 421.

[2] FUTCH, A.H., Jr., HOLDREN, J.P., KILLEEN, J., MIRIN, A.A., Plasma Phys. **14** (1972) 211.

[3] KILLEEN, J., MIRIN, A.A., RENSINK, M.E., "The solution of the kinetic equations for a multi-species plasma", Methods in Computational Physics **16**, Academic Press, New York (1976) 389–431.

[4] ROSENBLUTH, M.N., MacDONALD, W.M., JUDD, D.L., Phys. Rev. **107** (1957) 1.

[5] SPITZER, L., Physics of Fully Ionized Gases, 2nd edn, Wiley-Interscience, New York (1962).

[6] RICHTMYER, R.D., MORTON, K.W., Difference Methods for Initial-Value Problems, Wiley-Interscience, New York (1967).

[7] DAWSON, J.M., FURTH, H.P., TENNEY, F.H., Phys. Rev. Lett. **26** (1971) 1156.

[8] FURTH, H.P., JASSBY, D.L., Phys. Rev. Lett. **32** (1974) 1176.

EQUILIBRIUM RELATIONS IN THE PRESENCE OF ARBITRARY PLASMA DIFFUSION IN AXISYMMETRIC CONFIGURATIONS*

D. PFIRSCH
Max-Planck-Institut für Plasmaphysik,
Garching,
Federal Republic of Germany

Abstract

EQUILIBRIUM RELATIONS IN THE PRESENCE OF ARBITRARY PLASMA DIFFUSION IN AXISYMMETRIC CONFIGURATIONS.

A condition for general axisymmetric diffusive equilibria that relates the outward diffusion to the toroidal current density is derived. In an approximate version, it requires that some effective diffusion velocity v_D^* must not exceed the poloidal magnetic diffusion velocity v_m. Relevant consequences follow in the anomalous diffusion regime if diffusion is caused by an anomalous parallel electron viscosity instead of an anomalous perpendicular resistivity. In the former case, v_D^* equals the real diffusion velocity v_D, and an anomalous bootstrap current arises which leads to rather low upper limits for β_p. If the usual trapped ion or Bohm diffusion is assumed to be caused by enhanced viscosity, no stationary high-temperature equilibria would be possible in a system governed by the appropriate diffusion law.

1. INTRODUCTORY REMARKS

Essentially, this paper reviews a discussion by K. Borrass and D. Pfirsch (Max-Planck-Institut für Plasmaphysik Rep. IPP 6/155 of March 1977), and to elucidate the questions considered, two introductory remarks are made:

A. A self-consistent macroscopic theory using, among others, the following set of equations

$$\nabla p = \frac{1}{c}\, \underline{j} \times B, \quad \nabla \cdot j = 0, \quad \underline{E} + \frac{1}{c}\, \underline{v} \times B = \underline{\underline{\eta}} \cdot \underline{j}$$

leads uniquely to Pfirsch-Schlüter diffusion. Present-day tokamak transport codes are therefore only capable of describing neoclassical or anomalous diffusion within

* Work performed under the terms of the agreement on association between the Max-Planck-Institut für Plasmaphysik and Euratom.

TABLE I. ESTIMATES OF RATIO v_D/v_m FOR VARIOUS CASES

	a (cm)	R_0 (cm)	T (keV)	n (cm^{-3})	τ_p (seconds)	v_D/v_m
I	10	70	1	2×10^{13}	0.05	1
II	45	120	2.5	5×10^{13}	0.1	50
III	400	1000	15	5×10^{13}	2	5000

the framework of a macroscopic theory if some of the above equations are modified. Obviously, only Ohm's law allows major changes without influencing the fundamental pressure balance relation. To obtain the freedom of choosing any diffusion velocity, these codes therefore only use the toroidal component of Ohm's law and, furthermore, neglect the $\underline{v} \times \underline{B}$ term, i.e. Ohm's law is reduced in these codes to the equation

$$E_\phi = \eta_\parallel \; j_\phi$$

where ϕ is the toroidal angle.

Neglecting the $\underline{v} \times \underline{B}$ term means neglecting the convective transport of the magnetic field. A relation of the form $\underline{E} = \eta \underline{j}$ leads, on the other hand, to magnetic field diffusion with a velocity

$$v_m \approx \frac{\eta c^2}{4\pi a}$$

where a is a typical length for the magnetic field variation due to currents. In the following, a is chosen as the minor plasma radius.

Neglecting the $\underline{v} \times \underline{B}$ term is therefore only justified if the convective field transport is small compared to field diffusion. The convective transport occurs with the plasma diffusion velocity v_D. Hence we arrive at the condition $v_D \ll v_m$.

Table I contains estimates of the ratio v_D/v_m for various cases according to present-day code calculations. The resistivity is assumed to be normal. R_0 is the major radius, and τ_p is the particle confinement time. The values of v_D/v_m indicate that the present-day codes can probably only be applied to smaller experiments if anomalous diffusion is not caused by anomalous resistivity.

B. Plasma diffusion through a magnetic field means plasma motion relative to the magnetic field. Stationary conditions are therefore only possible if particle diffusion equals magnetic field diffusion. Since the scale length for the magnetic field can be larger than the plasma radius a, we must have

$$v_D \leqslant v_m = \frac{\eta c^2}{4\pi a}$$

With

$$v_m \approx \frac{2v_{cl}}{\beta}, \qquad \beta = \frac{8\pi p}{B^2}$$

where v_{cl} is the classical diffusion velocity in plane geometry, the inequality yields

$$\beta \leqslant \frac{2v_{cl}}{v_D}$$

A few examples may illustrate this result:

(a) Neoclassical banana diffusion is given by

$$v_D \sim q^2 A^{3/2}\, v_{cl}, \quad q = \text{safety factor}, \; A = R_o/a$$

The β relation leads to

$$\beta_{max} \sim \frac{1}{q^2 A^{3/2}} \quad \text{or} \quad \beta_{p\ max} \sim A^{1/2}$$

This result is known to be a consequence of the bootstrap current.

(b) Pseudoclassical diffusion is described by

$$v_D \sim q^2 A^2$$

leading to

$$\beta_{p\ max} \sim 1$$

(c) Bohm diffusion does not lead directly to a β limit. Because of the special form of v_D we obtain a temperature limit

$$T_{keV} \leqslant 0.02\, B_{kG}^{2/5}$$

If Bohm diffusion were to be formally inserted in the right-hand side of the β relation with T = 10 keV, B = 40 kg and n = 10^{14}, it would be found that

$$\beta_{max} \sim 10^{-7}$$

which, of course, is not a self-consistent result. It illustrates, however, in addition to the correct temperature limitation, the severe consequences of the relation $v_D \leqslant v_m$.

The next sections contain a rigorous treatment of these problems for axisymmetric configurations.

2. GLOBAL RELATIONS

We start with the stationary momentum equation for electrons, which can be derived from an exact Fokker-Planck equation:

$$-e\left(n\,\underline{E} + \frac{1}{c}\,\overline{n\underline{v}} \times B\right) = \underline{R}_{ee} + \underline{R}_{ei}$$

In this equation n, $\overline{n\underline{v}}$, $\underline{E}$, $\underline{B}$ can be considered as average quantities in the case of turbulent plasmas, $\underline{R}_{ee}$ and $\underline{R}_{ei}$ describe all kinds of momentum exchange among the electrons themselves and between electrons and ions, including pressure gradients, inertial terms or turbulent terms such as $\overline{\delta n\, \delta E}$ and $\overline{\delta\,(n\underline{v}) \times \delta \underline{B}}$. Subscripts e for electrons are omitted if not needed.

There can be many species of ions as well as neutrals. An important relation follows from the momentum equation by calculating the rate equation for the axial component of the total angular momentum of the electrons, which is

$$-\int\limits_{plasma} e\,R\left(nE_{\phi} + \frac{1}{c}\,\overline{nv_{\phi}}\,B_p\right) d^3x = \int\limits_{plasma} R\,(R_{ee\phi} + R_{ei\phi})\, d^3x$$

R is the major radius, p denotes the poloidal and ϕ the toroidal component and ψ the component perpendicular to poloidal flux surfaces ψ = const, when $2\pi\psi$ is the poloidal flux.

From axisymmetry it follows that

$$\int R\,R_{ee\phi}\, d^3x = 0$$

since momentum exchange among the electrons themselves cannot change their total angular momentum.

We now use the following co-ordinates:

ψ, 1 = length on ψ = const in the poloidal direction, and ϕ.

Observing that

$$\underline{B}_p = \nabla\psi \times \nabla\psi, \qquad |\nabla\psi| = \frac{1}{R}$$

we can express the volume integration by

$$\int d^3x \ldots = \int d\psi \oint_\psi dl \; 2\pi R \frac{1}{|\nabla\psi|} \ldots = \int d\psi \; 2\pi \oint_\psi \frac{dl}{B_p}$$

Here it is assumed for convenience that $B_p \geqslant 0$ holds, which does not mean a restriction for tokamak equilibria.

The angular momentum relation, divided by the charge e, then reads

$$\int_{\text{plasma}} d\psi \left[\oint_\psi dl \; 2\pi R \, n \frac{cE\phi}{B_p} + \oint_\psi dl \; 2\pi R \, \overline{nv_\phi} + \oint_\psi dl \; 2\pi R \frac{R_{ei\phi}}{eB_p} \right] = 0$$

The expression

$$\oint_\psi dl \; 2\pi R \; \overline{nv\psi} \equiv \Gamma_\psi$$

occurring in this relation is the total electron flux through a flux surface ψ = const.

Writing down similar relations for all kinds of particle species contained in the plasma, and multiplying each relation again by the corresponding charge and summing up all these relations, leads to

$$\int_{\text{plasma}} d\psi \sum e \, \Gamma_\psi = 0$$

This holds because the momentum exchange between different particle species cannot change the total angular momentum of the plasma, which makes the third term vanish when summed over all particle species. The first term cancels out if the plasma is neutral as a whole. The latter can be relaxed if wall interaction is taken into account. This result means that a necessary condition for ambipolar diffusion is always fulfilled which, however, might for self-consistency imply the existence of radial electric fields.

We now discuss in more detail the angular momentum relation for the electrons. We observe first that it exhibits the fact that there are only three causes for a change of the total angular momentum of the electrons: an external electric field E_ϕ leading to the first term; a poloidal field leading to the Γ_ϕ term, which means partly interaction with an external field and partly a non-stochastic interaction with the other particle species; and momentum exchange with other particle species given by the third term.

The first term can be considered as a generalized Ware effect. In equilibrium it must have the opposite sign of the $R_{ei\phi}$ term because the electron charge is negative. The same holds for Γ_ψ. Anomalous Γ_ψ therefore means that $R_{ei\phi}$ must also be anomalous.

Since $\underline{R_{ei}}$ describes the momentum exchange between electrons and other particles, it must somehow be related to the electrical resistance. Hence, if the concept of resistivity is still applicable, we can write

$$\underline{R_{ei}} = -e\,n\,\underline{\underline{\eta}}\cdot\underline{j} = -en(\eta_{\parallel}\,\underline{j}_{\parallel} + \eta_{\perp}\,\underline{j}_{\perp})$$

Anomalous Γ_ψ therefore means anomalous $(\eta\cdot j)_\phi$.

To draw conclusions from this, we note that experiments indicate that in the usual low electron drift case, $\eta_{\parallel}$ is close to its classical value; $\underline{j}_{\perp}$ follows from the pressure balance equation as

$$\underline{j}_{\perp} = -\frac{c\nabla p \times \underline{B}}{B^2}$$

Diffusion, being greater than classical Pfirsch-Schlüter diffusion, therefore requires either an enhanced perpendicular resistivity or a large bootstrap current. In the first case, experiments would demand $\eta_{\perp}$ to be a few thousand times $\eta_{\parallel}$. There are, however, arguments that an enhanced electron-electron collision frequency should be responsible for the observed diffusion, in which case $\eta_{\perp}$ should remain essentially normal and thus

$$(\underline{\underline{\eta}}\cdot\underline{j})_\phi \approx \eta_{\parallel}\,j_{\parallel\phi}$$

Assuming this to be true, we find, in the sense of mean values over the whole plasma,

$$v_\psi \approx \frac{c\,\eta_{\parallel}\,j_{\parallel\phi}}{B_p}$$

Here the Ware effect is neglected; with given v_ψ it would increase the necessary $j_{\parallel\phi}$. With this current, Maxwell's equation for the poloidal field reads, in the large aspect ratio limit,

$$\frac{1}{r}\frac{\partial}{\partial r}(r\,B_p) \approx \frac{4\pi}{c}\,j_{\parallel\phi} \approx \frac{4\pi}{c^2\eta_\parallel}\,B_p\,v_\psi$$

From this we obtain by order of magnitude

$$v_\psi \approx \frac{c^2\eta_\parallel}{4\pi a} = v_m \approx \frac{2\,v_{cl}}{\beta}$$

This is the relation stated in the second Introductory Remark.

3. LOCAL RELATIONS

If one can neglect the angular momentum transport through the magnetic surfaces compared with the angular momentum exchange between the electrons and the other particles, i.e. if

$$v_D/a \ll \nu_{ei}$$

holds, where ν_{ei} is an effective electron-ion collision frequency including turbulent effects, then we can drop the ψ integration in the global angular momentum relation. A first consequence of this local relation is then that ambipolarity holds locally. The approximation allows the presence of space charges in the sense of quasi-neutrality, i.e. $\left|\sum en\right| \ll en$ must be fulfilled.

If we express $\underline{R_{ei}}$ by $\underline{R_{ei}} = -en(\underline{\eta}\cdot\underline{j})$, the new relation reads

$$\oint_\psi dl\;2\pi Rn\,\frac{cE_\phi}{B_p} + \Gamma_\psi - \oint_\psi dl\;2\pi R\,n\,\frac{c(\underline{\eta}\cdot\underline{j})_\phi}{B_p} = 0$$

We shall now use this equation together with two other relations in order to derive an equation for Γ_ψ analogous to the formula for classical diffusion. To this end we have to eliminate $\underline{j}$ from the above formula, in which

$$(\underline{\underline{\eta}}\cdot\underline{j})_\psi = (\eta_\parallel - \eta_\perp)\,\frac{B_\phi B_p}{B^2}\,j_p + \left(\eta_\perp\,\frac{B_p^{\,2}}{B^2} + \eta_\parallel\,\frac{B_\phi^{\,2}}{B^2}\right) j_\phi$$

contains the two components j_p and j_ϕ of the current density.

The first additional relation we shall use is the equation for the pressure balance:

$$\nabla p = \frac{1}{c} \underline{j} \times \underline{B}$$

It is assumed here that all deviations of the total pressure tensor from its trace p can be neglected. The pressure tensor itself can be assumed to contain all kinds of inertial and turbulent contributions; p must then be a function of ψ alone, which leads to

$$cRB_p \frac{dp}{d\psi} = j_p B_\phi - j_\phi B_p$$

This equation allows us to eliminate j_ϕ, for instance. To eliminate j_p too we multiply the electron momentum equation by $\underline{B}$. Writing

$$\underline{R_{ee}} = \nabla p_e + \nabla \cdot \underline{\underline{\Pi}}_e$$

with $T_r \underline{\underline{\Pi}}_e = 0$, we obtain

$$0 = B_p \frac{\partial p_e}{\partial l} - en E_\phi B_\phi + en \frac{\partial \phi}{\partial l} B_p + \underline{B} \cdot \nabla \cdot \underline{\underline{\Pi}}_e$$

$$+ en \eta_{\parallel} (j_\phi B_\phi + j_p B_p)$$

The unknown electric potential ϕ for the poloidal field disappears if one operates on this equation by

$$2\pi \oint_\psi \frac{dl}{nB_p} \ldots$$

or, if one assumes that $n = n(\psi)$, just

$$2\pi \oint_\psi \frac{dl}{B_p} \ldots$$

in which latter case p_e also drops out.

If one again eliminates j_ϕ by using the pressure balance equation, a second equation is arrived at containing j_p in an integral over a magnetic surface. One can now eliminate j_p between these two equations by using the relation which holds for axisymmetric configurations, namely

$$\underline{B}_\phi = f(\psi)\ \nabla\phi$$

from which it follows that

$$\nabla \times \underline{B}_\phi = f'(\psi)\ \nabla_\psi \times \nabla_\phi = f'(\psi)\ \underline{B}_p = \frac{4\pi}{c}\ \underline{j}_p$$

Thus, j_p can be expressed by B_p times a function of ψ alone and this function can be eliminated between our two equations. This yields the envisaged relation for Γ_ψ. If, besides $n = n(\psi)$, one also assumes that $\eta_{\|} = \eta_{\|}(\psi)$ and writes

$$E_\phi = \frac{U}{2\pi R}$$

where U is the space-independent loop voltage, then one finds that

$$\frac{\Gamma_\psi}{V'(\psi)} = cn\,\frac{U}{2\pi}\left(\frac{(RB_\phi)^2}{\langle B^2\rangle}\ \langle\frac{1}{R^2}\rangle - 1\right) + \frac{c}{e}\ \frac{RB_\phi}{\langle B^2\rangle}\ \langle\ \underline{B}\cdot\nabla\cdot\underline{\underline{\Pi}}_e\rangle$$

$$- \eta_{\|}c^2 n\ \frac{dp}{d\psi}\left(R^2B_\phi^2\right)\left[\langle\frac{1}{B^2}\rangle - \frac{1}{\langle B^2\rangle}\right] - c^2 n\ \frac{dp}{d\psi}\ \langle\ \eta_\perp\ \frac{R^2B_p^2}{B^2}\rangle$$

In this equation it holds that

$$V'(\psi) = 2\pi\oint_\psi \frac{dl}{B_p}\ , \qquad \langle A\rangle = \frac{1}{V'(\psi)}\,2\pi\oint_\psi \frac{dl}{B_p}\ A$$

$\langle A\rangle$ is therefore a mean value of A over a magnetic surface.

The various terms in this equation have the following meaning:

> The last term is ‘classical’ diffusion; the last-but-one term is the Pfirsch-Schlüter correction; the first term is a pinch effect; and the second term is the possibly dominant term resulting from an effective parallel viscosity of the electrons. This term is responsible, for instance, for neoclassical diffusion.

Let us finally relate these results to the ϕ-component of Ohm’s law. To this end we eliminate Γ_ψ by introducing j_ϕ again. The relation we arrive at is:

$$\frac{1}{V'(\psi)}\ \frac{1}{c}\ \Gamma_\psi^* - \eta_{\|}n\ \langle Rj_\phi\rangle = -\ n\,\frac{U}{2\pi}\ \langle\frac{1}{R^2}\rangle\ \frac{(RB_\phi)^2}{\langle B^2\rangle} \leq 0$$

with

$$\frac{1}{V'(\psi)} \Gamma_\psi^* = \frac{c}{e} \frac{RB_\phi}{\langle B^2 \rangle} \langle \underline{B} \cdot \nabla \cdot \underline{\underline{\Pi}}_e \rangle$$

$$- \eta_\parallel c^2 n \frac{dp}{d\psi} (RB_\phi)^2 \left[\langle \frac{1}{B^2} \rangle - \frac{1}{\langle B^2 \rangle} \right]$$

$$- \eta_\parallel c^2 n \frac{dp}{d\psi} \langle \frac{R^2 B_p^{\ 2}}{B^2} \rangle$$

Γ_ψ^* is equal to Γ_ψ except for the terms containing $\eta_\perp$ and U. Thus, if electron viscosity dominates, we have $\Gamma_\psi^* \approx \Gamma_\psi$. The inequality then requires the existence of an anomalous bootstrap current, which leads to the former β limitation.

For large aspect ratio the new relation reads simply with $[j_\phi]: = \langle Rj_\phi \rangle / R_0$

$$E_\phi = \eta_\parallel [j_\phi] - \frac{1}{c} v_D^{\ *} B_p$$

where v_D^* corresponds to Γ_ψ^*. Thus the applicability of

$$E_\phi = \eta_\parallel j_\phi$$

in present-day transport codes depends on the validity of $v_D^* \ll v_m$ rather than $v_D \ll v_m$, as was heuristically stated in the first Introductory Remark.

In conclusion, it appears to be of great importance to get detailed information about the nature of anomalous diffusion: whether it is caused by anomalous electron viscosity or by anomalous perpendicular resistivity. Only in the latter case could interesting β-values be achieved.

COLLISIONAL TRANSPORT

D. PFIRSCH
Max-Planck-Institut für Plasmaphysik,
Garching,
Federal Republic of Germany

Abstract

COLLISIONAL TRANSPORT.

Collisional particle and heat transport is treated in plane and toroidal geometry. In particular, temperature gradient effects on impurity diffusion – so-called temperature screening – are considered for the different collisional regimes. The existence of quasistationary self-consistent tokamak equilibria with finite resistivity and a possible limitation of the maximum β caused by particle diffusion is discussed.

A. PLANE GEOMETRY

The case of plane geometry serves, on the one hand, as a reference case and, on the other, it can be of interest for tokamaks with highly elongated cross-sections.

1. AMBIPOLARITY

In this section the motion of single particles in a magnetic field $\underline{B}$ which is constant to lowest order in the gyro-radius is considered, and the influence of collisions on this motion is investigated. Let $\underline{B}$ have only a z-component in a rectangular coordinate system x,y,z, B = (0,0,B), then the unperturbed motion of a particle is described by

$$m\ddot{x} = + \frac{e}{c}\dot{y}B, \quad m\ddot{y} = - \frac{e}{c}\dot{x}B$$

For B $\approx$ const to lowest order in the gyro-radius we obtain with w = x + iy:

$$\ddot{w} = - i\,\omega_g \dot{w}, \quad \dot{w} = \dot{w}_o\, e^{-i\omega_g t}, \quad w = w_o - \frac{1}{i\omega_g}\dot{w}_o\, e^{-i\omega_g t}$$

where w_o is the complex gyro-centre position.

If the particle undergoes a collision with a sudden change in $\dot{w}$ but with w unchanged, one obtains a change in w_o given by

$$\Delta w_o = \frac{1}{i\omega_g} \Delta \dot{w}_o e^{-i\omega_g t} = \frac{1}{i} \frac{\Delta m\dot{w} \cdot c}{eB}$$

Using vector symbols $\underline{r}_o = (x_o, y_o)$, $\underline{v} = (\dot{x}, \dot{y})$, we obtain

$$\Delta \underline{r}_o = c \frac{\Delta m\underline{v} \times \underline{B}}{eB^2}$$

If we define the mean centre of charge of the colliding particles by

$$\Sigma e\underline{r}_o / \Sigma |e|$$

where the summation goes over the two colliding particles, then we get for its change during collision

$$\Sigma e \Delta \underline{r}_o / \Sigma |e| = c \frac{\Sigma \Delta m\underline{v} \times B}{B^2 \; \Sigma |e|} = 0 \quad \text{since } \Sigma \Delta m\underline{v} = 0$$

Thus, to lowest order in the gyro-radius there is ambipolar diffusion due to momentum conservation.

This holds very generally for all situations in which the change of the mean position of a particle during collision in the direction of the pressure gradient is proportional to the gyro-radius, i.e. proportional to the momentum.

2. CLASSICAL DIFFUSION

From $\nabla p = \frac{1}{c} \underline{j} \times \underline{B}$, $\eta \cdot \underline{j} = \underline{E} + \frac{1}{c} \underline{v} \times \underline{B}$, and $\underline{E} = 0$ one finds

$$\nabla p = \frac{1}{\eta c^2} (\underline{v} \times \underline{B}) \times \underline{B} = \frac{1}{\eta c^2} (\underline{B}(\underline{v} \cdot \underline{B}) - \underline{v} B^2)$$

from which

$$\underline{v} = - \frac{\eta c^2}{B^2} \nabla p \equiv \underline{v}_{cl}$$

results. This is so-called classical diffusion. For T = const, which, in reality, is assumed here from the beginning, this can be written as

$$\underline{v} = - \frac{\eta c^2}{B^2} T \nabla n \equiv - \frac{D}{n} \nabla n$$

with the diffusion coefficient

$$D = nT\eta c^2/B^2$$

The resistivity η can be expressed by the electron collision frequency

$$\eta = \nu_{ei} \frac{m}{e^2 n}$$

With this expression we obtain

$$D = \nu_{ei} \frac{mTc^2}{e^2 B^2} = \nu_{ei} r_{ge}^2$$

Thus, the diffusion can be described as a stochastic process with the gyro-radius as the step size of a single process.

The magnetic diffusion coefficient describing the skin effect is

$$D_m = \frac{\eta c^2}{4\pi}$$

which leads to the relation

$$D = \frac{4\pi nT}{B^2} D_m = \frac{1}{2} \beta D_m \quad , \quad \beta = \frac{nT}{B^2/8\pi}$$

This means that particle diffusion in plane geometry is small compared with magnetic diffusion as long as the local β is smaller than 1. Using the expression for the Spitzer resistivity

$$\eta = 1.15 \times 10^{-14} \frac{Z \ln\Lambda}{T_{eV}^{3/2}}$$

and taking $Z = 1$, $T_{eV} = 10^4$ (= 10 keV)

we obtain

$$D_m = 0{,}82 \ln\Lambda \ \mathrm{cm^2/sec}$$

$$\approx 15 \ \mathrm{cm^2/sec}$$

To get a confinement time of 1 sec for $\beta = 1$, it would be enough to have a half-thickness of the plasma of the order of 5 cm. A half-thickness of 5 m would lead to particle confinement times of 10^4 sec. For comparison, Bohm diffusion $D_B = \frac{cT}{16eB}$ for T = 10 keV and B = 40 kG would give $D_B = 1.56 \ 10^6 \ \mathrm{cm^2/sec}$.

3. CLASSICAL HEAT CONDUCTION PERPENDICULAR TO B

The transport of heat perpendicular to $\underline{B}$ in the direction of ∇T can be described by a stochastic process similar to diffusion. Thus

$$\underline{q} = -\kappa \nabla T = -D_T \, n \nabla T$$

where D_T now also contains contributions from equal particle collisions. Since $\nu_{ii} = \sqrt{\frac{m_i}{m_e}} \nu_{ie}$ and electrons and ions give the same contribution to D as far as collisions between these two species are concerned, only ion-ion collisions are of interest. Thus, the time-scale for heat conduction is shorter than for diffusion by a factor of $\sqrt{m_e/m_i}$, which is 1/61 in the case of deuterium and 1/74 for tritium.

4. IMPURITY DIFFUSION

Qualitatively, we have the following situation. Since the impurities interact much more strongly with the hydrogenic ions than with the electrons, the mean relative motion between impurities and ions is small compared with that between electrons and the heavy particles. We can therefore go into a system of reference in which all the heavy particles are at rest simultaneously. These particles are then not confined by a magnetic field but by an electrostatic space charge potential U(x) with, say, U(0) = 0, where x = 0 means the centre of the plasma. Only the electrons are confined magnetically. Then, since we have almost thermal equilibrium,

$$n_i(x) = n_i(0) \, e^{-\frac{e_i U(x)}{T_i}}$$

or

$$\frac{n_i(x)}{n_i(0)} = \left(\frac{n_k(x)}{n_k(0)} \right)^{\frac{e_i}{T_i} \frac{T_k}{e_k}}$$

For $e_k = e$ (hydrogenic ions) and $e_i = Z e$, $T_i = T_k = T$ we have

$$\frac{n_Z(x)}{n_Z(0)} = \left(\frac{n_H(x)}{n_H(0)} \right)^Z$$

a result first obtained by J.B. Taylor [1]. Since $Z \geq 2$, all kinds of impurities are more concentrated in the centre of the plasma than the hydrogenic ions as far as this formula is applicable. A remarkable change of this result occurs if one takes into account a temperature gradient. One can do a very simple perturbation theory with respect to a temperature gradient in a kinetic equation of the Bathnager-Gross-Krook type, leading to a perturbation of a Maxwellian distribution f_M caused by gradients and electric fields:

$$\delta f^{(Z)} = \frac{-\nu_Z v_x + \omega_Z v_y}{\nu_Z^2 + \omega_Z^2} \left[\frac{\partial}{\partial x} - \frac{ZeE_x}{T} \right] f_M^{(Z)} , \quad \omega_Z = \frac{ZeB}{m_Z c}$$

ν_Z is the collision frequency of impurities of charge Z with hydrogenic ions. Doing the corresponding calculation for the hydrogenic ions and imposing the condition of momentum conservation during collisions in the y direction, one can eliminate the electric field and obtain ambipolar diffusion in the x-direction. The flux of the impurities calculated from $\delta f^{(Z)}$ is then

$$\Gamma_Z = \langle n v_x \rangle_Z = - \frac{\nu_Z^{th} n_Z}{m_Z \omega_Z^2} \frac{3}{2\pi} TZ \left[\frac{1}{Z} \frac{1}{n_Z} \frac{\partial n_Z}{\partial x} - \frac{1}{n_H} \frac{\partial n_H}{\partial x} + \frac{1}{2} \left(1 - \frac{1}{Z}\right) \frac{1}{T} \frac{\partial T}{\partial x} \right]$$

If there were no temperature gradient, the stationary state would require that

$$\frac{1}{Z} \frac{1}{n_Z} \frac{\partial n_Z}{\partial x} = \frac{1}{n_H} \frac{\partial n_H}{\partial x}$$

or

$$n_Z \sim n_H^Z$$

as we had found before. To find out what the influence of the temperature gradient is, it is convenient to discuss temperature profiles related to the hydrogenic density profile by

$$T(x) \sim n_H^{\alpha}$$

For n_Z we write

$$n_Z \sim n_H^{\beta}$$

The stationary state is then obtained for

$$\frac{\beta}{Z} - 1 + \frac{\alpha}{2}\left(1 - \frac{1}{Z}\right) = 0$$

or

$$\beta = Z - \frac{\alpha}{2}(Z-1)$$

For $\alpha < 2$ we have $\beta > 1$, assuming $Z > 2$, and for $\alpha > 2$ we have $\beta < 1$. Thus, up to $\alpha = 2$ there is impurity influx. For $\alpha > 2$ there would be impurity outflux, an effect which is called temperature screening. The physics of this effect in plane geometry is, as can easily be seen from the calculations, the v-dependence of the collision frequency ν_Z. If ν_Z were independent of v, impurities could never be thrown out by temperature gradients.

B. TOROIDAL GEOMETRY

1. PFIRSCH-SCHLÜTER REGIME

In toroidal geometry we have to distinguish whether collisions are frequent or rare. The macroscopic theory applies if the mean free path of the particles is shorter than the connection length qR after which a field line has appreciably changed its meridional position on a magnetic surface. In this case, the only problem in establishing an equilibrium is that, unlike in the plane geometry, currents parallel to $\underline{B}$ also have to be driven against the plasma resistance. This cannot be done directly by diffusion through a $\frac{1}{c}\,\underline{v} \times \underline{B}$ term in Ohm's law since such a term acts only perpendicularly to $\underline{B}$. We therefore need, in addition, an electric field to drive the parallel currents. Such fields arise automatically by charge separation because the diamagnetic current is not divergence-free. But such space-charge fields have to be pure poloidal fields because of axial symmetry. The projection of these fields on the field lines must drive $j_{\parallel}$ and is therefore given by $\eta_{\parallel} j_{\parallel} \approx \eta_{\parallel} q j_{diam}$, which holds if the toroidal field is the main confining field. The poloidal electric field therefore has the strength $\eta_{\parallel} q j_{diam} \cdot \frac{qR}{r}$. The component of this field perpendicular to $\underline{B}$ has to be transformed away by a $\frac{1}{c} \underline{v} \times \underline{B}$ term, otherwise it would drive a current $\frac{\eta_{\parallel}}{\eta_{\perp}} q \frac{qR}{r} j_{diam} >> j_{diam}$.

The velocity, being of the $\underline{E} \times \underline{B}$ type, has to be in the poloidal and mainly outward direction (Fig. 1) and is obviously of the order

$$v_p \approx \frac{\eta_{\parallel}}{\eta_{\perp}} 2 \frac{q^2 R}{r} v_{cl}$$

To ensure an almost divergence-free motion of the plasma with a stationary density profile, the plasma has to flow back along the lines of force, the necessary velocity being

$$v_{\parallel} \approx \frac{qR}{r} v_p$$

But a fraction r/R of the poloidal particle flux leaks out, giving rise to a diffusion velocity[1]

$$v_{PS} \approx \frac{r}{R} v_p \approx \frac{\eta_{\parallel}}{\eta_{\perp}} 2q^2 v_{cl} = q^2 v_{cl}$$

This velocity means[2] that the expansive energy corresponding to v_{pS} not only goes into Joule dissipation of the diamagnetic current but also into Joule dissipation of the secondary current, which is a factor $\frac{\eta_{\parallel}}{\eta_{\perp}} 2q^2$ larger than the first one.

This so-called Pfirsch-Schlüter factor q^2 is also present in heat conduction. The physics there, however, is very different from that for diffusion. It has to do with a heat flow:

$$\underline{q_{k\perp}} = \frac{5}{2} \frac{n_k T_k}{e_k B^2} (\underline{B} \times \nabla T_k)$$

---

[1] First derived by D. Pfirsch and A. Schlüter [2].

[2] In agreement with a more general discussion by M.D. Kruskal and R.M. Kulsrud [3].

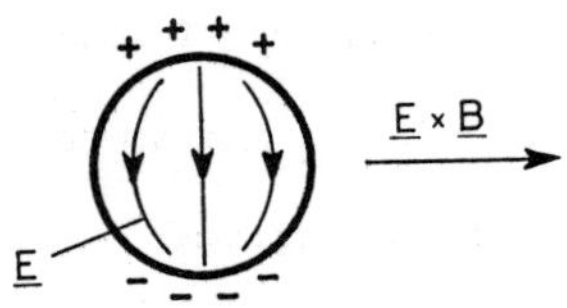

FIG.1. $\underline{E} \times \underline{B}$ drift caused by charge separation.

which is perpendicular to ∇T, which follows very easily from a formula like the one for δf above. Since it is not divergence-free, it leads to a heat flow parallel to $\underline{B}$ and consequently to a poloidal temperature gradient. With this gradient the above formula yields a radial heat flow which is then of the order of q^2 times the plane value.

The above discussion of particle diffusion and plasma convection does not lead to a condition that current reversal in the toroidal direction should not be allowed, as is sometimes said. Strictly speaking, however, the outlined theory applies only to low-ß stellarators. There might arise severe changes if one treats tokamak equilibria in a self-consistent way, and this shall now be done.

In a cylindrical coordinate system R,ϕ,Z one can describe an axisymmetric field configuration by

$$\underline{B}_p = \nabla\psi \times \nabla\phi \quad , \quad \underline{B}_t = F(\psi)\,\nabla\phi$$

The pressure p is a function of the flux function ψ alone, and because of the large heat conduction of the electrons parallel to $\underline{B}$ it is reasonable to assume that the temperature is also a function of ψ alone. For simplicity, we neglect anisotropy effects in the resistivity η . It thus holds that

$$p = p(\psi) \;, \quad \eta = \eta(\psi) \;, \quad n = n(\psi)$$

where n is the particle density.

The question is now what functions $p(\psi)$ and $F(\psi)$ are possible for self-consistent equilibria.

The answer follows from the question how the currents necessary for the equilibria can be driven. The current densities are

$$\frac{4\pi}{c} j_t = Rp' + \frac{1}{R}\frac{1}{2}(F^2)' \ , \quad \frac{4\pi}{c} j_p = F'\nabla\psi \times \nabla\phi = F'\underline{B}_p$$

The toroidal current can be driven partly by a loop voltage $2\pi U$ or the corresponding electric field U/R, partly by plasma convection. It thus follows that

$$\eta j_t = U/R + \frac{1}{c}(\underline{v} \times \underline{B})_\phi$$

With

$$(\underline{v} \times \underline{B})_\phi = (\underline{v}\times(\nabla\psi \times \nabla\phi))_\phi = \left[-\nabla\phi(\underline{v}\cdot\nabla\psi) + \nabla\psi(\underline{v}\cdot\nabla\psi)\right]_\phi = -\frac{1}{R}\underline{v}\cdot\nabla\psi$$

we find

$$\frac{1}{c}\frac{1}{R}\underline{v}\cdot\nabla\psi = U/R - \eta j_t = U/R - \eta(Rp' + \frac{1}{R}\frac{1}{2}(F^2)')\frac{c}{4\pi}$$

If $-\dot{N}(\psi)$ is the total number of particles diffusing through a magnetic surface ψ = const per unit time, then

$$n(\psi)\int_{\psi=\text{const}} \underline{v}\cdot d\underline{S} = -\dot{N}$$

It is

$$\int_{\psi=\text{const}} \underline{v}\cdot d\underline{S} = \oint_\psi 2\pi R\frac{\underline{v}\cdot\nabla\psi}{|\nabla\psi|}\,dl = 2\pi\oint_\psi \frac{\underline{v}\cdot\nabla\psi}{R}\,R\,\frac{dl}{B_p}$$

The volume enclosed by a flux surface is

$$V(\psi) = \int\!\!\int 2\pi R\,dl\frac{d\psi}{|\nabla\psi|} = \int^\psi d\psi \oint_\psi \frac{2\pi\,dl}{B_p}$$

and therefore

$$V'(\psi) = 2\pi\oint_\psi \frac{dl}{B_p}$$

Defining averages by

$$< G > \equiv \frac{2\pi \oint_{\psi} G \frac{dl}{B_p}}{2\pi \oint_{\psi} \frac{dl}{B_p}} = \frac{1}{V'(\psi)} \; 2\pi \oint_{\psi} G \frac{dl}{B_p}$$

we obtain

$$- \frac{\dot{N}}{nV'} = cU - \frac{\eta c^2}{4\pi} \left[p' <R^2> + \frac{1}{2} (F^2)' \right]$$

or

$$p' = \frac{1}{\eta <R^2>} \left[U + \frac{\dot{N}}{nV'} - \frac{\eta}{2} (F^2)' \right]$$

The poloidal current can be driven partly by a poloidal electric potential field $\underline{E} = - \nabla\Phi$ and partly again by convection. It thus follows that

$$\eta j_p = - \nabla\Phi + \frac{1}{c} (\underline{v} \times \underline{B})_p$$

For the last term here we find

$$(\underline{v} \times \underline{B})_p = (\underline{v} \times (\nabla\psi \times \nabla\phi))_p + (\underline{v} \times F\nabla\phi)_p = \nabla\psi \underline{v} \cdot \nabla\phi + F(\underline{v} \times \nabla\phi)$$

Therefore, it holds that

$$\frac{c\eta}{4\pi} F' \underline{B}_p = - \nabla\Phi + \nabla\psi \frac{1}{c} \underline{v} \cdot \nabla\phi + F \frac{1}{c} (\underline{v} \times \nabla\phi)$$

Uniqueness of the potential requires that

$$\oint_{\psi} \frac{\underline{B}_p \cdot \nabla\Phi}{B_p} dl = 0$$

With

$$\underline{B}_p \cdot (\underline{v} \times \nabla\phi) = (\underline{v} \times \nabla\phi) \cdot (\nabla\psi \times \nabla\phi) = \frac{1}{R^2} \underline{v} \cdot \nabla\psi$$

this condition yields

$$\oint dl \left[\frac{c\eta}{4\pi} F' B_p - \frac{1}{R^2} \frac{1}{B_p} \frac{1}{c} \underline{v} \cdot \nabla\psi \right] = \oint \frac{dl}{B_p} \left[\frac{c\eta}{4\pi} F' B_p{}^2 - \frac{1}{R^2} \frac{1}{c} \underline{v} \cdot \nabla\psi \right] = 0$$

Inserting the expression for $\underline{v} \cdot \nabla\psi$ and using the definition of averages, we get

$$\frac{c\eta}{4\pi} F' <B_p{}^2> - <\frac{F}{R}> \left[U/R - \frac{c\eta}{4\pi} (Rp' + \frac{1}{R} \frac{1}{2} (F^2)') \right] = 0$$

or

$$\frac{c\eta}{4\pi} F' \langle B_p^2 \rangle - FU \langle \frac{1}{R^2} \rangle + \frac{c\eta}{4\pi} p'F + \frac{c\eta}{4\pi} \frac{1}{2} (F^2)' F \langle \frac{1}{R^2} \rangle = 0$$

Multiplication with F and insertion of p' yields

$$\frac{c\eta}{4\pi} \frac{1}{2} (F^2)' \langle B_p^2 \rangle - F^2 \langle \frac{1}{R^2} \rangle \left[U - \frac{c\eta}{4\pi 2} (F^2)' \right] + F^2 \frac{1}{\langle R^2 \rangle} \left[U + \frac{\dot{N}}{nV'} \frac{c\eta}{4\pi 2} (F^2)' \right] = 0$$

or

$$\frac{c\eta}{4\pi 2} (F^2)' \langle B_p^2 \rangle = \frac{c\eta}{4\pi} p' F^2 \left[\langle R^2 \rangle \langle \frac{1}{R^2} \rangle - 1 \right] + \frac{\dot{N}}{cnV'} F^2 \langle \frac{1}{R^2} \rangle$$

From this we obtain with

$$F^2 \langle \frac{1}{R^2} \rangle = \langle \left(\frac{F}{R} \right)^2 \rangle = \langle B_t^2 \rangle$$

$$\dot{N} = \frac{nV'}{\langle B_t^2 \rangle} \frac{c^2 \eta}{4\pi} \left[\frac{1}{2} (F^2)' \langle B_p^2 \rangle - p' \langle B_t^2 \rangle \left[\langle R^2 \rangle - \langle \frac{1}{R^2} \rangle^{-1} \right] \right]$$

Multiplication by $\frac{\nabla\psi \cdot \nabla\psi}{|\nabla\psi|^2}$ yields, with $|\nabla\psi|^2 = R^2 B_p^2$,

$$nV' \nabla\psi = n\nabla V , \quad \dot{N} = - n\underline{v_D} \cdot \nabla V , \quad p' \nabla\psi = \nabla p, \text{etc.}$$

$$\underline{v_D} = - \frac{\eta}{\langle B_t^2 \rangle} \left[\frac{1}{2} \frac{\nabla R^2 B_t^2}{R^2} - \nabla p \frac{\langle B_t^2 \rangle}{B_p^2} \frac{1}{R^2} \left[\langle R^2 \rangle - \langle \frac{1}{R^2} \rangle^{-1} \right] \right]$$

For $\frac{1}{2} \frac{\nabla F^2}{R^2} \approx - \nabla p$ one has just the Pfirsch-Schlüter formula and it obviously does not matter how large $p/B_t^2/2$ is. Thus there is no condition that would not allow current reversal. There was only one relation between F and p which can be solved for p':

$$p' = - \frac{1}{2} (F^2)' \langle \frac{1}{R^2} \rangle \left[1 + \frac{\langle B_p^2 \rangle}{\langle B_t^2 \rangle} \right] + \frac{4\pi c U}{c^2 \eta} \langle \frac{1}{R^2} \rangle$$

Using this, the toroidal current density becomes

$$j_t = R \left(\frac{1}{2} (F^2)' \left(\frac{1}{R^2} - \langle \frac{1}{R^2} \rangle \right) - \frac{1}{2} (F^2)' \langle \frac{1}{R^2} \rangle \frac{\langle B_p^2 \rangle}{\langle B_t^2 \rangle} + \frac{4\pi c U}{c^2 \eta} \langle \frac{1}{R^2} \rangle \right)$$

U has to be chosen so as to guarantee q_o on axis greater than 1. If the magnetic surfaces close to the magnetic axis have circular cross-sections, then

$$\frac{4\pi}{c} j_{to} = \frac{2B_t^{\,o}}{R_o q_o} \quad \text{and therefore} \quad \left[\frac{4\pi cU}{c^2\eta} \left\langle\frac{1}{R^2}\right\rangle\right]_o = \frac{4\pi j_t^{\,o}}{cR_o} = \frac{2B_t^{\,o}}{R_o^{\,2} q_o}$$

Current reversal can occur if $\frac{4\pi cU}{c^2\eta}\left\langle\frac{1}{R^2}\right\rangle$ becomes of the order of $\frac{1}{2}(F^2)'\left(\frac{1}{R^2} - \left\langle\frac{1}{R^2}\right\rangle\right)$ somewhere. Since

$$\frac{1}{R^2} - \left\langle\frac{1}{R^2}\right\rangle \sim \frac{2r}{R}\frac{1}{R^2}, \quad \frac{1}{2}(F^2)' \approx R^2 p'$$

we get

$$\frac{1}{2}(F^2)'\left(\frac{1}{R^2} - \left\langle\frac{1}{R^2}\right\rangle\right) \approx p' \frac{2r}{R} = \frac{|\nabla p|\frac{2r}{R}}{RB_p} \approx \frac{\frac{2r}{a^2}p_o \frac{2r}{R}}{RB_p}$$

The ratio of this over the above quantity at r is therefore

$$\frac{\frac{2r\,p_o}{a^2}\frac{2r}{R}}{RB_p} \cdot \frac{R^2 q_o \eta}{2B_t \eta_o} = \frac{2r^2}{a^2}\frac{q_o p_o \eta}{B_p B_t \eta_o} = \frac{r^2}{a^2}\beta_p q_o \frac{B_p}{B_t}\frac{\eta}{\eta_o}$$

This is larger than 1 for $\beta_p > \frac{q}{q_o}\frac{R}{r}\frac{a^2}{r^2}\frac{\eta_o}{\eta}$.

Assuming $q/q_o = 2.5$, $T/T_o = \frac{1}{4}$, $r = a$, it holds that $\beta_p > 0.31 \frac{R}{a}$ and there is no problem with the existence of such solutions. To conclude this discussion, the relation for p is written by using vector quantities. We obtain after multiplying by $\nabla\psi$

$$\nabla p = -\left\langle\frac{1}{R^2}\right\rangle \frac{1}{2}\nabla(R^2 B_t^{\,2}) \left[1 + \frac{\langle B_p^{\,2}\rangle}{\langle B_t^{\,2}\rangle}\right] R \left\langle\frac{1}{R^2}\right\rangle \frac{4\pi cU}{c^2\eta}(R\nabla\phi \times \underline{B}_p)$$

2. BANANA REGIME

Even if there were no collisions at all, not all the particles in a tokamak plasma could move freely round the torus along the field lines. Since the magnetic field strength varies along a field

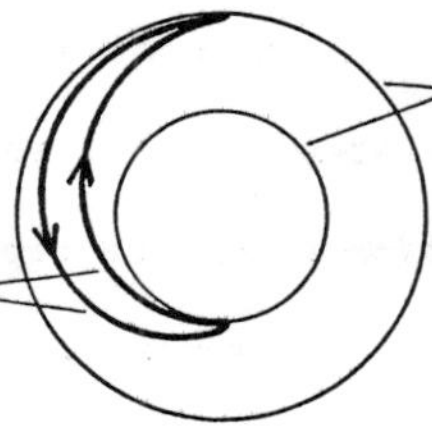

FIG. 2. Banana orbits.

line over a length after which a field line has appreciably changed its meridional position on a magnetic surface, i.e. after a length of the order qR, a particle sees magnetic mirrors at a distance qR. The strength of the mirrors $\Delta B/B$ is given by the inverse aspect ratio

$$\Delta B/B \approx a/R$$

There are therefore particles trapped between such mirrors according to the law of energy conservation:

$$\mu B + \frac{1}{2} m v_{\parallel}^2 = \text{const or } \mu \Delta B + \Delta \frac{1}{2} m v_{\parallel}^2 = 0$$

for which it holds that

$$|\Delta \frac{1}{2} m v_{\parallel}^2| = (\frac{1}{2} m v_{\parallel}^2)_{max}$$

Since $\mu = \frac{1}{2} m v_{\perp}^2/B$, this means that

$$\frac{v_{\parallel}^2}{v_{\perp}^2} = \frac{\Delta B}{B} = \frac{a}{R} << 1$$

Such particles drift essentially in the vertical direction with a velocity

$$v_{Drift} = c \frac{m v_{\perp}^2/R}{eB}$$

During the time the particle flies from one mirror to the other, that is, the time $qR/v_{\parallel}$, the particle moves a distance

$$\delta = v_{Drift}\, qR/v_{\parallel} = \frac{m v_{\perp} c}{eB} q \frac{v_{\perp}}{v_{\parallel}} = r_g q \sqrt{\frac{R}{a}}$$

out of a magnetic surface in the vertical direction. This is the thickness of the banana-like orbits (Fig. 2).

If there are few collisions, the first to occur is a reversal of $v_{\parallel}$, since $v_{\parallel}$ is much smaller than $v_{\perp}$. This means that an inner part of a banana orbit becomes an outer one or vice versa, i.e. the particle does steps of the order of the banana thickness δ . The banana thickness therefore replaces the gyroradius in plane geometry. The time to reverse $v_\parallel$ is, however, not the usual mean free time but is shorter than this by the factor $\frac{v_\parallel^2}{v^2}$. Thus we have to use a trapped particle collision frequency

$$\nu_t = \frac{v^2}{v_\parallel^2}\nu \approx \frac{R}{a}\nu$$

The number of such trapped particles is proportional to the $v_\parallel$ interval given by the trapping condition, i.e.

$$n_t = n\frac{v_\parallel}{v} = n\sqrt{\frac{a}{R}}$$

A stochastic process with δ as step size then yields the diffusion coefficient[3]

$$D_B = \delta^2\nu_t\frac{n_t}{n} = r_g{}^2q^2\frac{v_\perp{}^2}{v_\parallel{}^2}\frac{v^2}{v_\parallel{}^2}\nu\frac{v_\parallel}{v} = r_g{}^2\nu q^2\left(\frac{R}{a}\right)^{3/2}$$

$$= D_{PS}\ (R/a)^{3/2}$$

This derivation is valid as long as particle trapping is not inhibited by collisions, i.e. for $\nu_t qR/v_\parallel < 1$ or

$$\nu\frac{v^2}{v_\parallel{}^3}qR = \frac{qR}{\lambda}\left(\frac{v}{v_\parallel}\right)^3 = \frac{qR}{\lambda}A^{3/2} < 1 \quad \text{or } \lambda > A^{3/2}qR$$

where λ is the mean free path. Thus there is a regime left:

$$qR < \lambda < A^{3/2}\ qR$$

Since

$$D_B,\ D_{PS} \sim \frac{1}{\lambda}$$

one has

$$D_B(\lambda = A^{3/2}qR) = D_{PS}(\lambda = qR)$$

One therefore expects the following behaviour of D(shown in Fig. 3).

---

[3] First derived by A.A. Galeev and R.Z. Sagdeev [4].

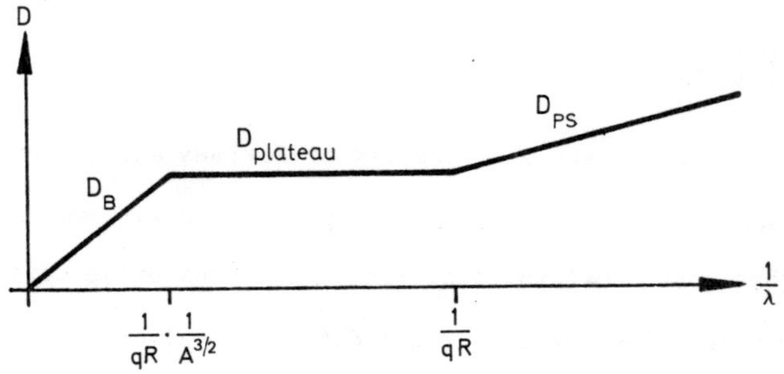

FIG.3. Diffusion coefficient as function of the inverse mean free path (idealized).

The inner part forms a plateau and is therefore called the plateau regime. In reality, there is a smooth transition from the banana to the Pfirsch-Schlüter regime.

Since the stochastic process including banana orbits is only modified by geometrical factors compared to that with gyro-orbits, ambipolar diffusion again holds.

There are two accompanying effects which are of importance:

1) the so-called bootstrap current,

2) the Ware effect.

There is, so to speak, an induction effect of the high diffusion velocity leading to a current density in the toroidal direction

$$j_B = \frac{1}{\eta} B_p \frac{v_B}{c} = - \frac{c}{B_p} \frac{dp}{dr} \left(\frac{r}{R}\right)^{1/2}$$

The corresponding current in the poloidal direction does not occur because of a special kind of viscosity. An important consequence of this current is as follows. Insertion of this current into Maxwell's equation yields

$$\frac{1}{r} \frac{d}{dr} r B_p = \frac{4\pi}{c} \frac{c}{B_p} \frac{dp}{dr} \left(\frac{r}{R}\right)^{1/2}$$

or

$$\beta_p = \frac{< p >}{B_p^2/8\pi} \approx A^{1/2}$$

I want to mention here a discussion of Borrass and the author [5]. In these considerations it is stated and proved for axisymmetric equilibria that under certain conditions the diffusion velocity should never exceed the magnetic field diffusion velocity in a plasma with finite resistivity. Assuming this to be true, one easily obtains a relation for an upper limit of ß in the following way. We had found above the relation valid in plane geometry

$$v_{cl} = \frac{1}{2}\,\beta\,v_{magn}$$

The magnetic diffusion described by v_{magn} does not depend on the geometry. Thus, from the condition

$$v_{magn} > v_D$$

one obtains

$$\beta < \frac{2v_{cl}}{v_D}$$

For the banana regime this yields $\beta < \frac{1}{A^{3/2}q^2}$ or with $\beta = \frac{1}{q^2A^2}\beta_{pol}$ one gets $\beta_{pol} < A^{1/2}$, in agreement with the result just derived. Pseudo-classical diffusion is given by $v_D \sim q^2A^2v_{cl}$, and therefore $\beta_{pol} < 1$, as is already well known. Pfirsch-Schlüter diffusion is expressed by $v_D \approx q^2v_{cl}$, and therefore $\beta_{pol} < A^2$, which would be much larger than often assumed, and which is probably not in contradiction to an exact theory as outlined in Section 1a. If this ß relation is correct, it would lead to extremely severe restrictions as to the permissible anomalous diffusion rates if these are not caused by a corresponding anomalous resistivity.

The Ware effect is the pendant to the bootstrap current in the sense of Onsager's relation and states that the usual E/B drift is replaced by

$$v_E = c \frac{E}{B_{pol}}$$

3. IMPURITY TRANSPORT

For impurities in the Pfirsch-Schlüter regime, the impurity transport is similar to that in plane geometry but is modified by a Pfirsch-Schlüter factor [6 - 8], i.e. there is impurity influx as long as the temperature profile is flatter than given by T n^2. Strictly speaking, this is true only if

$$U^2_{th_H} \tau_{D_H} \tau_{M_H} \gtrsim (qR)^2$$

as shown by Samain and Werkoff [8]. τ_{D_H} is the deflection time and τ_{M_H} the Maxwellization time for hydrogen ions. This is likely to be the case for the hydrogen ions in the banana or plateau regime. If the inequality is the other way around, temperature screening vanishes.

If $n_Z Z^2/n_i Z_i^2 > 1$, the ion flux is enhanced by a factor of m_i/m_e over the pure plasma case and the ion heat conduction becomes about $A^{3/2}$ times the ion heat conduction in the pure case [7], but the electron flux remains unchanged.

Calculations in which the impurities are also assumed to be in the banana or plateau regime were first made by Hinton and Moore [9].

If the impurities are in the banana regime, there is temperature screening as in the Pfirsch-Schlüter regime. If, however, the impurities are in the plateau regime, no temperature screening occurs. This might be a special problem for the α-particles. But for low Z and low mass impurities, long mean free path theories usually do not apply, and so one has to consider the situation as being open for such species.

Finally, I present a short description of the methods used:

1. Use of drift kinetic equation with Fokker-Planck collision term;

2. Zero-order solutions = local Maxwellian distributions;

3. Linear perturbation theory with respect to ∇n, ∇T, $\underline{E}$, $T_i - T_e$.

This is similar to the procedure decribed above for the plane case. The linear equations to be solved here are, however, much more complicated. A very elegant method for solution was used by Rosenbluth, Hazeltine and Hinton [10], who proved the equivalence of the linear equations with a principle of minimum entropy production for the banana regime. A modified variational principle was derived by Samain and Werkoff [8] and applied to plasmas with impurities in the collisional regime. The entropy production formula can be used, on the other hand, in the sense of Onsager's theory.

In the collisional regime one can, of course, do a momentum expansion [11], where it is generally necessary to go one order beyond Grad's 13-moment method.

REFERENCES

[1] TAYLOR, J.B., Phys. Fluids **4** (1961) 1142.

[2] PFIRSCH, D., SCHLÜTER, A., Max-Planck-Institut für Physik und Astrophysik, Munich, Rep. MPI/PA/7/62 (1962).

[3] KRUSKAL, M.D., KULSRUD, R.M., Phys. Fluids **1** (1958) 265.

[4] GALEEV, A.A., SAGDEEV, R.Z., Zh. Ehksp. Teor. Fiz. **53** (1967) 348.

[5] BORASS, K., PFIRSCH, D., Max-Planck-Institut für Plasmaphysik, Garching, Rep. IPP 6/155 (1977).

[6] RUTHERFORD, P.H., Phys. Fluids **17** (1974) 1782.

[7] HIRSHMAN, S.P., MIT Rep. PRR-7513 (1975).

[8] SAMAIN, A., WERKOFF, F., Nucl. Fusion **17** (1977) 53.

[9] HINTON, F.L., MOORE, T.B., Nucl. Fusion **14** (1974) 639.

[10] ROSENBLUTH, M.N., HAZELTINE, R.D., HINTON, F.L., Phys. Fluids **15** (1972) 116.

[11] HIRSHMAN, S.P., Princeton Plasma Physics Lab. Rep. PPPL-1291 (1976).

MHD theory

NON-LINEAR NUMERICAL ALGORITHMS FOR STUDYING TEARING MODES*

B.V. WADDELL**, M.N. ROSENBLUTH
Institute for Advanced Study,
Princeton, New Jersey

D.A. MONTICELLO, R.B. WHITE
Princeton Plasma Physics Laboratory,
Princeton, New Jersey

B. CARRERAS***
Oak Ridge National Laboratory,
Oak Ridge, Tennessee,
United States of America

Abstract

NON-LINEAR NUMERICAL ALGORITHMS FOR STUDYING TEARING MODES.

The numerical methods that have recently been developed to study the non-linear evolution of tearing modes in tokamaks are summarized. The essential features of tearing modes can be described by the resistive MHD equations. The numerical algorithms described here are based on a reduced set of two-dimensional resistive MHD equations that are numerically tractable. Two distinct types of numerical methods are described in detail. In the first method, referred to as the MASSLESS algorithm, the inertia is neglected. On the other hand, in the second method, referred to as the MASS algorithm, the inertia is retained and consequently the scheme is capable of handling a larger variety of problems. Codes based on these two algorithms give similar results for the non-linear evolution of the m = 2 tearing mode.

I. INTRODUCTION

The research efforts of the last decade clearly show that the tokamak is the most promising magnetic containment device for attaining controlled nuclear fusion. Furthermore, during this period, experimental and theoretical evidence has been accumulating in support of the hypothesis that resistive instabilities or "tearing" modes [1] are an important aspect of tokamak discharges. For example, tearing modes could be responsible for the long-wavelength temperature oscillations in the plasma interior[2]; these oscillations are characterized by poloidal mode numbers m = 1 and m = 2. The essential features of

---

* Research sponsored in part by USERDA under contract with the Union Carbide Corporation.

** Present address: Oak Ridge National Laboratory, Oak Ridge, Tennessee, USA.

*** Visitor from Junta de Energía Nuclear, Madrid, Spain.

tearing modes can be described by the relatively simple set of resistive MHD equations.[1] These equations are so complex, however, that in general they must be solved numerically [3]. Furthermore, an analytic treatment of the nonlinear theory of the tearing mode for $m > 1$ shows that once its amplitude is large enough to be observable, then its time evolution is highly nonlinear [4]. Consequently, a desirable requirement of a numerical algorithm for studying tearing modes is that it be nonlinear.

In this article, two distinct numerical algorithms for studying the nonlinear evolution of tearing modes are presented. Both algorithms are based on a simplified (or reduced) set of MHD equations that make the problem tractable numerically [5]. The reduced equations are obtained by making a series of sensible approximations that reduces the number of independent and dependent variables and eliminates the fastest MHD time scale. The fastest MHD time scale is $\tau_H = r_w/V_A$, where r_w is the minor radius of the torus and V_A is the Alfvén speed. The elimination of this time scale is particularly important for resistive instabilities because they evolve slowly on a time that is much longer than the MHD time. In one of the numerical algorithms, the MHD time scale is really completely eliminated because the inertia term in the momentum equation is ignored. This algorithm, which is referred to as the MASSLESS algorithm, is capable of following the nonlinear evolution of a single tearing mode with $m > 1$ on the resistive time scale. The other algorithm, referred to as the MASS algorithm, retains the inertia term and is capable of following the evolution of any mode on the slowest MHD time scale, i.e. the time scale $\tau_{Hp} = R/V_A$, where R is the major radius of the torus.

In Section II, the approximations required to derive the reduced set of resistive MHD equations are discussed. These reduced equations are sufficient to describe the evolution of perturbations of a given pitch in a low-β fluid; toroidal effects are neglected, i.e. cylindrical geometry is employed. In Section III, the equations are reduced further by ignoring the inertia, and a numerical algorithm for solving these MASSLESS equations is described. The inertia is retained, however, in Section IV, and the numerical stability properties of various algorithms for solving the full set of MASS equations are discussed. Finally, the results obtained using the MASSLESS and MASS algorithms are compared in Section V.

II. REDUCTION OF THE RESISTIVE MHD EQUATIONS

The analysis begins with the standard set of resistive magnetohydrodynamic equations:

$$\underline{J} = \frac{1}{\mu_0} \underline{\nabla} \times \underline{B} \qquad \text{(Ampère's Law)}$$

$$\frac{\partial \underline{B}}{\partial t} = -\underline{\nabla} \times \underline{E} \qquad \text{(Faraday's Law)}$$

$$\underline{\nabla} \cdot \underline{B} = 0$$

$$\rho \frac{d\underline{V}}{dt} = \rho \frac{\partial \underline{V}}{\partial t} + \underline{V} \cdot \underline{\nabla V} = -\underline{\nabla} P + \underline{J} \times \underline{B} \qquad \text{(Momentum Balance)}$$

$$\frac{\partial}{\partial t} \rho + \underline{\nabla} \cdot (\rho \underline{V}) = 0 \qquad \text{(Continuity Equation)}$$

$$\underline{E} + \underline{V} \times \underline{B} = \eta \underline{J} \qquad \text{(Ohm's Law)}$$

$$\frac{d}{dt} (P\rho^{-\gamma}) = 0 \qquad \text{(Example of an Equation of State)}$$

Here $\underline{J}$ is the current density, μ_o is the magnetic permeability (rationalized MKS units are employed), $\underline{B}$ is the magnetic field, $\underline{E}$ is the electric field, ρ is the mass density, $\underline{V}$ is the fluid velocity, P is the pressure, and γ is the adiabatic constant. The resistivity η is usually a specified function of space and time.

In order to reduce this set of equations to a tractable form, the following approximations are made:

a) Cylindrical geometry. Toroidal effects are ignored. The standard cylindrical coordinate system is employed with r, θ, and z designating, respectively, the radial, poloidal and toroidal coordinates.

b) $\varepsilon \equiv 2\pi r_W/L << 1$. This is the standard tokamak ordering where L = length of the cylinder = $2\pi R$ and r_w and R are respectively the minor and major radii. Since the pitch of the field lines is of order unity, this ordering corresponds to $B_\theta/B_z \sim \varepsilon$.

c) B_z is constant in space and time. This approximation is consistent with the tokamak ordering since $\tilde{B}_z \lesssim |\tilde{B}_\perp| \lesssim B_\theta \sim \varepsilon B_z$, where the tildes denote perturbations and $\perp$ means perpendicular to $\hat{z}$. The main purpose of this approximation is to eliminate the fast time scale $\tau_H = r_W/V_A$, the characteristic time for Alfvén waves propagating across the magnetic field, where V_A is the Alfvén speed. Then, the fastest time scale remaining in the equations is R/V_A, the characteristic time scale for Alfvén waves propagating along the magnetic field.

d) Helical symmetry. All quantities are assumed to be initially spatial functions of only r and $\tau = m\theta + kz$, where m and k are respectively the poloidal and toroidal mode numbers. In terms of the commonly used toroidal mode number n, k = n/R. Clearly, this symmetry is preserved as the system evolves.

e) $\beta \sim \varepsilon^2$. Here, of course, β is the ratio of the plasma pressure and the magnetic pressure. This ordering is valid for present tokamak parameters.

In addition to simplifying the algebra, approximations a) and b) are necessary for the consistency of approximations c) and d). Approximations c) and d) permit the introduction of the helical flux function ψ:

$$\underline{B} = \underline{\nabla}\psi \times \hat{z} + B_z\hat{h}$$

Here, $\hat{h} = \hat{z} - \delta\hat{\theta}$ is the unit vector along the helix defined by τ and $\delta \equiv kr/m$. It can be shown that $\psi = \hat{h} \cdot \underline{A}$, where $\underline{A}$ is vector potential of the magnetic field. In addition to allowing the magnetic field to be expressed in terms of a single scalar quantity, the introduction of ψ ensures that $\underline{\nabla} \cdot \underline{B} = 0$, which means that field lines should not wander spuriously. Orderings b) and e) and the z-component of the momentum balance equation imply that V_z is of order $\varepsilon|\underline{V}_\perp|$ and hence can be neglected. Furthermore, approximation c) and the z component of Faraday's law imply that $\underline{\nabla} \cdot \underline{V}$ is of order $\varepsilon^2|\underline{V}_\perp|/r_W$ and thus can also be neglected. Consequently, we can introduce the stream function A for the velocity:

$$\underline{V} = \underline{\nabla}A \times \hat{z}$$

Notice that the necessary equation of state in this ordering is $\underline{\nabla} \cdot \underline{V} = 0$. Then Faraday's and Ohm's laws imply that

$$\frac{d\psi}{dt} = \frac{\partial\psi}{\partial t} + \underline{V} \cdot \underline{\nabla}\psi = -\eta J_z/S \tag{1}$$

Here, and in the following, all derivatives are in the plane perpendicular to the z-axis. (Although we take into account helical symmetry, it is convenient to write the equations in the z = 0 plane.) All lengths are normalized to r_W; the time is normalized to $\tau_{Hp} = m(\mu_o\bar{\rho})^{1/2}/kB_z$, where $\bar{\rho}$ is the characteristic mass density; the flux function ψ is normalized to $kB_z r_W^2/m$; J_z is normalized to $kB_z/(\mu_o m)$; and η is normalized to the characteristic value $\bar{\eta}$. The quantity $S \equiv \tau_R/\tau_{Hp}$, where $\tau_R \equiv \mu_o r_W^2/\bar{\eta}$ is the resistive or "skin" time. The toroidal component of the current density can be obtained from Ampère's law:

$$J_z = -\nabla^2\psi - 2 \tag{2}$$

If the mass density is constant, taking the z-component of the momentum balance equation yields

$$\frac{d}{dt} U = \hat{z} \cdot [\underline{\nabla}\psi \times \underline{\nabla}J_z] \tag{3}$$

where

$$U \equiv \nabla^2 A \tag{4}$$

is the vorticity. Equations (1) - (4) constitute a set of reduced equations [5] which is sufficient to describe the fluid when approximations a) - e) are valid. The MHD equations have been reduced to two equations in two scalar unknowns with the conditions $\underline{\nabla} \cdot \underline{B} = 0$ and $\underline{\nabla} \cdot \underline{V} = 0$ automatically satisfied.

III. MASSLESS ALGORITHM

In some cases, it is possible to reduce the resistive MHD equations even further. Since Eq. (1) implies that $\partial/\partial t \sim \eta$, we might expect from Eq. (3) that $\partial \nabla^2 A/\partial t \sim \eta^2$, so that the inertia can be neglected. This is valid approximation in the nonlinear regime of a single magnetic island if $m > 1$. By the term "single magnetic island" it is meant that for a given ratio of m and n there is only one magnetic island because the safety factor q is a single-valued function of r. If q is double-valued, there can be two islands each of which is localized about a radius where $q = m/n$; we refer to this case as a "double" magnetic island. In order to describe the linear regime, an $m = 1$ mode, or a double magnetic island, the inertia must be retained.

If the inertia is ignored, Eq. (3) implies that the current density must be a function only of ψ. Ignoring the inertia, however, requires that the velocity $\underline{V}$ must also be eliminated from Eq. (1). Consequently, we average over flux contours:

$$\langle F \rangle_\psi \equiv \oint_\psi \frac{d\ell}{\nabla\psi} F \quad \oint_\psi \frac{d\ell}{\nabla\psi}$$

where F is an arbitrary function. In resistive time units, the equations become [6]

$$\nabla^2\psi = -J_z(\psi) - 2 \equiv f(\psi)$$

and

$$\left\langle \frac{\partial\psi}{\partial t} \right\rangle_\psi = -\langle\eta\rangle_\psi J_z(\psi) + E_{zw}$$

Here, we have replaced ψ by $\psi - tE_{zw}$ where E_{zw} is the electric field at the wall, so that the boundary condition is $\psi(r_w) = 0$. Clearly, these equations evolve only on the resistive time scale.

The primary numerical difficulty in solving these equations is ensuring that $\nabla^2\psi = f(\psi)$. In order to accomplish this, we employ an iteration scheme that involves expanding $\nabla^2\psi$ in a polynomial series:

$$\nabla^2\psi^{t+\Delta t}_{N+1} = \sum_k C_k P_k(\psi_N^{t+\Delta t})$$

$$= -2 - J_z^{t+\Delta t}(\psi_N^{t+\Delta t})$$

Here, ψ is known at time t and must be found at time $t + \Delta t$. The iteration index is denoted by N, and P_k is some polynomial (for example, Legendre). The coefficients C_k depend on both the time and iteration number. Thus, if we define

$$G_k(\psi_N^{t+\Delta t}) \equiv \nabla^{-2} P_k(\psi_N^{t+\Delta t})$$

then

$$\psi_{N+1}^{t+\Delta t} = \sum_k C_k G_k(\psi_N^{t+\Delta t})$$

In practice, ψ is a multi-valued fucntion and different polynomial expansions are required for different regions of the plasma.

The coefficients C_k for each iteration are determined from the flux evolution equation, which is advanced implicitly in time. We have

$$\langle \psi_{N+\ell}^{t+\Delta t} \rangle_{\psi_{N,\ell}^{t+\Delta t/2}} = \langle \psi^t \rangle_{\substack{t+\Delta t/2\\ N,\ell}} - \Delta t \eta(\psi_{N,\ell}^{t+\Delta t/2}) \, J_{zN,\ell}^{t+\Delta t/2} + \Delta t E_{zw}$$

where ℓ denotes the particular flux contour,

$$\psi_N^{t+\Delta t/2} = \frac{1}{2} (\psi_N^{t+\Delta t} + \psi^t)$$

$$\eta(\psi_{N,\ell}^{t+\Delta t/2}) = \langle \eta \rangle_{\psi_{N,\ell}^{t+\Delta t/2}}$$

and

$$J_{zN,\ell}^{t+\Delta t/2} = \langle \frac{1}{2} [J_z^{t+\Delta t} (\psi_N^{t+\Delta t}) + J_z^t(\psi^t)] \rangle_{\psi_{N,\ell}^{t+\Delta t/2}}$$

The corresponding matrix equation for the coefficients is

$$\sum_k H(\ell|k) C_k = R(\ell)$$

where

$$H(\ell|k) = \langle G_k(\psi_N^{t+\Delta t}) - \frac{\Delta t}{2} \eta(\psi_{N,\ell}^{t+\Delta t/2}) \, P_k(\psi_N^{t+\Delta t}) \rangle_{\psi_{N,\ell}^{t+\Delta t/2}}$$

$$R(\ell) = \langle \psi^t \rangle_{\psi_{N,\ell}^{t+\Delta t/2}} - \frac{\Delta t}{2} \eta(\psi_{N,\ell}^{t+\Delta t/2}) [\langle J_z^t(\psi_N^t) \rangle_{\psi_{N,\ell}^{t+\Delta t/2}} -2] + \Delta t E_{zw}$$

Rather than satisfying the preceding equation for each value of ℓ, it is desirable to satisfy it in a least-squares sense by requiring

$$V = \sum_{\ell}[R(\ell) - \sum_{k} H(\ell|k)C_k]^2$$

to be a minimum. Of course, the number of contours must be greater than or equal to the number of functions. Then, the equation to be solved is

$$\sum_{k} M(k'|k)C_k = S(k')$$

where

$$M(k'|k) \equiv \sum_{\ell} H(\ell|k')H(\ell|k)$$

and

$$S(k') \equiv \sum_{\ell} H(\ell|k')R(k')$$

In principle the above scheme is extremely efficient because the time step can be quite large. In practice, however, a large number of polynomials is often necessary in order to represent ψ accurately and, consequently, a large amount of time is required to repeatedly invert Poisson's equation.

IV. MASS ALGORITHMS

In this section, we analyse various numerical algorithms for solving the reduced equations including the inertia. If we use a time advancement scheme that is totally explicit, we expect to encounter the standard diffusion limitation on the time step, i.e. $\Delta t < (\Delta x)^2/(S\eta)$, where Δx is the normalized grid spacing and Δt is the time step in MHD units. Because Δx must be small in order to resolve the singularities in the equations and because in cylindrical geometry $\Delta x \sim r\Delta\theta$ becomes very small near the origin, this restriction is unacceptable. In order to eliminate the restriction, we must employ an implicit scheme.

In the following paragraphs, we consider both a partially implicit scheme [7] and a fully implicit scheme. For the purposes of comparison, we also consider the details of a fully explicit scheme. In each scheme, once U is obtained at a particular time, Eq. (4) must be inverted in order to obtain A and thus $\underline{V}$. Also, we employ a two-step algorithm in each scheme in order to center the time advancement. To simplify the notation, we define

$$Z \equiv Z_r + Z_\theta = \hat{z} \cdot [\underline{\nabla}\psi \times \underline{\nabla} J_z]$$

$$Z_r \equiv -\left(\frac{1}{r}\frac{\partial}{\partial\theta}\nabla^2\psi\right)\frac{\partial\psi}{\partial\theta}$$

and

$$Z_\theta \equiv (\frac{\partial}{\partial r} \nabla^2\psi)(\frac{1}{r}\frac{\partial\psi}{\partial\theta}).$$

We also denote the finite difference forms of $\frac{\partial}{\partial r}$, $\frac{1}{r}\frac{\partial}{\partial\theta}$, $\frac{1}{r}\frac{\partial}{\partial r} r \frac{\partial}{\partial r}$ and $\frac{1}{r^2}\frac{\partial^2}{\partial\theta^2}$ by, respectively, $\frac{\delta}{\delta r}$, $\frac{\delta}{\delta\theta}$, $\frac{\delta^2}{\delta r^2}$ and $\frac{\delta^2}{\delta\theta^2}$.

a) Partially Implicit

Step 1:

$$U^{t+\Delta t/2} = U^t + \frac{\Delta t}{2}[Z - \underline{V} \cdot \nabla\psi]^t$$

$$\psi^{t+\Delta t/2} = \psi^t - \frac{\Delta t}{2}[V_r\frac{\delta\psi}{\delta r} - \eta\frac{\delta^2\psi}{\delta r^2}]^{t+\Delta t/2} - \frac{\Delta t}{2}[V_\theta\frac{\delta\psi}{\delta\theta} - \eta\frac{\delta^2\psi}{\delta\theta^2} - 2\eta]^t$$

Step 2:

$$U^{t+\Delta t} = U^t + \Delta t[Z - \underline{V} \cdot \underline{\nabla}U]^{t+\Delta t/2}$$

$$\psi^{t+\Delta t} = \psi^{t+\Delta t/2} - \frac{\Delta t}{2}[V_r\frac{\delta\psi}{\delta r} - \eta\frac{\delta^2\psi}{\delta r^2}]^t - \frac{\Delta t}{2}[V_\theta\frac{\delta\psi}{\delta\theta} - \eta\frac{\delta^2\psi}{\delta\theta^2} - 2\eta]^{t+\Delta t/2}$$

Here, the vorticity is stepped explicitly. The flux function, however, is stepped implicitly by the alternating direction method; in the first step the r derivatives are implicit and in the second step the θ derivatives are implicit. In both steps, ψ at the forward time can be found by inverting a tridiagonal matrix.

b) Fully Implicit

Step 1:

$$U_{N+1}^{t+\Delta t/2} = U^t + \frac{\Delta t}{2}[Z_r - V_r\frac{\delta\psi}{\delta r}]_N^{t+\Delta t/2} + \frac{\Delta t}{2}[Z_\theta - V_\theta\frac{\delta\psi}{\delta\theta}]^t$$

$$\psi_{N+1}^{t+\Delta t/2} = \psi^t - \frac{\Delta t}{2}[V_r\frac{\delta\psi}{\delta r} - \eta\frac{\delta^2\psi}{\delta r^2}]_{N+1}^{t+\Delta t/2} - \frac{\Delta t}{2}[V_\theta\frac{\delta\psi}{\delta\theta} - \eta\frac{\delta^2\psi}{\delta\theta^2} - 2\eta]^t$$

Step 2:

$$U_{N+1}^{t+\Delta t} = U^{t+\Delta t/2} + \frac{\Delta t}{2}[Z_r - V_r\frac{\delta\psi}{\delta r}]^{t+\Delta t/2} + \frac{\Delta t}{2}[Z_\theta - V_\theta\frac{\delta\psi}{\delta\theta}]_N^{t+\Delta t}$$

$$\psi_{N+1}^{t+\Delta t} = \psi^{t+\Delta t/2} - \frac{\Delta t}{2}[V_r\frac{\delta\psi}{\delta r} - \eta\frac{\delta^2\psi}{\delta r^2}]^{t+\Delta t/2} - \frac{\Delta t}{2}[V_\theta\frac{\delta\psi}{\delta\theta} - \eta\frac{\delta^2\psi}{\delta\theta^2} - 2\eta]_{N+1}^{t+\Delta t}$$

Here, by iterating, both the vorticity and the flux are stepped implicitly using the alternating direction method; N denotes the iteration index. In the first step, the r derivatives are implicit; in the second step, the θ derivatives are implicit.

c) Fully Explicit

Step 1:	*Step 2:*
$U^{t+\Delta t/2} = U^t + \frac{\Delta t}{2}[Z - \underline{V} \cdot \underline{\nabla} U]^t$	$U^{t+\Delta t} = U^t + \Delta t[Z - \underline{V} \cdot \underline{\nabla} U]^{t+\Delta t/2}$
$\psi^{t+\Delta t/2} = \psi^t - \frac{\Delta t}{2}[\underline{V} \cdot \underline{\nabla}\psi + \eta J_z]^{t+\Delta t/2}$	$\psi^{t+\Delta t} = \psi^t + \Delta t[\underline{V} \cdot \underline{\nabla}\psi + \eta J_z]^{t+\Delta t/2}$

Here, both the vorticity and flux equations are stepped explicitly.

We employ the standard von Neumann numerical stability analysis to study the stability of these time advancement schemes. We linearize Eqs (1) and (3) and neglect all small terms in order to make the analysis tractable. Since only short wavelength instabilities are of interest, the analysis can be performed in slab rather than cylindrical geometry. Furthermore, for short wavelength instabilities, ψ_o' is approximately constant and hence Eq. (4) can be trivially inverted to give A. Here, ψ_o is the unperturbed flux and the prime denotes the radial derivative. In finite difference form, the equations become

$$\frac{\partial \tilde{A}}{\partial t} = -\psi_o' \frac{\delta \tilde{\psi}}{\delta y}$$

$$\frac{\partial \tilde{\psi}}{\partial t} + \frac{\delta \tilde{A}}{\delta y} \psi_o' = \frac{\eta}{S}\left[\frac{\delta^2 \tilde{\psi}}{\delta x^2} + \frac{\delta^2 \tilde{\psi}}{\delta y^2}\right]$$

where "o" subscripts denote unperturbed quantities, tildes denote numerical perturbations, and the rectangular coordinates x and y correspond respectively to r and θ. We take the numerical perturbations to have the form

$$\tilde{f} = f(t)\ e^{i(jk_x\Delta x + kk_y\Delta y)}$$

where j and k are the grid indices, k_x and k_y are the wave numbers, and Δx and Δy are the grid spacings. Then the equations reduce to

$$\frac{\partial A}{\partial t} = -i\,\frac{\gamma}{\Delta t}\,\psi$$

$$\frac{\partial \psi}{\partial t} = -i\,\frac{\gamma}{\Delta t}\,A - \frac{1}{\Delta t}(\alpha_x + \alpha_y)\psi$$

where

$$\gamma \equiv \frac{\Delta t \psi_o'}{\Delta y} \sin k_y \Delta y$$

$$\alpha_x \equiv 4\frac{\Delta t \eta}{S(\Delta x)^2} \sin^2(k_x \Delta x/2)$$

$$\alpha_y \equiv 4\frac{\Delta t \eta}{S(\Delta y)^2} \sin^2(k_y \Delta y/2)$$

First, we consider the partially implicit method. The linearized version of the time advancement equations becomes

$$A^{t+\Delta t/2} = A^t - \frac{1}{2} i\gamma\psi^t$$

$$\psi^{t+\Delta t/2} = \psi^t - \frac{1}{2}[i\gamma A + \alpha_x\psi]^{t+\Delta t/2} - \frac{1}{2}\alpha_y\psi^t$$

$$A^{t+\Delta t} = A^t - i\gamma\psi^{t+\Delta t/2}$$

$$\psi^{t+\Delta t} = \psi^{t+\Delta t/2} - \frac{1}{2}[i\gamma A + \alpha_x\psi]^{t+\Delta t/2} - \frac{1}{2}\alpha_y\psi^{t+\Delta t}$$

After some algebra, these equations can be written in matrix form as

$$\begin{bmatrix} A^{t+\Delta t} \\ \\ \psi^{t+\Delta t} \end{bmatrix} = \begin{bmatrix} \dfrac{2+\alpha_x-\gamma^2}{2+\alpha_x} & -\dfrac{i\gamma(2-\alpha_y-\gamma^2/2)}{2+\alpha_x} \\ \\ -\dfrac{4i\gamma}{(2+\alpha_x)(2+\alpha_y)} & \dfrac{(2-\alpha_x)(2-\alpha_y)-2\gamma^2}{(2+\alpha_x)(2+\alpha_y)} \end{bmatrix} \begin{bmatrix} A^t \\ \\ \psi^t \end{bmatrix} = \lambda \begin{bmatrix} A^t \\ \\ \psi^t \end{bmatrix}$$

Thus, the amplification factor λ is a solution of the eigenvalue equation

$$a\lambda^2 + b\lambda + c = 0$$

where

$$a \equiv (2 + \alpha_x)(2 + \alpha_y)$$

$$b \equiv -8 - 2\alpha_x\alpha_y + \gamma^2(4 + \alpha_y)$$

$$c \equiv -\gamma^2\alpha_y + (2 - \alpha_x)(2 - \alpha_y)$$

The scheme is stable if and only if $|\lambda| \leq 1$. We wish to determine the restriction imposed on Δt by this inequality. We must consider two possibilities: either $b^2 - 4ac < 0$ or $b^2 - 4ac > 0$. If $b^2 - 4ac < 0$, then $|\lambda| \leq 1$ implies that $a \geq c$; this inequality is always satisfied. On the other hand, if $b^2 - 4ac > 0$, then $|\lambda| \leq 1$ implies that $a + c \geq |b|$; i.e.

$$\gamma^2 \leq \frac{8 + 2\alpha_x\alpha_y}{2 + \alpha_y}$$

Since the maximum value of γ^2 is $(\Delta t\psi_0'/\Delta y)^2$, the restriction on the size of the time step is

$$\Delta t \leq \frac{\Delta y}{|\psi_0'|}\left[\frac{8}{2 + \alpha_y^m}\right]^{\frac{1}{2}}$$

where α_y^m is the maximum of α_y. In cylindrical geometry, $r\Delta\theta$ corresponds to Δy.

The linearized equations for the fully implicit scheme are

$$A^{t+\Delta t/2} = A^t - \frac{i}{2}\gamma\psi^{t+\Delta t/2}$$

$$\psi^{t+\Delta t/2} = \psi^t - \frac{1}{2}[i\gamma A + \alpha_x\psi]^{t+\Delta t/2} - \frac{1}{2}\alpha_y\psi^t$$

$$A^{t+\Delta t} = A^{t+\Delta t/2} - \frac{i}{2}\gamma\psi^{t+\Delta t/2}$$

$$\psi^{t+\Delta t} = \psi^{t+\Delta t/2} - \frac{1}{2}[i\gamma A + \alpha_x\psi]^{t+\Delta t/2} - \frac{1}{2}\alpha_y\psi^{t+\Delta t}$$

provided the iteration scheme converges. The eigenvalue equation for the amplification factor is

$$a\lambda^2 + b\lambda + c = 0$$

where

$$a \equiv (2 + \alpha_y)(2 + \alpha_x + \gamma^2/2)$$

$$b \equiv -8 - 2\alpha_x\alpha_y + 2\gamma^2$$

$$c \equiv (2 - \alpha_y)(2 - \alpha_x + \gamma^2/2)$$

In this case, both conditions $a \geq c$ and $a + c \geq |b|$ are trivially satisfied and there is no restriction on Δt.

Now, however, the convergence of the iteration scheme for the fully implicit method must be studied. For the first part of the time step, the equations in linearized form are

$$A_{N+1}^{t+\Delta t/2} = A^t - \frac{i}{2}\gamma\,\psi_N^{t+\Delta t/2}$$

$$\psi_{N+1}^{t+\ t/2} = \psi^t - \frac{1}{2}[i\gamma A + \alpha_x\psi]_N^{t+\Delta t/2} - \frac{1}{2}\alpha_y\psi^t$$

Then,

$$\psi_{N+1}^{t+\Delta t/2} = [\frac{2 - \alpha_y}{2 + \alpha_x}\ \psi^t - i\frac{\gamma}{2 + \alpha_x}A^t] - \frac{1}{2}\frac{\gamma^2}{2 + \alpha_x}\psi_N^{t+\Delta t/2}$$

We define $\delta\psi_N^{t+\Delta t/2} = \psi^{t+\Delta t/2} - \psi_N^{t+\Delta t/2}$, where $\psi^{t+\Delta t/2}$ is the actual value of ψ at the forward time. Then,

$$\delta\psi_{N+1}^{t+\Delta t/2} = -\frac{\gamma^2}{2(2 + \alpha_x)}\delta\psi_N^{t+\Delta t/2}$$

and to ensure convergence, we must require $\gamma^2 \leq 2(2 + \alpha_x)$. Therefore,

$$\Delta t \leq 2\frac{\Delta y}{|\psi_o'|}$$

Thus, the time step restriction for the iteration to converge is similar to the one for the partially implicit time advancement to be numerically stable.

The von Neumann stability analysis of the explicit scheme yields the requirements $\alpha_x + \alpha_y \leq 2$ and $\gamma^2 \leq 3$. The requirement that $\alpha_x + \alpha_y \leq 2$ is the standard diffusion limitation on the time step, and for typical parameters it is more severe than the requirement that $\gamma^2 \leq 3$. The corresponding restriction on the size of the time step is

$$\Delta t \leq \text{Min}\left\{ \frac{S}{2\eta}\left[\frac{(\Delta x)^2(\Delta y)^2}{(\Delta x)^2 + (\Delta y)^2}\right] , \frac{\sqrt{3}\,\Delta y}{|\psi_o'|} \right\}$$

Typical time steps for the two schemes are compared in Table I for various values of $\Delta \equiv \Delta x = \Delta y$ and S. We observe that usually the time step for the explicit method is much smaller than for the implicit methods. The time step for the fully implicit method can be larger than the time step for the partially implicit method by about a factor of four. However, we expect that the time step for the fully implicit method must be reduced by about a factor of two in order to ensure rapid convergence of the iteration. Then, since iterating requires at least twice as much computer time per step, the fully implicit method probably has little, if any, advantage over the partially implicit method. Consequently, the partially implicit method is the one that we have tested and employed.

TABLE I. TIME STEPS FOR THE MASS ALGORITHMS

	Fully implicit		Partially implicit		Fully explicit	
S	$\Delta = 10^{-2}$	$\Delta = 10^{-3}$	$\Delta = 10^{-2}$	$\Delta = 10^{-3}$	$\Delta = 10^{-2}$	$\Delta = 10^{-3}$
10^3	2	2	0.60	0.13	0.025	2.5×10^{-4}
10^4	2	2	1.1	0.27	0.25	2.5×10^{-3}
10^5	2	2	2.0	0.60	$\sqrt{3}$	0.025
10^6	2	2	2.0	1.1	$\sqrt{3}$	0.25

V. COMPARISON OF THE ALGORITHMS

We have built and used two codes based on the two algorithms described in this article. In Fig. 1 are plotted the results from both codes for the m = 2 magnetic island width as a function of time. The safety factor profile is single valued with q(o) = 1.6 and $q(r_w) = 4.1$. The resistivity is constant in time and proportional to the reciprocal of the unperturbed current density. The MASS code result is for S = 1.25×10^5. In typical runs, the number of θ grid points employed was approximately 12 and the number of radial grid points was between 60 and 120. It was necessary to localize the radial grid about the singular surface, particularly for small island widths. In the MASSLESS code, the radial grid could be expanded during the course of the run in order to more accurately follow the evolution of the

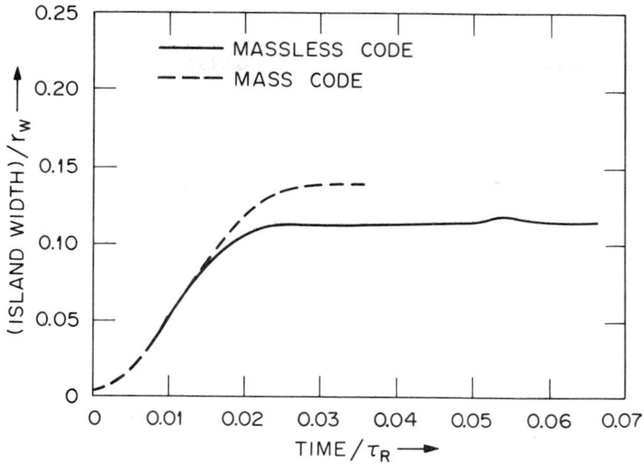

FIG.1. Comparison of the MASS and MASSLESS codes: m = 2 magnetic island width as a function of time.

growing island. We observe that in the nonlinear regime where the island width is larger than the tearing layer width and hence both codes should apply, the results are in agreement. Specifically, during the period when the island is growing linearly with time, the growth rates obtained from the two codes are essentially identical. The island widths at which saturation occurs differ somewhat. The magnitude of the difference is reasonable considering the delicate nature of the steady state that exists when the island width saturates.

The main advantage of the MASSLESS Code is that the cost per run is about a factor of (at least) four less than the MASS Code. However, unlike the MASSLESS Code, the MASS Code can be used to study the linear regime, the m = 1 tearing mode, and complicated island structures such as those produced by a double m = 2 tearing mode. More detailed results from these two codes have been presented elsewhere [8,9].

REFERENCES

[1] FURTH, H.P., KILLEEN, J., ROSENBLUTH, M.N., Phys. Fluids **6** (1963) 459.
[2] VON GOELER, S., STODIEK, W., SAUTHOFF, N., Phys. Rev. Lett. **33** (1974).
[3] FURTH, H.P., RUTHERFORD, P.H., SELBERG, H., Phys. Fluids **16** (1973) 1054.
[4] RUTHERFORD, P.H., Phys. Fluids **16** (1973) 1903.
[5] ROSENBLUTH, M.N., MONTICELLO, D.A., STRAUSS, H., WHITE, R.B., Phys. Fluids **19** (1976) 1987.
[6] GRAD, H., HU, P.M., STEVENS, E.C., Proc. Natl. Acad. Sci. **72** (1975) 3789.
[7] BISKAMP, D., WELTER, H., "Numerical studies of resistive instabilities", Plasma Physics and Controlled Nuclear Fusion Research 1976 (Proc. 6th Int. Conf. Berchtesgaden, 1976) **1**, IAEA, Vienna (1977) 579.
[8] WADDELL, B.V., et al., Nucl. Fusion **16** (1976) 528.
[9] WHITE, R.B., et al., "Non-linear tearing modes in tokamaks", Plasma Physics and Controlled Nuclear Fusion Research 1976 (Proc. 6th Int. Conf. Berchtesgaden, 1976) **1**, IAEA, Vienna (1977) 569.

MAGNETIC RECONNECTION IN A SPACE PLASMA

A.A. GALEEV, L.M. ZELENY
Space Research Institute,
Academy of Sciences of the USSR,
Moscow, USSR

Abstract

MAGNETIC RECONNECTION IN A SPACE PLASMA.

A review is presented of reconnection in a space plasma of collisionless magnetic field lines with special regard to the earth's magnetosphere. The collisionless tearing mode is considered to be the main mechanism for the merging of magnetic field lines. It is shown that the difference between the nose and tail reconnection is related to the difference in the geometries of magnetic field lines. Non-linear analysis of the collisionless tearing mode provides an estimate of the reconnection rate of interplanetary and geomagnetic field lines. Magnetospheric tail instability during substorms is discussed in terms of the tearing-mode development.

INTRODUCTION

It is well known that even small magnetic fields in large space scales can be the main energy source that drives many dramatic events in a space plasma. That is why the reconnection processes, such as the mechanism of the free magnetic field energy release, have received so much attention during recent decades. The importance of magnetic field line reconnection was first recognized for solar [1] and magnetospheric [2] physics. These days, it is widely discussed among astrophysicists and even appeared to be the mechanism to drive the disruptive instability in the thermonuclear device, the tokamak [3]. As in the case of collisionless shock structure, there is no unified theory of reconnection. Here we restrict ourselves to the simplest case of a very diffuse plasma slab immersed in a sheared magnetic field when the current velocity in a plasma is well below the thresholds of current instabilities. That is really the case for magnetic reconnection in the earth's magnetosphere and interplanetary space where the thickness of the plasma slabs is greater than the ion Larmor radius.

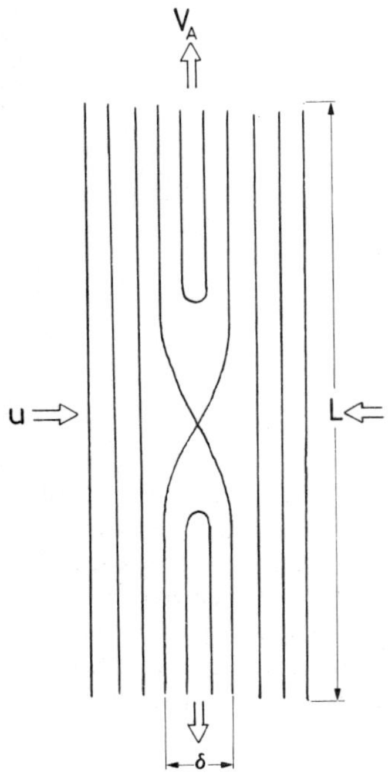

FIG.1. Parker-Sweet model of reconnection.

1. MHD MODELS OF RECONNECTION

The simplest MHD model of the magnetic field line reconnection in a plane plasma slab was developed by Parker and Sweet [4, 5] (see Fig. 1). It is based on the resistive dissipation of the plane current layer resulting in the magnetic field merging. Plasma entering the merging region and being frozen in the magnetic field is then streaming along the layer.

Momentum conservation demands that the streaming velocity is equal to the Alfvén velocity. A continuity equation defines one of the parameters that had remained unknown: δ is the width of the current layer, L the length of the plasma slab along the field lines, and u the merging rate:

$$uL \sim v_A \, \delta \tag{1}$$

In the resistive plasma, the merging rate and the layer width are related through the magnetic field diffusion coefficient:

$$u \sim \frac{D}{\delta}, \quad D \cong \frac{c^2}{4\pi\sigma} \tag{2}$$

Here σ is the plasma conductivity.

As a result we obtain from Eqs (1) and (2)

$$u = \left(\frac{c^2}{4\pi\sigma v_A L}\right)^{1/2} v_A \sim \frac{v_A}{Re_m^{1/2}} \tag{3}$$

One can think that the application of this MHD model to a cosmic plasma (i.e. solar flares, magnetospheric substorms, etc.) is only a question of the anomalous resistivity of the highly collisionless space plasma. But we should always bear in mind that in order to develop the plasma turbulence causing the anomalous resistivity it is necessary to exceed the current instability threshold (i.e. the current layer should be thin enough). But according to Eq. (1), since the length L is usually large, we shall get a very low power of the processes considered (see also Ref. [6]).

Let us consider in greater detail the earth's magnetosphere problems. According to Dungey's early model of the magnetosphere,there are two neutral lines (merging regions): one is at the magnetosphere nose and the other in the tail [2].

The power generated during the magnetospheric substorm comes from the energy of the magnetic field stored in the magnetospheric tail. According to the Parker-Sweet merging model,this power can be estimated as a rate of the magnetic energy inflow into the merging region:

$$\mathscr{P} = u_z L_y L_x \frac{B_x^2}{8\pi} \tag{4}$$

where $u_z = v_A \delta_z / L_x$, and δ_z is defined by the threshold current velocity V_{cr}:

$$\delta_z \cong \frac{c B_x}{4\pi e n_0 V_{cr}} \tag{5}$$

Since ions in the magnetospheric tail are much hotter than electrons, the critical current velocity can be estimated as [7]

$$v_{cr} \cong \left(\frac{T_\pm}{m}\right)^{\frac{1}{2}} \tag{6}$$

where ion temperature can be found from the pressure balance:

$$T_+ \sim \frac{B_x^2}{8\pi n_0} \tag{7}$$

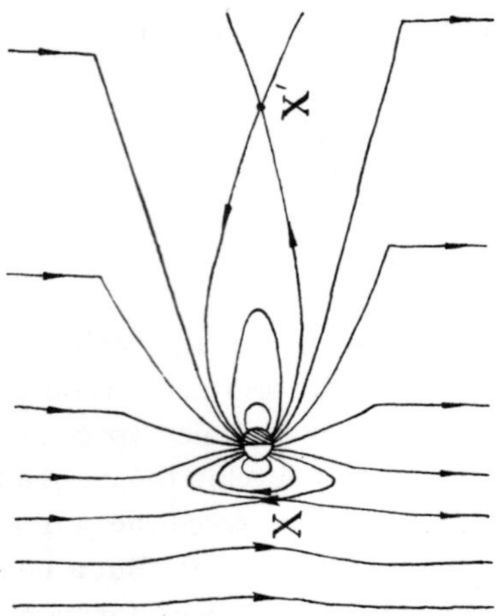

FIG.2. Dungey model of reconnection in the earth's magnetosphere.

For typical magnetospheric plasma parameters:

$$B_x \sim 1.5 \times 10^{-4} \text{ gauss}$$

$$n_0 \sim 0.1 \text{ cm}^{-3}$$

$$L_x \sim L_y \sim 10^{10} \text{ cm}$$

we obtain from Eqs (4) – (7):

$$\mathscr{P} = v_A (c/\omega_p) L_y \frac{B_x^2}{4\pi} \sim 10^8 \text{ watts}$$

This is three orders of magnitude less than the well-known observed power of auroral activity. In the same way, the Parker-Sweet model with anomalous resistivity cannot provide a high enough reconnection rate at the magnetosphere nose to explain the measured electric field within the magnetosphere. The latter is the simple mapping along the highly conductive magnetic field lines of the electric field induced by plasma flow near the merging region (see Fig. 2):

$$E_y = \frac{1}{c} u_x B_z \tag{8}$$

A solution of these problems was proposed by Petschek [8], who drew a picture of the hydrodynamic flow of a plasma with frozen-in magnetic field, with the singularity along the neutral line. The essential features of this flow are shown in Fig. 3.

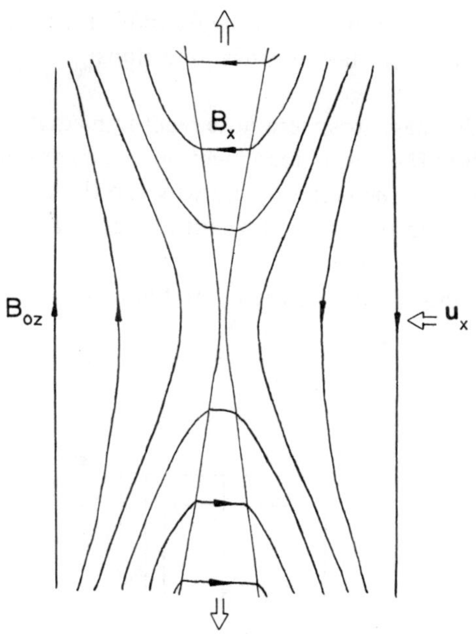

FIG.3. Petschek model of reconnection.

We restrict ourselves to only this oversimplified picture and refer the reader for more detailed consideration to Vasyliunas' survey [9]. In Fig. 3 the standing Alfvén wave separates the region of constant B_X magnetic field from the external hydrodynamic flow described by the curl-free magnetic field:

$$B_Z \cong B_0 - \frac{\mathcal{M} B_0}{\Theta_0} \ln \frac{r}{z_*}$$

$$B_X = \frac{\mathcal{M} B_0 \theta}{\Theta_0} \tag{9}$$

where Θ_0 is the position of the standing Alfvén wave in polar co-ordinates (r, θ); $\mathcal{M} = u_X/v_A$ is the Alfvén Mach number of the plasma flowing in with the Alfvén speed $u_X = B_X/\sqrt{4\pi n_0 M}$ into the standing Alfvén wave; and z_* is the size of the diffusion region where the above Parker-Sweet picture of the reconnection takes place. We see that the maximum reconnection rate depends on the diffuse region sizes and therefore on the anomalous resistivity only logarithmically:

$$u_{X,\max} \sim \frac{\pi v_A}{\ln Re_m} \tag{10}$$

Petschek's model provides, in principle, the desirable reconnection rate but it has never answered two important physical questions:

(1) Why does the magnetosphere nose react immediately on the interplanetary magnetic field direction (i.e. 'nose' reconnection occurs at once after the appearance of free magnetic energy in the system), and why can the plasma sheet with oppositel directed magnetic fields in the north and south lobes of the magnetospheric tail be stable for long periods of time?

(2) Why, where and when does the neutral line in the magnetospheric tail appear?

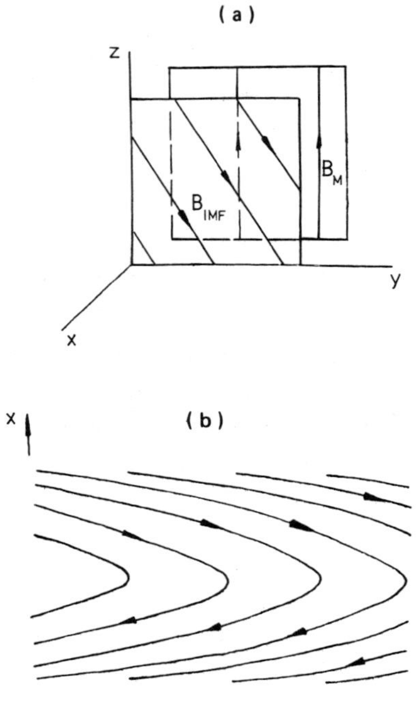

FIG.4. Magnetic field geometries near (a) the 'nose' and (b) the tail merging regions.

2. TEARING-MODE STABILITY ANALYSIS

We shall now show that the tearing instability analysis gives the answer to both questions. We should mention that the difference between the 'nose' and 'tail' reconnection comes out from the difference in the magnetic field geometry near the 'neutral' plane (see Fig. 4).

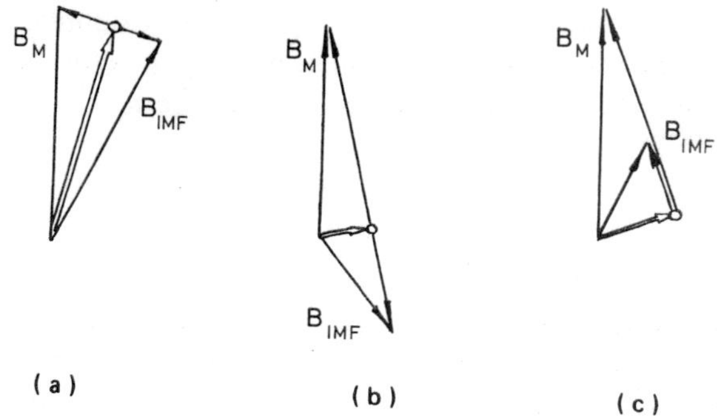

FIG.5. Different types of geometry of interplanetary (B_{IMF}) and geomagnetic (B_M) field interaction [13, 14]. *Symmetrical configuration (a); unsymmetrical configurations with (b) and without (c) free magnetic energy.*

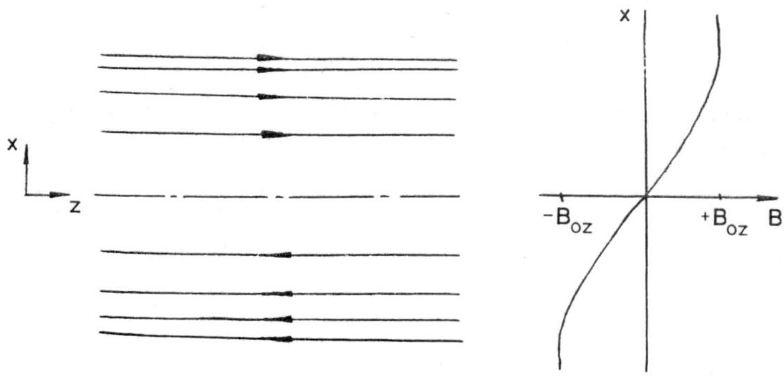

FIG.6. Model of plane neutral layer.

To describe these geometries we use the simplest models of magnetic configurations:

$$\vec{B} = B_{oz}\,\mathrm{th}\left(\frac{x}{\Delta}\right)\hat{e}_z + B_y\,\hat{e}_y \tag{11a}$$

$$\vec{B} = B_{oz}\,\mathrm{th}\left(\frac{x}{\Delta}\right)\hat{e}_z + B_x\,\hat{e}_x \tag{11b}$$

where B_{oz}, B_y, B_x are constants. The magnetic field profile is specified by Harris' choice of the distribution function [10]:

$$f_{oj}(x,\vec{v}) = n_o\left(\frac{m_j}{2\pi T_j}\right)^{\frac{3}{2}} \exp\left\{-\frac{m_j v^2}{2\,T_j} + \frac{u_j}{T_j}\left(m_j v_y + \frac{e_j}{c} A_{oy}(x)\right) - \frac{m_j^2 u_j^2}{2\,T_j}\right\} \tag{12}$$

$$u_j = -2\,c\,T_j/e_j\,B_{oz}\,\Delta$$

This choice is, of course, sufficiently specific. For example, for the magnetic field configuration (11a) at the nose of the magnetosphere there is one more invariant-canonical momentum along the z-axis:

$$P_{zj} = m_j v_z + \frac{e_j}{c} A_{oz}(x)$$

Therefore we can easily change the density profile, introducing the arbitrary function $n_0(P_{zj}(x, v_z))$. Moreover the symmetric profile of $n(x)$ corresponds only to the rather special case of equal magnetic field strength on both sides of the reconnection region [11, 12] (see also Fig. 5, taken from Refs [13] and [14]). For the magnetospheric tail configuration (11b) Harris equilibrium is slightly distorted in the close vicinity of the B_z-field reversal plane but, as in the case of trapped-particle drift instabilities, here again the main contribution to the dispersion relation comes from the change of particle trajectory topology, and the distribution function can be considered as invariable and Maxwellian.

Let us recall first the basic idea by Coppi et al. [11] of the stability analysis of the simplest magnetic field configuration (see Fig. 6):

$$B = B_{0z} \,\mathrm{th} \left(\frac{x}{\Delta}\right) \hat{e}_z \tag{13}$$

with the distribution function defined by Eq. (12). Tearing instability of a plane neutral sheet results in pinching a plane current sheet, as shown in Fig. 7. Such a type of perturbation can be described by vector potential in a current direction:

$$\vec{A}_1 = A_{1y}(x) \exp[-i\omega t + ik_z z] \hat{e}_y \tag{14}$$

Everywhere but the closest vicinity of the neutral layer ($x < \sqrt{\rho_{jz} \Delta}$, ρ_{jz} is the Larmor radius of particles of j-specie in B_{0z} magnetic field), the Larmor radius is smaller than the characteristic scale length of inhomogeneity and therefore low-frequency oscillations ($\omega \ll \omega_{cj}(x)$, $\omega_{cj}(x) = e_j B_z(x)/m_j c$) can be considered as adiabatic. In this case the distribution function still depends on (z, t) only through the y-component of the particle canonical momentum. Considering the perturbation as small, we obtain, to a linear approximation:

$$f_j = f_{0j}(x, \vec{v}) + f_{1j}^{ad}(x, \vec{v}, z, t)$$

$$f_{1j}^{ad} = \frac{e_j u_j}{c T_j} A_{1y}(x) f_{0j}(x, \vec{v}) \exp[-i\omega t + ik_z z] \tag{15}$$

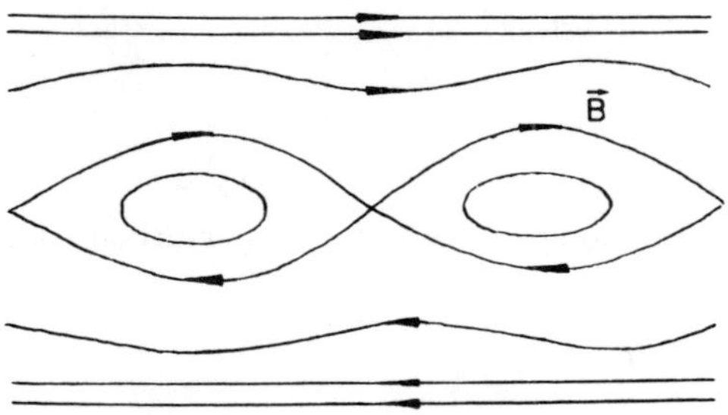

FIG. 7. Tearing mode in a plane neutral layer.

Combining this result with the Maxwell equation:

$$\text{rot}\,\vec{B}_1 = (4\pi/c)\sum_j e_j \int \vec{v}\, f_{1j}\, d^3\vec{v} \tag{16}$$

we finally come to the Schrödinger-type equation for the y-component of the vector potential:

$$\frac{d^2 A_{1y}}{dx^2} - [k_z^2 + V_0(x)]\, A_{1y}(x) = 0 \tag{17}$$

where

$$V_0(x) = -\sum_j \frac{4\pi e_j^2\, u_j^2}{c^2 T_j}\, n(x) = -\frac{2}{\Delta^2\, \text{ch}^2\left(\dfrac{x}{\Delta}\right)}$$

As is very well known, this equation has only one eigenfunction, vanishing at infinity:

$$A_{1y}(x) = \frac{1}{2\Delta}\, \text{ch}^{-1}\left(\frac{x}{\Delta}\right) \tag{18}$$

corresponding to the eigenvalue $k_z^2\,\Delta^2 = 1$.

Now we should turn back to the nonadiabatic contribution from the vicinity of the neutral plane $x < \sqrt{\rho_{jz}\,\Delta}$, where $\rho_{jz}(x) > x$. Neglecting the magnetic field in this region, we can obtain the linearized kinetic equation in the form:

$$\frac{\partial\, \delta f_{1j}}{\partial t} + \vec{v} \cdot \nabla \delta f_{1j} = \frac{e_j}{T_j}\, \frac{i\omega}{c}\, A_{1y}(x)\, v_y\, f_{0j}\quad \exp\,[-i\omega t + ik_z z] \tag{19}$$

where

$$\delta f_{1j} = f_{1j} - f_{1j}^{ad}$$

Integrating this equation along the particle's unperturbed trajectory, we obtain

$$\delta f_{1j} = \frac{e_j f_{oj}}{c T_j} \left[i\omega \int_{-\infty}^{0} A_{1y}(x)\, v_y \;\exp\left[-i\omega\tau + ik_z v_z \tau\right] d\tau \right] \tag{20}$$

Then the current perturbation causes an additional contribution to the 'effective potential' in Eq. (17):

$$V_1 = \sum_j \frac{4\pi e_j^2}{c^2 T_j} \int d^3\vec{v}\, f_{oj}\, v_y \left[-i\omega \int_{-\infty}^{0} A_{1y}\, v_y \;\exp\left[-i(\omega - k_z v_z)\tau\right] d\tau \right] \tag{21}$$

For low-frequency perturbations, the main contribution to the integral comes from the half-residual part:

$$V_1(x, \omega, k) = \sum_j \frac{\omega_{pj}^2}{c^2} \frac{-i\pi^{1/2}\omega}{|k_z|\, v_{thj}} \tag{22}$$

where

$$\omega_{pj}^2 = \frac{4\pi e_j^2 n(x)}{m_j}, \qquad v_{thj} = \frac{2\, T_j}{m_j}$$

For purely imaginary frequency of perturbations ($-i\omega \equiv \gamma > 0$), the nonadiabatic contribution to the effective potential has the form of a narrow potential barrier in the centre of the Teller potential well $V_0(x)$ (see Fig. 8). For, in the integral sense, small V_1, we can use the quantum-mechanical perturbation theory [16] to find the small eigenvalue change:

$$1 - k_z^2 \Delta^2 = \frac{\Delta}{2} \int_{-\infty}^{+\infty} V_1(x)\, \mathrm{ch}^{-2}\left(\frac{x}{\Delta}\right) dx \tag{23}$$

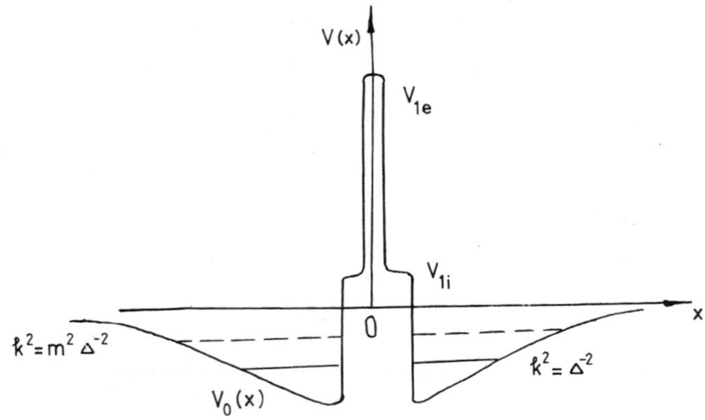

FIG.8. Effective potentials for Eq. (17).

This equation serves as the dispersion relation for obtaining the growth rate. For the case of a plane neutral sheet, the main contribution to the integral (23) comes from electrons and finally

$$\gamma = \left(1 + \frac{T_i}{T_e}\right) \pi^{1/2} k_z v_{the} \left(\frac{\rho_{ez}}{\Delta}\right)^{3/2} (1 - k_z^2 \Delta^2) \tag{24}$$

3. RECONNECTION OF THE INTERPLANETARY FIELD LINES AND EARTH'S MAGNETIC FIELD LINES IN TERMS OF THE TEARING MODES

Since the interplanetary magnetic field vector and the earth's magnetic field vector are not collinear, we usually have the magnetic field component within the field reversal layer (see Fig. 4 and Eq. (11a)):

$$\vec{B} = B_{oz}\,\mathrm{th}\left(\frac{x}{\Delta}\right) \hat{e}_z + B_y \hat{e}_y \tag{25}$$

If this y-component is strong enough,

$$b_y \equiv \frac{B_y}{B_{oz}} \gg \epsilon_j^{1/2} \quad , \qquad \epsilon_j = \frac{\rho_{jz}}{\Delta} \tag{26}$$

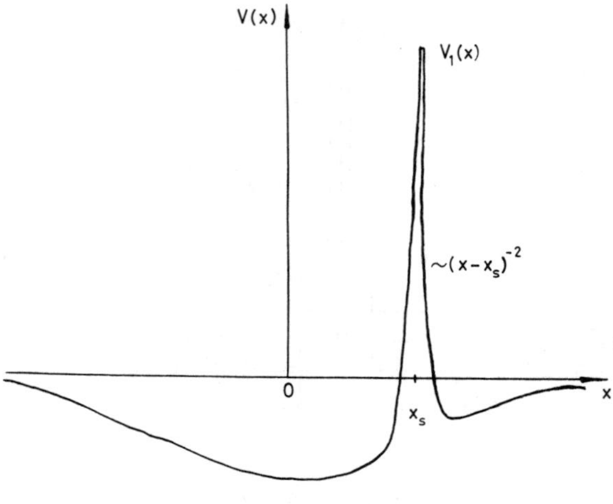

FIG.9. Effective potential barrier in Eq. (30) for the model of a magnetic field in the presence of the component along the equilibrium current (Eq. (11a)).

We can use the drift approximation everywhere, including the B_z-field reversal layer, both for electrons and ions. Then we have the following expression for the distribution function instead of Eq. (12):

$$f_{0j}(x, v_\parallel) = \frac{n_0}{ch^2\left(\frac{x}{\Delta}\right)} \left(\frac{m_j}{2\pi T_j}\right)^{1/2} \exp\left[-\frac{m_j}{2T_j}\left(v_\parallel - u_j \frac{B_y}{B}\right)^2\right] \qquad (27)$$

Here we restrict ourselves to the simplest case of large $b_y \gtrsim 1$. For two-dimensional perturbation of the form:

$$\vec{A}_1(\vec{r}, t) = [A_{1y}(x)\hat{e}_y + A_{1z}(x)e_z] \exp[-i\omega t + ik_y y + ik_z z]$$

$$\vec{k}\cdot\vec{A}_1(\vec{r}, t) = 0 \qquad (28)$$

the wave-particle interaction resulting in the mode growth takes place mainly near the 'singular' surface defined by the condition

$$k_\parallel(x_s) = \frac{k_z B_z(x_s) + k_y B_y}{B} \cong 0 \qquad (29)$$

Wave-particle interaction can influence the mode development only if this surface is inside the effective potential well $V_0(x)$, i.e. only if $k_y/k_z \lesssim B_{0z}/B_y \ll 1$. In this case we can still use Eq. (17) for the adiabatic part of the potential $V_0(x)$. To find the non-adiabatic correction $V_1(x)$, resulting from wave-particle interaction, we should make the following changes in Eq. (21):

(a) The frequency outside the trajectory integral in Eq. (21) should be Doppler-shifted by the quantity $k_y u_j$ according to the Lorentz transformation of the perturbed electric field.

(b) In the drift approximation, the particle trajectory is given by

$$\vec{r} = \frac{\vec{B}}{|\vec{B}|} v_\parallel \tau$$

and the distribution function on $v_\parallel$ is shifted by the drift velocity component along the magnetic field (see Eq. (27)). Then we rewrite Eq. (21) for $V_{1j}(x)$ in the form:

$$V_{1j}(x, \omega, \vec{k}) = -\frac{4\pi e_j^2}{c^2 T_j} \int_{-\infty}^{+\infty} dv_\parallel\, f_{0j}(x, v_\parallel)\, v_\parallel$$

$$\times \left[i(\omega - k_y u_j) \int_{-\infty}^{0} v_\parallel \exp\left[-i\omega\tau + ik_\parallel v_\parallel \tau\right] d\tau \right]$$

$$\cong \frac{2\omega_{pj}^2}{c^2} \frac{\omega_{*j} - \omega_j}{k_\parallel v_{thj}} Z_2\left(\frac{\omega_j}{k_\parallel v_{thj}}\right)$$

where

$$\omega_j = \omega - k_\parallel u_j \frac{B_y}{B}, \qquad \omega_{*j} \cong -\frac{c\vec{k} \cdot [T_j \nabla n \times \vec{B}]}{e_j B^2 n(x)}$$

$$Z_2(\zeta) = \pi^{-1/2} \int_{-\infty}^{+\infty} \frac{z^2 e^{-z^2} dz}{z - \zeta - i\theta\, \mathrm{sign}\, k_\parallel}, \quad \theta \to 0$$

We see that the main contribution comes from the narrow vicinity of the singular surface (see Fig. 9):

$$|x - x_s| \lesssim \frac{|\omega|\, b_y\, \Delta}{k_z\, v_{thj}}$$

For the low-energy level shift in an effective potential well $V_0(x)$ due to the non-adiabatic wave-particle interaction, we can again use the quantum-mechanical perturbation theory (see Eq.(23)). The integrals in Eq. (23) can be taken exactly and we obtain the dispersion relation [17–19]:

$$\omega = \omega_{*e}(x_s) + i\,\frac{k\, v_{the}}{\pi^{1/2} b_y} \left(\frac{c}{\omega_{pe}\,\Delta}\right)^2 (1 - k^2 \Delta^2) \tag{30}$$

In the limit of the weakly magnetized neutral layer ($b_y \sim \epsilon_e^{1/2}$), this result, as can be easily shown, is consistent with the result of Laval et al. [18] for the case $b_y = 0$. At the opposite limit of a very strong magnetic field, when the ion Larmor radius becomes smaller than the singular region width near $x = 0$,

$$\left(\frac{T_e}{T_i}\right)^{1/2} \rho_{iy} < \frac{\gamma\, b_y\, \Delta}{k\, v_{the}} \sim \epsilon_e^{\frac{2}{3}}$$

$$b_y > \frac{\sqrt{\dfrac{2\, m_i}{m_e}} \left(\dfrac{k\,\Delta}{1 - k^2 \Delta^2}\right)}{\eta_e\, \epsilon_e}$$

i.e.

$$\eta_j = \frac{(T_i + T_e)}{T_j}$$

We should take into account the electric potential perturbations and ion contribution in Eq. (17). This procedure enables us to obtain on a common kinetic basis the well-known MHD results of Coppi [20]:

$$\gamma \cong \epsilon_e^{\frac{3}{2}} \frac{v_{thi}}{\Delta} \left(\frac{T_i}{T_e}\right)^{1/2} \frac{\eta_e^{\frac{2}{3}}}{I^2} (1 - k^2 \Delta^2)^2, \quad I \equiv 2\pi\, \frac{\Gamma(3/4)}{\Gamma(1/4)}$$

and demonstrates the continuous transition from the kinetic regime (30) to the MHD regime of inertial tearing instability in collisionless plasma, as shown in Fig. 10.

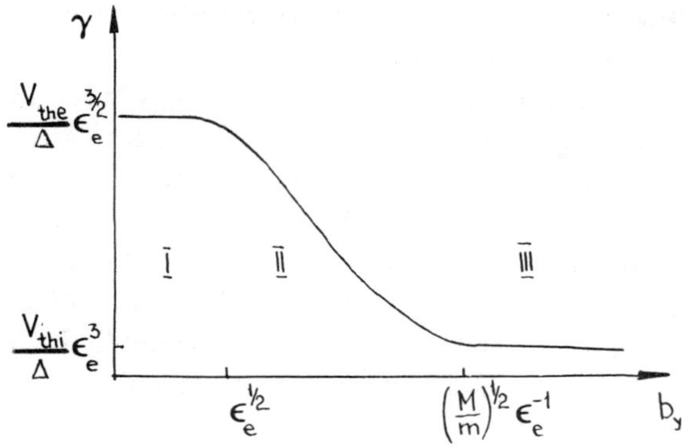

FIG.10. The dependence of tearing-instability development regimes in a collisionless plasma from the value of the magnetic field component along the current (i.e. from the angle of rotation of the magnetic field in a transition layer).
I: kinetic resonant tearing mode; II: kinetic inertial tearing mode; III: MHD inertial tearing mode.

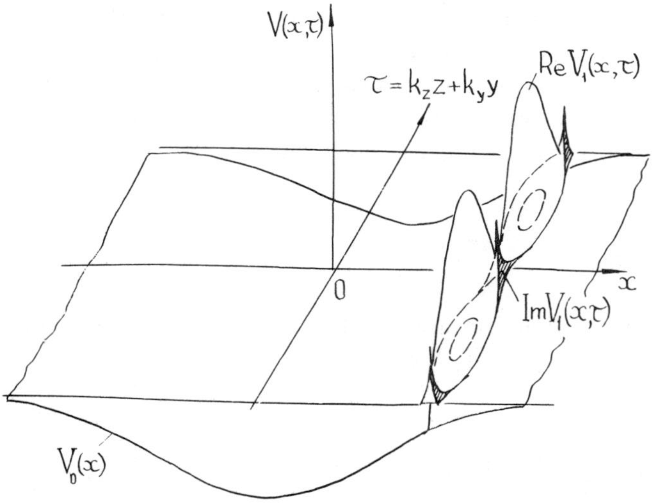

FIG.11. Adiabatic $V_0(x)$ and non-adiabatic $V_1(x)$ contributions to the effective potential $V(x)$ for the single-mode regime.

As mentioned earlier, this kind of analysis is incomplete since we restricted ourselves to the symmetrical $B_z(x)$ profile only. The case of arbitrary $B_z(x)$ profiles was considered by Schindler and Soop [21], but only on a plane model with $B_y = 0$. They have shown that any profile is unstable if and only if it has a field reversal (i.e. when $B_z(x)$ changes sign at some point $x = x_0$). This conclusion may be extended to a more general class of configurations in the presence of a constant uniform component of the magnetic field perpendicular to the plane considered, and this is compatible with results of Sonnerup [13], Gonzalez and Mozer [14], shown in Fig. 5.

To describe the magnetic field lines'reconnection, we need to consider the non-linear stage of the tearing instability. This problem is greatly simplified in two limiting cases: single-mode growth and random-phase approximation for the growth of a large number of modes. These two cases give qualitatively the same result, though the numbers are of course different. Since we are not going to follow precise calculations, we consider here only the case of the single-mode regime. Single-mode growth for the magnetic field model by Eq. (25) results in the formation of a magnetic island near the singular surface $k_{\parallel}(x) = 0$.

The width of the magnetic island increases with the wave amplitude:

$$w = \Delta\left(\frac{2\,B_{1x}}{k_z\,\Delta\,B_{oz}}\right)^{1/2} \tag{31}$$

When we are calculating the contribution to $V_1(x, \omega, \vec{k})$ from the particles moving along the closed magnetic surfaces (i.e. within the islands), we should take into account that $k_{\parallel}$ oscillates along the particle trajectory. Therefore, for such particles $|\omega_j| > \overline{k_{\parallel}\, v_{\parallel}}$. Particles moving along the unclosed magnetic surfaces still contribute to the wave-particle resonant interaction (see Fig. 11). For an island width larger than the singular layer width,

$$w > \frac{\omega_{*e}\, b_y\, \Delta}{k\, v_{thj}} \sim \frac{T_e}{T_j}\, \rho_{jz} \tag{32}$$

the resonant interaction decreases rapidly. As a result, the dispersion relation takes the form (for $\rho_{iz}\, T_e/T_i > w > \rho_{ez}$):

$$1 - k^2\Delta^2 = \left(\frac{\omega_{pe}}{c}\,\Delta\right)^2 (\omega - \omega_{*e})\left[\frac{w}{\Delta\,\omega} - i\frac{b_y}{|k_z|\, v_{the}}\,\frac{\rho_{ez}^2}{w^2}\right]$$

$$- i\left(\frac{\omega_{pi}}{c}\,\Delta\right)^2 \frac{\omega - \omega_{*i}}{|k_z|\, v_{thi}}\, b_y \tag{33}$$

In principle, particles moving along the unclosed magnetic surfaces make a slightly larger contribution to wave-particle interactions near the singular surface, where $k_{\parallel}(x) = 0$. So the contribution of the wave-particle interaction effects to Eq. (33) decreases a little slower (as $w^{-3/2}$ instead of w^{-2}). But since this effect differs for the single-mode regime and random-phase approximation we shall not pay it much attention and restrict ourselves to simple estimates.

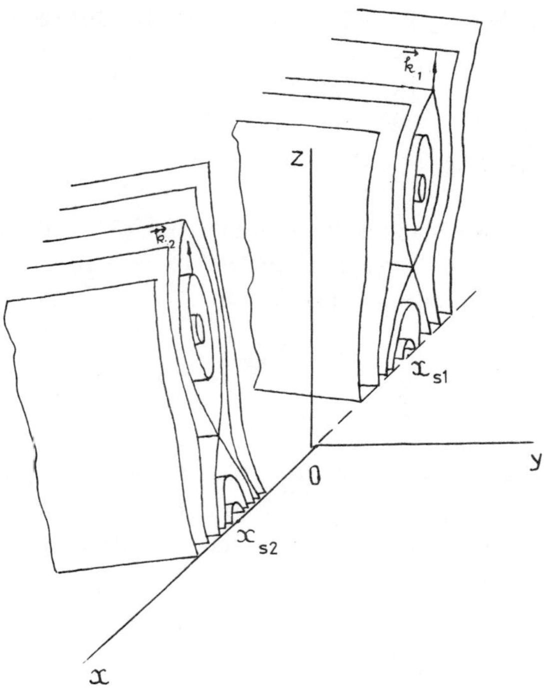

FIG.12. The growth of magnetic islands due to development of tearing modes with different angles of propagation with respect to magnetic field (for simplicity only the case of two independent modes is shown).

We see that in the non-linear regime, when $w > \rho_{ez}$, the growth rate for a diffuse neutral layer ($\epsilon_i < 1$) can be larger than the linear one:

$$\gamma \sim \eta_e \epsilon_e^2 \frac{|k_z| v_{the}}{b_y} \frac{\rho_{ez}}{\epsilon_i w} \tag{34}$$

Further growth of the single mode results in the rapid decrease of the growth rate ($\gamma \sim w^{-4}$ for $w > \rho_{zi} T_e/T_i$). But, as with the hydrodynamic theory of the tearing mode, the growth of the magnetic island could be stopped only by quasilinear relaxation [22].

In contrast to the toroidal systems for the infinite-plane plasma slab, the perturbation wave numbers have a continuous spectrum and therefore the singular surfaces can continuously change their position (see Fig. 12). As a result, the different magnetic islands can overlap and lead to stochastic diffusion of the magnetic field lines. Following Ref. [23], we can write the diffusion coefficient in the form:

$$D \cong v_{thi}^2 \frac{B_{1x}^2}{B^2} \frac{1}{|k_{\parallel}(w)|\, v_{thi}}$$

Taking into account that the island width is given by Eq. (31) and that the resonant interaction with ions decreases as $(\rho_{iz}/w)^3$ for $w > \rho_{iz}\, T_e/T_i$, we find that the diffusion coefficient saturates in this limit:

$$D \cong v_A\, \Delta\epsilon_i^3 \left(\frac{T_e}{T_i}\right)^3 \eta_i^{-1/2}\, b_y^{-2} \tag{35}$$

where v_A is the Alfvén speed in the field $B_y \gtrsim B_{oz}$. Combining this result with the Parker-Sweet theory of magnetic reconnection for a plane slab of length L discussed above (see Eq.(31)), we can find the estimate of maximum reconnection rate for collisionless plasma:

$$u \sim v_A \left(\frac{\rho_{iz}}{L}\right)^{3/4} b_y^{-5/4} \left(\frac{T_e}{T_i}\right)^{3/4} \eta_i^{-1/8} \tag{36}$$

4. TEARING INSTABILITY AS A MECHANISM OF THE MAGNETOSPHERIC TAIL EXPLOSION

We now turn to the stability analysis of two-dimensional plasma equilibria relevant to the magnetospheric tail problems [24, 25]. We consider here the simple model of the magnetic field:

$$\vec{B} = B_{oz}\, \mathrm{th}\left(\frac{x}{\Delta}\right) \hat{e}_z + B_x\, \hat{e}_x \tag{37}$$

The particle distribution function corresponding to this model, for the purpose of stability analysis, can be chosen approximately the same as for Harris's equilibria (see Eq. (12)).

The difference in the study of the stability questions comes from the change in the particle trajectory topology near the plane x = 0. Here the x-component of the magnetic field tends to magnetize particle motion under the condition

$$\omega_{xj} = \frac{e_j B_x}{m_j c} \gtrsim \gamma \tag{38}$$

For approximately circular orbits of particles in the normal magnetic field we can easily calculate the non-adiabatic contribution to the effective potential (compare Eq. (21)):

$$V_{1j}(x, \omega, k_z) = \frac{4\pi e_j^2}{c^2 T_j} \int d^3\vec{v}\, f_{0j}(x, \vec{v})\, v_\perp \cos\theta$$

$$\times \left\{ -i\omega \int_{-\infty}^{0} d\tau\, v_\perp \cos(\omega_{xj}\tau + \theta) \right.$$

$$\left. \times \exp\left[-i\omega\tau - i\frac{k_z v_\perp}{\omega_{xj}} (\cos(\omega_{xj}\tau + \theta) - \cos\theta) \right] \right\}$$

$$\cong 2\frac{\omega_{pj}^2}{c^2}\frac{d}{d\Lambda_j}[\Lambda_j e^{-\Lambda_j} I_1(\Lambda_j)] \tag{39}$$

where $\Lambda_j = 0.5\, k_z^2\, \rho_{xj}^2$ and $I_n(\Lambda_j)$ is the modified Bessel function. In expressions (39) we have introduced polar co-ordinates in the velocity space with the x-axis as the polar axis and have restricted ourselves to the most important mode propagating in the z-direction. This contribution (39) comes from particles within the thin layer $|x| < \sqrt{\rho_{jz}\,\Delta}$ where we can describe the trajectories of particles from the bulk of plasma distribution as slow Larmor rotation in the (y, z)-plane (in the B_x field) and (in rough approximation) fast oscillations in the $|x| < \sqrt{\rho_{jz}\,\Delta}$ region. However, when the x-component of the magnetic field becomes stronger,

$$B_x > \sqrt{\frac{\rho_{jz}}{\Delta}}\, B_{0z}$$

we can use the drift approximation for particle orbits and find that the result given by Eq. (39) is still approximately valid but already for a wider range of space, i.e. for $|x| < (B_x/B_{0z})\,\Delta$.

The stabilizing influence of the x-component of the magnetic field consists of two effects: (1) prohibition of Landau wave-particle interaction under the condition of Eq. (38); and (2) the shift of the 'energy level', $E = -k^2$, in the Teller potential well $V_0(x)$ due to the change of height of the potential barrier $V_1(x)$ in the centre of this well (see Figs 8 and 13). To describe the second effect, it is

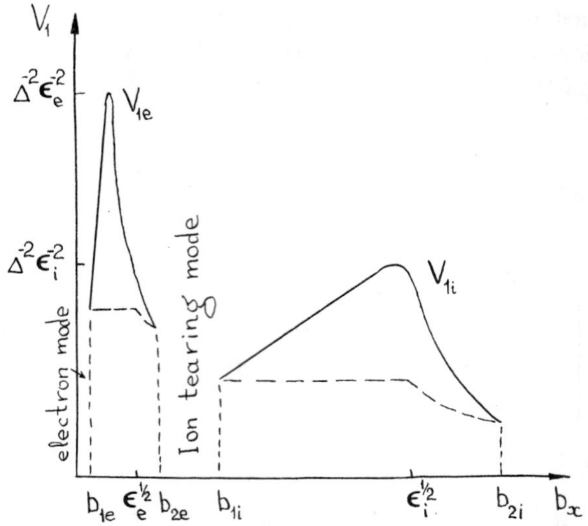

FIG.13. Potential barrier height as a function of the magnetic field component normal to the neutral layer.

convenient to plot the potential barrier height $V_{1j}(0)$ as a function of the value of the x-component of the magnetic field; this is done in Fig. 13. We see that for the given mode number $m = k\Delta$, electron Landau resonance is destroyed for very small values of normal magnetic field (see Eq. (38)):

$$b_x \equiv \frac{B_x}{B_z} \gtrsim b_{1e} \cong \eta_e(1-m^2)\,\epsilon_e^{5/2} \qquad (40)$$

At the same time, the height of the potential barrier increases and it becomes non-transparent. Since the energy level does not exist in either half of the divided Teller well, the tearing mode does not exist until the potential barrier decreases with the B_x-field increase and again becomes transparent for perturbations for values $b_x > b_{2e}$. Then the development of the tearing mode again appears to be possible but its growth can be stipulated now only by the resonance interaction with ions and therefore this mode is called the ion-tearing mode [26].

Since, in the same way, the B_x-field can quench the ion Landau resonance for $b_x > b_{1i}$, when the ion-cyclotron frequency is larger than the growth rate of the ion tearing mode, the ion-tearing mode can in principle exist only within a limited range of B_x-field strengths:

$$b_{2e} < b_x < b_{1i} \qquad (41)$$

and only if the plasma parameters permit the b_{1i} to be larger than b_{2e}.

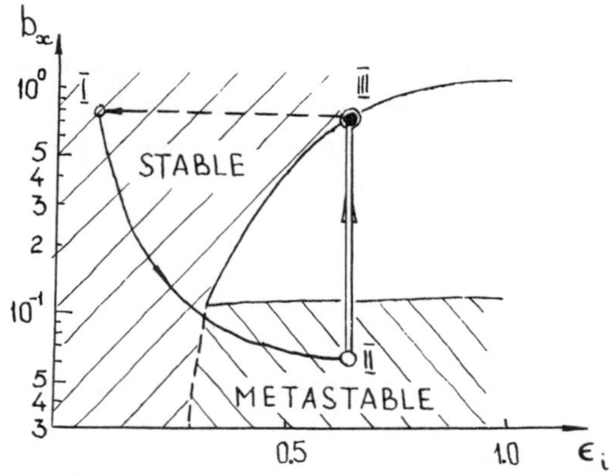

FIG.14. Tearing mode stability diagram and changes of the magnetospheric tail state during substorms.
I–II: growth phase; II–III: break-up and expansion phase; and III–I: recovery phase.

In Fig. 14 we have drawn the marginal stability curve in the plane (b_x, ϵ_i), which is obviously described analytically by the equation

$$b_{ii}(m, \epsilon_i) = b_{2e}(m, \epsilon_i) \tag{42}$$

Plasma with parameters (b_x, ϵ_i) to the left of this curve (low plasma current and high magnetic field component normal to the neutral layer) is stable against the tearing mode. The lower solid curve on the stability diagram (Fig. 14) is the curve $b_{2e}(m_{min}, \epsilon_i)$, corresponding to the maximum wavelength $2\pi\Delta/m_{min}$ of perturbations that can develop in the system with finite characteristic length. Plasma states below this curve are metastable in the sense that only a finite disturbance can lead to instability.

Coppi, Laval and Pellat [15] were the first to propose the tearing mode as the mechanism of magnetic field reconnection in the magnetospheric tail during substorms. Later this idea was worked out in detail by Schindler [26], who understood that the electron-tearing mode is practically always suppressed by the field component normal to the neutral layer and therefore turned his attention to the ion-tearing mode. But until more detailed analyses by Galeev and Zeleny [24, 25] it was not known why the magnetospheric tail stays quiet for a long time and then suddenly explodes.

With the help of the stability diagram, Fig. 14, we can give in principle a more quantitative description of the UCLA picture of a substorm [27, 28] (for a more detailed discussion see Ref. [29]). In Fig. 14 the curve I-II shows qualitatively the changes of plasma states during the growth phase when the plasma sheet is becoming thinner (dimensionless parameter ϵ_i is increasing) and the value of the B_x-component is decreasing (tail field lines become more stretched in an antisolar direction). To draw this curve with real numbers we should solve the non-stationary problem for slow evolution of plasma equilibria in the magnetospheric tail. Plasma in the state labelled II in Fig. 14 can be quiet for a long time, until the break-up phase of the substorm is triggered by some finite external disturbances, e.g. by a sudden northward increase of the interplanetary magnetic field, as described by Russell [30]. At this moment the neutral line (the merging region) is formed in the magnetospheric tail and reconnection can proceed further, approaching the steady-state regime. As already described, most of the magnetic field energy will be transferred to the fast plasma flows and only a relatively small part of the energy may be dissipated immediately into the heat by anomalous resistivity and can be used accelerating the particles either in the induced electric fields or due to wave-particle interactions.

As was recently shown by simultaneous particle and field observations [31], there is a close association between the formation of neutral lines in the magnetotail and bursts of energetic electrons whose spectrum extends to MeV range. The electric field responsible for this acceleration seems to be of an impulsive, inductive nature, as the characteristic time of magnetic field changes (related to the merging region formation) evaluated as an inverse increment of the ion-tearing mode does not exceed a few tens of seconds. And such fast times for field changes are quite sufficient [31] for the desired inductive acceleration of electrons up to the observed energies.

5. DISCUSSION

We have shown that the collisionless tearing instability of different hot plasma ($\beta \sim 1$) configurations with inverse magnetic fields that are realized in the earth's magnetosphere may be responsible for the macroscopic picture of the reconnection of interplanetary and geomagnetic fields and also for magnetospheric tail dynamics during the development of substorms.

However, this instability and related reconnection processes have a much wider range of applications. Especially promising seems to be the use of the results obtained here for the case of large-scale structures that form under the 'winding' of magnetic field lines on the central magnetized body, immersed in a high-conductivity plasma. The enormous width of the layer, where at least one of the components of the magnetic field changes its sign, excluded the possibility of collisionless dissipation on the base of current instabilities (having the finite current velocity threshold). Thus, the tearing mode is practically the only mechanism of field-line merging in this case.

An interesting example of such large-scale configuration represents the sector structure of the interplanetary magnetic fields [32]. Mariner-5 experiments [33] have shown that the width of the transition layer between sectors is of the order of 10^3 to 10^4 km, which considerably exceeds the ion Larmor radius.

More detailed discussion is to be found in Ref. [34] where the observations are interpreted in terms of a resistive tearing mode that is scarcely of importance for solar wind conditions. The collisionless tearing results (see § 3), on the other hand, for typical interplanetary plasma parameters ($T_e \cong 15$ eV; $T_i \cong 6$ eV; $n \cong 5\ cm^{-3}$; $v_{sw} \cong 4 \times 10^7$ cm/s; $B \cong 5\ \gamma$) show that smearing of the layer between sectors up to the observed width takes place at distances of the order of 1 astron. unit = 2×10^{13} cm. The complete disappearance of sector structure due to this highly efficient process can evidently be achieved only at distances of the order of

$$L \sim \frac{\Delta_s^2 v_{sw}}{D(\Delta_s)}$$

where

$$D(\Delta_s) \cong v_A \Delta_s \left(\frac{\rho_i}{\Delta_s}\right)^3 \left(\frac{T_e}{T_i}\right)^3 \left(1 + \frac{T_e}{T_i}\right)^{-1/2}$$

is the diffusion coefficient of the magnetic field for the characteristic layer width equal to the transverse dimension of the sector structure:

$$\Delta_s \cong v_{sw}/\Omega_\odot \sim 1 \text{ astron. unit}$$

Using the real parameter of solar wind in the expression above, and taking into account the diminishing of the values of density and magnetic field with the distance from the sun, we can draw a conclusion that the interplanetary magnetic field conserves its sector structure up to the distances ~ 100 astron. units where the interaction of solar wind and the interstellar medium begins [35]. However, for astrophysical objects such as the Crab Nebula, having much greater cover [36], the field line reconnection process due to the development of tearing instability plays an even more crucial role in releasing the energy of the magnetic field and corresponding plasma heating.

REFERENCES

[1] GIOVANELLI, R.G., Mon. Not. R. Astron. Soc. **107** (1947) 338.

[2] DUNGEY, J.W., Phys. Rev. Lett. **6** (1961) 47.

[3] KADOMTSEV, B.B., Fiz. Plasmy **1** (1975) 710.

[4] PARKER, E.N., J. Geophys. Res. **62** (1963) 509; Astrophys. J., Suppl. Ser. **8** (1963) 177.

[5] SWEET, P.A., in Electromagnetic Phenomena in Cosmic Plasma (Proc. 6th Int. Astronomical Union, Stockholm, 1958), Cambridge Univ. Press, New York (1958) 123.

[6] PIKELNER, S.B., TSYTOVICH, V.N., Astron. Zh. **52** (1975) 738.

[7] CORONITY, F.V., EVIATOR, A., Astrophys. J., Suppl. **33** (1977) 189.
[8] PETSCHEK, H.E., in Solar Wind (Proc. Jet Propulsion Lab. Conf.), (MACKIN, R.J., NEUGEBAUER, M., Eds), Pergamon, Oxford (1966).
[9] VASYLIUNAS, V.M., Rev. Geophys. Space Phys. **13** (1975) 303.
[10] HARRIS, E.G., Nuovo Cim. **23** (1962) 115.
[11] HAERENDEL, G., Microscopic Plasma Processes related to Reconnection, Max-Planck-Inst. für Phys. und Astrophys. Garching, Rep. MPI-PAE/Extraterr. 129, Dec. 1976.
[12] PASCHMANN, G., et al., J. Geophys. Res. **81** (1976) 2883.
[13] SONNERUP, B.U.O., J. Geophys. Res. **79** (1974) 1546.
[14] GONZALEZ, W.D., MOZER, F.S., J. Geophys. Res. **79** (1974) 4186.
[15] COPPI, B., LAVAL, G., PELLAT, R., Phys. Rev. Lett. **16** (1966) 1207.
[16] LANDAU, L.D., LIFSHITZ, E.M., Quantum Mechanics, Nauka, Moscow (1974).
[17] DRAKE, J.F., LEE, Y.S., Phys. Fluids **20** (1977) 1341.
[18] LAVAL, G., PELLAT, R., VUILLEMIN, M., "Instabilités électromagnétiques des plasmas sans collisions", Plasma Physics and Controlled Nuclear Fusion Research 1965 (Proc. 2nd Int. Conf. Culham, 1965) **2**, IAEA, Vienna (1966) 259.
[19] GALEEV, A.A., ZELENY, L.M., Pis'ma Zh. Ehksp. Teor. Fiz. **25** (1977) 407.
[20] COPPI, B., Phys. Fluids **8** (1965) 2273; Phys. Lett. **11** (1964) 226.
[21] SCHINDLER, K., SOOP, M., Phys. Fluids **11** (1968) 1192.
[22] WHITE, R.B., MONTICELLO, D., ROSENBLUTH, M.N., WADDELL, B.V., Phys. Fluids **20** (1977) 800.
[23] ROSENBLUTH, M.N., et al., Nucl. Fusion **6** (1966) 297.
[24] GALEEV, A.A., ZELENY, L.M., JETP Lett. **22** (1975) 170.
[25] GALEEV, A.A., ZELENY, L.M., Zh. Ehksp. Teor. Fiz. **69** (1975) 882; **70** (1976) 2135.
[26] SCHINDLER, K., J. Geophys. Res. **79** (1974) 2803.
[27] RUSSELL, C.T., McPHERRON, R.L., Space Sci. Rev. **15** (1973) 205.
[28] CORONITY, F.V., KENNEL, C.F., J. Geophys. Res. **78** (1973) 2837.
[29] GALEEV, A.A., ZELENY, L.M., "The role of plasma processes in the dynamics of magnetospheric substorms", Paper TS-3-39, Int. Symp. Solar and Terrestrial Physics, Boulder, 1976; see also preprint D-231 of Space Research Institute, Moscow (1976).
[30] RUSSELL, C.T., in Physics of Solar Planetary Environments (Proc. Int. Symp. Solar and Terrestrial Physics, Boulder, 1976) **2**, Am. Geophys. Union, Boulder (1976) 526.
[31] TERASAWA, T., NISHIDA, N., Planet. Space Sci. **24** (1976) 1.
[32] WILCOX, J.M., NESS, N.F., J. Geophys. Res. **70** (1965) 5793.
[33] SMITH, E.J., in Solar Wind, NASA Spec. Publ., 308 (1972), Discussion, p. 469.
[34] BAVASANO, B., DOBROWOLNY, M., MARIANI, F., J. Geophys. Res. **81** (1976) 1.
[35] BARANOV, V.B., KRASNOBAYEV, K.V., RUDERMAN, M.S., Astrophys. Space Sci. **41** (1976) 491.
[36] PIDDINGTON, J.H., Cosmic Electrodynamics, Wiley, New York – London (1969) 237.

SIMULATION OF COMPACT BREAKEVEN AND IGNITION EXPERIMENTS

E. BITTONI, B. COPPI*,
M. HAEGI,** A. TARONI
Centro di Calcolo,
Comitato Nazionale per l'Energia Nucleare,
Bologna, Italy

Abstract

SIMULATION OF COMPACT BREAKEVEN AND IGNITION EXPERIMENTS.
An analysis is presented that explores the range of parameters for which breakeven or ignition conditions can be reached in compact D-T burning devices, called Ignitors.

1. INTRODUCTION

The objective of the analysis presented is to explore the range of parameters for which breakeven or ignition conditions can be reached in compact D-T burning devices [1] which we refer to as Ignitors [2].

The main feature of these devices is to be able to produce, simultaneously, relatively high values of the poloidal magnetic field, the plasma current density and the plasma current. The minimum value considered for the latter parameter is 2.2 MA, so that an appreciable fraction of the α-particles produced in D-T reactions can be contained.

Here we analyse two reference types of device. One, called Ignitor-Au, for which the confinement configuration adopted is close to that of experiments such as the Frascati Torus or the Alcator C, in which the auxiliary heating system plays a primary role in the heating cycle towards breakeven or ignition conditions. The other, called Ignitor-Oh, adopts a tight aspect ratio configuration and can sustain higher currents and current densities than Ignitor-Au, so that direct Ohmic heating plays a primary role in the heating cycle. As will be indicated later, we have also explored a limited number of intermediate cases and shall report our results in subsequent papers.

We start the heating cycle with a plasma density that is below the limit for which the centre of the plasma column can be brought to a high enough temperature for radiation bremsstrahlung emission to be compensated by the α-particle

---

* Permanent address: Massachusetts Institute of Technology, Cambridge, Mass., USA.
** Permanent address: Euratom-CNEN Association, Frascati, Italy.

heating [2]. The initial density is chosen in such a range that the ion temperature is close to the electron temperature up to the desired pre-ignition values while the time needed to reach ignition is kept to a minimum. Then the central plasma density is increased by injection of cold plasma at the edge of the plasma column, the heating of the cold plasma being provided both by the α-particles produced in the inner part of the plasma column and by the external heating systems adopted.

2. SIMULATION MODEL

2.1. Transport code

Our numerical analysis has been performed by the same code and basically the same plasma model as was used to simulate numerically the high-density discharges in Alcator [3].

We recall that the electron thermal energy transport is assumed to be anomalous, with a diffusion coefficient

$$D_{an}^{T} = \frac{cT_{eff}}{eB} \frac{E_{\parallel}}{E_R^c} \tag{1}$$

where E_R^c is the collisional runaway electric field defined as $E_R^c = \eta_{cl} v_{the} en$; η_{cl} being the classical electrical resistivity; T_{eff} is an effective temperature that is assumed to be constant and is estimated empirically; and $E_{\parallel}$ is the applied longitudinal field. We recall that T_{eff} is a measure of the fluctuating potential of the microinstabilities that are considered to induce the observed anomalous transport.

The particle loss is also considered to be anomalous and is represented by

$$\Gamma_{an} = -\alpha D_{an}^{T} \frac{\partial n}{\partial r} \tag{2}$$

where the parameter α is obtained by estimates based on existing experiments. The ion thermal energy transport is taken to be purely collisional [3] and the same assumption is made on the electrical resistivity, as the characteristic values of the parameters $\xi_{\parallel} = v_{D\parallel}/v_{the}$ and $E_{\parallel}/E_R^c$ are very low for the regime of interest. Notice that $E_{\parallel}/E_R^c = \xi_{\parallel}\eta_{\parallel}/\eta_{cl}$, where $\eta_{\parallel}$ is the actual collisional resistivity that is relevant to the regime of interest. Thus in the trapped electron regime $\eta_{\parallel}/\eta_{cl} \geqslant 1$ and is an increasing function of the radius.

We recall at this point that a commonly considered scaling for the energy replacement time is $\tau_E \propto na^2$, a being the plasma radius. We may argue that the corresponding D_{an}^T would scale as $D_{an}^T \propto 1/n$, while Eq.(1) would give $D_{an}^T \propto (J_{\parallel}/B)/(nGT^{1/2})$ where $G = \eta_{cl}/\eta_{\parallel}$ and $J_{\parallel}$ is the longitudinal current density. Thus, in evaluating the anomalous transport for Ignitor we use a more pessimistic scaling than $\tau \propto na^2$ in that G becomes as small as $1/6.3$ at the edge of the plasma column and the considered values for $J_\parallel/B$ are larger than those typical of the experiments to which the scaling $\tau \propto na^2$ has been applied, while the increase in $T^{1/2}$ does not compensate that of $J_\parallel/(GB)$ for the regimes of interest.

2.2. Alpha-particle heating

We take into account within the electron energy balance equation, for simplicity, the power density produced by the slowing-down of α-particles due to fusion reactions. In particular we adopt the following expression:

$$P_F = 5.15 \times 10^{-25}\, n^2 T_i^{-2/3} [\exp(-199.4/T_i^{1/3})] \times \epsilon_h \tag{3}$$

where T_i is in eV, n in cm^{-3}, P_F in $W \cdot cm^{-3}$ and equal tritium and deuterium densities, $n_D = n_T = n/2$, are assumed. The parameter ϵ_h is introduced in order to take into account the losses of α-particles due to diffusion. A more complete model replacing Eq.(3) is being formulated and will be included in a subsequent report.

2.3. Neutral beam and microwave injection

The power input resulting from the injection of energetic neutrals or of microwave power is taken into account by a very rough model, expressed by a source term with an appropriate but simple radial profile in the electron and ion thermal energy balance equations. We may expect that even a very simple model can give a sufficiently correct estimate of the effect of neutral beam or microwave injection if the value of the prescribed injected power is realistic.

In the case of Ignitor-Au, which is analysed in Section 4, the relevant source term is proportional to the plasma density as it assumes a uniform distribution of the injected particles over the plasma cross-section. Considering an injection energy of 80 keV and the value of the electron temperature during injection, and assuming classical slowing-down, we estimate that about 20% of the injected power goes to the electrons and about 80% to the ions.

2.4. Gas injection

No satisfactory transport model relevant to high-density regimes ($n \gtrsim 2 \times 10^{14}\ cm^{-3}$) has been developed to reproduce the plasma density increase observed in the Alcator experiment when neutral gas is injected into the plasma chamber [3]. For this reason we cannot compute consistently the development of the density profile during the phase of the discharge when the density is increased. Thus, if $t = t^*$ is the time at which a density increase is prescribed, we model this by

$$n(r,t) = f(t)\, n(r,t^*)$$

$$f(t) = 1 + c_n \frac{t - t^*}{\tau_n} \quad \text{for} \quad t^* \leqslant t < t^* + \tau_n' \tag{4}$$

$$f(t) = 1 + c_n \quad \text{for} \quad t > t^* + \tau_n'$$

where c_n and τ_n are chosen ad hoc and $n(r,t^*)$ is the density profile obtained at $t = t^*$. This profile is computed from the initial time up to the time t from the particle diffusion equation, taking into account neutral recycling at the boundary [2, 4].

We do not consider here the energy loss due to ionization and line radiation, as this is considerably smaller than the energy necessary to bring the injected gas up to the existing plasma temperature. Therefore we include the following terms in the thermal energy balance equation for the electrons and the ions:

$$P_{eff}^{i} = -\frac{3}{2} T_e \left(\frac{dn}{dt}\right)_{ex} \tag{5}$$

and

$$P_{eff}^{e} = -\frac{3}{2} T_i \left(\frac{dn}{dt}\right)_{ex} \tag{6}$$

where $(dn/dt)_{ex}$ is the rate of density rise imposed by the rate of gas injection.

2.5. Surface heating and current-density distribution

To produce a better distribution of the current density and thus decrease the value of $\iota(r = 0)$ we consider the case where an auxiliary heating system is applied. The property of this is to raise the electron temperature at the periphery of the plasma column preferentially. Thus we assume

$$P_A = P_A^0 \frac{r^2}{a^2} \tag{7}$$

for the auxiliary injected power density into the electron population; the corresponding total power is

$$W_A = P_A^0 \pi^2 R_0 a^2$$

and, if we assume $P_A^0 \cong 5\ W \cdot cm^{-3}$, we obtain $W_A \cong 1$ MW, for $R_0 \cong 50$ cm and $a \cong 20$ cm.

The current-density distribution is strongly influenced by the form assumed for the electrical resistivity. When we assume that no trapped particle scattering is produced by the modes that can be excited in the relevant regime we adopt the form derived from Hirschman et al. in Ref.[5]:

$$\eta_\parallel = \frac{\eta_{cl\parallel}}{G(r)} \quad (8)$$

where

$$G(r) = (1-f_T)(1-0.28\, f_T)$$

$$f_T = \{1-(1-\Delta)^2/[(1-\Delta^2)^{1/2}(1+1.46\,\Delta^{1/2})]\}/(1+0.78\nu_*^e)$$

$$\Delta = r/R_0, \quad \nu_*^e = \frac{\sqrt{2}\, q(r) R_0 \nu_e}{v_{the}\ \Delta^{3/2}}$$

$$v_{the} = (2\, T_e/me)^{1/2}$$

$$q(r) = 2\pi/\iota(r)$$

$\iota(r)$ is the rotational transform and ν_e the average electron collision frequency.

When we assume that the rate of trapped electron scattering is high enough for all electrons to contribute to carry the current, as in the case of a constant magnetic field, we adopt $\eta_\parallel = \eta_{cl}$, i.e. $G(r) = 1$.

Note that in the former case, when $\nu_*^e < 1$, the current-density distribution acquires a well peaked profile as $G(r)$ departs rapidly from unity near $r = 0$. On the basis of the results reported in Ref.[6] on the drastically improved stability of resistive modes with poloidal wavenumber $m^0 = 1$, we consider a given current-density distribution as acceptable if the area over which $q(r) \leqslant 1$ is less than 15% of the plasma cross-section area. Finally, we notice that, given the relatively high density regimes in which we are interested, we have assumed $Z_{eff} \cong 1$.

3. THE IGNITOR-OH

The Ignitor-Oh series of devices is characterized by a tight and compact magnet structure designed to maximize the contribution of Ohmic heating toward achieving ignition temperatures in a D-T plasma. The auxiliary heating systems that can be adopted are chosen with the criterion that they do not have such requirements of access to the plasma chamber as to involve alterations and decreased efficiency of the basic magnet structure. Besides contributing to the total injected power, an auxiliary heating system can raise the plasma temperature on the periphery of the plasma column and therefore lead to a current-density profile that is relatively broad. This in turn can permit stable plasma discharges with relatively low values of the safety factor:

$$q_s = \frac{B_T(r=0)\ a}{R_0 B_p (r=a)}$$

An additional means by which this can be accomplished is to decrease the aspect ratio to relatively low values such as 2.5 or less. In this case we can have $q(r=a) \cong 2.5$ and $q(r=0) \cong 0.8$ for reasonable [2] current distributions and $q_s \cong 2$, where $q(r) = 2\pi/\iota(r)$ and $\iota(r)$ is the rotational transform, i.e. the rate of magnetic shear is enhanced in a tight aspect ratio configuration. In assessing the acceptable values of $q(r=0)$ and q_s we take into account that in the high density and temperature regimes of interest the effects of ion-ion collisions tend to improve the stability of internal resistives modes.

In this connection we shall evaluate the parameter ϵ_q, which is the ratio of the area over which $q(r) < 1$ to the total cross-section area, and consider $\epsilon_q \cong 0.15$ an acceptable value, as indicated earlier. We also notice that the values of

$$\beta_p = \frac{8\pi}{B_p^2 (r=a)} \langle n(T_e + T_i) \rangle$$

for which ignition temperatures can be achieved are well below unity, and experiments carried out on the Alcator device have indicated that stable discharges can be produced for values of $q(r=a)$ as low as 2. We shall therefore consider $q_s \cong 2.15$ as a reference value so that $q(r \cong a) \cong 2.7$ for a current-density profile somewhat more peaked than a parabolic one.

In particular, we have carried out our analysis for a set of reference parameters relevant to a device for which a preliminary design study is being undertaken [7]:

$R_0 = 54$ cm	$a = 21.5$ cm
$B_T = 160$ kG	$I \cong 3.22$ MA
$q_s \cong 2.15$	$B_p \cong 30$ kG

We adopt relatively high densities, close to the limit for which bremsstrahlung emission would prevent the system from reaching the ideal ignition temperature of 4 keV. The criterion for choosing a given density value is, in fact, that of decreasing the ignition time to a minimum. This is defined as the time at which the α-particle heating exactly compensates all losses. Thus, after a number of trials, we have chosen optimally $n^0 \cong 2 \times 10^{15}\ cm^{-3}$ as the peak initial plasma density.

Referring to Eq.(1) giving an expression for the anomalous diffusion coefficient for the thermal energy, we take $T_{eff} \cong 50$ eV in most of the cases that we analyse and, referring to Eq.(2), $\alpha \cong 10^{-2}$ as for the case treated in Ref.[3].

We prescribe parabolic profiles of density and temperatures as initial values with $n_0 = 2 \times 10^{15}\ cm^{-3}$, $T_{e0} = T_{i0} = 1$ keV at the centre. The prescribed value $B_p = 30$ kG and the requirement of a parabolic profile of the current density J determine the initial values of B_p. Boundary values are at all times: $n_a = 10^{14}\ cm^{-3}$; $T_{ea} = T_{ia} = 10$ eV; $B_p = 30$ kG. Impurities are not considered and we take $Z_{eff} = 1$. This assumption seems reasonable given the high densities considered here, as shown by the Alcator results [8].

We define t_{i4} to be the time at which the peak ion temperature reaches 4 keV. This is the ideal ignition temperature where the power loss by bremsstrahlung emission is equal to the α-particle fraction of the power produced in D-T reactions. For the density range we consider, t_{i4} is also about the time at which the peak electron temperature is 4 keV. In addition, we define the ignition time t_I at which the α-particle heating compensates all types of power loss from the plasma column.

Case (a): In the case where we assume $\epsilon_h = 1$ we obtain, in the absence of auxiliary heating,

$$t_{i4} \cong 0.19\ s;\ \beta_p(t = t_{i4}) \cong 0.3;\ \ \epsilon_q(t \cong t_{i4}) \cong 0.1$$

$$t_I \cong 0.67\ s;\ \beta_p(t = t_I) \cong 0.53;\ \ \epsilon_q(t \cong t_I) \cong 0.21$$

Case (b): In the case where we take $\epsilon_h = 0.7$ we have correspondingly

$$t_{i4} \cong 0.20\ s;\ \beta_p(t_{i4}) \cong 0.3;\ \epsilon_q(t_{i4}) \cong 0.11$$

$$t_I \cong 1.12\ s;\ \beta_{pI} \cong 0.66;\ \epsilon_{qI} \cong 0.25$$

where $\beta_{pI} = \beta_p(t = t_I)$ and $\epsilon_{qI} = \epsilon_q(t = t_I)$. The peak ion temperature evolution as a function of time for these and two other cases is represented in Fig.1. In addition, Figs 2–4 show the profiles of density, temperatures and current density at $t = t_{i4}$ for case (a). At the same time, the anomalous electron conductivity confinement time is $\tau_e \cong 0.37$ s, and the classical ion conductivity confinement time is $\tau_i \cong 2.3$ s,

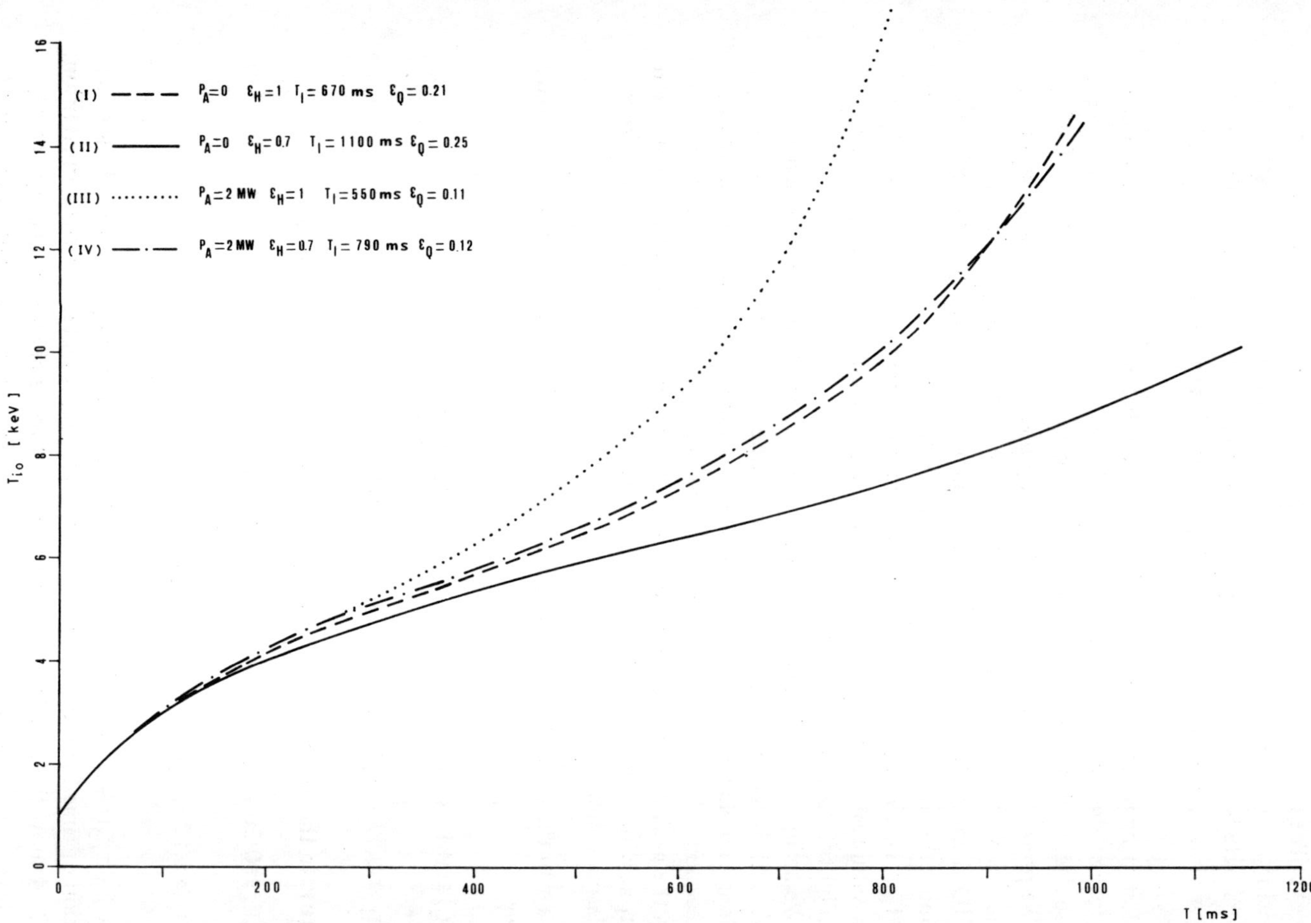

FIG.1. Ignitor-Oh. Reference parameters: $R = 54$ cm; $a = 21.5$ cm; $B_t = 160$ kG; $I = 3.22$ MA; $T_{eff} = 50$ eV; $\alpha = 0.01$.
Peak ion temperature evolution $T_i(r = 0)$ as a function of time.
Curve I: $\epsilon_h = 1$ and $P_A = 0$. Curve II: $\epsilon_h = 0.7$ and $P_A = 0$. Curve III: $\epsilon_h = 1$ and $P_A = 2$ MW. Curve IV: $\epsilon_h = 0.7$ and $P_A = 2$ MW.

indicating that classical conductivity losses are negligible in comparison to the anomalous losses.

Nearly equal results are obtained for case (b) at $t \cong t_i$ and we observe that this set of results corresponds to two of the many cases we have investigated by varying the dimensions of the device, the magnetic field, the plasma current, etc.

To decrease $\epsilon_q(t=t_I)$ to acceptable values, we have considered the possibility of raising the electron temperature at the periphery of the plasma column by means of an external surface heating as described at the end of Section 2. We assume that the external surface heating is provided from the initial phase of the discharge onwards. In our computations we have considered several values of the external power P_A; in particular, we find that $P_A \cong 2$ MW is required to obtain acceptable values of ϵ_q at $\cong t = t_I$.

Typical results for the standard Ignitor-Oh parameters are:

Case (c): $P_A = 2$ MW; $\epsilon_h = 1$:

$$t_{i4} \cong 0.18 \text{ s}; \quad \beta_p(t_{i4}) \cong 0.35; \quad \epsilon_q(t_{i4}) \cong 0.088$$

$$t_I \cong 0.55 \text{ s}; \quad \beta_p(t_I) \cong 0.66; \quad \epsilon_q(t_I) \cong 0.11$$

The electron and ion thermal conductivity energy confinement times at $t=t_{i4}$ are: $\tau_e \cong 0.26$ s; $\tau_i \cong 1.8$ s. At $t=t_I$ the effective power input from α-particles is $P_\alpha = \epsilon_h P_{\alpha\,total} \cong 3.3$ MW and the total power of fusion-produced neutrons is $P_n \cong 13.2$ MW. Figures 5–7 show the profiles of density, temperature and current density at $t=t_I$.

Case (d): $P_A = 2$ MW; $\epsilon_h = 0.7$

$$t_{i4} \cong 0.18 \text{ s}; \quad \beta_p(t_{i4}) \cong 0.35; \quad \epsilon_q(t_{i4}) \cong 0.088$$

$$t_I \cong 0.79; \quad \beta_p(t_I) \cong 0.8; \quad \epsilon_q(t_I) \cong 0.12$$

The energy confinement times at $t=t_{i4}$ are practically the same as in case (c). At $t=t_I$ we have $P_\alpha \cong 3.6$ MW; $P_n \cong 20.6$ MW. In addition, Figs 8–10 show the profiles of density, temperature and current density at $t=t_I$.

The evolution of the peak ion temperatures as a function of time in cases (c) and (d) is represented in Fig.1. We notice that we have varied T_{eff} in order to find the upper value for which the ignition phase can be reached. We find this value to be $T_{eff} \cong 150$ eV. In addition, an increase in T_{eff} corresponds to an increase in t_{i4}, t_I, ϵ_{qI} and t_0.

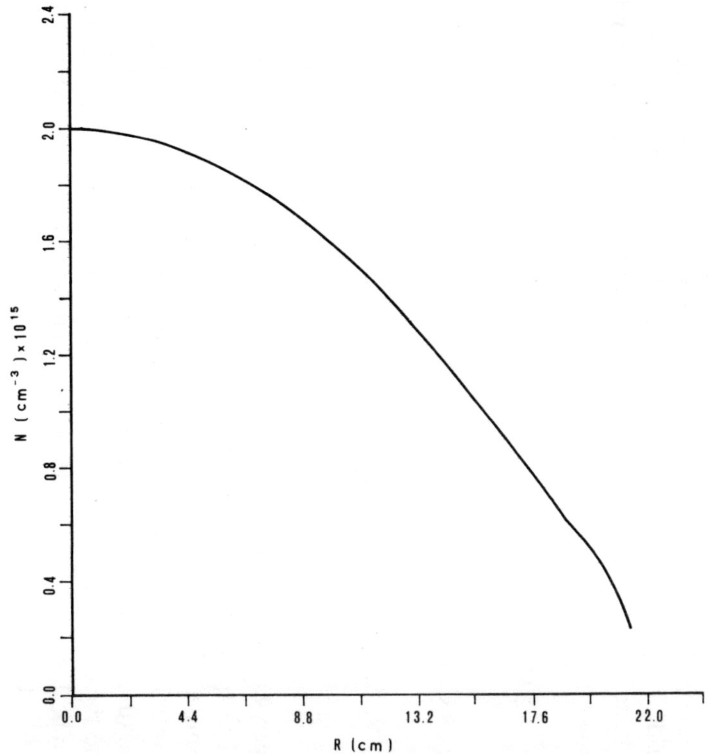

FIG.2. Ignitor-Oh. Case $\epsilon_h = 1$; $P_A = 0$. Density profile at $t = t_{i4} = 0.19$ s.

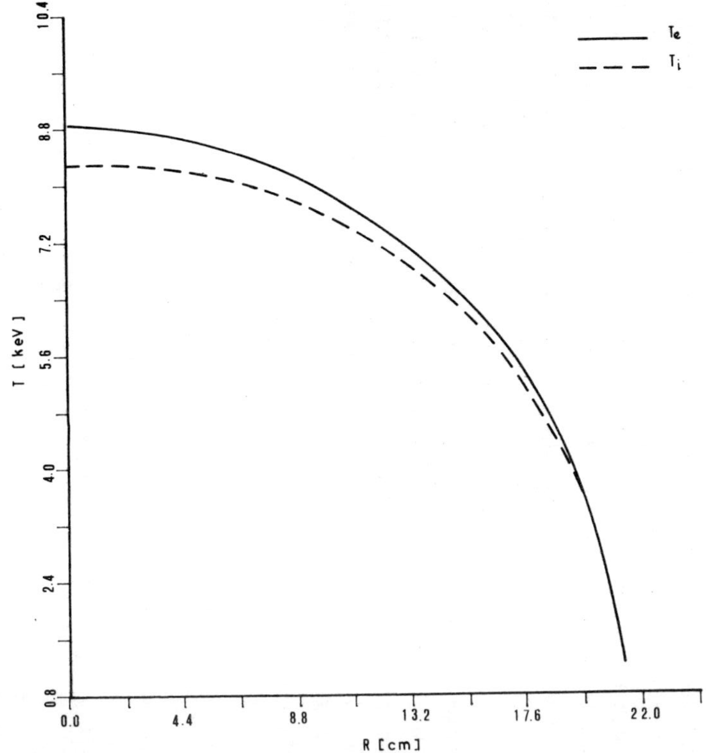

FIG.3. Ignitor-Oh. Case $\epsilon_h = 1$; $P_A = 0$. Electron and ion temperature profiles at $t = t_{i4} = 0.19$ s.

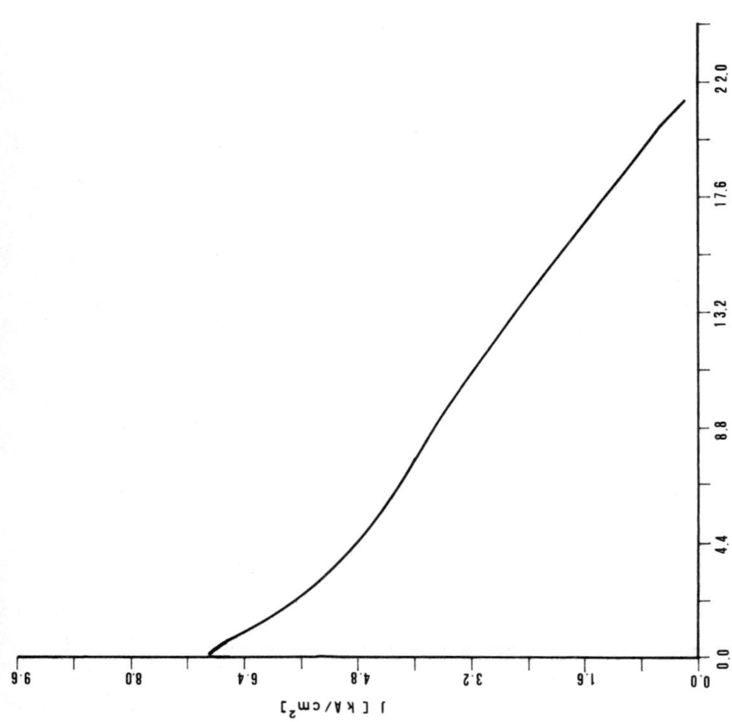

FIG.4. Ignitor-Oh. Case $\epsilon_h = 1$; $P_A = 0$. Current density profile at $t = t_{i4} = 0.19$ s.

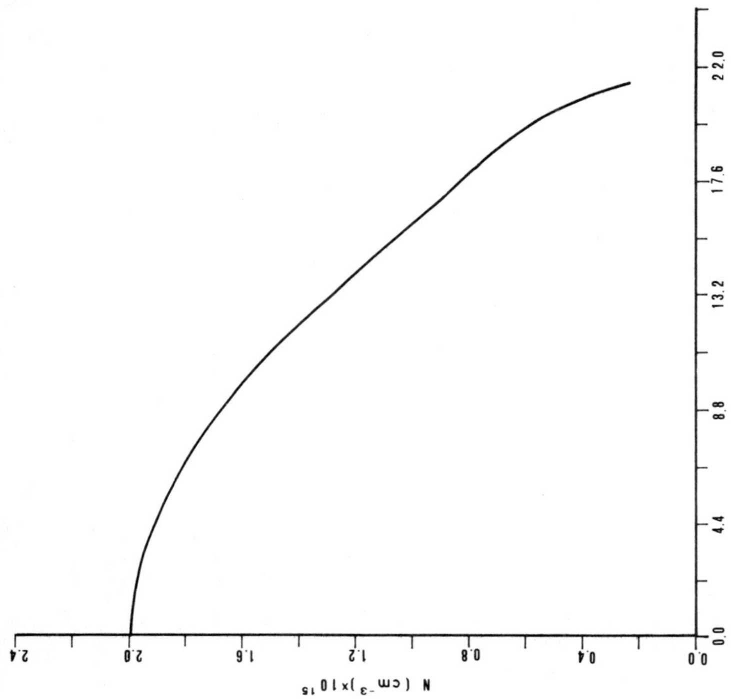

FIG.5. Ignitor-Oh. Case $\epsilon_h = 1$; $P_A = 2$ MW. Density profile at the ignition time $t_I = 0.55$ s.

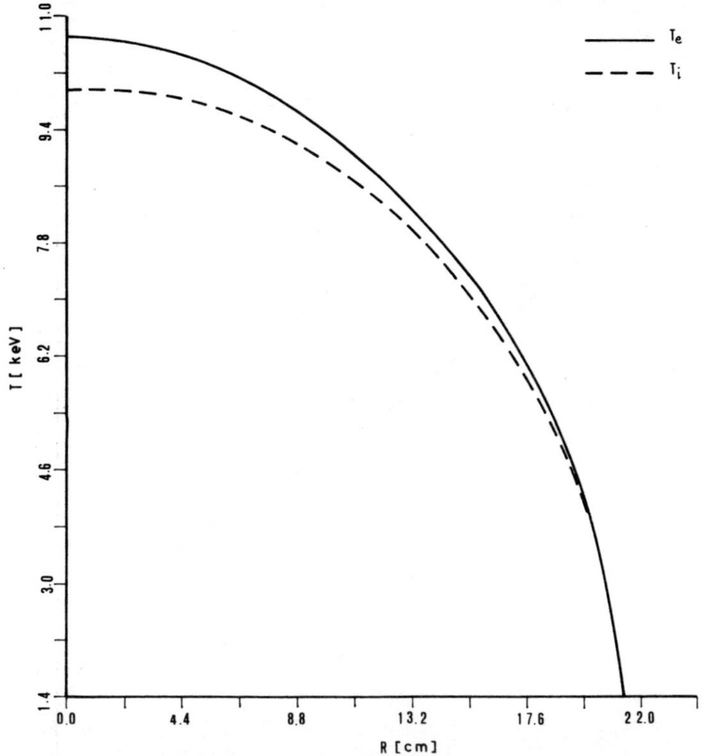

FIG.6. Ignitor-Oh. Case $\epsilon_h = 1$; $P_A = 2$ MW. Electron and ion temperature profiles at the ignition time $t_I = 0.55$ s.

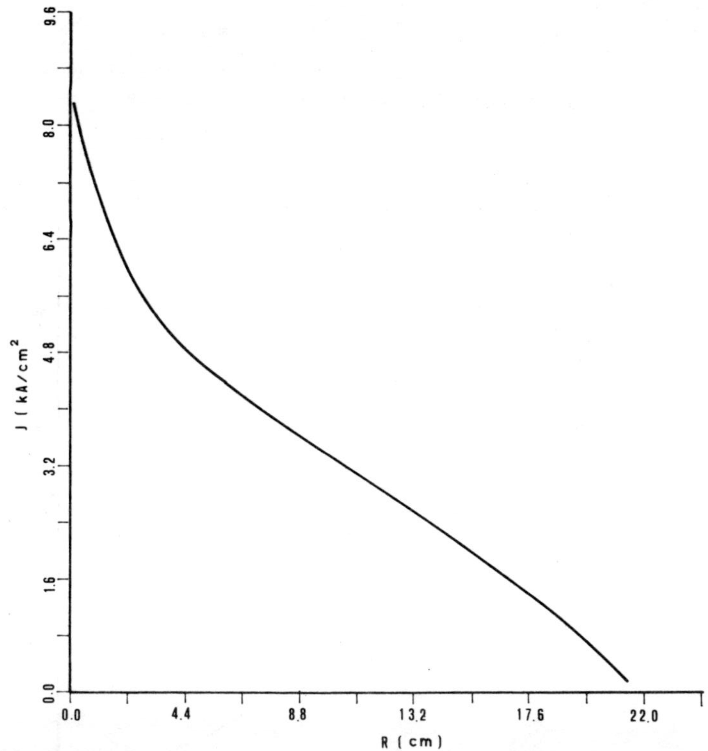

FIG.7. Ignitor-Oh. Case $\epsilon_h = 1$; $P_A = 2$ MW. Current density profile at the ignition time $t_I = 0.55$ s.

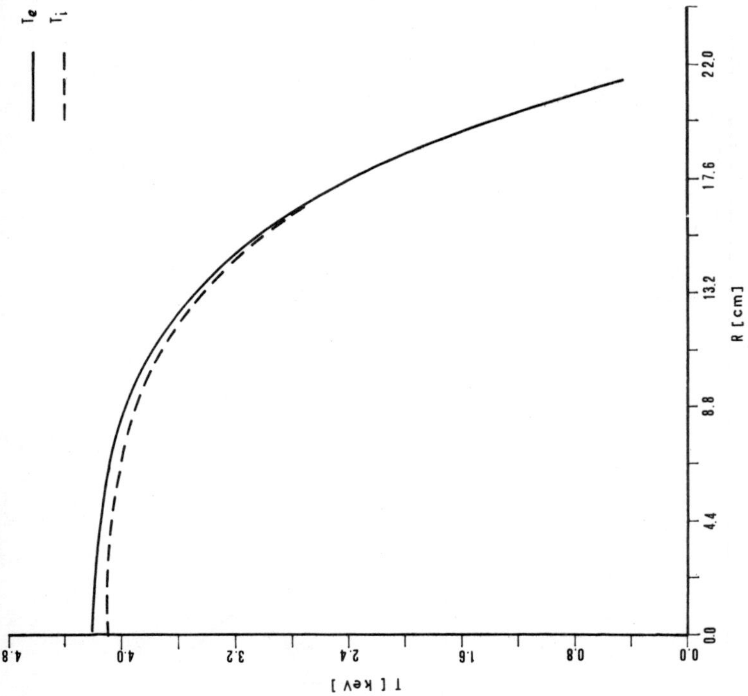

FIG.9. Ignitor-Oh. Case $\epsilon_h = 0.7$; $P_A = 2$ MW. Electron and ion temperature profiles at the ignition time $t_I = 0.79$ s.

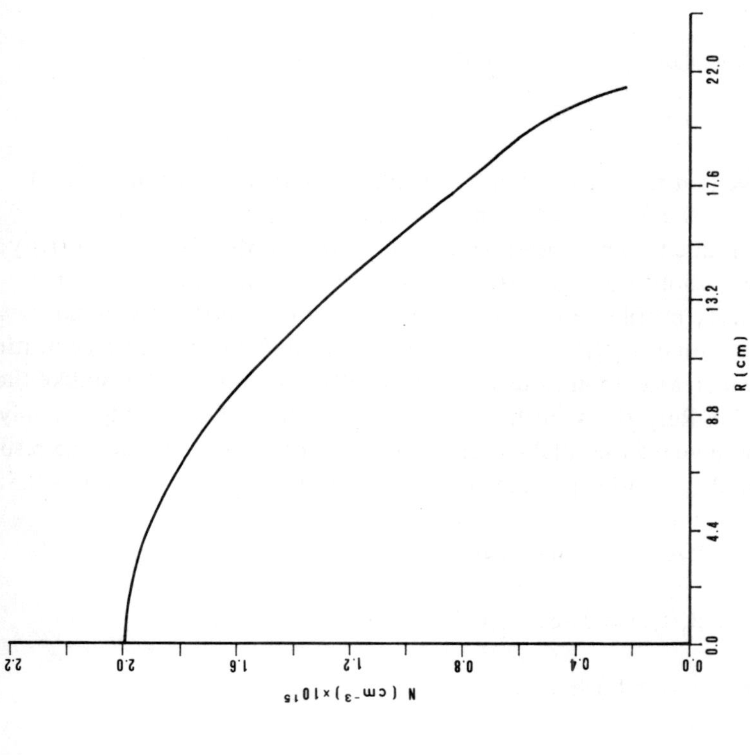

FIG.8. Ignitor-Oh. Case $\epsilon_h = 0.7$; $P_A = 2$ MW. Density profile at the ignition time $t_I = 0.79$ s.

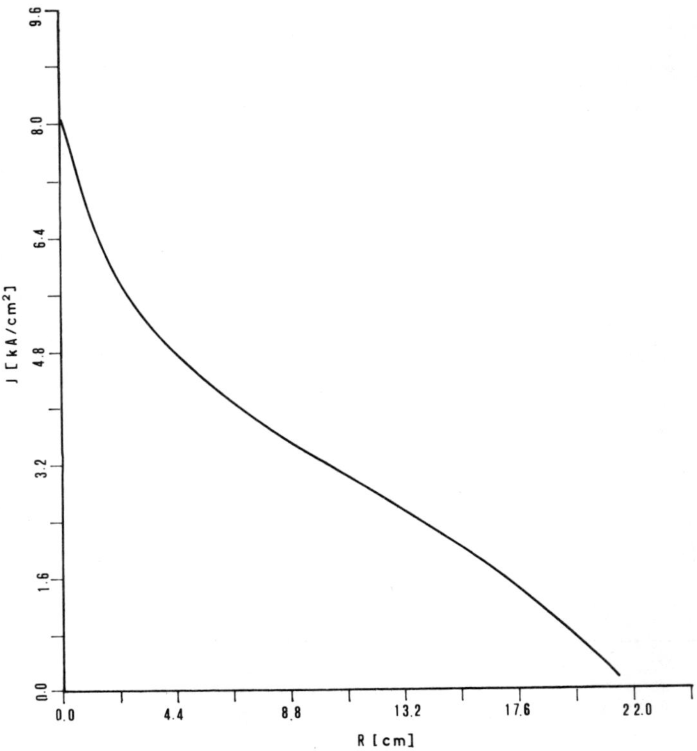

FIG.10. Ignitor-Oh. Case $\epsilon_h = 0.7$; $P_A = 2$ MW. Current density profile at the ignition time $t_I = 0.79$ s.

We observe that the current density profiles are rather peaked in cases (c) and (d) in spite of the fact that the temperature profiles are very flat or even inverted. This is due to the trapped-particle correction to the classical resistivity. Computer runs without trapped-particle correction to the resistivity give very flat current density profiles in these cases and $\epsilon_q(t_I) = 0$. For this reason no significant reduction of $\epsilon_q(t_I)$ is achieved, with or without external surface heating, by considering increased anomalous transport coefficients when $q < 1$, unlike the case considered in Ref.[9]. A further reduction of $\epsilon_q(t_I)$ and t_0 at $t = t_I$ can only be obtained by reducing the total current I. We have obtained the following results with the standard Ignitor-Oh parameters, but $I = 2.8$ MA, $P_A = 2$ MW and $\epsilon_h = 0.5$:

Case (e): ($I = 2.8$ MA; $P_A = 2$ MW; $\epsilon_h = 0.5$)

$t_{i4} \cong 0.27$ s; $\beta_p(t_{i4}) \cong 0.48$; $\epsilon_q(t_{i4}) \cong 0.03$

$t_I \cong 1.5$ s; $\beta_p(t_I) \cong 1.18$; $\epsilon_q(t_I) \cong 0.04$

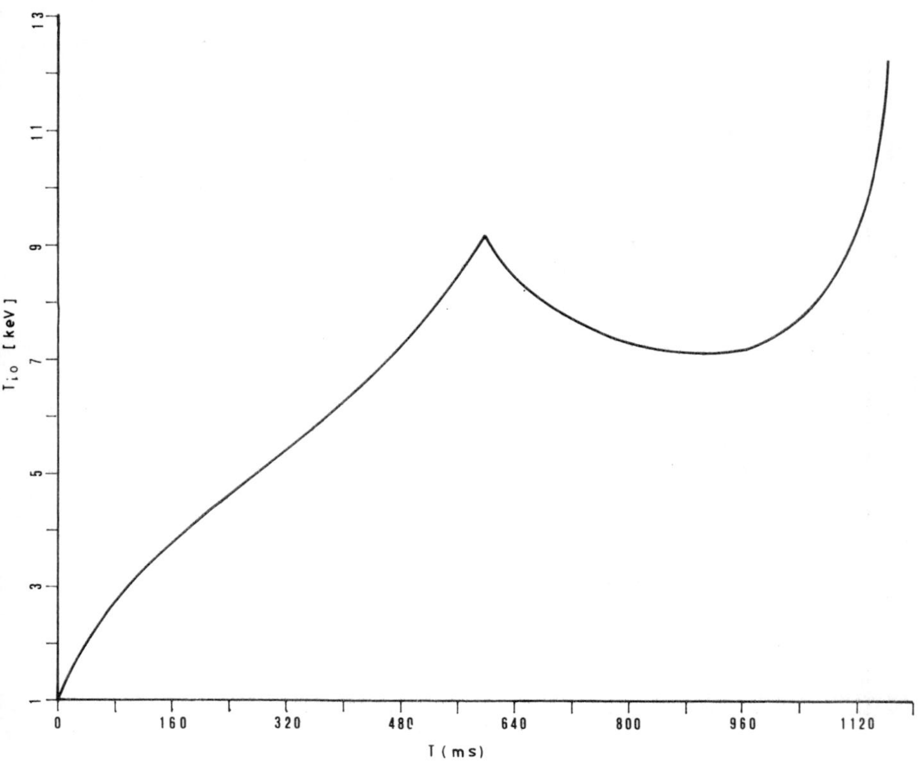

FIG.11. Ignitor-Oh. Case $\epsilon_h = 1$, $P_A = 2$ MW, and density increase by a factor 4. Peak ion temperature evolution $T_i(r = 0)$ as a function of time.

The electron and ion thermal conductivity energy confinement times at $t = t_{i4}$ are $\tau_e \cong 0.29$ s; $\tau_i \cong 1.35$ s. At $t = t_I$ we have $P_\alpha \cong 3.5$ MW and $P_n \cong 28$ MW, as the corresponding average ion temperature is $\bar{T}_i \cong 6.7$ keV. The high value of t_I obtained in this case is due to the rather pessimistic choice of $\epsilon_h = 0.5$.

To increase the power produced by fusion reaction, we programme a density increase such that the α-particle heating contributes significantly to raising the temperature of the newly injected plasma. We have seen that it is not convenient in the case of Ignitor-Oh to increase the plasma density before the ignition time t_I. In fact, to decrease t_I it is convenient to exploit the properties of Ohmic heating by beginning the heating cycle with the maximum permissible density and start to increase the density only when $t \gtrsim t_I$. Our computations give an estimate of the maximum density increase that can be achieved in a given time τ. We choose $\tau = 0.6$ s after a series of trials, assuming that the considered discharge can last up to 1.4 s, including the pre-ignition phase.

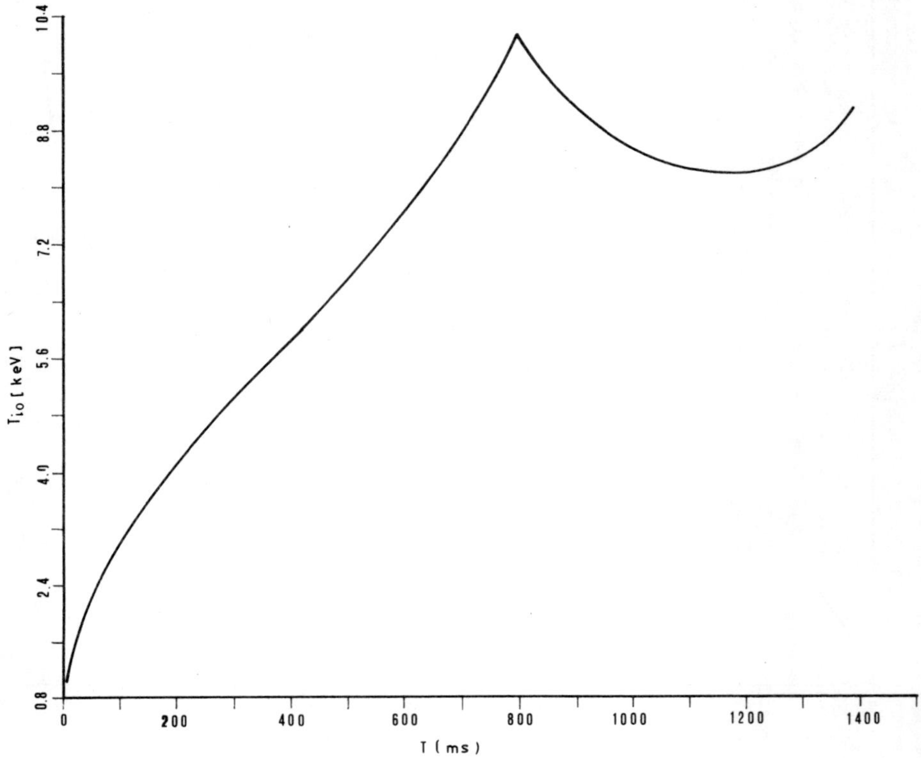

FIG.12. Ignitor-Oh. Case $\epsilon_h = 0.7$, $P_A = 2$ MW, and density increase by a factor 3. Peak ion temperature evolution $T_i(r = 0)$ as a function of time.

Our numerical results show that in case (c) the density can be enhanced by a factor 4 in a time $\tau = 0.6$ s, starting at $t = 0.6$ s, slightly after the ignition time is reached. At $t = 1.2$ s we obtain $P_\alpha \cong 100$ MW; $P_n \cong 400$ MW. At the same time, the following average values are obtained:

$$\bar{n} = 4.2 \times 10^{15}\ \text{cm}^{-3};\ \bar{T}_e = 5.91\ \text{keV};\ \bar{T}_i = 5.53\ \text{keV};\ \beta_p \cong 2.6$$

Figure 11 shows the evolution in time of the peak ion temperature for the considered case.

In case (d) the density can be increased by a factor 3, starting at $t = 0.8$ s ($t_I = 0.79$ s). At $t = 1.4$ s we obtain:

$$P_\alpha \cong 37\ \text{MW};\ P_n \cong 211\ \text{MW};\ \bar{n} = 3.3 \times 10^{15}\ \text{cm}^{-3}$$

$$\bar{T}_e = 5.04\ \text{keV};\ \bar{T}_i = 4.96\ \text{keV};\ \beta_p \cong 1.7$$

Figure 12 shows the evolution in time of the peak ion temperature for the considered case.

4. THE IGNITOR-AU DEVICE

The purpose of the Ignitor-Au device is to make use of magnet technologies that have already been developed and tested in order to realize a D-T toroidal plasma in which breakeven conditions are reached. The adoption of an auxiliary heating system compatible with the limited access allowed by the high-field magnets employed is an integral part of the Ignitor-Au design. One of the uses foreseen for a device of this type is that of a Toroidal Material Testing Reactor as indicated in Ref.[1]. In this context it is proposed that the magnetic core of the device be treated as disposable, with the plasma chamber itself serving as the material to be tested under a variety of thermal and mechanical loading conditions. The reference parameters we consider for an Ignitor-Au device are [1]:

$$R_0 = 78 \text{ cm}; \quad a = 24 \text{ cm}; \quad q_s = 2.5$$

$$B_T = 150 \text{ kG}; \quad I = 2.22 \text{ MA}; \quad B_p = 18.5 \text{ kG}$$

$$S_0 \cong 7.4 \times 10^4 \text{ cm}^2; \quad V_0 = 0.89 \times 10^6 \text{ cm}^3$$

Here S_0 and V_0 indicate the surface and the volume of the plasma surface respectively.

Initial and boundary values of n, T_e, T_i and B_p are given as in the case of Ignitor-Oh. Only n_0, n_a and B_p are changed, as we take $n_0 = 3 \times 10^{14} \text{ cm}^{-3}$; $n_a = 3 \times 10^{13} \text{ cm}^{-3}$; $B_{pa} = 18.5$ kG.

In the present case we foresee the use of neutral injection for auxiliary heating of the plasma as described in Section 5. According to the results of Section 5, we assume that an effective power input $P_{NI} = 4$ MW is available from the beginning of the discharge onwards and distribute it to electrons and ions as described at the end of Section 2. The initial density is much smaller than in Ignitor-Oh because the Ohmic heating is now smaller, and in order to deal with reasonable values of the energy for the injected neutrals we adopt $n_0 = 3 \times 10^{14} \text{ cm}^{-3}$ and reach the stage of the discharge when the power input due to fusion becomes larger than the power loss due to bremsstrahlung, within a time $t_{Br} \cong 0.06$ s.

In this case it is convenient to start the density increase phase as soon as possible, taking advantage of the external power input. Thus we can avoid operating in low-collisionality regimes, with relatively high temperatures and low densities. We follow the 'radiation tunnelling' procedure described in Ref.[2] and choose $t = t_{Br}$ as the initial time of the density increase phase of the discharge.

A series of computations relating to Ignitor-Au show that ignition can be reached in a time $t_I \cong 1.0 - 1.4$ s for different choice of the parameters T_{eff}, ϵ_h, P_{NI} and α. Our final assumptions for the values of these parameters are

$$T_{eff} = 100 \text{ eV}; \quad \alpha = 0.01; \quad \epsilon_h = 0.7; \quad P_{NI} = 4 \text{ MW}$$

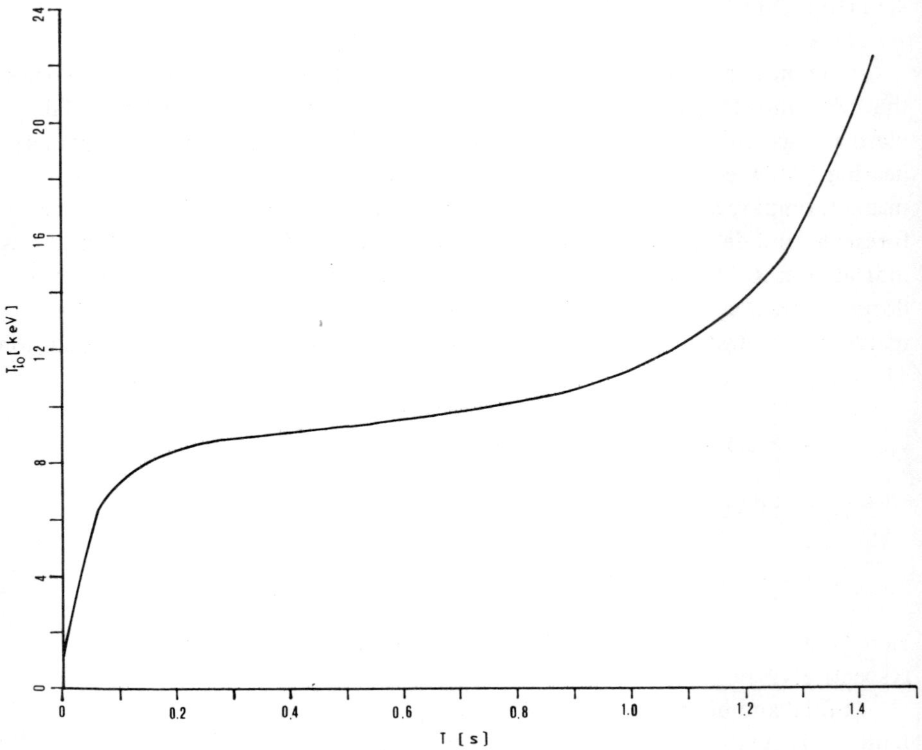

FIG.13. Ignitor-Au. Reference parameters: $R = 78$ cm; $a = 24$ cm; $B_T = 150$ kG; $I = 2.22$ MA; $P_{NI} = 4$ MW; $T_{eff} = 100$ eV; $\alpha = 0.01$; $\epsilon_h = 0.7$.
Peak ion temperature evolution $T_i(r = 0)$ as a function of time. In the considered case the density is increased by a factor 7.

We choose the duration of the density-increase phase of the discharge to be τ=1.2 s, and assume that a typical discharge in Ignitor-Au can last 1.5 s. With the parameters given above, the density can be increased by a factor 7, reaching a peak of 2.1×10^{15} cm^{-3} and we obtain:

$$t_{Br} \cong 0.06 \text{ s}; \ \beta_p(t_{Br}) \cong 0.17; \ \epsilon_q(t_{Br}) \cong 10^{-3}; \ T_e(t_{Br}) \cong 0.1 \text{ s}$$

$$T_i(t_{Br}) \cong 1.5 \text{ s}; \ t_I \cong 0.93 \text{ s}; \ \beta_p(t_I) \cong 1.7; \ \epsilon_q(t_I) \cong 0.04$$

The electron and ion thermal conductivity energy confinement times at time t_{Br} are $\tau_e \cong 0.1$ s, $\tau_i \cong 1.5$ s. At $t = t_I$ we have $P_\alpha \cong 5$ MW and $P_n \cong 28.5$ MW.

The evolution of the peak ion temperature as a function of time in the considered case is represented in Figure 13.

5. NEUTRAL INJECTION HEATING FOR IGNITOR-AU

The method of neutral injection heating which, for this device, provides the highest efficiency in a wide range of operating plasma densities involves vertical or quasivertical injection into the upper confinement disc. The confinement discs are regions of space facing the injection windows where the magnetic well (i.e. the local mirror in which particles with small $v_{\parallel}/v_{\perp}$ can be trapped and drift out of the plasma in the $\vec{B} \times \nabla B$ direction) along a field line is suppressed by the pitch of the helical toroidal field. Thus confinement discs exist in regions where the slope of the field intensity, along a given helical line of force, is higher than the slope of the ripple field, i.e.

$$\left| \frac{\partial B_{\text{helical}}}{\partial s} \right| \gtrsim \left| \frac{\partial B_{\text{ripple}}}{\partial s} \right| \tag{9}$$

where s is a parameter along the field line.

This condition can be satisfied in the central part of the cross-section, for symmetric coils where the ripple is low, and in the upper and lower half of the plasma cross-section where field along a magnetic line varies most rapidly. In the case of symmetric coils the confinement discs look like two ellipses roughly centred at the upper and lower half [10, 11] of the cross-section.

Other methods of injection at high densities would require very high energies for the neutrals to penetrate towards the centre of the plasma. In addition: (i) tangential injection would require a complicated coil structure for the type of device under consideration; and (ii) perpendicular horizontal injection in the confinement discs would lead to substantial losses because of the beam attenuation from the external border of the plasma up to the confinement discs, which are displaced toward the symmetry axis, given the compact structure of the magnet.

We consider two possible regimes of operation for the case of vertical injection:

(a) Low density

For $\bar{n} \cong 10^{14}$ cm^{-3} and a neutral deuteron energy of $W_{od} = 80$ keV, the total ionization mean free path is $\lambda \sim a$. Thus the penetration of the beam into the plasma is not a problem. If the perpendicularly injected neutral is ionized before it encounters the confinement disc, it will drift upwards until it intersects the disc and then be confined. The size of the confinement disc is such that all drifting particles produced by the beam are intercepted.

(b) High density

For $\bar{n} \cong 10^{15}$ cm^{-3} the 80-keV deuteron is ionized near the surface of the plasma as $\lambda \sim a/10$. Further penetration has to rely on ∇B drift. Therefore, the

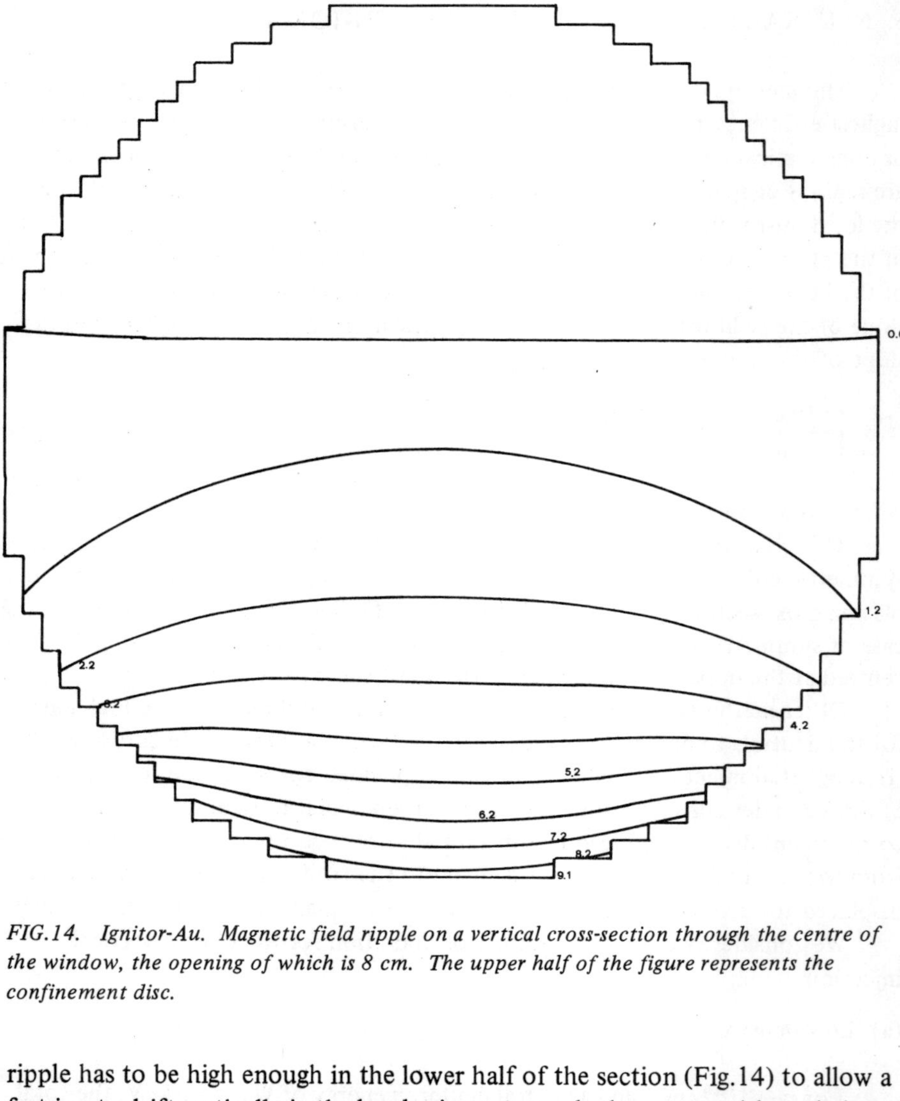

FIG.14. Ignitor-Au. Magnetic field ripple on a vertical cross-section through the centre of the window, the opening of which is 8 cm. The upper half of the figure represents the confinement disc.

ripple has to be high enough in the lower half of the section (Fig.14) to allow a fast ion to drift vertically in the local mirror towards the centre without being trapped into a banana trajectory (Fig.15). On the upper half of the cross-section, the ripple has to be low in order to have, as for the low-density case, a relatively large confinement disc [10,12].

Numerical calculations on single-particle trajectory including collisions are under way, using the three-dimensional FAR code (originally developed by the Laboratory of Fontenay-aux-Roses, France) with an improved small-angle scattering mechanism, in order to identify the proper design parameters of the relevant window coil. Meanwhile, to formulate some basic requirements for the ripple, we shall give simple estimates.

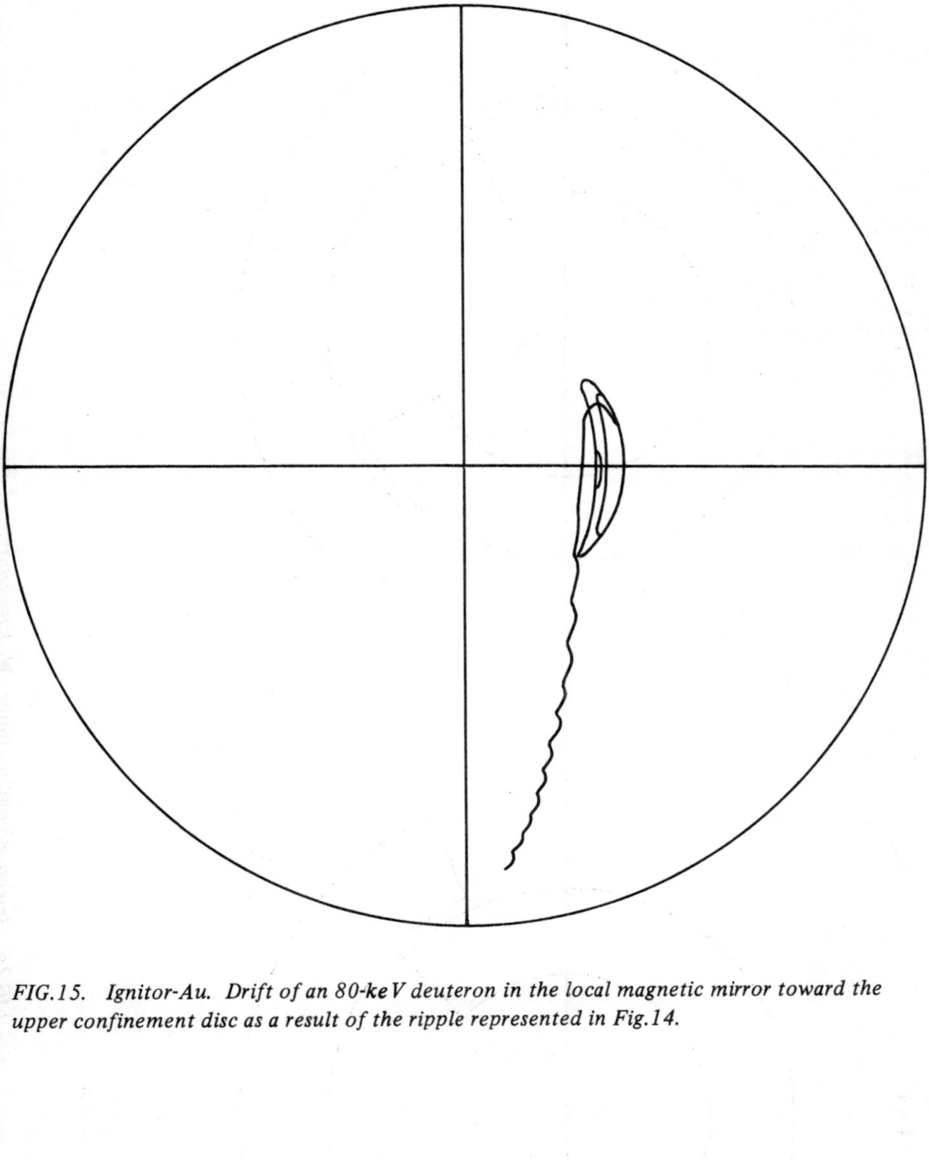

FIG.15. Ignitor-Au. Drift of an 80-keV deuteron in the local magnetic mirror toward the upper confinement disc as a result of the ripple represented in Fig.14.

The mean field particle density, necessary to give sufficient $v_{\parallel}/v_{\perp}$ to the purely $v_{\perp}$ injected deuteron during its $\vec{B} \times \nabla B$ drift from the border to the centre of the plasma, so that it can escape from the $\delta B/B$ ripple mirror, is given by

$$\bar{n} \cong 5.3 \times 10^{17} \frac{\delta B}{B} \text{ cm}^{-3}$$

for $A_i = 2$; $W_{od} = 80$ keV; $R_0 = 78$ cm; $a = 24$ cm; $B_0 = 150$ kG.

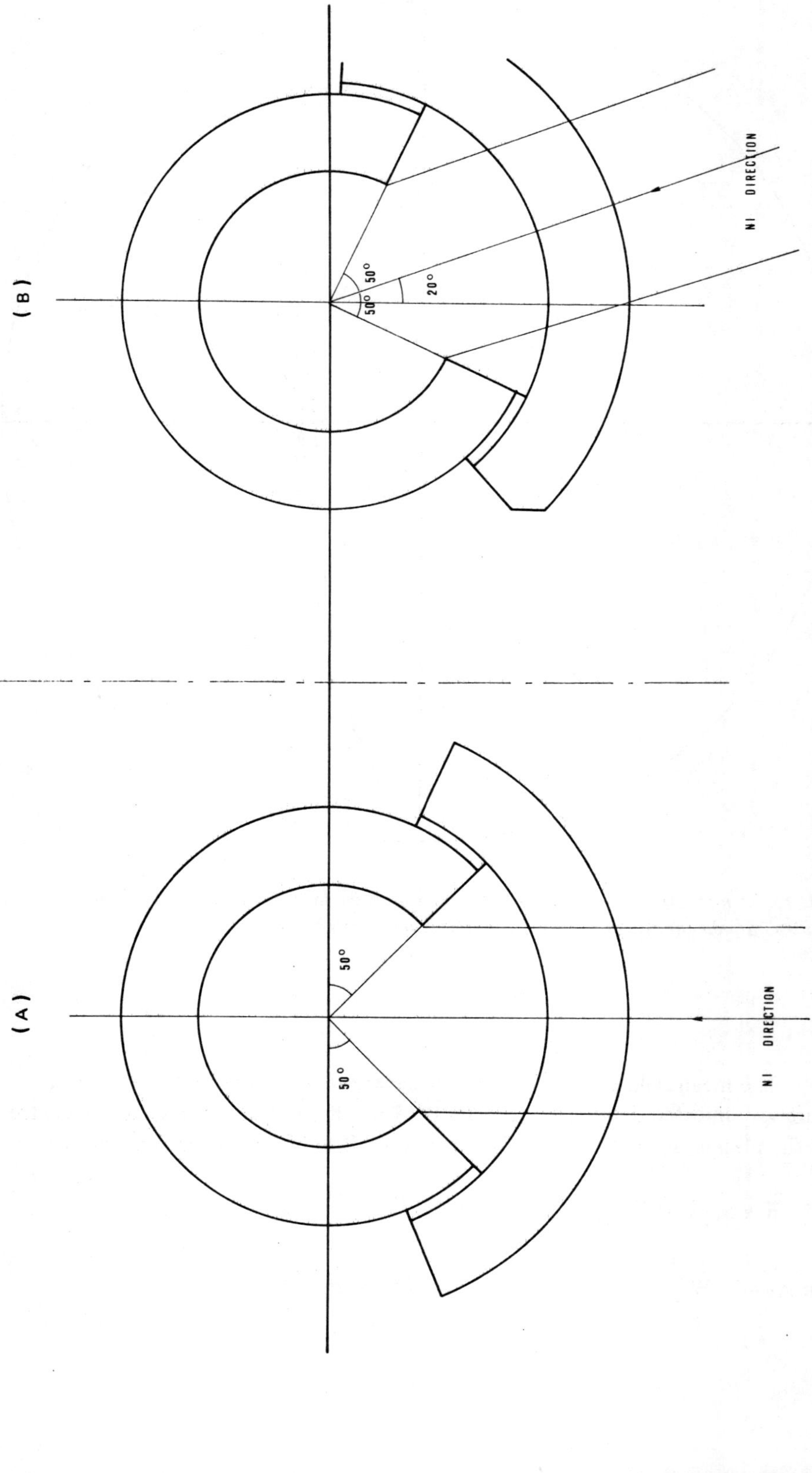

FIG.16. Injection configurations. We have considered case (a) for our calculations; case (b) features an improved access.

Then for a maximum operating mean density of $\bar{n}_{max} \sim 2 \times 10^{15}\ cm^{-3}$, the ripple has to be

$$\left(\frac{\delta B}{B}\right)_{\substack{lower\\half}} > 3.7 \times 10^{-3} \qquad (10)$$

to avoid collisional trapping before reaching the central core. In addition, to allow the drifting particle to reach the central region, the confinement disc in the lower half of the section has to be almost suppressed. Assuming a sinusoidal function for $B_{helical}$ and B_{ripple}, the condition (9) which defines the confinement disc, is

$$\left| \frac{\partial}{\partial s} \left(\frac{B_{or}}{R_0} \cos \phi/q(r) \right) \right| \geqslant \left| \frac{\partial}{\partial s} \left(\frac{\delta B(r)}{2} \cos(N\phi) \right) \right| \qquad (11)$$

where ϕ is the angle around the axis of the torus, $\delta B_{(r)}$ the peak-to-peak average ripple at the radius r, and N the number of coils.

Writing the poloidal angle $\theta = \phi/q$, Eq.(11) becomes

$$\frac{\delta B(r)}{B_0} \lesssim \frac{2r\,|\sin\theta\,|}{N R_0 q(r) f}$$

where f is a correction factor dependent on the actual behaviour of the field.

A numerical computation of the field for the considered Ignitor-Au configuration gives $f \cong 3$.

Thus the ripple configuration of Ignitor-Au can be schematically summarized as follows. In the centre of the lower half of the cross-section the ripple should be

$$\left(\frac{\delta B}{B}\right)_{\substack{lower\\half}} > 4\% \qquad (12)$$

in order to suppress the lower confinement disc. One sees that condition (12) is stronger than condition (10). In the centre of the upper half of the section the ripple has to be

$$\left(\frac{\delta B}{B}\right)_{\substack{upper\\half}} \lesssim 0.5\% \qquad (13)$$

in order to have a large confinement disc there.

Among the possible coil systems which can provide such a ripple configuration, a solution is shown in Fig.16(A). To realize the window, an arc of the turns of the toroidal field coil is removed, lifted outwards over the remaining circular turns, and pushed on both sides of the window itself. The ripple in the vertical cross-section through the centre of the window and the confinement disc in the upper half of the plasma cross-section are shown in Fig.14. The relevant coil system meets the requirements (10–13).

Neutral injection in high-field devices with narrow windows like Ignitor-Au obviously require low-divergence beams. This divergence, which can now reach $\pm 0.3°$, is expected to reach $\pm 0.1°$ in the near future [13, 14] and we shall assume the latter value for the following calculations. Assuming a beam steering, we shall take at the window a bulk-equivalent current density of D^0 of $0.25\ A \cdot cm^{-2}$, which corresponds to an extracted D^+ current density of $0.40\ A \cdot cm^{-2}$ neutralized with an efficiency of 0.62 corresponding to 80 keV.

Setting the injector at 2 m from the window, assuming an effective aperture width of 4 cm (the coil gap being larger) and a diverging envelope of $\pm 0.2°$, the bulk of the beam has a width 2.6 cm and a length $24 \times \sqrt{2}$ cm = 34 cm. For four windows the surface is $354\ cm^2$. The total neutral equivalent intensity is then $0.25\ A \cdot cm^{-2} \times 354\ cm^2 = 88\ A_{eq.D^0}$. Thus the total power entering the plasma is $88\ A \times 80\ keV \cong 7$ MW or 1.75 MW per window. The largest losses are presumably shared by the ripple diffusion of the banana orbits and the charge exchange of the residual thermal neutral population. No clear evaluation of these losses can be made at present, but in our opinion they should not exceed 50%. Therefore the effective additional heating power is $3.5\ MW < P_{NI} < 7\ MW$.

In this connection we notice that a real injection system does not have monochromatic beams and this lowers its effective total heating power. On the other hand, if we inject with an angle of ~ 20° with respect to the vertical direction we can have easier and larger access to the plasma, and increase the possible heating power (see Fig.16(B)).

Finally, we estimate that a contribution of about 10% of the neutral injected power will be given through direct beam-plasma D-T reactions. However, the presence of a ripple in the confining magnetic field will spoil the confinement of a fraction of the α-particle orbits and we have not evaluated here the consequent degradation of their heating power. If we also take into account the fact that the assumed current for the reference Ignitor-Au parameters is not sufficient to contain all the trapped α-particle orbits, we may argue that the assumed value of 0.7 for ϵ_h is probably too optimistic.

ACKNOWLEDGEMENTS

The work reported here was performed in part under the sponsorship of the CNEN-Euratom Association, and in part with the support of USERDA.

REFERENCES

[1] COPPI, B., Massachusetts Institute of Technology, RLE Rep. PRR-7518 (1975).
[2] COPPI, B., Massachusetts Institute of Technology, RLE Rep. PRR-76/31 (1976); and in Proc. 1976 Int. School on Fusion Reactor Technology, Erice, Sept. 1976, Pergamon, Oxford (1977) 303.
[3] COPPI, B., TARONI, A., Massachusetts Institute of Technology, RLE Rep. PRR-76/12 (1976); and in Proc. 1976 Varenna Symp. on the Heating of Thermonuclear Plasmas, Editrice Compositori, Bologna (1977) 192.
[4] DÜCHS, D.F., FURTH, H.P., RUTHERFORD, P.M., Princeton Plasma Physics Lab. Rep. TM-265 (1973).
[5] HIRSHMAN, S.P., HAWRYLUK, R.J., BIRGEN, B., Princeton Univ. Rep. PPPL-1325 (1977).
[6] BASU, B., et al., "Transport processes and instabilities in magnetically confined plasmas with $T_{i\perp} \approx T_{e\perp}$", Plasma Physics and Controlled Nuclear Fusion Research 1976 (Proc. 6th Int. Conf. Berchtesgaden, 1976) **2**, IAEA, Vienna (1977) 455.
[7] BRUNELLI, B., COPPI, B., DE MENNA, L., ESPOSITO, F., Centro di Calcolo CNEN, Bologna, Rep. Doc. CEC(77), IGN-77/2 (1977).
[8] APGAR, E., et al., "High-density and collisional plasma regimes in the Alcator programme", Plasma Physics and Controlled Nuclear Fusion Research 1976 (Proc. 6th Int. Conf. Berchtesgaden, 1976) **1**, IAEA, Vienna (1977) 247, and Massachusetts Institute of Technology, RLE Rep. PRR-76/36 (1976).
[9] RUTHERFORD, P.H., DÜCHS, D.F., Princeton Plasma Physics Lab. Rep. MATT-1272 (1976).
[10] HAEGI, M., BITTONI, E., in Proc. 1976 Varenna Symp. on the Heating of Thermonuclear Plasmas, Editrice Compositori, Bologna (1977) 293.
[11] HAEGI, M., BITTONI, E., Internal Report, CNEN-Euratom Association, Frascati (Dec. 1975).
[12] JASSBY, D.L., GOLDSTON, R.J., Princeton Plasma Physics Lab. Rep. MATT-1202 (1976).
[13] SWEETMAN, D., private communication (Sep. 1976).
[14] MORGAN, O.B., private communication (Sep. 1976).

Kinetic theory

KINETIC THEORY OF SURFACE WAVES IN SEMIBOUNDED PLASMAS

A.G. SITENKO
Institute for Theoretical Physics,
Academy of Sciences of the Ukrainian SSR,
Kiev, USSR

Abstract

KINETIC THEORY OF SURFACE WAVES IN SEMIBOUNDED PLASMAS.

A kinetic theory of non-linear wave interaction in semibounded plasmas is developed for the specular reflection model. A kinetic field equation is derived on the basis of which resonant interaction of surface waves, causing decay and explosive instabilities, is investigated. Surface and volume fluctuations in semibounded non-equilibrium plasmas and time development of fluctuation spectra due to non-linear wave interaction are considered. A kinetic equation for surface waves is derived and its possible applications to the description of wave scattering and radiation in semibounded plasmas are discussed.

1. INTRODUCTION

The electrodynamic properties of spatially homogeneous plasma are described by linear and non-linear electric susceptibilities. Electromagnetic fields are determined from non-linear equations by given charge and current distributions in plasma [1]; they also depend on boundary conditions if plasma is confined in some finite volume. A characteristic feature of bounded plasma is that there exist both volume electromagnetic oscillations with the spectrum of the same type as in infinite plasma (the wavelength of such oscillations is small compared with characteristic plasma dimensions) and collective electromagnetic oscillations of a new type, the surface waves, which propagate along the boundary and damp inside plasmas.

The structure of surface waves essentially depends on the shape of the surface and on boundary conditions. The latter are determined by the type of interaction between plasma particles and the boundary. The simplest description of surface waves may be given within the framework of the specular reflection model, i.e. when it is assumed that the charged particles, reaching the surface, are reflected specularly [2–4].

The properties of different surface waves in the simplest semibounded plasma case were investigated in detail in Refs [5–10] (see also [11, 12]). The existence of surface Langmuir oscillations in semibounded plasma was mentioned for the first time in Refs [13, 14]. Excitation of surface Langmuir waves in semibounded plasma due to the motion of charged particles along the boundary was investigated in Ref. [15]. A kinetic theory of high-frequency surface Langmuir and low-frequency ion sound waves was developed by Yu.A. Romanov [5, 6]. Thermal fluctuations due to volume and surface eigenoscillation excitation in plasma halfspace were studied in Ref. [16]. Surface fluctuations in semibounded non-equilibrium plasma and spontaneous radiation from such plasmas were studied in Refs [17–19]; in particular, it was shown in [19] that the surface excitations can cause critical fluctuations. An instability, arising from the growth of ion sound surface waves in the presence of ion-electron relative motion in semibounded plasma, was considered in Ref. [10].

When surface-wave intensities become large enough, non-linear effects have to be taken into account. Non-linear wave interaction in semibounded plasma was investigated in hydrodynamic approximation in Refs [20–22]; in particular, the decay instability of the surface wave was considered in Ref. [20]. Non-linear wave interaction in semibounded plasma, causing echo surface oscillations, was investigated in the kinetic approach in Refs [23, 24].

In the present paper a kinetic theory of non-linear wave interaction in semibounded plasma is developed for the specular reflection model. A general non-linear equation for the electromagnetic field is derived on the basis of which different non-linear effects are discussed. Resonant interaction of three surface waves, causing decay and explosive instabilities, is considered. Surface and volume fluctuations in semibounded non-equilibrium plasma are investigated, and fluctuation spectra time evolution due to non-linear wave interaction is studied. A kinetic equation for surface waves is derived and its possible applications to the description of wave scattering, transformation and radiation from semibounded plasma are considered.

2. NON-LINEAR EQUATION FOR POTENTIAL FIELD IN SEMIBOUNDED PLASMAS

It is convenient to investigate surface and volume eigenoscillations and their non-linear interaction in semibounded plasma in the same way as in infinite plasmas, based on non-linear equations for the electromagnetic field, which may be derived from kinetic equations for particle distribution functions and for the selfconsistent field. We shall consider stationary spatially homogeneous plasma filling halfspace $z > 0$. Assume that $z < 0$ halfspace is filled with some insulator, characterized by the dielectric constant ϵ_0. We first restrict ourselves to consideration of the electrostatic interaction between charged particles (the self-consistent electric field is potential in this case).

The kinetic equations for electron and ion distribution functions and the equation for the selfconsistent field in $z > 0$ halfspace are as follows:

$$\frac{\partial f}{\partial t} + \vec{v}\frac{\partial f}{\partial \vec{r}} + \frac{e}{m}\vec{E}\frac{\partial}{\partial \vec{v}}(f_0 + f) = 0 \tag{1}$$

$$\mathrm{div}\,\vec{E} = 4\pi\left(\sum e\int d\vec{v}\, f + \rho^0\right) \tag{2}$$

where f is the deviation of the electron or ion distribution function from the unperturbed distribution f_0 (in equilibrium plasma one must take the Maxwellian distribution function for f_0), $\vec{E}$ is the selfconsistent electric field, and ρ^0 is the external charge density. (The sign Σ in Eq.(2) means summation over the electron and ion components). The electric field $\vec{E}$ in $z < 0$ halfspace satisfies the equation

$$\mathrm{div}\,\vec{E} = 0 \tag{3}$$

(we assume there are no external charges outside the plasma).

The kinetic equation (1) must be completed by conditions for the distribution function f on the boundary $z = 0$. Those for specular particle reflection can be taken in the form

$$f(x, y, z = 0; v_x, v_y, v_z) = f(x, y, z = 0; v_x, v_y, -v_z) \tag{4}$$

The electric field on the surface $z = 0$ must satisfy the usual boundary conditions, which are reduced to the continuity of the electric field tangential component and the electric induction normal component.

To solve the set of equations (1), (2) in the halfspace $z > 0$, we shall use the following formal procedure. We perform even and odd continuations of electric field components $\vec{E}_\perp$ and E_z correspondingly into $z < 0$ halfspace (denoting such continued field $\vec{E}^+$) and suppose that kinetic equations determine the distribution functions in the whole space (we label them f^+):

$$\frac{\partial f^+}{\partial t} + \vec{v}\frac{\partial f^+}{\partial \vec{r}} + \frac{e}{m}\vec{E}^+\frac{\partial}{\partial \vec{v}}(f_0 + f^+) = 0 \tag{5}$$

The differential operator in these equations is at such electric field continuation invariant relative to the substitution $(z, v_z) \to (-z, -v_z)$, so the solutions must have the same property:

$$f^{+}(x,y,z;\ v_x,v_y,v_z) = f^{+}(x,y,-z;\ v_x,v_y,-v_z) \tag{6}$$

(it is assumed that the unperturbed distributions f_0 are even functions of v_z). Distribution functions f and f^+ must coincide in $z > 0$ halfspace, because relations (6) at $z = 0$ lead directly to the boundary conditions (4).

The electric field $\vec{E}^+$ satisfies the condition

$$\operatorname{div}\vec{E}^{+} = 4\pi\left(e\int d\vec{v}\, f^{+} + \rho^{0}\right) + 2E_z^{+}(x,y,z=0)\,\delta(z) \tag{7}$$

which differs from (2) by taking into account additional surface charge, providing for field normal component discontinuity on the boundary. (External charge density ρ^0 is assumed to be continued into $z < 0$ in an even way.) We continue the electric field, governed at $z < 0$ by Eq.(3), into halfspace $z > 0$ in a similar way. This field (denoted by $\vec{E}^-$) satisfies the equation

$$\operatorname{div}\vec{E}^{-} = -2E_z^{-}(x,y,z=0)\,\delta(z) \tag{8}$$

The solution of Eq.(7) at $z > 0$ determines the electric field in plasmas ($\vec{E}^+(r) = \vec{E}(r)$), and the solution of (8) at $z < 0$ is the electric field outside the plasma ($\vec{E}^-(r) = \vec{E}(r)$). The following boundary conditions must be satisfied on the plasma surface:

$$\vec{E}_{\perp}^{+}(x,y,z=0) = \vec{E}_{\perp}^{-}(x,y,z=0) \tag{9}$$

$$E_z^{+}(x,y,z=0) = \varepsilon_0 E_z^{-}(x,y,z=0) \tag{10}$$

The space Fourier transformation of Eq.(7) is

$$ikE_{\vec{k}}^{+} = 4\pi\left(e\int d\vec{v}\, f_{\vec{k}}^{+} + \rho_{\vec{k}}^{0}\right) + 2E_{z\vec{k}_\perp}^{+}(0) \tag{11}$$

and, due to the longitudinal character of the field,

$$E_{z\vec{k}_\perp}^{+}(0) \equiv \frac{1}{2\pi}\int dk_z\, \frac{k_z}{k} E_{\vec{k}}^{+} \tag{12}$$

The Fourier-transformed Eq.(8) takes the form

$$ikE^-_{\vec{k}} = -2E_{z\vec{k}_\perp}(0) \tag{13}$$

The value $E_{z\vec{k}_\perp}(0)$, entering (11), may be found from the boundary conditions. Due to (10),

$$E^+_{z\vec{k}_\perp}(0) = \varepsilon_0 E^-_{z\vec{k}_\perp}(0)$$

Using (13), $E^-_{z\vec{k}_\perp}(0)$ may be easily expressed in terms of $E^-_{\perp\vec{k}_\perp}(0)$. Indeed,

$$E^-_{\perp k_\perp}(0) = \frac{k_\perp}{2\pi}\int dk_z \frac{E^-_k}{k} = i\frac{k_\perp}{\pi}E^-_{zk_\perp}(0)\int\frac{dk_z}{k_\perp^2}$$

so

$$E^-_{z\vec{k}_\perp}(0) = -iE^-_{\perp\vec{k}_\perp}(0) \tag{14}$$

And as, from (9), it follows that

$$E^+_{\perp\vec{k}_\perp}(0) = E^-_{\perp\vec{k}_\perp}(0)$$

then Eq.(11) may be written as

$$ikE^+_{\vec{k}} = 4\pi\left(e\int d\vec{v}\, f^+_{\vec{k}} + \rho^0_{\vec{k}}\right) - 2i\varepsilon_0 E^+_{\perp\vec{k}_\perp}(0) \tag{15}$$

where

$$E^+_{\perp k_\perp}(0) = \frac{k_\perp}{2\pi}\int dk_z \frac{E^+_k}{k} \tag{16}$$

To simplify the notation we shall omit the plus sign, i.e. we shall write E instead of E^+.

Performing space-time Fourier transformation of the kinetic equation (5) and solving by the successive approximation method the equation obtained, we can write $f^{\pm}_{\vec{k}\omega}$ as an expansion in a series of field intensity $E_{\vec{k}\omega}$. Substituting this expansion into (15), we obtain the following non-linear equation, which entirely determines the electric field in the part of the space filled with plasma:

$$\varepsilon(\omega,\vec{k})\,E_{\vec{k}\omega} + \sum_{\substack{\omega_1+\omega_2=\omega \\ \vec{k}_1+\vec{k}_2=\vec{k}}} \varkappa^{(2)}(\omega_1,\vec{k}_1;\omega_2,\vec{k}_2)\,E_{\vec{k}_1\omega_1}E_{\vec{k}_2\omega_2}$$

$$+ \sum_{\substack{\omega_1+\omega_2+\omega_3=\omega \\ \vec{k}_1+\vec{k}_2+\vec{k}_3=\vec{k}}} \varkappa^{(3)}(\omega_1,\vec{k}_1;\omega_2,\vec{k}_2;\omega_3,\vec{k}_3)\,E_{\vec{k}_1\omega_1}E_{\vec{k}_2\omega_2}E_{\vec{k}_3\omega_3} + \ldots + 2\varepsilon_0\frac{k_\perp}{k}\sum_{k_z'}\frac{E_{\vec{k}'\omega}}{k'} = -\frac{4\pi i}{k}\rho^{\circ}_{\vec{k}\omega} \tag{17}$$

where $\epsilon(\omega, \vec{k})$ is the dielectric permittivity; $\kappa^{(2)}(\omega_1, \vec{k}_1; \omega_2, \vec{k}_2)$ and $\kappa^{(3)}(\omega_1, \vec{k}_1; \omega_2, \vec{k}_2; \omega_3, \vec{k}_3)$ are the non-linear susceptibilities for infinite homogeneous plasma. The electric field outside the plasma can, according to Eqs (13), (14) and (16), be expressed directly in terms of the solution of Eq.(17):

$$E^{-}_{\vec{k}\omega} = 2\frac{k_\perp}{k}\sum_{k_z'}\frac{E_{\vec{k}'\omega}}{k'} \tag{18}$$

Thus determination of the field in the plasma halfspace reduces to the solution of the non-linear equation (17).

3. DISPERSION EQUATIONS FOR VOLUME AND SURFACE WAVES (LINEAR APPROXIMATION)

Neglecting non-linear terms in Eq.(17) and taking external charge density equal to zero, we obtain the basic equation of the linear approximation, which describes eigenoscillations in semibounded plasma:

$$\varepsilon(\omega,\vec{k})\,E_{\vec{k}\omega} + 2\frac{\varepsilon_0}{k}E_{\vec{k}_\perp\omega} = 0 \tag{19}$$

Here and further on we shall use the following notation:

$$E_{\vec{k}_\perp\omega} \equiv E_{\perp\vec{k}_\perp\omega}(0) = \frac{k_\perp}{2\pi}\int dk_z\,\frac{E_{\vec{k}\omega}}{k} \tag{20}$$

i.e. $E_{\vec{k}_\perp\omega}$ is the value of the field intensity tangential component on the plasma surface. It is easy to obtain from (19) the following equation for $E_{\vec{k}_\perp\omega}$:

$$\zeta(\omega,\vec{k}_\perp)\, E_{\vec{k}_\perp\omega} = 0 \qquad (21)$$

where

$$\zeta(\omega,\vec{k}_\perp) = 1 + \frac{\varepsilon_0 k_\perp}{\pi}\int dk_z \frac{1}{k^2\varepsilon(\omega,\vec{k})} \qquad (22)$$

It follows from (19) that eigenoscillations of two types – volume and surface – can exist in semibounded plasma.

The dispersion equation for volume eigenoscillations ($E_{\vec{k}\omega} \neq 0$, $E_{\vec{k}_\perp\omega} = 0$) is determined by the same condition as in infinite plasmas:

$$\varepsilon(\omega,\vec{k}) = 0 \qquad (23)$$

We use the notation $\omega_{\vec{k}}$ for the eigenfrequencies, which are solutions of (23) at fixed values of $\vec{k}$, and present the eigenoscillation field in the form

$$E_{\vec{k}\omega} = \pi E_{\vec{k}}\left\{e^{-i\phi_{\vec{k}}}\delta(\omega-\omega_{\vec{k}}) + e^{i\phi_{\vec{k}}}\delta(\omega+\omega_{\vec{k}})\right\} \qquad (24)$$

where $E_{\vec{k}}$ and $\phi_{\vec{k}}$ are initial amplitude and phase. Using boundary conditions (9) and (10) and relation (14), it is easy to show that

$$E_{\vec{k}_\perp\omega} = \frac{i}{2\pi\varepsilon_0}\int dk_z \frac{k_z E_{\vec{k}\omega}}{k}$$

and as $E_{\vec{k}}$, $\phi_{\vec{k}}$ and $\omega_{\vec{k}}$ are even functions of k_z, the following condition is satisfied for the volume oscillations:

$$E_{\vec{k}_\perp\omega} = 0$$

The dispersion equation for the surface eigenwave ($E_{\vec{k}_\perp\omega} \neq 0$) is determined by the condition

$$\zeta(\omega,\vec{k}_\perp) = 0 \qquad (25)$$

We denote the eigenfrequencies of surface oscillations by $\omega_{\vec{k}_\perp}$ and present the surface oscillation field in the form

$$E_{\vec{k}_\perp\omega} = \pi E_{\vec{k}_\perp} \left\{ e^{-i\phi_{\vec{k}_\perp}} \delta(\omega - \omega_{\vec{k}_\perp}) + e^{i\phi_{\vec{k}_\perp}} \delta(\omega + \omega_{\vec{k}_\perp}) \right\} \tag{26}$$

As $\epsilon(\omega,\vec{k}) \neq 0$ for surface waves, it is easy to find the total space component of the field for the surface oscillations from (19):

$$E_{\vec{k}\omega} = -\frac{2\varepsilon_0}{k} \frac{E_{\vec{k}_\perp\omega}}{\varepsilon(\omega,\vec{k})} \tag{27}$$

It is not difficult to show that, when moving away from the boundary, the surface oscillation field diminishes exponentially.

4. SURFACE WAVES

We now consider different types of surface waves in semibounded plasma, determined by the dispersion equation (25).

The dispersion of both surface and volume waves in the high-frequency domain is determined by the plasma electron component. The eigenfrequency and damping coefficient of high-frequency surface waves in the long-wave limit $a^2 k_\perp^2 \ll 1$ (a is the Debye radius) are described by the formulas:

$$\omega_{\vec{k}_\perp} = \frac{\Omega}{\sqrt{1+\varepsilon_0}}\left(1 + \frac{\sqrt{3\varepsilon_0}}{2} a k_\perp + \ldots\right), \quad \gamma_{\vec{k}_\perp} = \sqrt{\frac{2}{3\pi}}\, k_\perp s \tag{28}$$

Surface wave dispersion in the low-frequency domain essentially depends on both electrons and ions. Suppose that the electron temperature is much greater than the ion one $T_e \gg T_i$ (strongly non-isothermal plasma) and consider the frequency domain, satisfying the condition

$$s \gg \frac{\omega}{k_\perp} \gg s_i$$

In this case we can use the following approximate expression for the dielectric permittivity:

$$\varepsilon(\omega,\vec{k}) = 1 + \frac{1}{a^2 k^2} - \frac{\Omega_i^2}{\omega^2}$$

and then it is easy to find the value (22):

$$\zeta(\omega,\vec{k}_\perp) = 1 + \frac{\varepsilon_0}{\varepsilon_i}\frac{1}{\sqrt{1+\frac{1}{a^2k_\perp^2\varepsilon_i}}}, \qquad \varepsilon_i = 1 - \frac{\Omega_i^2}{\omega^2} \tag{29}$$

Equating $\zeta(\omega, \vec{k}_\perp)$ to zero, we find two roots of the dispersion equation:

$$\omega^2 = \frac{\Omega_i^2}{1 + \frac{1}{2a^2k_\perp^2}\left(1 \pm \sqrt{1+4a^4k_\perp^4\varepsilon_0^2}\right)} \tag{30}$$

Taking the plus sign, we obtain the eigenfrequencies of the surface waves in the long-wave and short-wave limits correspondingly:

$$\left.\begin{aligned} \omega_{\vec{k}_\perp} &= k_\perp v_s\left(1 - \frac{1}{2}a^2k_\perp^2\right), & a^2k_\perp^2 \ll 1 \\ \omega_{\vec{k}_\perp} &= \frac{\Omega_i}{\sqrt{1+\varepsilon_0}}\left(1 - \frac{1}{2a^2k_\perp^2(1+\varepsilon_0)}\right), & a^2k_\perp^2 \gg 1 \end{aligned}\right\} \tag{31}$$

If the minus sign is taken in (30), the surface wave eigenfrequencies are correspondingly equal to:

$$\left.\begin{aligned} \omega'_{\vec{k}_\perp} &= \Omega_i\left(1 + \frac{1}{2}a^2k_\perp^2\varepsilon_0^2\right), & a^2k_\perp^2 \ll 1 \\ \omega'_{\vec{k}_\perp} &= \frac{\Omega_i}{\sqrt{1-\varepsilon_0+\frac{1}{2a^2k_\perp^2}}}, & a^2k_\perp^2 \gg 1 \end{aligned}\right\} \tag{32}$$

these expressions are valid for $\epsilon_0 = 0$. We mention that the eigenfrequencies (31) are less than the ion Langmuir frequency, while eigenfrequencies (32) exceed it.

The high-frequency electron and low-frequency ion sound and ion surface waves considered are characterized by positive energy. In semibounded non-equilibrium plasma, surface waves with negative energy can also exist. For example, we consider plasma with velocity $\vec{u}$, parallel to the boundary. Neglecting ion thermal motion, we can use the following approximate expression for plasma dielectric permittivity:

$$\varepsilon(\omega,\vec{k}) = 1 + \frac{1}{a^2k^2} - \frac{\Omega_i^2}{\omega^2}\left[1 + \eta\frac{\omega^2}{(\omega-\vec{k}_\perp\vec{u})^2}\right], \qquad \eta \equiv \frac{n_0'}{n_0} \ll 1 \tag{33}$$

valid for frequencies $\omega \ll k_\perp s$. Substituting (33) into the general formula (22) and integrating over k_z, we again obtain for $\zeta(\omega, \vec{k}_\perp)$ expression (29), in which ϵ_i should be understood as

$$\varepsilon_i = 1 - \frac{\Omega_i^2}{\omega^2}\left[1 + \eta \frac{\omega^2}{(\omega - \vec{k}\vec{u})^2}\right]$$

Equating then $\zeta(\omega, \vec{k}_\perp)$ to zero, we can write the dispersion equation in the form:

$$1 + \frac{1}{2a^2k_\perp^2}\left(1 \pm \sqrt{1 + 4a^4k^4\varepsilon_0^2}\right) = \frac{\Omega_i^2}{\omega^2}\left[1 + \eta \frac{\omega^2}{(\omega - \vec{k}\vec{u})^2}\right] \tag{34}$$

Consider the long-wave limit $a^2k^2 \ll 1$, and assume that, approximately,

$$\sqrt{1 + 4a^4k_\perp^4\varepsilon_0^2} \simeq 1 + 2a^4k_\perp^4\varepsilon_0^2$$

Taking the plus sign in (34), we can rewrite the dispersion equation as

$$1 + \frac{1}{a^2k_\perp^2} + a^2k_\perp^2\varepsilon_0^2 = \frac{\Omega_i^2}{\omega^2}\left[1 + \eta \frac{\omega^2}{(\omega - \vec{k}_\perp\vec{u})^2}\right]$$

Supposing that the beam density is small enough ($\eta \ll 1$), it is easy to find the roots of this equation, corresponding to the eigenfrequencies of surface waves:

$$\left.\begin{aligned} \omega_{\vec{k}_\perp}^{(1)} &= \frac{k_\perp v_s}{\sqrt{1 + a^2k_\perp^2}} \\ \omega_{\vec{k}_\perp}^{(2)} &= \vec{k}_\perp\vec{u} - \sqrt{\eta}\,\frac{\Omega_i}{\sqrt{1 + \frac{1}{a^2k_\perp^2} - \frac{\Omega_i^2}{(\vec{k}_\perp\vec{u})^2}}} \\ \omega_{\vec{k}_\perp}^{(3)} &= \vec{k}\vec{u} + \sqrt{\eta}\,\frac{\Omega_i}{\sqrt{1 + \frac{1}{a^2k_\perp^2} - \frac{\Omega_i^2}{(\vec{k}_\perp\vec{u})^2}}} \end{aligned}\right\} \tag{35}$$

Waves corresponding to eigenfrequencies $\omega^{(1)}_{\vec{k}_\perp\omega}$ and $\omega^{(3)}_{\vec{k}_\perp\omega}$ are characterized by positive energy, while the wave corresponding to eigenfrequency $\omega^{(2)}_{\vec{k}_\perp\omega}$ is characterized by negative energy ($\partial\xi/\partial\omega < 0$).

Taking the minus sign in (34), we obtain the dispersion relation in the form

$$1 - a^2 k_\perp^2 \varepsilon_0^2 = \frac{\Omega_i^2}{\omega^2}\left[1 + \eta \frac{\omega^2}{(\omega - \vec{k}_\perp \vec{u})^2}\right], \qquad \varepsilon_0 < 0$$

The roots of this equation determine the following surface wave eigenfrequencies:

$$\left.\begin{aligned} \omega_{\vec{k}_\perp}^{(1)'} &= \Omega_i \left(1 + \frac{1}{2} a^2 k_\perp^2 \varepsilon_0^2\right) \\ \omega_{\vec{k}_\perp}^{(2)'} &= \vec{k}_\perp \vec{u} - \sqrt{\eta}\, \frac{\Omega_i}{\sqrt{1 - a^2 k_\perp^2 \varepsilon_0^2 - \frac{\Omega_i^2}{(\vec{k}_\perp \vec{u})^2}}} \\ \omega_{\vec{k}_\perp}^{(3)'} &= \vec{k}_\perp \vec{u} + \sqrt{\eta}\, \frac{\Omega_i}{\sqrt{1 - a^2 k_\perp^2 \varepsilon_0^2 - \frac{\Omega_i^2}{(\vec{k}_\perp \vec{u})^2}}} \end{aligned}\right\} \tag{36}$$

(expressions for $\omega_{\vec{k}_\perp}^{(2)'}$ and $\omega_{\vec{k}_\perp}^{(3)'}$ are valid for $(\vec{k}_\perp \vec{u})^2 \gg \Omega_i^2$). Waves with frequencies $\omega_{\vec{k}_\perp}^{(1)'}$ and $\omega_{\vec{k}_\perp}^{(3)'}$, as in the previous case, are characterized by positive energy, and the wave with frequency $\omega_{\vec{k}_\perp}^{(2)'}$ by negative energy.

Consider now surface waves in semibounded plasmas in an external constant and a homogeneous magnetic field. Assume for simplicity that the magnetic field $\vec{B}_0$ is perpendicular to the boundary. The dielectric permittivity of magnetoactive plasma in the high-frequency domain is determined by the expression [25]:

$$\varepsilon(\omega, \vec{k}) = 1 + \frac{1}{a^2 k^2}\left\{1 - \sum_n \frac{\omega}{\omega - n\omega_B}\left[\varphi(z_n) - i\sqrt{\pi}\, z_n e^{-z_n^2}\right]\Lambda_n(\beta)\right\} \tag{37}$$

where

$$\Lambda_n(\beta) = e^{-\beta} I_n(\beta), \quad \beta = \frac{k_\perp^2 s^2}{3\omega_B^2}, \quad \omega_B = \frac{eB_0}{mc}, \quad z_n = \sqrt{\frac{3}{2}}\, \frac{\omega - n\omega_B}{|k_z| s}$$

Consider waves with frequencies close to multiple electron cyclotron frequencies $\omega \simeq n\omega_B$. If $n\omega_B \gg |\omega - n\omega_B| \gg |k_z| s$,

$$\varepsilon(\omega, \vec{k}) = 1 + \frac{1}{a^2 k^2}\left\{1 - \Lambda_0(\beta) - \frac{\omega}{\omega - n\omega_B}\Lambda_n(\beta)\right\}$$

and for $\zeta(\omega, \vec{k}_\perp)$ we obtain

$$\zeta(\omega,\vec{k}_\perp) = 1 + \frac{\varepsilon_0}{\sqrt{1+\frac{1}{a^2k_\perp^2}\left\{1-\Lambda_0(\beta)-\frac{\omega}{\omega-n\omega_B}\Lambda_n(\beta)\right\}}} \tag{38}$$

So the dispersion equation for surface waves with frequencies close to multiple electron cyclotron frequency may be written as follows:

$$1+\frac{1}{a^2k_\perp^2}\left\{1-\Lambda_0(\beta)-\frac{\omega}{\omega-n\omega_B}\Lambda_n(\beta)\right\} = \varepsilon_0^2 \quad , \qquad \varepsilon_0 < 0$$

and we find surface wave eigenfrequencies in magnetoactive plasma:

$$\omega_{\vec{k}_\perp} = (1+\Delta)\, n\omega_B \tag{39}$$

where

$$\Delta \equiv \frac{\Lambda_n(\beta)}{(1-\varepsilon_0^2)a^2k_\perp^2 + 1 - \Lambda_0(\beta) - \Lambda_n(\beta)} \ll 1 \tag{40}$$

Taking the imaginary part of $\zeta(\omega, \vec{k}_\perp)$ into account, we obtain for the damping coefficient $(\gamma_{\vec{k}_\perp} \ll \omega_{\vec{k}_\perp})$

$$\gamma_{\vec{k}_\perp} = \sqrt{\frac{6}{\pi}}\left\{(1+\theta)e^{\theta}\int_\theta^\infty dy\,\frac{e^{-y}}{y} - 1\right\}\Delta^2\,\frac{(n\omega_B)^2}{k_\perp s} \tag{41}$$

where

$$\theta \equiv \frac{3}{2\varepsilon_0^2}\left(\frac{n\omega_B}{k_\perp s}\right)^2\Delta^2$$

Thus a weakly damping surface wave with frequencies close to multiple electron cyclotron frequency can propagate in semibounded plasma in a normal-to-the-boundary external magnetic field.

5. NON-LINEAR INTERACTION OF SURFACE WAVES

Non-linear interaction of volume and surface waves in semibounded plasma is described by the general non-linear equation (17). Assuming that there are no external charges, we rewrite this equation in the form:

$$\varepsilon(\omega,\vec{k})E_{\vec{k}\omega} + \sum_{\substack{\omega_1+\omega_2=\omega \\ \vec{k}_1+\vec{k}_2=\vec{k}}} \varkappa^{(2)}(\omega_1,\vec{k}_1;\omega_2,\vec{k}_2)E_{\vec{k}_1\omega_1}E_{\vec{k}_2\omega_2}$$

$$+\sum_{\substack{\omega_1+\omega_2+\omega_3=\omega \\ \vec{k}_1+\vec{k}_2+\vec{k}_3=\vec{k}}} \varkappa^{(3)}(\omega_1,\vec{k}_1;\omega_2,\vec{k}_2;\omega_3,\vec{k}_3)E_{\vec{k}_1\omega_1}E_{\vec{k}_2\omega_2}E_{\vec{k}_3\omega_3} + \ldots + \frac{2\varepsilon_0}{k}E_{\vec{k}_\perp\omega} = 0 \tag{42}$$

Using the multiple time-scale expansion method, one can derive from (42) a hierarchy of equations which determine the time dependence of amplitudes due to non-linear resonant wave interaction.

The simplest example of non-linear resonant wave interaction is the three-wave resonance, which takes place when the frequencies of interacting waves satisfy the condition

$$\omega_{\vec{k}_1} + \omega_{\vec{k}_2} = \omega_{\vec{k}} \tag{43}$$

It is obvious that the three-wave resonance in semibounded plasmas is possible in the case of interaction between three volume waves, two volume waves and one surface wave (two cases are possible: a surface wave is created as a result of the interaction of two volume waves, and interaction between a volume wave and a surface wave leads to the creation of a volume wave), and three surface waves. When there are no three-wave resonances, the four-wave resonant interaction is the most essential, and it is possible under the condition that

$$\omega_{\vec{k}_1} + \omega_{\vec{k}_2} + \omega_{\vec{k}_3} = \omega_{\vec{k}} \tag{44}$$

We restrict ourselves to a detailed consideration of the resonant interaction of surface waves. Multiply Eq.(42) by $k_\perp/(k\epsilon(\omega,\vec{k}))$ and integrate over k_z, and then, taking the surface character of interacting waves into account, express the fields $E_{\vec{k}_1\omega_1}$, $E_{\vec{k}_2\omega_2}$, ... , using (27), in terms of the surface components $E_{\vec{k}_{1\perp}\omega_1}$, $E_{\vec{k}_{2\perp}\omega_2}$, ... As a result the basic equation, describing the non-linear interaction of surface waves in semibounded plasmas, may be written as follows:

$$\zeta(\omega,\vec{k}_\perp)E_{\vec{k}_\perp\omega} + \sum_{\substack{\omega_1+\omega_2=\omega \\ \vec{k}_{1\perp}+\vec{k}_{2\perp}=\vec{k}_\perp}} \tilde{\varkappa}^{(2)}(\omega_1,\vec{k}_{1\perp};\omega_2,\vec{k}_{2\perp})E_{\vec{k}_{1\perp}\omega_1}E_{\vec{k}_{2\perp}\omega_2}$$

$$+\sum_{\substack{\omega_1+\omega_2+\omega_3=\omega \\ \vec{k}_{1\perp}+\vec{k}_{2\perp}+\vec{k}_{3\perp}=\vec{k}_\perp}} \tilde{\varkappa}^{(3)}(\omega_1,\vec{k}_{1\perp};\omega_2,\vec{k}_{2\perp};\omega_3,\vec{k}_{3\perp})E_{\vec{k}_{1\perp}\omega_1}E_{\vec{k}_{2\perp}\omega_2}E_{\vec{k}_{3\perp}\omega_3}+\ldots=0 \tag{45}$$

where $\tilde{\mathfrak{X}}^{(2)}$ and $\tilde{\mathfrak{X}}^{(3)}$ are plasma non-linear surface susceptibilities:

$$\tilde{\mathfrak{X}}^{(2)}(\omega_1,\vec{k}_{1\perp};\omega_2,\vec{k}_{2\perp}) = \frac{\varepsilon_0^2 k_\perp}{\pi^2}\int dk_{1z}\int dk_{2z}\frac{\mathfrak{X}^{(2)}(\omega_1,\vec{k}_1;\omega_2,\vec{k}_2)}{k_1k_2k\,\varepsilon(\omega_1,\vec{k}_1)\varepsilon(\omega_2,\vec{k}_2)\varepsilon(\omega,\vec{k})} \tag{46}$$

$$\tilde{\mathfrak{X}}^{(3)}(\omega_1,\vec{k}_{1\perp};\omega_2,\vec{k}_{2\perp};\omega_3,\vec{k}_{3\perp}) = -\frac{\varepsilon_0^3 k_\perp}{\pi^3}\int dk_{1z}\int dk_{2z}\int dk_{3z}\frac{\mathfrak{X}^{(3)}(\omega_1,\vec{k}_1;\omega_2,\vec{k}_2;\omega_3,\vec{k}_3)}{k_1k_2k_3k\,\varepsilon(\omega_1,\vec{k}_1)\varepsilon(\omega_2,\vec{k}_2)\varepsilon(\omega_3,\vec{k}_3)\varepsilon(\omega,\vec{k})} \tag{47}$$

Now we apply the multiple time-scale expansion method to the non-linear equation (45). The fields of surface waves in the first approximation are determined, as before, by expression (26), but due to non-linear wave interaction, the amplitudes $E_{\vec{K}_\perp}$ and phases $\phi_{\vec{K}_\perp}$ should be considered as slowly varying functions of time. Equations for time dependence of amplitudes $E_{\vec{K}_\perp}$ and phases $\phi_{\vec{K}_\perp}$ may be found from the condition that the secular parts of higher-order approximations of Eq.(45) should turn to zero. Under the condition of three-surface wave resonance,

$$\omega_{\vec{k}_{1\perp}} + \omega_{\vec{k}_{2\perp}} = \omega_{\vec{k}_\perp} \tag{48}$$

The equation for time dependence of the linear approximation amplitude of surface waves is

$$\frac{\partial}{\partial t} E_{\vec{k}_\perp} e^{-i\phi_{\vec{k}_\perp}} = \frac{i}{2}\,\frac{1}{\dfrac{\partial\zeta(\omega_{\vec{k}_\perp},\vec{k}_\perp)}{\partial\omega_{\vec{k}_\perp}}} \times \sum_{\vec{k}_{1\perp}+\vec{k}_{2\perp}=\vec{k}_\perp} \tilde{\mathfrak{X}}^{(2)}(\omega_{\vec{k}_{1\perp}},\vec{k}_{1\perp};\omega_{\vec{k}_{2\perp}},\vec{k}_{2\perp})\,E_{\vec{k}_{1\perp}}E_{\vec{k}_{2\perp}}\,e^{-i(\phi_{\vec{k}_{1\perp}}+\phi_{\vec{k}_{2\perp}})} \tag{49}$$

If condition (48) is not satisfied, then the correction to the field in second approximation is expressed in terms of fields in first approximation in the following way:

$$E^{(2)}_{\vec{k}_\perp\omega} = -\frac{1}{\zeta(\omega,\vec{k}_\perp)}\sum_{\substack{\omega_1+\omega_2=\omega\\ \vec{k}_{1\perp}+\vec{k}_{2\perp}=\vec{k}_\perp}} \tilde{\mathfrak{X}}^{(2)}(\omega_1,\vec{k}_{1\perp};\omega_2,\vec{k}_{2\perp})\,E_{\vec{k}_{1\perp}\omega_1}E_{\vec{k}_{2\perp}\omega_2} \tag{50}$$

Amplitude and phase-time dependence may be found from the removal of the secularity in the third approximation equation. Resonant interaction takes place if

$$\omega_{\vec{k}_{1\perp}} + \omega_{\vec{k}_{2\perp}} + \omega_{\vec{k}_{3\perp}} = \omega_{\vec{k}_\perp} \tag{51}$$

The equation for surface-wave field time dependence is, in the case of four-wave resonance, as follows:

$$\frac{\partial}{\partial t} E_{\vec{k}_\perp} e^{-i\phi_{\vec{k}_\perp}} = -\frac{i}{4} \frac{1}{\frac{\partial \zeta(\omega_{\vec{k}_\perp}, \vec{k}_\perp)}{\partial \omega_{\vec{k}_\perp}}}$$

$$\times \sum_{\vec{k}_{1\perp}+\vec{k}_{2\perp}+\vec{k}_{3\perp}=\vec{k}_\perp}{}' \left\{ 2 \frac{\tilde{\varkappa}^{(2)}(\omega_{\vec{k}_{1\perp}}, \vec{k}_{1\perp}; \omega_{\vec{k}_{2\perp}}+\omega_{\vec{k}_{3\perp}}, \vec{k}_{2\perp}+\vec{k}_{3\perp})\, \tilde{\varkappa}^{(2)}(\omega_{\vec{k}_{2\perp}}, \vec{k}_{2\perp}; \omega_{\vec{k}_{3\perp}}, \vec{k}_{3\perp})}{\zeta(\omega_{\vec{k}_{2\perp}}+\omega_{\vec{k}_{3\perp}}, \vec{k}_{2\perp}+\vec{k}_{3\perp})} \right.$$

$$\left. - \tilde{\varkappa}^{(3)}(\omega_{\vec{k}_{1\perp}}, \vec{k}_{1\perp}; \omega_{\vec{k}_{2\perp}}, \vec{k}_{2\perp}; \omega_{\vec{k}_{3\perp}}, \vec{k}_{3\perp}) \right\} E_{\vec{k}_{1\perp}} E_{\vec{k}_{2\perp}} E_{\vec{k}_{3\perp}} e^{-i(\phi_{\vec{k}_{1\perp}} + \phi_{\vec{k}_{2\perp}} + \phi_{\vec{k}_{3\perp}})} \tag{52}$$

The prime near the sum symbol on the right-hand side of Eq.(52) means that it is necessary to take into account all possible combinations of waves which are in accordance with the resonance condition (51) for different signs of frequencies.

6. THREE-WAVE DECAY OF SURFACE WAVES

Consider resonant interaction of three surface waves with frequencies $\omega_{\vec{k}_\perp}$, $\omega_{\vec{k}_{1\perp}}$ and $\omega_{\vec{k}_{2\perp}}$ and fixed wave vectors $\vec{k}_\perp$, $\vec{k}_{1\perp}$ and $\vec{k}_{2\perp}$, satisfying the resonance conditions:

$$\omega_{\vec{k}_{1\perp}} + \omega_{\vec{k}_{2\perp}} = \omega_{\vec{k}_\perp}, \qquad \vec{k}_{1\perp} + \vec{k}_{2\perp} = \vec{k}_\perp \tag{53}$$

Each of the interacting waves is characterized by the energy

$$W_{\vec{k}_\perp} = -\frac{\varepsilon_0}{8\pi k_\perp} \omega_{\vec{k}_\perp} \frac{\partial \zeta(\omega_{\vec{k}_\perp}, \vec{k}_\perp)}{\partial \omega_{\vec{k}_\perp}} \left| E_{\vec{k}_\perp} \right|^2 \tag{54}$$

Energies of separate waves may be both positive and negative (the character of wave energy is determined by the sign of the derivative $\zeta'_{\vec{k}_\perp} \equiv (\partial\zeta(\omega_{\vec{k}\perp}, \vec{k}_\perp))/\partial\omega_{\vec{k}_\perp}$. We introduce for convenience amplitudes $A_{\vec{k}_\perp}$ and sign factors $s_{\vec{k}_\perp}$:

$$A_{\vec{k}_\perp} \equiv \frac{1}{\sqrt{16\pi}} \sqrt{|\zeta'_{\vec{k}_\perp}|}\, E_{\vec{k}_\perp} e^{-i\phi_{\vec{k}_\perp}}, \qquad s_{\vec{k}_\perp} \equiv \operatorname{sgn} \zeta'_{\vec{k}_\perp} \tag{55}$$

Then the expressions for energy and momentum of surface waves take the form:

$$W_{\vec{k}_\perp} = s_{\vec{k}_\perp} \omega_{\vec{k}_\perp} |A_{\vec{k}_\perp}|^2, \qquad \vec{P}_{\vec{k}_\perp} = s_{\vec{k}_\perp} \vec{k}_\perp |A_{\vec{k}_\perp}|^2 \tag{56}$$

Using definitions (55), we can write Eq.(49) in the form of a Schrödinger equation in interaction representation:

$$i \frac{\partial A_{\vec{k}_\perp}}{\partial t} = s_{\vec{k}_\perp} V_{\vec{k}_\perp; \vec{k}_{1\perp}, \vec{k}_{2\perp}} A_{\vec{k}_{1\perp}} A_{\vec{k}_{2\perp}} \tag{57}$$

where $V_{\vec{k}_\perp; \vec{k}_{1\perp}, \vec{k}_{2\perp}}$ is the interaction matrix element

$$V_{\vec{k}_\perp; \vec{k}_{1\perp}, \vec{k}_{2\perp}} \equiv -\sqrt{4\pi}\, \frac{\tilde{\kappa}^{(2)}(\omega_{\vec{k}_{1\perp}}, \vec{k}_{1\perp}; \omega_{\vec{k}_{2\perp}}, \vec{k}_{2\perp})}{\sqrt{|\zeta'_{\vec{k}_\perp} \zeta'_{\vec{k}_{1\perp}} \zeta'_{\vec{k}_{2\perp}}|}} \tag{58}$$

Taking into account the symmetry properties of plasma non-linear susceptibilities $\tilde{\kappa}^{(2)}(\omega_{\vec{k}1\perp}, \vec{k}_{1\perp}; \omega_{\vec{k}_{2\perp}}, \vec{k}_{2\perp})$, it is easy to show that amplitude time dependence for surface waves with frequencies $\omega_{\vec{k}_{1\perp}}$ and $\omega_{\vec{k}_{2\perp}}$ is described by the equations:

$$\left.\begin{aligned} i \frac{\partial A_{\vec{k}_{1\perp}}}{\partial t} &= s_{\vec{k}_{1\perp}} V^*_{\vec{k}_\perp; \vec{k}_{1\perp}, \vec{k}_{2\perp}} A_{\vec{k}_\perp} A^*_{\vec{k}_{2\perp}} \\ i \frac{\partial A_{\vec{k}_{2\perp}}}{\partial t} &= s_{\vec{k}_{2\perp}} V^*_{\vec{k}_\perp; \vec{k}_{1\perp}, \vec{k}_{2\perp}} A^*_{\vec{k}_{1\perp}} A_{\vec{k}_\perp} \end{aligned}\right\} \tag{59}$$

which contain the same matrix element as in (57). The set of coupled equations (57) and (59) gives a full description of three interacting surface-wave dynamics. This set may be solved exactly [1, 26, 27].

When three surface waves with one and the same sign of energy (positive or negative) are interacting, i.e. when

$$s_{\vec{k}_{1\perp}} = s_{\vec{k}_{2\perp}} = s_{\vec{k}_\perp} \tag{60}$$

then a decay instability arises in the system. Consider a wave with frequency $\omega_{\vec{k}_\perp}$, which at zero time t = 0 has large amplitude

$$|A_{\vec{k}_\perp}|^2 \gg |A_{\vec{k}_{1\perp}}|^2 \quad \text{and} \quad |A_{\vec{k}_\perp}|^2 \gg |A_{\vec{k}_{2\perp}}|^2$$

As a result of resonant interaction,amplitude $A_{\vec{k}_\perp}$ during the initial stage of time evolution changes slightly, while amplitudes $A_{\vec{k}_{1\perp}}$ and $A_{\vec{k}_{2\perp}}$ grow exponentially with time. The growth rates of waves with frequencies $\omega_{\vec{k}_{1\perp}}$ and $\omega_{\vec{k}_{2\perp}}$ depend on the amplitude of the wave with frequency $\omega_{\vec{k}_\perp}$:

$$|\gamma| = \sqrt{|V_{\vec{k}_\perp;\vec{k}_{1\perp},\vec{k}_{2\perp}} A_{\vec{k}_\perp}|^2} \tag{61}$$

A value, inverse to (61), determines the decay time. As an example of surface-wave decay interaction, one can indicate the decay of a surface Langmuir wave into surface Langmuir and ion sound waves (in strongly non-isothermal plasma), decay of a surface cyclotron wave with frequency $\cong 2\omega_B$ into two surface cyclotron waves with frequencies $\cong\omega_B$ (in magnetoactive plasma), and so on.

In the case of resonant interaction of three surface waves with energies of different sign, e.g. when

$$S_{\vec{k}_{1\perp}} = S_{\vec{k}_{2\perp}} = -S_{\vec{k}_\perp} \tag{62}$$

an explosive instability arises in the system, i.e. the amplitudes of interacting waves turn to infinity at some finite time t_∞. The wave with negative energy gives energy to the waves with positive energy (or the waves with negative energy give energy to the wave with positive energy), and the amplitudes of interacting waves grow to infinity in spite of total energy conservation in the system. By means of appropriate choice of initial condition, we can provide the amplitude of the most intensive wave (at zero time) to develop in time according to the law:

$$A_{\vec{k}_\perp}(t) = \frac{A_{\vec{k}_\perp}(0)}{1 - \frac{t}{t_\infty}} \tag{63}$$

The explosion time t_∞ is determined by the initial amplitude and the non-linear interaction matrix element:

$$t_\infty = \frac{1}{\sqrt{|V_{\vec{k}_\perp;\vec{k}_{1\perp},\vec{k}_{2\perp}} A_{\vec{k}_\perp}(0)|^2}} \tag{64}$$

Explosive instability can take place in semibounded plasma with compensated ion beam due to resonant interaction between three surface waves, dispersion of which is determined by (35) (or (36)).

7. FLUCTUATIONS

When considering fluctuations in non-equilibrium semibounded plasma it is convenient, as in the infinite plasma case, to use a non-linear equation for a field with fluctuation sources, which may be derived from Maxwell equations and the equation for the microscopic density, describing particle motion in plasmas. We restrict ourselves, for simplicity, to consideration of the potential field and assume that the plasma in the halfspace is homogeneous and stationary. We separate the fluctuation part δf of the microscopic density and present it as a sum of the microscopic density fluctuation part in the absence of particle interaction δf^0 and the difference between the exact microscopic density and the microscopic density for non-interacting particles f. The value f and the microscopic field $\vec{E}$ are described by equations which differ from (1) and (2) only by the additional terms, respectively,

$$\frac{e}{m}\vec{E}\frac{\partial}{\partial \vec{v}}\delta f^0 \quad \text{and} \quad 4\pi e\int d\vec{v}\,\delta f^0$$

Assuming that specular reflection conditions are satisfied on the surface, and continuing the electric field outside the plasma as before, we obtain for f^+ an equation determined in the whole space. Presenting the solution of this equation in the form of an expansion in a series of field amplitude and substituting it into the expression for induced charge, we thus obtain the following non-linear equation for the fluctuation field:

$$\varepsilon(\omega,\vec{k})E_{\vec{k}\omega} + \sum_{\substack{\omega_1+\omega_2=\omega\\ \vec{k}_1+\vec{k}_2=\vec{k}}} \varkappa^{(2)}(\omega_1,\vec{k}_1;\omega_2,\vec{k}_2)E_{\vec{k}_1\omega_1}E_{\vec{k}_2\omega_2}$$

$$+ \sum_{\substack{\omega_1+\omega_2+\omega_3=\omega\\ \vec{k}_1+\vec{k}_2+\vec{k}_3=\vec{k}}} \varkappa^{(3)}(\omega_1,\vec{k}_1;\omega_2,\vec{k}_2;\omega_3,\vec{k}_3)E_{\vec{k}_1\omega_1}E_{\vec{k}_2\omega_2}E_{\vec{k}_3\omega_3} + \ldots$$

$$+ \sum_{\omega',\vec{k}'}\delta\varepsilon(\omega,\vec{k};\omega',\vec{k}')E_{\vec{k}'\omega'} + \sum_{\substack{\omega_1+\omega_2=\omega\\ \vec{k}_1+\vec{k}_2=\vec{k}\\ \omega',\vec{k}'}}\delta\varkappa^{(2)}(\omega_1,\vec{k}_1;\omega_2,\vec{k}_2;\omega',\vec{k}')E_{\vec{k}_1\omega_1}E_{\vec{k}'\omega'} + \ldots$$

$$+ \frac{2\varepsilon_0}{k}E_{\vec{k}_\perp\omega} = -\frac{4\pi i}{k}\rho^0_{\vec{k}\omega} \qquad (65)$$

where $\rho^0_{\vec{k}\omega}$ is the fluctuation charge density due to the random motion of the separate charged particles:

$$\rho^0_{\vec{k}\omega} = \sum e \int d\vec{v}\, \delta f^0_{\vec{k}\omega} \tag{66}$$

($\rho^0(z)$ is continued into the halfspace outside the plasmas in an even way); $\delta\epsilon$ and $\delta\kappa^{(2)}$ are fluctuation variations of dielectric permittivity and non-linear susceptibility.

If the field intensity is small and non-linear effects are negligible we obtain from (65), in linear approximation,

$$\varepsilon(\omega,\vec{k})\, E_{\vec{k}\omega} + \frac{2\varepsilon_0}{k} E_{\vec{k}_\perp\omega} = -\frac{4\pi i}{k}\rho^0_{\vec{k}\omega} \tag{67}$$

and the surface fluctuation field $E_{\vec{k}_\perp\omega}$ satisfies the following equation:

$$\zeta(\omega,\vec{k}_\perp) E_{\vec{k}_\perp\omega} = -4\pi i k_\perp \sum_{k_z} \frac{\rho^0_{\vec{k}\omega}}{k^2\varepsilon(\omega,\vec{k})} \tag{68}$$

Relation (67) connects the fluctuation field in plasmas with distribution function fluctuations in the absence of particle interaction. Using this relation, we can express field correlation functions in plasmas directly in terms of correlation functions for the system of non-interacting particles.

Spectral distribution of distribution function fluctuations neglecting particle interaction (but taking into account specular reflection from the boundary) is determined by the formula:

$$\langle \delta f^0_{k_z}(\vec{v})\, \delta f^0_{k'_z}(\vec{v}') \rangle_{\vec{k}_\perp\omega} = (2\pi)^2 \delta(\vec{v}_\perp - \vec{v}'_\perp) \left\{ \delta(v_z - v'_z)\, \delta(k_z - k'_z) \right.$$
$$\left. + \delta(v_z + v'_z)\, \delta(k_z + k'_z) \right\} \delta(\omega - \vec{k}\vec{v})\, f_0(\vec{v}) \tag{69}$$

The first term on the right-hand side of (69) describes fluctuations in an unbounded system in the absence of particle interaction; the second term is due to particle specular reflection from the boundary. Integrating (69) over velocities, we find spectral distribution of charge density fluctuations neglecting particle interaction (but taking into account the reflection from the boundary):

$$\langle \rho_{k_z} \rho_{k'_z} \rangle^0_{\vec{k}_\perp\omega} = 4\pi^2 \sum e^2 \int d\vec{v}\, \delta(\omega - \vec{k}\vec{v})\, f_0(\vec{v}) \left\{ \delta(k_z - k'_z) + \delta(k_z + k'_z) \right\} \tag{70}$$

Using Eq.(68), we determine the electric field surface fluctuation spectral distribution:

$$\langle E^2\rangle_{\vec{k}_\perp\omega} = \frac{16\pi^2 k_\perp^2}{|\zeta(\omega,\vec{k}_\perp)|^2}\sum_{k_z,k_z'}\frac{\langle\rho_{k_z}\rho_{k_z'}\rangle^0_{\vec{k}_\perp\omega}}{k^2k'^2\varepsilon(\omega,\vec{k})\varepsilon^*(\omega,\vec{k}')} \tag{71}$$

Expressing the field surface component $E_{\vec{k}_\perp\omega}$ in (67) in terms of the fluctuation charge density $\rho^0_{\vec{k}\omega}$, we obtain, as before, the total field fluctuation spectral distribution in semibounded plasma:

$$\langle E_{k_z}E_{k_z'}\rangle_{\vec{k}_\perp\omega} = \frac{16\pi^2}{kk'\varepsilon(\omega,\vec{k})\varepsilon^*(\omega,\vec{k}')}\left\{\langle\rho_{k_z}\rho_{k_z'}\rangle^0_{\vec{k}_\perp\omega} - \frac{2\varepsilon_0 k_\perp}{\zeta(\omega,\vec{k}_\perp)}\sum_{k_z''}\frac{\langle\rho_{k_z''}\rho_{k_z'}\rangle^0_{\vec{k}_\perp\omega}}{k''^2\varepsilon(\omega,\vec{k}'')}\right.$$

$$\left. - \frac{2\varepsilon_0 k_\perp}{\zeta^*(\omega,\vec{k}_\perp)}\sum_{k_z''}\frac{\langle\rho_{k_z}\rho_{k_z''}\rangle^0_{\vec{k}_\perp\omega}}{k''^2\varepsilon^*(\omega,\vec{k}'')} + \frac{4\varepsilon_0^2k_\perp^2}{|\zeta(\omega,\vec{k}_\perp)|^2}\sum_{k_z,k_z'}\frac{\langle\rho_{k_z}\rho_{k_z'}\rangle^0_{\vec{k}_\perp\omega}}{k^2k'^2\varepsilon(\omega,\vec{k})\varepsilon^*(\omega,\vec{k}')}\right\} \tag{72}$$

The first term on the right-hand side of (72) describes the fluctuations due to volume field oscillations in semibounded plasma. As this term contains delta-functions of the difference or the sum of k_z and k_z', it exceeds all other terms in (72) in the corresponding region of k_z and k_z' magnitudes (volume fluctuation domain). So the electric field volume fluctuation spectral distribution in semi-bounded plasma may be written approximately in the form:

$$\langle E_{k_z}E_{k_z'}\rangle_{\vec{k}_\perp\omega} = \frac{16\pi}{kk'}\frac{\langle\rho_{k_z}\rho_{k_z'}\rangle^0_{\vec{k}_\perp\omega}}{\varepsilon(\omega,\vec{k})\varepsilon^*(\omega,\vec{k}')} \tag{73}$$

We note that, by integrating (72) over k_z and k_z' components, we immediately obtain surface fluctuation spectral distributions (71).

Expressions (71) and (73) are valid for the description of the electric field surface and volume fluctuations both in thermodynamic equilibrium and non-equilibrium (but stationary and stable) plasmas. Expressions (71) and (73) may be essentially simplified for thermodynamic equilibrium plasma. Noting that, in equilibrium plasma,

$$\langle\rho_{k_z}\rho_{k_z'}\rangle^0_{\vec{k}_\perp\omega} = k^2\{\delta(k_z-k_z')+\delta(k_z+k_z')\}\frac{T}{\omega}\,\mathrm{Im}\,\varepsilon(\omega,k) \tag{74}$$

we can present surface and volume fluctuation spectral distributions in the form:

$$\langle E^2\rangle_{\vec{k}_\perp\omega} = \frac{8\pi k_\perp}{\varepsilon_0}\,\frac{T}{\omega}\,\frac{\mathrm{Im}\,\zeta^*(\omega,k_\perp)}{|\zeta(\omega,k_\perp)|^2} \tag{75}$$

$$\langle E_{k_z}E_{k_z'}\rangle_{\vec{k}_\perp\omega} = 16\pi^2\,\frac{T}{\omega}\,\frac{\mathrm{Im}\,\varepsilon(\omega,k)}{|\varepsilon(\omega,k)|^2}\left\{\delta(k_z-k_z') + \delta(k_z+k_z')\right\} \tag{76}$$

In the spectra of surface and volume fluctuations, in addition to a wide maximum in the low-frequency domain due to the random motion of charged particles, there are also sharp maxima, corresponding to collective surface or volume field fluctuation oscillations. Spectral distributions of surface and volume fluctuations near eigenfrequencies are determined by the following expressions:

$$\langle E^2\rangle_{\vec{k}_\perp\omega} = \pi I_{\vec{k}_\perp}\left\{\delta(\omega-\omega_{\vec{k}_\perp}) + \delta(\omega+\omega_{\vec{k}_\perp})\right\}, \quad I_{\vec{k}_\perp} = \frac{8\pi}{\varepsilon_0}\,\frac{T}{\omega_{\vec{k}_\perp}\zeta'_{\vec{k}_\perp}} \tag{77}$$

$$\langle E_{k_z}E_{k_z'}\rangle_{\vec{k}_\perp\omega} = 2\pi^2 I_{\vec{k}}\left\{\delta(\omega-\omega_{\vec{k}})+\delta(\omega+\omega_{\vec{k}})\right\}\left\{\delta(k_z-k_z')+\delta(k_z+k_z')\right\}, \quad I_{\vec{k}} = 8\pi\frac{T}{\omega_{\vec{k}}\varepsilon'_{\vec{k}}} \tag{78}$$

We note that intensities of fluctuation oscillations in non-equilibrium plasma may essentially differ from the thermal level. Intensities may increase greatly compared with (77) and (78) if the plasma state is near the kinetic stability threshold. Non-linear interaction of fluctuation oscillations must be taken into account in such a case.

We separate in (65) the part corresponding to surface wave non-linear interaction. As a result the equation may be rewritten in the form:

$$\zeta(\omega,\vec{k}_\perp)E_{\vec{k}_\perp\omega} + \sum_{\substack{\omega_1+\omega_2=\omega\\ \vec{k}_{1\perp}+\vec{k}_{2\perp}=\vec{k}_\perp}}\tilde{\varkappa}^{(2)}(\omega_1,\vec{k}_{1\perp};\omega_2,\vec{k}_{2\perp})E_{\vec{k}_{1\perp}\omega_1}E_{\vec{k}_{2\perp}\omega_2} + \sum_{\substack{\omega_1+\omega_2+\omega_3=\omega\\ \vec{k}_{1\perp}+\vec{k}_{2\perp}+\vec{k}_{3\perp}=\vec{k}_\perp}}\tilde{\varkappa}^{(3)}(\omega_1,k_{1\perp};\omega_2,k_{2\perp};\omega_3,k_{3\perp})E_{\vec{k}_{1\perp}\omega_1}E_{\vec{k}_{2\perp}\omega_2}E_{\vec{k}_{3\perp}\omega_3}$$

$$+\ldots+\sum_{\omega',\vec{k}_\perp'}\delta\zeta(\omega,\vec{k}_\perp;\omega',\vec{k}_\perp')E_{\vec{k}_\perp'\omega'} + \sum_{\substack{\omega_1+\omega_2=\omega\\ \vec{k}_{1\perp}+\vec{k}_{2\perp}=\vec{k}_\perp\\ \omega',\vec{k}_\perp'}}\delta\tilde{\varkappa}^{(2)}(\omega_1,\vec{k}_{1\perp};\omega_2,\vec{k}_{2\perp};\omega',\vec{k}_\perp')E_{\vec{k}_{1\perp}\omega_1}E_{\vec{k}_\perp'\omega'}+\ldots$$

$$= -4\pi i k_\perp\sum_{k_z'}\frac{\rho^0_{\vec{k}\omega}}{k^2\varepsilon(\omega,\vec{k})} \tag{79}$$

where

$$\delta\zeta(\omega,\vec{k}_\perp;\omega',\vec{k}_\perp') = -\sum_{k_z,k_z'}\frac{2\varepsilon_0 k_\perp}{kk'}\,\frac{\delta\varepsilon(\omega,\vec{k};\omega',\vec{k}')}{\varepsilon(\omega,\vec{k})\varepsilon(\omega',\vec{k}')}$$

and so on. In order to describe the fluctuation field, taking the non-linear interaction into account, it is necessary, generally speaking, to find not only the quadratic correlation function $\langle E^2\rangle_{\vec{k}_\perp\omega}$, but also the higher-order correlation functions (i.e. the third and fourth orders in our approximation). Multiplying successively the left- and right-hand parts of Eq.(79) by themselves, we can obtain a set of non-uniform integral equations which determine a consequence of these correlation functions. In particular, we obtain the following equation for the second-order correlation function:

$$\varsigma(\omega,\vec{k}_\perp)\langle E^2\rangle_{\vec{k}_\perp\omega} - \frac{2}{\varsigma^*(\omega,\vec{k}_\perp)}\sum_{\omega',\vec{k}'_\perp}|\tilde{\varkappa}^{(2)}(\omega-\omega',\vec{k}_\perp-\vec{k}'_\perp;\omega',\vec{k}'_\perp)|^2\langle E^2\rangle_{\vec{k}'_\perp\omega'}\langle E^2\rangle_{\vec{k}_\perp-\vec{k}'_\perp,\omega-\omega'}$$

$$-\sum_{\omega',\vec{k}'_\perp}\tilde{a}(\omega,\vec{k}_\perp;\omega',\vec{k}'_\perp)\langle E^2\rangle_{\vec{k}'_\perp\omega'}\langle E^2\rangle_{\vec{k}_\perp\omega}$$

$$=\frac{1}{\varsigma^*(\omega,\vec{k}_\perp)}\left\{\sum_{\omega',\vec{k}'_\perp}\tilde{b}(\omega,\vec{k}_\perp;\omega',\vec{k}'_\perp)\langle E^2\rangle_{\vec{k}'_\perp\omega'}+\tilde{q}_{\vec{k}_\perp\omega}\right\} \tag{80}$$

where the value $\tilde{a}$ is expressed in terms of non-linear susceptibilities,

$$\tilde{a}(\omega,\vec{k}_\perp;\omega',\vec{k}'_\perp)=4\frac{\tilde{\varkappa}^{(2)}(\omega-\omega',\vec{k}_\perp-\vec{k}'_\perp;\omega',\vec{k}'_\perp)\tilde{\varkappa}^{(2)}(\omega,\vec{k}_\perp;-\omega',-\vec{k}'_\perp)}{\varsigma(\omega-\omega',\vec{k}_\perp-\vec{k}'_\perp)}$$

$$+2\frac{\tilde{\varkappa}^{(2)}(\omega,\vec{k}_\perp;0,0)\tilde{\varkappa}^{(2)}(\omega',\vec{k}'_\perp;-\omega',-\vec{k}'_\perp)}{\varsigma(0,0)}-3\tilde{\varkappa}^{(3)}(\omega',\vec{k}'_\perp;\omega,\vec{k}_\perp;-\omega',-\vec{k}'_\perp) \tag{81}$$

and values $\tilde{b}$ and $\tilde{q}$ are expressed in terms of correlation functions of fluctuation sources,

$$\tilde{q}_{\vec{k}_\perp\omega}=16\pi^2k_\perp^2\sum_{k_z,k'_z}\frac{\langle\rho_{k_z}\rho_{k'_z}\rangle^0_{\vec{k}_\perp\omega}}{k^2k'^2\varepsilon(\omega,\vec{k})\varepsilon(\omega,\vec{k}')}+\ldots \tag{82}$$

Using Eq.(80), we can find fluctuation field spectral distribution, taking into account non-linear surface wave interaction. This interaction, in particular, causes additional maxima in the spectrum at combination frequencies, as well as saturation of critical surface fluctuations in non-equilibrium plasma by surface oscillation eigenfrequency non-linear shifts.

Equation (80), which determines the spectral distribution of electric field surface fluctuations, was derived on the assumption that this distribution is stationary. But in real conditions, taking into account non-linear interaction of the surface waves together with linear damping or growth of oscillation leads to a possibility for the field fluctuation spectral distribution to change in time. If particle distributions are stationary, the equation for surface fluctuation spectral distribution time evolution may be derived from (80) by substituting

$$\zeta(\omega,\vec{k}_\perp)\langle E^2\rangle_{\vec{k}_\perp\omega} \rightarrow \left\{\zeta(\omega,\vec{k}_\perp) + \frac{i}{2}\frac{\partial \operatorname{Re}\zeta(\omega,\vec{k}_\perp)}{\partial\omega}\frac{\partial}{\partial t}\right\}\langle E^2\rangle_{\vec{k}_\perp\omega} \tag{83}$$

and taking the imaginary part. As a result we obtain

$$\begin{aligned}\frac{1}{2}\frac{\partial \operatorname{Re}\zeta(\omega,\vec{k}_\perp)}{\partial\omega}\frac{\partial}{\partial t}\langle E^2\rangle_{\vec{k}_\perp\omega} &= -\operatorname{Im}\zeta(\omega,\vec{k}_\perp)\langle E^2\rangle_{\vec{k}_\perp\omega}\\ &+2\operatorname{Im}\frac{1}{\zeta^*(\omega,\vec{k}_\perp)}\sum_{\omega',\vec{k}'_\perp}|\tilde{\varkappa}^{(2)}(\omega-\omega',\vec{k}_\perp-\vec{k}'_\perp;\omega',\vec{k}'_\perp)|^2\langle E^2\rangle_{\vec{k}'_\perp\omega'}\langle E^2\rangle_{\vec{k}_\perp-\vec{k}'_\perp,\omega-\omega'}\\ &+\operatorname{Im}\sum_{\omega',\vec{k}'_\perp}\tilde{a}(\omega,\vec{k}_\perp;\omega',\vec{k}'_\perp)\langle E^2\rangle_{\vec{k}'_\perp\omega'}\langle E^2\rangle_{\vec{k}_\perp\omega}\\ &+\operatorname{Im}\frac{1}{\zeta^*(\omega,\vec{k}_\perp)}\left\{\sum_{\omega',\vec{k}'_\perp}\tilde{b}(\omega,\vec{k}_\perp;\omega',\vec{k}'_\perp)\langle E^2\rangle_{\vec{k}'_\perp\omega'}+\tilde{q}_{\vec{k}_\perp\omega}\right\}\end{aligned} \tag{84}$$

This equation describes the time dependence of the spectral density $\langle E^2\rangle_{\vec{k}_\perp\omega}$ due to linear dissipation and non-linear wave interaction. It is not difficult to derive a kinetic equation for surface waves in semibounded plasmas on the basis of (84).

8. KINETIC EQUATION FOR SURFACE WAVES

The general solution of Eq.(84) in linear approximation is

$$\langle E^2\rangle_{\vec{k}_\perp\omega} = \langle E^2\rangle^{\circ}_{\vec{k}_\perp\omega} + \pi I_{\vec{k}_\perp}(t)\left\{\delta(\omega-\omega_{\vec{k}_\perp})+\delta(\omega+\omega_{\vec{k}_\perp})\right\} \tag{85}$$

where the first term, which is determined by the non-uniform part of (84), characterizes the stationary level of surface fluctuations, and the second term describes surface eigenoscillations of the electric field due to the initial conditions. The stationary surface fluctuation level in equilibrium plasma is determined by temperature; therefore we can neglect thermal oscillations if the induced oscillation level is high enough.

Substituting spectral distribution (85) into (84), we obtain the following equation for the intensity of oscillations of a certain type $I_{\vec{k}_\perp}$:

$$\frac{\partial I_{\vec{k}_\perp}}{\partial t} = -2\gamma_{\vec{k}_\perp} I_{\vec{k}_\perp} + \frac{2\pi}{(\zeta'_{\vec{k}_\perp})^2} \sum_{\vec{k}'_\perp} \left| \tilde{\varkappa}^{(2)}(\omega_{\vec{k}_\perp} \mp \omega_{\vec{k}'_\perp}, \vec{k}_\perp - \vec{k}'_\perp; \pm\omega_{\vec{k}'_\perp}, \vec{k}'_\perp) \right|^2$$

$$\times\, \delta(\omega_{\vec{k}} \mp \omega_{\vec{k}'_\perp} \mp \omega_{\vec{k}_\perp - \vec{k}'_\perp})\, I_{\vec{k}'_\perp} I_{\vec{k}_\perp - \vec{k}'_\perp}$$

$$+ \frac{1}{\zeta'_{\vec{k}_\perp}} \operatorname{Im} \sum_{\vec{k}'_\perp} \tilde{a}(\omega_{\vec{k}_\perp}, \vec{k}_\perp; \pm\omega_{\vec{k}'_\perp}, \vec{k}'_\perp)\, I_{\vec{k}'_\perp} I_{\vec{k}_\perp}$$

$$+ \frac{2}{\zeta'_{\vec{k}_\perp}} \operatorname{Im} \sum_{\omega', \vec{k}'_\perp} \tilde{a}(\omega_{\vec{k}_\perp}, \vec{k}_\perp; \omega', \vec{k}'_\perp) \langle E^2 \rangle^0_{\vec{k}'_\perp \omega'} I_{\vec{k}_\perp}$$

$$+ \frac{1}{(\zeta'_{\vec{k}_\perp})^2} \sum_{\vec{k}'_\perp} \Big\{ \tilde{b}(\omega_{\vec{k}_\perp}, \vec{k}_\perp; \pm\omega_{\vec{k}'_\perp}, \vec{k}'_\perp)$$

$$+ 4 \left| \tilde{\varkappa}^{(2)}(\omega_{\vec{k}_\perp} \mp \omega_{\vec{k}'_\perp}, \vec{k}_\perp - \vec{k}'_\perp; \pm\omega_{\vec{k}'_\perp}, \vec{k}'_\perp) \right|^2 \langle E^2 \rangle^0_{\vec{k}_\perp - \vec{k}'_\perp, \omega_{\vec{k}_\perp} - \omega_{\vec{k}'_\perp}} \Big\} I_{\vec{k}_\perp} \qquad (86)$$

This equation describes surface wave dynamics, taking into account both interaction between themselves and with surface fluctuations in plasmas. If we neglect in (86) the fluctuation oscillation intensity

$$\langle E^2 \rangle^0_{\vec{k}_\perp \omega}$$

and the value

$$\tilde{b}(\omega_{\vec{k}_\perp}, \vec{k}_\perp; \omega_{\vec{k}'_\perp}, \vec{k}'_\perp)$$

which is connected with particle distribution fluctuations, we obtain a kinetic equation for surface waves:

$$\frac{\partial I_{\vec{k}_\perp}}{\partial t} = -2\gamma_{\vec{k}_\perp} I_{\vec{k}_\perp} + \frac{2\pi}{(\zeta'_{\vec{k}_\perp})^2} \sum_{\vec{k}'_\perp} \left| \tilde{\varkappa}^{(2)}(\omega_{\vec{k}_\perp} \mp \omega_{\vec{k}'_\perp}, \vec{k}_\perp - \vec{k}'_\perp; \pm\omega_{\vec{k}'_\perp}, \vec{k}'_\perp) \right|^2 \delta(\omega_{\vec{k}_\perp} \mp \omega_{\vec{k}'_\perp} \mp \omega_{\vec{k}_\perp - \vec{k}'_\perp}) I_{\vec{k}'_\perp} I_{\vec{k}_\perp - \vec{k}'_\perp}$$

$$+ \frac{1}{\zeta'_{\vec{k}_\perp}} \operatorname{Im} \sum_{\vec{k}'_\perp} \tilde{a}(\omega_{\vec{k}_\perp}, \vec{k}_\perp; \pm\omega_{\vec{k}'_\perp}, \vec{k}'_\perp) I_{\vec{k}'_\perp} I_{\vec{k}_\perp} \tag{87}$$

The kinetic equation (87) describes changes of surface wave spectral density $I_{\vec{k}_\perp}$ due to linear dissipation and non-linear wave-wave and wave-particle interaction. The three-wave decay processes, i.e. transformation of two waves into one and decay of the wave into two others, are taken into account in (87). In addition, induced scattering of waves on particles is taken into account, which causes additional damping of waves – non-linear Landau damping [28].

As already mentioned, the kinetic equation (87) is valid only when wave intensities are large enough and fluctuation oscillations in plasma may be neglected. That is why scattering and transformation of waves due to interaction with fluctuation fields are not described by (87). In contrast to (87), the kinetic equation (86) takes into account interaction between waves and fluctuation fields and so may be used for the description of wave scattering and transformation on fluctuations in semibounded plasmas. Equation (84) also allows the description of scattering of charged particles on electric field fluctuations, accompanied by bremsstrahlung. That is why, using the generalization (taking into account not only potential but also vorticity electric fields) of Eq.(84), we can investigate the spontaneous radiation from plasma into the surrounding medium. This radiation may differ essentially from thermal emission in non-equilibrium plasma.

9. NON-LINEAR EQUATION FOR THE FIELD (GENERAL CASE)

We now consider a non-linear equation for the electric field in semibounded plasma without the assumption about potentiality of the field. In such a case, besides the self-consistent electric field, the self-consistent magnetic field also enters the kinetic equation. We continue the self-consistent electric and magnetic fields into space outside plasmas (the components $\vec{E}_\perp$ and B_z, respectively, in the even and E_z and $\vec{B}_\perp$ in the odd relative substitution $z \to -z$) and supplement the definition of the distribution function f as in the case of the potential electric field. Induced current density in semibounded plasma is expressed in terms of linear and non-linear susceptibilities of infinite uniform plasma, and space limitation leads to additional surface current in the field equation. The non-linear equation for the electric field in semibounded plasma (in the absence of external sources) may be written as

$$\Lambda_{ij}(\omega,\vec{k})E_{j\vec{k}\omega} + \sum_{1+2=\omega,\vec{k}} \varkappa^{(2)}_{ijk}(1,2)E_{j1}E_{k2} + \sum_{1+2+3=\omega,\vec{k}} \varkappa^{(3)}_{ijkl}(1,2,3)E_{j1}E_{k2}E_{l3}+\ldots$$

$$= 2i\,\frac{c}{\omega}\,\varepsilon_{zij}\,B_{j\vec{k}_\perp\omega}(0) \qquad (88)$$

where $B_{j\vec{k}_\perp\omega}(0)$ is the magnetic field on the boundary

$$B_{j\vec{k}_\perp\omega}(0) \equiv \frac{1}{2\pi}\int dk_z\,B_{j\vec{k}\omega} \qquad (89)$$

Supposing that there is an insulator with dielectric permittivity outside the plasma and, using the boundary conditions for fields, it is not difficult to show that

$$\varepsilon_{zij}\,B_{j\vec{k}_\perp\omega}(0) = i\alpha_{ij}(\omega,\vec{k}_\perp)\,E_{j\vec{k}_\perp\omega}(0) \qquad (90)$$

where

$$\alpha_{ij} = \begin{pmatrix} \alpha_1 & 0 & 0 \\ 0 & \alpha_2 & 0 \\ 0 & 0 & 0 \end{pmatrix}, \quad \alpha_1 = \frac{\varepsilon_0}{\sqrt{\eta_\perp^2-\varepsilon_0}}, \quad \alpha_2 = -\sqrt{\eta_\perp^2-\varepsilon_0}$$

Now we can rewrite the basic non-linear equation for the electric field in semi-bounded plasma in a closed form:

$$\Lambda_{ij}(\omega,\vec{k})E_{j\vec{k}\omega} + \sum_{1+2=\omega,\vec{k}} \varkappa^{(2)}_{ijk}(1,2)E_{j1}E_{k2} + \sum_{1+2+3=\omega,\vec{k}} \varkappa^{(3)}_{ijkl}(1,2,3)E_{j1}E_{k2}E_{l3}$$

$$+ \ldots + \frac{2c}{\omega}\alpha_{ij}(\omega,\vec{k}_\perp)\sum_{k_z} E_{j\vec{k}\omega} = 0 \qquad (91)$$

The linear approximation equation for the field in semibounded plasma is

$$\Lambda_{ij}(\omega,\vec{k})E_{j\vec{k}\omega} + \frac{2c}{\omega}\alpha_{ij}(\omega,\vec{k}_\perp)\,E_{j\vec{k}_\perp\omega} = 0 \qquad (92)$$

where $E_{j\vec{k}_\perp\omega}$ is the electric field on the boundary

$$E_{j\vec{k}_\perp\omega} \equiv E_{j\vec{k}_\perp\omega}(0) = \frac{1}{2\pi}\int dk_z E_{j\vec{k}\omega} \tag{93}$$

It is not difficult to derive from (92) the following dispersion equation for the surface waves:

$$\left| \delta_{ik} + \frac{c}{\pi\omega}\int dk_z \Lambda_{ij}^{-1}(\omega,\vec{k})\alpha_{jk}(\omega,\vec{k}_\perp) \right| = 0 \tag{94}$$

When particle distribution is isotropic, the dispersion equation (94) decomposes into two independent equations:

$$1 - \frac{c}{\pi\omega}\sqrt{\eta_\perp^2 - \varepsilon_0}\int dk_z \frac{1}{\varepsilon_t(\omega,k) - \eta^2} = 0 \tag{95}$$

$$1 + \frac{c}{\pi\omega}\frac{\varepsilon_0}{\sqrt{\eta_\perp^2 - \varepsilon_0}}\int \frac{dk_z}{k^2}\left\{\frac{k_\perp^2}{\varepsilon_\ell(\omega,k)} + \frac{k_z^2}{\varepsilon_t(\omega,k) - \eta^2}\right\} = 0 \tag{96}$$

corresponding to surface oscillations with different polarizations of the electric field (s and p are the polarizations).

The non-linear equation (91) makes it possible to investigate three-wave decays in semibounded plasma in the general case and is basic for the derivation of the kinetic equation for waves, which also describes the transverse electromagnetic waves.

REFERENCES

[1] SITENKO, A.G., Fluctuations and Nonlinear Wave Interaction in Plasmas, Naukova Dumka, Kiev (1977) (in Russian).

[2] LANDAU, L.D., Zh. Ehksp. Teor. Fiz. **16** (1946) 574.

[3] REUTER, G.E.H., SONDHEIMER, E.H., Proc. R. Soc. (London) **A195** (1948) 336.

[4] SHAFRANOV, V.D., Zh. Ehksp. Teor. Fiz. **34** (1958) 1475.

[5] ROMANOV, Yu.A., Sov. Phys. – Radiophys. Quantum Electron. **7** (1964) 242.

[6] ROMANOV, Yu.A., Sov. Phys. – Solid State **7** (1965) 970.

[7] KONDRATENKO, A.N., Nucl. Fusion **5** (1965) 267.

[8] KONDRATENKO, A.N., Zh.Tekh. Fiz. **36** (1966) 1943.

[9] GUERNSEY, R.L., Phys. Fluids **12** (1969) 1852.

[10] FUSE, M., ICHIMARU, S., J. Phys. Soc. Jpn. **38** (1975) 559.

[11] GINSBURG, V.L., RUKHADZE, A.A., Waves in Magnetoactive Plasmas, Nauka. Moscow (1975) (in Russian).

[12] ABA-ASALI, E.A., ALTERKOP, B.A., RUKHADZE, A.A., Plasma Phys. **17** (1975) 218.

[13] SHUMAN, W.O., Z. Naturforsch. A. **5** (1950) 181.

[14] FAINBERG, Ya.B., Proc. Symp. CERN **1** (1956) 84.

[15] SITENKO, A.G., TKALICH, V,S., Zh. Tekh. Fiz. **21** (1959) 1074.

[16] SITENKO, A.G., YAKIMENKO, I.P., in Advances in Plasma Physics **5**, Wiley, New York (1974).

[17] YAKIMENKO, I.P., ZAGORODNY, A.G., Phys. Scr. **10** (1974) 244.

[18] ZAGORODNY, A.G., YAKIMENKO, I.P., Kiev Preprint ITP-75-124R (1975) (in Russian).

[19] POPOV, V.S., YAKIMENKO, I.P., Zh. Tekh. Fiz. **45** (1975) 1381.

[20] KARPĹYUJK, K.S., ORAEVSKY, V.N., Zh. Tekh. Fiz. **38** (1968) 1214.

[21] KARPLYUJK, K.S., KOLESNICHENKO, Ya.I., ORAEVSKY, V.N., Nucl. Fusion **10** (1970) 3.

[22] KONDRATENKO, A.N., SHAPTALA, V.G., Ukr. Fiz. Zh. **14** (1969) 1092.

[23] SITENKO, A.G., PAVLENKO, V.N., ZASENKO, V.I., Ukr. Fiz. Zh. **20** (1975) 324.

[24] SITENKO, A.G., PAVLENKO, V.N., ZASENKO, V.I., Sov. J. Plasma Phys. **2** (1976) 804.

[25] SITENKO, A.G., STEPANOV, K.N., Zh. Ehksp. Teor. Fiz. **31** (1956) 642.

[26] COPPI, B., ROSENBLUTH, M.N., SUDAN, R.N., Ann. Phys. **55** (1969) 207, 248.

[27] SAGDEEV, R.Z., GALEEV, A.A., Nonlinear Plasma Theory, Benjamin, New York (1969).

[28] KADOMTSEV, B.B., Plasma Turbulence, Academic Press, New York (1965).

IAEA-SMR-31/27-A

PARAMETRIC EXCITATION

K. NISHIKAWA
Faculty of Science,
Hiroshima University,
Hiroshima,
Japan

Abstract

PARAMETRIC EXCITATION.
General aspects of the theory of parametric instabilities are surveyed. First, the basic concept of the parametric excitation is explained and its characteristic features are described with the use of the Mathieu equation model. Two specific types of instabilities – resonant type and non-resonant type purely growing instabilities – are specifically discussed. The theory is then extended to a general three-mode coupling problem in a uniform medium and a physical mechanism of the instability is presented. A brief classification of various instabilities is also presented. Finally, some geometrical effects on the resonant decay instability are discussed based on physical rather than mathematical arguments, with particular reference to the absolute versus convective types of instabilities.

1. INTRODUCTION

Parametric instability is an instability of natural oscillations due to the periodic modulation of a parameter which characterizes the oscillation. Consider the equation of a harmonic oscillator:

$$\frac{d^2 X(t)}{dt^2} + \Omega^2 X(t) = 0 \tag{1}$$

This equation is characterized by a single parameter Ω, which is the frequency of the oscillator. If this parameter is modulated periodically in the form

$$\Omega^2 = \Omega_0^2 [1 + Q(t)], \quad Q(t+T) = Q(t) \tag{2}$$

where Ω_0 is the frequency in the absence of modulation and T is the modulation period, the oscillation of X(t) amplifies when certain matching conditions are satisfied. For the case of weak modulation, i.e. $|Q| \ll 1$, matching conditions are given by

$$\Omega_0 T \cong n\pi \quad (n = 1, 2, \ldots) \tag{3}$$

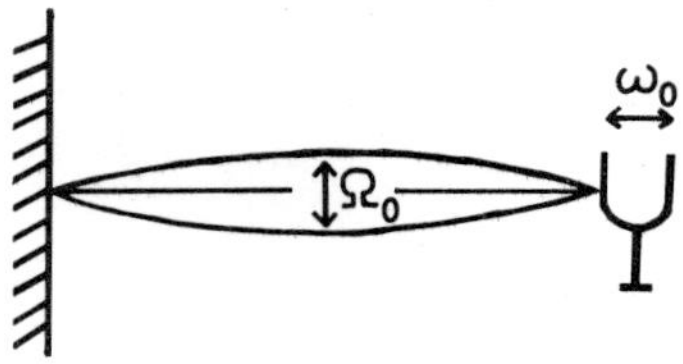

FIG.1. Stretched string (natural transverse vibration frequency Ω_0) with one end fixed to a wall and the other end attached to a tuning fork of frequency ω_0.

If one of these conditions is met, a strong coupling is produced between the two natural oscillations, $X \sim \exp[-i\Omega_0 t]$ and $X \sim \exp[i\Omega_0 t]$, due to the modulation, and this coupling results in the amplification of the natural oscillations by subtracting energy from the modulator. We note a similarity of the matching conditions (3) to the Bragg reflection conditions for the electron in a periodic potential in solids [1].

The first systematic investigation of the parametric excitation was made by Lord Rayleigh [2]. He considered a stretched string with one end fixed and the other end attached to a tuning fork (see Fig.1). When the frequency of the tuning fork, ω_0, is adjusted to twice the natural frequency of the transverse vibration of the string, Ω_0, i.e. $\omega_0 \cong 2\Omega_0$, then the transverse vibration is amplified. In this case, the oscillation of the tuning fork modulates the tension of the string, which determines the frequency of the transverse vibration, and the condition $\omega_0 \cong 2\Omega_0$ corresponds to the condition (3) for $n = 1$.

A more familiar example is a child's swing. The child moves downward every time the swing comes to the bottom and the child's motion produces a periodic modulation of the effective length of the swing and hence its frequency at twice its value, again corresponding to the case $n = 1$ of the condition (3).

In plasmas, we often encounter a large-amplitude monochromatic oscillation which is excited by some external sources, such as an electromagnetic wave, electron or ion beam, etc. Such a large-amplitude oscillation can act as a modulator of plasma parameters which characterize the dispersion relation of the natural oscillations in the plasma. Parametric instabilities resulting from the modulation then strongly affect the efficiency of the energy deposition rate of the external source on the plasma. The effect is not simply an excitation of some discrete number of natural oscillations, but it often results in an evolution of a strongly turbulent situation or in a complete modification of the plasma profile. Phenomena of parametric instabilities are therefore important not only in fusion research but also in the basic study of non-linear problems.

I shall start by a simple model, the Mathieu equation model, to demonstrate the basic features of the parametric instability, and then generalize the argument to more complicated situations where a modulator excites a set of high- and low-frequency oscillations, showing some examples in actual plasmas, and finally discuss some geometrical effects on the instability characteristics by physical rather than mathematical arguments. All arguments will be restricted to the linear stage of the instability; for non-linear behaviour arising from the parametric instability, I refer to the paper in these Proceedings by Dr. Tsytovich.

2. THE MATHIEU EQUATION MODEL

In this section, we consider Eqs (1) and (2) with Q(t) given by

$$Q(t) = -2\epsilon \cos \omega_0 t, \quad \omega_0 = 2\pi/T \tag{4}$$

where ϵ is the modulation amplitude which is assumed to be constant. In this case, Eq.(1) is reduced to a Mathieu equation whose periodic solutions are given by $ce_n(\omega_0 t/2, \epsilon\Omega_0^2/8)$ (n = 0, 1, 2, ...) and $se_n(\omega_0 t/2, \epsilon\Omega_0^2/8)$ (n = 1, 2, ...), $ce_n(x,q)$ and $se_n(x,q)$ being the nth order Mathieu functions of modulus q. The properties of these functions are well known and can be found in the usual textbooks on applied mathematics.

Here, instead of using these functions, I shall restrict myself to the weak modulation case, $|\epsilon| \ll 1$, and analyse the solutions by a perturbation method.

We use the Fourier representation,

$$X(t) = \int \frac{d\omega}{2\pi} e^{-i\omega t} X(\omega)$$

in which Eq.(1) is written as

$$D(\omega) X(\omega) = -\epsilon\Omega_0^2 \{X(\omega - \omega_0) + X(\omega + \omega_0)\} \tag{5}$$

where

$$D(\omega) = \omega^2 - \Omega_0^2 \tag{6}$$

The left-hand side of Eq.(5) is the linear contribution, the vanishing of D(ω) giving the linear dispersion relation, while the right-hand side stands for the mode-coupling to the responses at frequencies shifted by $\pm\omega_0$ due to the modulation. We can write similar equations for $X(\omega \pm \omega_0)$, obtaining

$$D(\omega \pm \omega_0) X(\omega \pm \omega_0) = -\epsilon\Omega_0^2 \{X(\omega) + X(\omega \pm 2\omega_0)\} \tag{7}$$

This equation now contains the responses at frequencies $\omega \pm 2\Omega_0$, in addition to that at the original frequency ω, and therefore we must proceed to write the equations for $X(\omega \pm 2\omega_0)$. If we continue this procedure, we obtain a hierarchy of equations which is never closed, because of the subsequent appearance of new responses at frequencies shifted by a greater integral multiple of ω_0 from the original frequency ω.

At this point, we recall that we are considering the case $|\epsilon| \ll 1$; then Eq.(5) implies that unless ω is close to either Ω_0 or $-\Omega_0$, the response $X(\omega)$ becomes very small (of order ϵ). The same consideration can be applied to Eq.(7), where the response $X(\omega \pm \omega_0)$ is very small unless $(\omega \pm \omega_0)$ is close to either Ω_0 or $-\Omega_0$. The obvious approximation which we can immediately think of is to retain only those responses which have frequencies near $\pm\Omega_0$ and neglect all others. This approximation leads to our first example, the case of $n = 1$ in Eq.(3).

2.1. The case of $\omega_0 \cong 2\Omega_0$

Let ω be close to Ω_0 in Eq.(5); then $\omega - \omega_0 \cong -\Omega_0$, which yields a large response, but $\omega + \omega_0 \cong 3\Omega_0$, which is off-resonant from the natural frequency. We therefore keep only $X(\omega - \omega_0)$ and neglect $X(\omega + \omega_0)$ in Eq.(5). Similarly in Eq.(7), we keep $X(\omega)$ but neglect $X(\omega \pm 2\omega_0)$. In this way we obtain a closed set of equations for $X(\omega)$ and $X(\omega - \omega_0)$. A non-trivial solution can be obtained by setting the determinant of the coefficients equal to zero, i.e.

$$D(\omega)\,D(\omega - \omega_0) = \epsilon^2 \Omega_0^4 \tag{8}$$

which gives the dispersion relation.

Now we recall that we are considering the case $\omega \cong \Omega_0$ and $\omega - \omega_0 \cong -\Omega_0$, so we can make the so-called *resonance approximation* which amounts to the approximation:

$$D(\omega) = (\omega - \Omega_0)(\omega + \Omega_0) \cong 2\Omega_0\,(\omega - \Omega_0)$$

$$D(\omega - \omega_0) = (\omega - \omega_0 - \Omega_0)(\omega - \omega_0 + \Omega_0) \cong -2\Omega_0\,(\omega - \Omega_0 - \Delta) \tag{9}$$

where we introduced the *frequency mismatch* defined by

$$\Delta = \omega_0 - 2\Omega_0 \tag{10}$$

Use of the approximation (9) reduces Eq.(8) to a quadratic form which can immediately be solved as

$$\omega = \Omega_0 + \frac{1}{2}\,[\Delta \pm \sqrt{\Delta^2 - \epsilon^2 \Omega_0^2}\,] \tag{11}$$

In the limit of small $|\epsilon|$, this equation yields two real solutions, $\omega \cong \Omega_0$ and $\omega \cong \omega_0 - \Omega_0$, the former being the natural oscillation and the latter the driven oscillation. The modulation yields a coupling of these two oscillations and produces an instability when

$$\epsilon^2 > \frac{\Delta^2}{\Omega_0^2} \tag{12}$$

For the given mismatch Δ, the right-hand side gives the *threshold* intensity for the modulation to cause instability. The threshold becomes *zero* when $\Delta = 0$. The growing solutions have real frequency given by

$$\omega_r = \Omega_0 + \frac{\Delta}{2} = \frac{\omega_0}{2} \tag{13}$$

which is independent of the original natural frequency, indicating a *frequency locking*. The growth rate above threshold is given by

$$\gamma = \frac{1}{2}\,[\epsilon^2 \Omega_0^2 - \Delta^2]^{1/2} \leqslant \frac{|\epsilon|\Omega_0}{2} \tag{14}$$

where the right-hand expression gives the *maximum growth rate,* $\gamma_{max} = |\epsilon|\Omega_0/2$, which obtains at exact matching $\Delta = 0$.

2.2. The case of $\omega_0 \cong \Omega_0$

The foregoing approximation of neglecting all the off-resonant responses is no longer useful for the cases $n > 1$ in Eq.(3). In these cases, one has to retain some non-resonant responses which connect the two resonant responses at $\pm\Omega_0$. Here we consider the case of $n = 2$, i.e. $\omega_0 \cong \Omega_0$; then, if $\omega \cong \Omega_0$ in Eq.(5), neither of $X(\omega \pm \omega_0)$ becomes resonant. Of these, the mode $X(\omega - \omega_0)$ couples directly to $X(\omega - 2\omega_0 \cong -\Omega_0)$ which is another resonant mode, while the mode $X(\omega + \omega_0)$ couples back to the original resonant mode $X(\omega)$. Although both couplings have to be retained for a precise quantitative argument, we shall keep here only the one which couples to $X(\omega - 2\omega_0)$ and neglect $X(\omega + \omega_0)$. The latter simply yields a frequency shift which does not contribute to instability.

To make the equations symmetric, we choose ω to be close to zero; then $X(\omega)$ becomes the one which connects the two resonant responses $X(\omega \pm \omega_0 \cong \pm\Omega_0)$. Then our approximation amounts to neglecting $X(\omega \pm 2\omega_0)$ in Eq.(7), retaining all terms in Eq.(5). The equations are closed among $X(\omega \pm \omega_0)$ and $X(\omega)$ and we get the dispersion relation in the form

$$1 = \frac{\epsilon^2 \Omega_0^4}{D(\omega)} \left[\frac{1}{D(\omega + \omega_0)} + \frac{1}{D(\omega - \omega_0)} \right] \tag{15}$$

As before, we make the resonance approximation for $D(\omega \pm \omega_0 \cong \pm \Omega_0)$. Introducing the *frequency mismatch* δ by

$$\delta = \omega_0 - \Omega_0 \tag{16}$$

we have

$$D(\omega \pm \omega_0) \cong \pm 2\Omega_0 \, (\omega \pm \delta) \tag{17}$$

If we further approximate $D(\omega)$ by $D(0) = -\Omega_0^2$, Eq.(15) again becomes quadratic in ω and can be solved as

$$\omega^2 = \delta \left[\epsilon^2 \Omega_0 + \delta \right] \tag{18}$$

This equation has a growing solution when

$$-\epsilon^2 \Omega_0 < \delta < 0 \tag{19}$$

i.e. when the modulation frequency is slightly less than the natural frequency. The *threshold* modulation intensity for a given frequency mismatch is given by

$$\epsilon^2 > -\delta/\Omega_0 \tag{20}$$

which vanishes at $\delta = 0$. There, however, the growth rate γ also vanishes. For given ϵ^2, the *maximum growth rate* obtains at $\delta = -\epsilon^2 \Omega_0/2$ with

$$\gamma_{max} = \frac{\epsilon^2 \Omega_0}{2} \tag{21}$$

Note that it is smaller than γ_{max} for the case of $n = 1$ by a factor ϵ. This is because we had to invoke a non-resonant response in order to produce the coupling in the present case.

An interesting feature of the present instability is that the non-resonant growing mode has zero frequency ($\omega_r = 0$). Therefore, this instability is called the *purely growing mode instability*. It also implies that the accompanying resonant responses $X(\omega \pm \omega_0)$ have frequency exactly equal to the modulation frequency $\pm\omega_0$, again indicating a *frequency locking*. Note that the instability excites all three modes, $X(\omega)$ and $X(\omega \pm \omega_0)$, with the same growth rate, but their relative intensities are different:

$$|X(\omega)| \sim |\epsilon|\,|X(\omega \pm \omega_0)| \ll |X(\omega \pm \omega_0)|$$

as can be seen from Eq.(5).

2.3. Effect of damping

So far, we have considered the parametric excitation of undamped natural oscillations. It is easy to investigate the effect of damping by introducing a phenomenological damping Γ in the present problem, i.e. instead of Eq.(1), we use

$$\frac{d^2}{dt^2} X(t) + 2\Gamma \frac{dX(t)}{dt} + (\Omega^2 + \Gamma^2)\, X(t) = 0 \tag{22}$$

which in the absence of modulation yields the two damped natural oscillations, $X(t) \sim \exp[\pm i\Omega_0 t - \Gamma t]$.

Now, Eq.(22) can be reduced to Eq.(1) by the transformation

$$\tilde{X}(t) \equiv X(t)\, e^{\Gamma t} \tag{23}$$

The analyses given above can then be used for $\tilde{X}(t)$. The only modification that arises from the damping is to reduce the growth rate from γ to $(\gamma - \Gamma)$. As a result, we now get a *finite threshold* for instability independently of the frequency mismatch, since we need a growth rate γ for $\tilde{X}(t)$ to be greater than Γ. The *minimum threshold* is obtained by setting the maximum growth rate γ_{max} calculated above equal to the damping Γ, i.e. $\gamma_{max} = \Gamma$, which gives

$$\epsilon^2 > \frac{4\Gamma^2}{\Omega_0^2} \quad \text{for} \quad \omega_0 \cong 2\Omega_0 \tag{24}$$

$$\epsilon^2 > \frac{2\Gamma}{\Omega_0} \quad \text{for} \quad \omega_0 \cong \Omega_0 \tag{25}$$

The former obtains at exact matching, i.e. $\Delta = 0$, while the latter obtains at $\delta = -\epsilon^2 \Omega_0 / 2$.

3. COUPLED-MODE PARAMETRIC EXCITATION

The foregoing example shows that the parametric instability occurs as a result of the coupling of natural oscillations at different frequencies. In the above example it is assumed that all the excited oscillations have the same natural frequency $|\Omega_0|$. An obvious extension of this analysis is to consider a coupling of oscillations or wave modes which have different natural frequencies, the difference being due either to different wavenumbers or to different branches. Cases which have many applications in plasma problems are those in which the modulator, which we call the *pump,* produces a coupling of high-frequency waves with a low-frequency wave. We therefore consider such a situation in this section.

We first derive a simple dispersion relation which is a generalization of Eq.(15) based on a model set of equations. Then we consider two special cases, that of a spatially uniform pump and that in which the process describes a stimulated scattering of the pump wave. Some terminologies which are often used in the literature to classify various instabilities are briefly summarized. Finally, examples in plasma physics problems are described together with physical mechanisms causing the parametric coupling.

3.1. Dispersion relation

We consider two branches of normal modes: the low-frequency mode represented by $X_L(\vec{k}, \omega)$ and the high-frequency mode represented by $X_H(\vec{k}, \omega)$. In the absence of a pump, they are assumed to satisfy the dispersion relations:

$$D_L(\vec{k}, \omega) \equiv \omega^2 - \omega_L^2(\vec{k}) = 0 \quad \text{for} \quad X_L \tag{26}$$

$$D_H(\vec{k}, \omega) \equiv \omega^2 - \omega_H^2(\vec{k}) = 0 \quad \text{for} \quad X_H \tag{27}$$

where for simplicity we have neglected damping.

The pump, or modulator, which we denote by

$$Z(\vec{r}, t) = 2Z_0 \cos[\vec{k}_0 \cdot r - \omega_0 t], \quad Z_0 = \text{const} \tag{28}$$

produces a coupling of these two modes. In analogy to the Mathieu equation model, we assume the following form of the coupling:

$$D_L(\vec{k}, \omega) X_L(\vec{k}, \omega) = Z_0 \{\lambda_+ X_H(\vec{k} + \vec{k}_0, \omega + \omega_0) + \lambda_- X_H(\vec{k} - \vec{k}_0, \omega - \omega_0)\} \tag{29}$$

$$D_H(\vec{k} \pm \vec{k}_0, \omega \pm \omega_0)\, X_H(\vec{k} \pm \vec{k}_0, \omega \pm \omega_0) = Z_0 \mu_\pm X_L(\vec{k}, \omega) \qquad (30)$$

where the coupling coefficients $\lambda_\pm$ and $\mu_\pm$ are constants.[1] Solving Eq.(30) for X_H and substituting the result in Eq.(29) yields the dispersion relation:

$$1 = \frac{Z_0^2}{D_L(\vec{k}, \omega)} \left\{ \frac{\lambda_+ \mu_+}{D_H(\vec{k} + \vec{k}_0, \omega + \omega_0)} + \frac{\lambda_- \mu_-}{D_H(\vec{k} - \vec{k}_0, \omega - \omega_0)} \right\} \qquad (31)$$

which resembles Eq.(15).

As before, we shall restrict ourselves to the case of weak pumping, i.e. $Z_0^2 \ll 1$. Then Eq.(31) can be satisfied only when one of the dispersion functions, $D_{L,H}$, becomes nearly equal to zero. This is, however, not sufficient to cause instability; instability occurs when at least two of the zeros of the dispersion functions merge into each other. There are two situations:

(a) $D_L(\vec{k}, \omega) \cong 0$ and $D_H(\vec{k} + \vec{k}_0, \omega + \omega_0) \cong 0$ or

$D_H(\vec{k} - \vec{k}_0, \omega - \omega_0) \cong 0$

(b) $D_H(\vec{k} + \vec{k}_0, \omega + \omega_0) \cong 0$ and $D_H(\vec{k} - \vec{k}_0, \omega - \omega_0) \cong 0$

From the analogy to the Mathieu equation model, case (a) corresponds to the *resonant type* coupling, i.e. n = 1 case, and case (b) to the *non-resonant type* coupling, i.e. n = 2 case.

In either case, $|\omega|$ is assumed to be much less than ω_0 and hence $|\omega \pm \omega_0| \sim \omega_0 \sim \omega_H$. We can then make the resonance approximation (9) for D_H to obtain

$$D_H(\vec{k} \pm \vec{k}_0, \omega \pm \omega_0) \cong \pm 2\omega_0 \left[\omega \pm \omega_0 \mp \omega_H(\vec{k} \pm \vec{k}_0)\right]$$

$$= \pm 2\omega_0 \left[(\omega - \alpha) \pm \delta\right] \qquad (32)$$

where we introduced two parameters:

$$\alpha = \left[\omega_H(\vec{k} + \vec{k}_0) - \omega_H(\vec{k} - \vec{k}_0)\right]/2 \qquad (33)$$

$$\delta = \omega_0 - \left[\omega_H(\vec{k} + \vec{k}_0) + \omega_H(\vec{k} - \vec{k}_0)\right]/2 \qquad (34)$$

---

[1] Strictly speaking, we should also include in Eq.(30) the terms proportional to $Z_0^2 X_H(\vec{k} \mp \vec{k}_0, \omega \mp \omega_0)$ which we neglect here for simplicity.

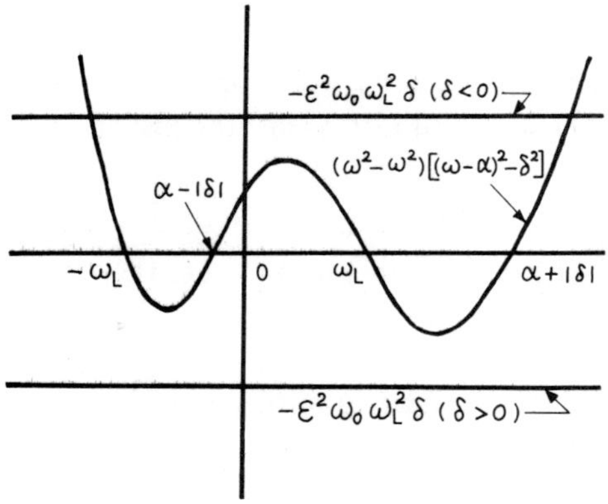

FIG.2. Curves showing both sides of Eq.(36) for the case in which $\alpha > 0$ and the extrema of the left-hand side are located between $\epsilon^2\omega_0\omega_L^2|\delta|$ and $-\epsilon^2\omega_0\omega_L^2|\delta|$.

Here δ is the *mismatch* of the pump frequency from the average of the two natural frequencies of the high-frequency modes, analogous to the mismatch introduced by Eq.(16), and α is their frequency difference which arises from the finiteness of the pump wavenumber $\vec{k}_0$. Note that we are choosing ω_H to be positive. Substitution of Eqs (26) and (32) into Eq.(31) yields a biquadratic equation for ω.

In many cases of interest, particularly when the growth rate assumes a maximum value, the coupling coefficients $\lambda_+\mu_+$ and $\lambda_-\mu_-$ in Eq.(31) become real and identical [3]. For simplicity, we therefore restrict ourselves to such situations and introduce a dimensionless small parameter ϵ by the relation

$$Z_0^2\lambda_+\mu_+ = Z_0^2\lambda_-\mu_- \equiv \epsilon^2\omega_0^2\omega_L^2(\vec{k}) \tag{35}$$

The dispersion relation is then reduced to the following simple form:

$$(\omega^2-\omega_L^2)[(\omega-\alpha)^2-\delta^2] = -\epsilon^2\omega_0\omega_L^2\delta \tag{36}$$

Figure 2 shows curves for both sides of this equation for the case $\alpha > 0$. It can be clearly seen from this figure that there are four complex solutions or two growing solutions for the case $\delta > 0$ and two complex or one growing solution for the case $\delta < 0$. Setting

$$\omega = \omega_r + i\gamma \tag{37}$$

and taking the imaginary part of Eq.(36), we find that these growing solutions arise in the following frequency ranges:

$$\delta > 0 : \quad \omega_r \geqslant \alpha \qquad \text{(Mode I)}$$

$$0 \geqslant \omega_r \qquad \text{(Mode II)} \tag{38}$$

$$\delta < 0 : \quad \alpha \geqslant \omega_r \geqslant 0 \qquad \text{(Mode III)}$$

In § 3.2 and § 3.3, these three types of growing solution are illustrated by consideration of special cases.

3.2. Uniform or dipole pump: $\vec{k}_0 = 0$

In this case, α vanishes and δ becomes identical to Eq.(16) with $\omega_H(\vec{k}) = \Omega_0$. The dispersion relation (36) becomes quadratic in ω^2 and can hence be solved immediately.

Let us first consider the case of a very weak pump, $\epsilon^2 \ll 1$. In this case we have two types of solution similar to those obtained in § 2:

(a) Resonant type for $\delta \cong \omega_L(\vec{k}) > 0$ with the growth rate

$$\gamma \cong [\gamma_{max}^2 - \Delta^2/4]^{1/2} \tag{39}$$

where Δ $(= \delta - \omega_L(\vec{k}))$ is the mismatch and γ_{max} is the maximum growth rate given by

$$\gamma_{max} = |\epsilon| (\omega_0 \omega_L)^{1/2}/2 \tag{40}$$

which can be compared with Eq.(14). The excited modes have frequencies close to $\omega_H(\vec{k})$ and $\omega_L(\vec{k})$, having the sum frequency exactly equal to the pump frequency ω_0. This type of instability is called the *resonant decay instability*. Note that in this special case of $\vec{k}_0 = 0$, modes I and II are degenerate, i.e. $\omega_r \cong \pm \omega_L(\vec{k})$.

(b) Non-resonant type for the case $\delta < 0$ (mode III) with the same growth characteristics as those derived in § 2.2, provided that the growth rate is much smaller than ω_L. This non-resonant type instability is called the *oscillating two-stream instability (OTSI)*.

In both cases, one can derive the minimum threshold for instability if one introduces phenomenological damping rates, Γ_L and Γ_H, by the relations:

$$D_{H,L}(\vec{k}, \omega) = [\omega - \omega_{H,L} + i\Gamma_{H,L}][\omega + \omega_{H,L} + i\Gamma_{H,L}] \tag{41}$$

The results can be written as

$$\gamma_{max} > (\Gamma_H \Gamma_L)^{1/2} \qquad \text{(resonant type)} \tag{42}$$

$$\gamma_{max} > \Gamma_H \qquad \text{(non-resonant type)} \tag{43}$$

which are to be compared with Eqs (24) and (25).

A new type of solution arises when the pump intensity becomes sufficiently large to satisfy the inequality:

$$\epsilon^2 \gg \omega_L/\omega_0 \tag{44}$$

In this case, the growth rate becomes greater than ω_L and therefore the growth characteristics substantially deviate from those obtained in §2. Maximization of the growth rate with respect to δ then yields

$$\omega = \begin{cases} \frac{1}{2}\left(\frac{\epsilon^2 \omega_0 \omega_L^2}{4}\right)^{1/3} [\pm\sqrt{5} + i\sqrt{3}] & \text{for } \delta > 0 \\ i\left(\frac{\epsilon^2 \omega_0 \omega_L^2}{2}\right)^{1/3} & \text{for } \delta < 0 \end{cases} \tag{45}$$

We note the one cube root dependence of the maximum growth rate on the pumping power ϵ^2 in both cases. These modes are sometimes called the *quasi-reactive modes*. For further details of the solutions see Ref. [4].

In concluding this subsection, we mention that in the case of a uniform pump the dispersion relation can be derived correct to all orders in ϵ. This is because all particles of the same species oscillate in unison in the presence of a uniform pump field and hence the pump field can be eliminated from the Klimontovich equation by using the oscillating frame of reference. We can then use the usual linear theory analysis for the fluctuations in the absence of the pump field, with one difference, that the reference frame now depends on the particle species. By introducing an appropriate transformation operator, one can easily manipulate it to derive the dispersion relation correct to all orders in ϵ (see e.g. [5]).

3.3. Stimulated scattering

We next consider the situation where the pump wave satisfies the same dispersion relation as the excited high-frequency wave, i.e.

$$\omega_0 = \omega_H(\vec{k}_0) \tag{46}$$

This corresponds to the situation where the pump wave is coherently scattered by the low-frequency wave. In this case, the mismatch (34) becomes

$$\delta = \omega_H(\vec{k}_0) - [\omega_H(\vec{k}_0 + \vec{k}) + \omega_H(\vec{k}_0 - \vec{k})]/2 \tag{47}$$

where we used the relation $\omega_H(\vec{k}) = \omega_H(-\vec{k})$, since we are choosing ω_H to be positive. In the particular case where the linear group dispersion of the high-frequency wave, i.e. $d^2\omega_H/dk^2$, calculated along the line connecting the three wavenumber vectors $\vec{k}_0$ and $\vec{k}_0 \pm \vec{k}$ has a fixed sign, the mismatch δ also has a fixed sign:

$$\begin{aligned} &\delta > 0 \quad \text{if} \quad \frac{d^2\omega_H}{dk^2} < 0 \\ &\delta < 0 \quad \text{if} \quad \frac{d^2\omega_H}{dk^2} > 0 \end{aligned} \tag{48}$$

Combining this relation with relation (38), we then find that in this case we can have only modes I and II or mode III, depending on the linear dispersion characteristics of the high-frequency mode. This situation is quite similar to the modulational instability which can be obtained from the non-linear Schrödinger equation (see e.g. [6]), and therefore one can give a simple physical interpretation for the mechanism of the instability.

For clarity, let us consider the case $d^2\omega_H/dk^2 > 0$. In this case only mode III, which has a real frequency in the region $\alpha \geqslant \omega_r \geqslant 0$, can be excited. Now from the definition (33), α/k denotes the propagation speed of the beat mode (low-frequency mode) as calculated from the *linear* dispersion relation of the high-frequency wave. Therefore $(\omega_r - \alpha)/k$ denotes the *non-linear shift* in the beat propagation speed. For mode III, this non-linear shift is negative, so that the beat tends to move faster in the low-amplitude region than in the high-amplitude region (Fig.3(a)). As a result, the beat tends to be antisteepened (Fig.3(b)). Now for $d^2\omega_H/dk^2 > 0$, the group velocity, $d\omega_H/dk \equiv v_{gH}$, is larger in the steep region (i.e. behind the crest of the beat) than in the smooth region (i.e. in front of the crest). This results in the amplification of the beat as shown in Fig.3(c). It is obvious that mode III is stable if $d^2\omega_H/dk^2 < 0$. Instability of modes I and II for the case $d^2\omega_H/dk^2 < 0$ can be explained in a similar fashion.

In all cases, the growth rate becomes largest and the minimum threshold resulting from damping becomes lowest when the resonance condition

$$\omega_H(\vec{k}_0) = \omega_H(\vec{k}_0 - \vec{k}) + \omega_L(\vec{k}) \tag{49}$$

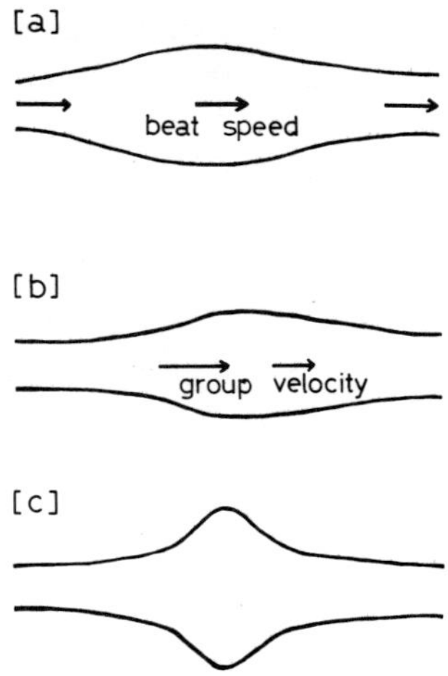

FIG.3. Physical mechanism of instability for the case $\alpha \geqslant \omega_r \geqslant 0$ *and* $d^2\omega_H/dk^2 > 0$: *(a) initial modulation; (b) antisteepening due to the beat motion; and (c) amplified modulation due to the group dispersion effect.*

is satisfied. This is in general called the *resonant decay instability.* In an isotropic plasma this condition is satisfied when $|\vec{k}_0 - \vec{k}| \cong k_0$ with the maximum growth rate under the *backscattering* condition

$$\vec{k}_0 \cong \vec{k} - \vec{k}_0, \quad \vec{k} \cong 2\vec{k}_0 \tag{50}$$

Figure 4 shows an example of the real frequency of the excited low-frequency oscillation as a function of α for a relatively weak pump case. The term *modified decay instability* is used for the non-resonant region of mode III instability. When α vanishes due to a particular vectorial relation of $\vec{k}_0$ and $\vec{k}$, a purely growing mode can be excited for the case $\delta < 0$, which corresponds to the *filamentation instability*. Finally, the term *modulational instability* is used for the case where $k_0 \gg k$, although in some literature it is used for a wider class of instabilities including the oscillating two-stream instability and the modified decay instability.

3.4. Examples in plasma physics problems

The most extensive studies of parametric instabilities in plasmas have been made in connection with laser plasma interactions (see e.g. [7]). Here we

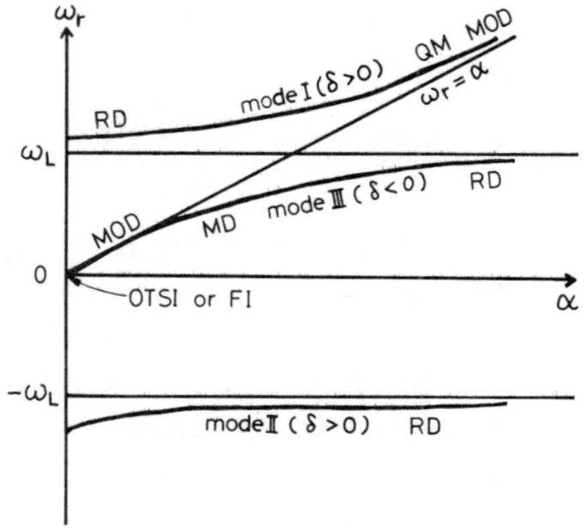

FIG.4. Classification of various parametric instabilities.
RD: resonant decay instability
QM: quasi-mode instability
MOD: modulational instability
MD: modified decay instability
OTSI: oscillating two-stream instability
FI: filamentation instability

consider a situation where an intense laser light is incident on a solid target and creates a high-density plasma (n ~ $10^{21}/cm^3$ for a Nd glass laser and n ~ $10^{19}/cm^3$ for a CO_2 laser) which in turn strongly interacts with the laser light. The plasma is unmagnetized and the typical scale length of the plasma is much greater than the Debye length, so that, to lowest order, we can treat the plasma as isotropic and uniform.

Naturally, the incident laser light can act as a pump for various parametric instabilities, of which the following are typical examples:

(a) Stimulated Raman scattering, which is a scattering of the laser light by a Langmuir wave and is important when ω_0 is close to $2\omega_{pe}$, where ω_{pe} is the electron plasma frequency which is proportional to the square root of the plasma density.

(b) Stimulated Brillouin scattering, which is a scattering of laser light by an ion-acoustic wave and is important in causing a backscattering of the incident light.

(c) Two-plasmon decay, which is a decay of the laser light at frequency near $\omega_0 \doteq 2\omega_{pe}$ into two Langmuir waves and is important for both laser light absorption and re-emission at frequencies $\omega_0/2$ and $3\omega_0/2$.

(d) Resonant decay into a couple of Langmuir and ion-acoustic waves, which occurs near $\omega_0 \cong \omega_{pe}$.

(e) Oscillating two-stream instability, which also occurs near $\omega_0 = \omega_{pe}$.

In addition to these, an important parametric instability takes place owing to a linear-mode conversion of the electromagnetic wave into a Langmuir wave near the critical density, i.e. near $\omega_0 = \omega_{pe}$. That is, an electromagnetic wave propagating obliquely to the plasma density gradient can excite an electrostatic oscillation which peaks at the resonance frequency $\omega_0 = \omega_{pe}$. This oscillation in turn acts as a pump for a modified decay instability at the critical density region, and the excited high-frequency wave can in turn be linearly transformed to an electromagnetic wave. This instability is called the *radiating decay instability* and is found to have an extremely low threshold [8].

When the pump intensity is increased, there can be various types of non-resonant instabilities. In addition to the reactive quasi-mode discussed above, there exist *resistive quasi-mode instabilities* in which one of the excited modes is highly damped due to wave-particle interactions. In the case of stimulated scattering, this corresponds to the *non-linear Landau damping* of the pump wave. When the pump wave is electromagnetic, it is often called *stimulated Compton scattering.* A situation also exists where a *stimulated mode conversion* is produced due to excitation of a highly damped mode. The *kinetic instability* discussed by Silin [9] is the case where an electromagnetic wave is converted into a short-wavelength ion plasma wave via scattering on electrons. The *transitional instability* which is considered to limit the Langmuir wave collapse [10] is the case where a long-wavelength Langmuir wave (condensed plasmon) is converted into a short-wavelength ion oscillation via electrons.

Parametric instabilities are also important in r.f. heating of a magnetized plasma, such as lower hybrid heating, ion-cyclotron heating and so on, and we refer to Refs [11, 12] for these applications.

We conclude this section by describing some of the physical mechanisms that cause the parametric coupling of waves. In most cases, the driving force of the low-frequency wave by the high-frequency waves, which include both pump and excited high-frequency waves, is the *ponderomotive force*. In the fluid approximation, it can be written as

$$\vec{\Phi}_p = \sum_s m_s \left\{ -\langle \vec{v}_{Hs} \cdot \nabla \vec{v}_{Hs} \rangle + \langle \vec{v}_{Hs} \times \vec{\Omega}_{Hs} \rangle \right\}$$

$$= \sum_s \left\{ -\frac{m_s}{2} \nabla \langle |\vec{v}_{Hs}|^2 \rangle + m_s \langle \vec{v}_{Hs} \times (\vec{\Omega}_{Hs} + \nabla \times \vec{v}_{Hs}) \rangle \right\} \qquad (51)$$

where m_s is the mass, $\vec{v}_{Hs}$ is the high-frequency component of the fluid velocity; $\vec{\Omega}_{Hs}$ $(= q_s \vec{B}_H/m_s c)$ is the high-frequency component of the cyclotron frequency vector (q_s being the charge); the suffix s denotes the particle species, and the angular bracket denotes the time average. The component of this ponderomotive force parallel to the static magnetic field causes a modification of the average plasma density δn_0, whereas the component perpendicular to the static magnetic field causes a drift current which in turn modifies the average magnetic field and the average flow velocity, yielding $\delta\vec{B}_0$ and $\delta\vec{v}_0$. These perturbations for the averaged quantities correspond to the low-frequency mode and modify the dispersion relation for the high-frequency wave. In this way, a coupling is produced between the high- and low-frequency waves. In some cases we also need other non-linear effects; these include the kinetic pressure effect, the second harmonic generation effect, non-linearity between density and potential, etc.

4. GEOMETRICAL EFFECTS ON RESONANT DECAY INSTABILITY

As pointed out in §3, the resonant decay instability has the lowest threshold and the maximum growth rate for a relatively weak pump. In this section, we restrict ourselves to the resonant decay instability and discuss some geometrical effects, namely the effects of spatial non-uniformity of the medium and finite spatial extent of the pump. We present these effects based on 'physical' arguments which ought to be unrigorous. However, all the results presented here can be derived by a more rigorous mathematical treatment which you can find in Ref. [13].

4.1. Interaction region

The most important geometrical effect is the appearance of a finite spatial extent of the interaction region. This can be seen most easily for the case where the pump has a finite spatial extent. Without a pump there is no parametric coupling of waves, so the parametric interaction is limited to a certain spatial region where the pump exists with a sufficient amplitude (greater than the threshold value).

Now, even if the pump has an infinite spatial extent with constant amplitude, the region for a given resonant decay instability can be limited if the plasma itself is spatially non-uniform. To show this, we first note that for resonant decay instability the following matching condition is required:

$$\omega_0 \cong \omega_L(\vec{k}) + \omega_H(\vec{k}_0 - \vec{k}) \tag{52}$$

If we denote the real part of the frequency of the excited low-frequency wave by ω_r, we can alternatively write this condition as

$$\vec{k}_0(\omega_0) \cong \vec{k}_L(\omega_r) + \vec{k}_H(\omega_0 - \omega_r) \tag{53}$$

where $\vec{k}_{0,L,H}(\omega)$ stand for the solutions of the linear dispersion relations for the pump wave (suffix 0), the low-frequency wave (suffix L) and the excited high-frequency wave (suffix H) when their frequencies are given. In a weakly inhomogeneous plasma, the wavenumbers $\vec{k}_{0,L,H}$ vary as the waves propagate, and their variation can be described by the WKB approximation. Suppose at a certain point, say at $\vec{r} = 0$, the wavenumber matching condition is exactly satisfied, i.e. $\vec{k}_0 = \vec{k}_L + \vec{k}_H$ at $\vec{r} = 0$. Then at $\vec{r} \neq 0$, a mismatch

$$\vec{\kappa}(\vec{r}) \equiv \vec{k}_H(\omega_0 - \omega_r, \vec{r}) + \vec{k}_L(\omega_r, \vec{r}) - \vec{k}_0(\omega_0, \vec{r}) \tag{54}$$

is produced. When this mismatch exceeds a certain critical value, the parametric instability ceases. Therefore, a finite spatial extent for the interaction region is produced.

As shown by Eqs (14) and (39), the maximum allowable frequency mismatch for a given pump amplitude is given by $\Delta^2_{max} = 4\gamma^2_{max}$. For the case of a one-dimensional propagation of waves, the relation between the frequency mismatch Δ and the wavenumber mismatch κ is given by

$$\Delta = |\kappa| \, |v_{gH} v_{gL}|^{1/2} \tag{55}$$

where v_{gH} and v_{gL} are the group velocity of the high- and low-frequency waves. Therefore, the interaction region is determined by the condition

$$\kappa^2(x) < \frac{4\gamma^2_{max}}{|v_{gH} v_{gL}|} \tag{56}$$

4.2. Absolute and convective instabilities

When the interaction region is limited in space, the effect of instability becomes entirely different depending on whether it is an absolute instability or a convective instability. A convective instability is one in which the wave amplitude increases as the wave packet propagates through the medium. In this case, amplification of the wave packet is observed only when we move with the group velocity of the wave. In other words, amplification ceases when the wave packet moves outside the interaction region. Therefore, as far as the spatial

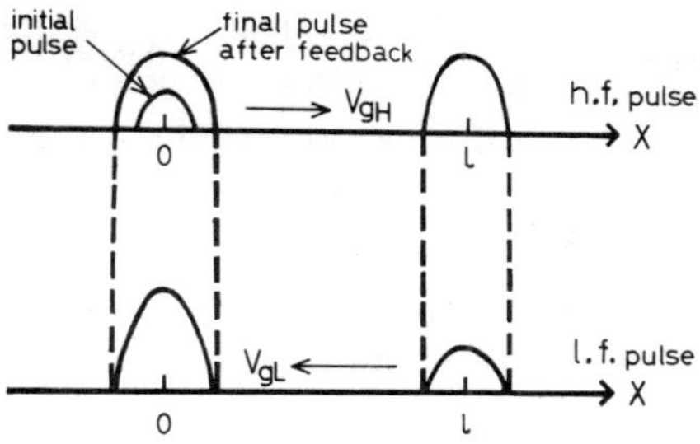

FIG.5. Physical mechanism of absolute instability in uniform plasma. Initial high-frequency pulse excited at x = 0 amplifies as it propagates with group velocity v_{gH}, *then produces a low-frequency pulse at* $x = \ell$ *by beating with the pump; the excited low-frequency pulse propagates back to x = 0 while being amplified and produces the high-frequency pump which has greater amplitude than the initial pulse.*

extent of the interaction region is sufficiently small, no substantial amplification can be produced. On the other hand, for the case of an absolute instability, the wave amplitude increases exponentially at the point where the wave packet was initiated. This instability is therefore little affected by the finite size of the interaction region and hence causes a strong non-linear effect in a local region.

Obviously when the excited high- and low-frequency waves move in the same direction, i.e. $v_{gH} v_{gL} > 0$, the instability is always convective. All the wave packets initiated at a local point move away in the same direction. On the other hand, if $v_{gH} v_{gL} < 0$, there can be an absolute instability. We show this by considering the case of uniform pump and uniform plasma. Let there be an initial pulse of high-frequency wave at x = 0. It amplifies as it moves with group velocity v_{gH} (see Fig.5). At the same time, it beats with the pump and creates the low-frequency wave which in turn propagates back to the opposite direction with group velocity v_{gL}. It is also amplified and produces the high-frequency wave by beating with the pump. If the high-frequency wave excited by beating at the original point x = 0 has greater amplitude than the initial pulse, this process results in an absolute instability. We can see from this argument that the absolute instability in this case results from a *feedback* loop of the wave energy. In the non-damping case, this feedback process always causes an absolute instability in a uniform medium.

When damping exists in the natural oscillations, the energy dissipation during the excursion of the wave packets must be taken into account. Then a finite threshold is needed for absolute instability. It is easy to show that the spatial amplification rate of the wave packet in the non-damping propagation is given by

$$q \equiv \frac{\gamma}{|v_{gH} v_{gL}|^{1/2}} \tag{57}$$

where γ is the temporal growth rate calculated in the preceding sections. Suppose the wave packet initiated at $x = 0$ made an excursion of length ℓ and came back to the original position $x = 0$ through the feedback path shown in Fig.5. Then in the non-damping case it amplifies with the factor $\exp[2q\ell]$. On the other hand, the packets are damped during their excursion at the rate $\Gamma_H/|v_{gH}|$ on the way from $x = 0$ to $x = \ell$ and at the rate $\Gamma_L/|v_{gL}|$ on the way back from $x = \ell$ to $x = 0$. Therefore, after the entire excursion the net amplification is obtained when

$$\exp\left[2q\,\ell - \frac{\Gamma_H \ell}{|v_{gH}|} - \frac{\Gamma_L \ell}{|v_{gL}|}\right] > 1 \tag{58}$$

Using Eq.(57), we then find the minimum threshold for the absolute instability as

$$\gamma_{max} > \sqrt{|v_{gH} v_{gL}|}\left[\frac{\Gamma_H}{|v_{gH}|} + \frac{\Gamma_L}{|v_{gL}|}\right]/2 \tag{59}$$

Note that the right-hand side is greater than (or equal to) the minimum threshold value, $\sqrt{\Gamma_H \Gamma_L}$, for the convective instability (see (42)). Therefore, in the presence of damping in a uniform medium, the resonant decay instability with $v_{gH} v_{gL} < 0$ is convective when the pump amplitude is relatively weak, and becomes absolute when it exceeds the threshold determined by (59).

4.3. Effects of spatial non-uniformity

We now discuss how this feedback process is affected by the spatial non-uniformity of the medium. Suppose at $x = 0$ the exact matching condition is satisfied, i.e. $k_H(x=0) = k_0(x=0) - k_L(x=0)$. When the high-frequency wave packet propagates up to the point $x = \ell$ and produces a beat low-frequency mode, it acquires the wavenumber $k_0(x=\ell) - k_H(x=\ell)$. In coming back to the point $x = 0$, this low-frequency mode will receive a wavenumber shift $k_L(0) - k_L(\ell)$. Therefore, when this low-frequency wave beats with the pump at $x = 0$, the high-frequency wave produced will have the wavenumber given by

$$k_0(0) - [k_0(\ell) - k_H(\ell) + k_L(0) - k_L(\ell)] = k_H(0) + \kappa(\ell) \tag{60}$$

where κ is the mismatch at $x = \ell$ defined by Eq.(54). This means that the spatial non-uniformity produces a mismatch $\kappa(\ell)$ during the excursion. If this mismatch exceeds the right-hand side of (56), the feedback is completely destroyed. In other words, we have a feedback length, ℓ_{FB}, determined by the inequality

$$\kappa^2(\ell_{FB}) < \frac{4\gamma_{max}^2}{|v_{gH} v_{gL}|} \tag{61}$$

The feedback for the mode excited at the exact matching point is maintained only within the region $|x| < \ell_{FB}$.

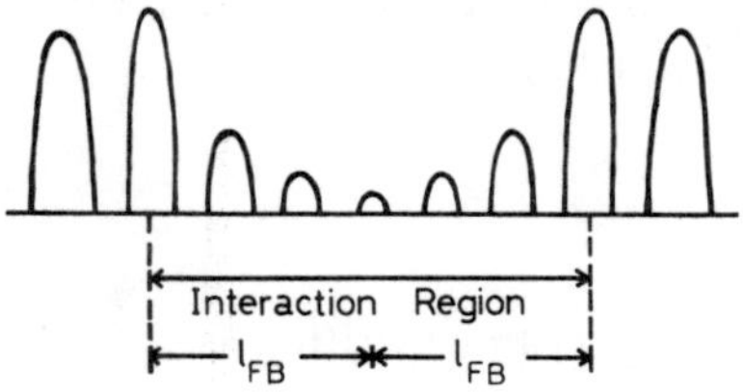

FIG.6. Schematic picture showing the suppression of the feedback loop in an inhomogeneous medium.

Now for the case where the interaction region is determined by the spatial non-uniformity alone, i.e. by the inequality (56), the feedback length is just one-half the width of the interaction region. As can be seen from Fig.6, this means that the main part of the wave energy does not contribute to the feedback and, therefore, the feedback process is almost completely destroyed. Indeed, for the case where $[d\kappa/dx]_{x=0} \neq 0$, we can rigorously show that there is no absolute instability, whatever the sign of $v_{gH}v_{gL}$ [14]. The instability in this case becomes convective.

For the case of convective instability, we can estimate the maximum amplification factor by neglecting damping and the non-linear effect as follows:

$$\left|\frac{X_H(\ell_{int}/2)}{X_H(x=0)}\right| = \exp\left[\int_0^{\ell_{int}/2} dx\, q(x)\right]$$

$$= \exp\left\{\int_0^{\ell_{int}/2} dx\left[\frac{\gamma_{max}^2}{|v_{gH} v_{gL}|} - \frac{\kappa(x)^2}{4}\right]^{1/2}\right\}$$

$$= \exp\left[\frac{\pi \gamma_{max}^2}{|\kappa' v_{gH} v_{gL}|}\right] \tag{62}$$

where ℓ_{int} is the width of the interaction region and in the last line we approximated $\kappa(x)$ by $\kappa' x$, in which case the interaction width is given by the relation

$$(\ell_{int}/2)^2 = 4\gamma_{max}^2/\kappa'^2 \, | v_{gH} v_{gL} | \tag{63}$$

The maximum amplification factor (62) was first derived by Rosenbluth [14].

The above result can alternatively be looked upon as giving a *threshold* for instability in the presence of 'damping' due to convection loss of wave energy, i.e. Eq.(62) indicates that for a sufficient amplification we need a pump power which satisfies the inequality

$$\gamma_{max}^2 > | \kappa' v_{gH} v_{gL} | \tag{64}$$

If we denote $|\kappa'| \equiv L^{-2}$, Eq.(64) can be written as

$$\gamma_{max}^2 > \left| \frac{v_{gH}}{L} \right| \left| \frac{v_{gL}}{L} \right| \tag{65}$$

Comparing this inequality with (42), we can interpret $|v_g/L|$ as an effective damping rate due to the convection loss of the wave energy. Note that $|L/v_g|$ is the time needed for the wave energy convecting across the distance L.

Although there is no absolute instability for the case where the pump has an infinite spatial extent and $\kappa' \neq 0$, one can have an absolute instability either when the pump has a finite spatial extent or when $\kappa' = 0$. First, if the pump duration length L_p is less than the feedback length ℓ_{FB}, then the feedback is maintained in the entire interaction region the extent of which is determined by L_p. Consequently, the absolute instability in a uniform system is recovered. In addition to the condition that $L_p < \ell_{FB}$, we need an additional condition for a sufficient growth, i.e.

$$\int_0^{L_p} dx \, q(x) > 1$$

which yields

$$\gamma_{max} > \frac{| v_{gH} v_{gL} |^{1/2}}{L_p} \tag{66}$$

For the case when $\kappa' = 0$, we can also have an absolute instability even if the pump has an infinite extent. In this case, we can write the mismatch as

$\kappa(x) \cong \kappa'' x^2/2$, which is symmetric around the exact matching point. Because of this symmetry, the wave initiated at $x = -\ell$ can couple strongly to the wave initiated at $x = \ell$ and this strong coupling enhances the feedback effect, resulting in an absolute instability. In this case, the interaction region is given by

$$(\ell_{int}/2)^4 = 16\gamma_{max}^2/\kappa''^2 |v_{gH} v_{gL}|$$

and the condition for sufficient amplification,

$$\int_0^{\ell_{int}/2} dx\, q(x) > 1$$

yields

$$\gamma_{max} > 4^{-1/3} |\kappa''|^{1/3} \sqrt{|v_{gH} v_{gL}|} \qquad (67)$$

In general, when an absolute instability takes place, a strong non-linear effect occurs such as the formation of a strong density gradient. Such a profile modification in turn strongly affects the subsequent parametric processes. These strongly non-linear problems have recently been investigated by several computer simulations, while analytical investigations are still at a primitive stage.

REFERENCES

[1] KITTEL, C., Introduction to Solid State Physics, 5th Edn, Wiley, New York, London, Sydney, Toronto (1976) Ch.7.
[2] RAYLEIGH, Lord, Philos. Mag. **15** (1883) 229.
[3] DRAKE, J.F., et al., Phys. Fluids **17** (1974) 778.
[4] NISHIKAWA, K., J. Phys. Soc. Jpn. **24** (1967) 467.
[5] GALEEV, A.A., SAGDEEV, R.Z., Nucl. Fusion **13** (1973) 603.
[6] KADOMTSEV, B.B., KARPMAN, V.I., Usp. Fiz. Nauk **103** (1971) 193.
[7] BRUECKNER, K.A., JORNA, S., Rev. Mod. Phys. **46** (1974) 325.
[8] FORSLUND, D.W., et al., Phys. Rev. Lett. **34** (1975) 193.
[9] SILIN, V.P., Zh. Ehksp. Teor. Fiz. **51** (1966) 1842.
[10] GALEEV, A.A., et al., JETP Lett. **21** (1975) 539; Sov. J. Plasma Phys. **1** (1975) 10.
[11] KAW, P.K., in Advances in Plasma Physics (SIMON, A., THOMPSON, W.B., Eds) **6**, Wiley, New York, London, Sydney, Toronto (1976) 179.
[12] PORKOLAB, M., Phys. Fluids **17** (1974) 1432.
[13] NISHIKAWA, K., LIU, C.S., in Advances in Plasma Physics (SIMON, A., THOMPSON, W.B., Eds) **6**, Wiley, New York, London, Sydney, Toronto (1976) 29.
[14] ROSENBLUTH, M.N., Phys. Rev. Lett. **29** (1972) 565.

HYBRID SIMULATIONS OF QUASINEUTRAL PHENOMENA IN MAGNETIZED PLASMA*

J.A. BYERS, B.I. COHEN,
W.C. CONDIT, J.D. HANSON
Lawrence Livermore Laboratory,
University of California,
Livermore, California,
United States of America

Abstract

HYBRID SIMULATIONS OF QUASINEUTRAL PHENOMENA IN MAGNETIZED PLASMA.

A new class of numerical algorithms is presented for computer simulation of low-frequency electromagnetic and elctrostatic phenomena in magnetized plasma. Maxwell's equations are solved in the limits of quasineutrality and negligible transverse displacement current (Darwin's model). Electrons are modelled as a fluid with polarization effects ignored. Ions are described as particles.

We present a new class of numerical algorithms for computer simulation of low-frequency ($\omega \ll \omega_{ce}, \omega_{pe}$) electromagnetic and electrostatic phenomena in magnetized plasma. Maxwell's equations are solved in the limits of quasineutrality and negligible transverse displacement current (Darwin's model): $\nabla \cdot \underline{J} = 0$, $\nabla \times \underline{B} = 4\pi c^{-1}\underline{J}$, and $-\partial \underline{B}/\partial t = c\nabla \times \underline{E}$. Electrons are modeled as a fluid with polarization effects ignored: $0 \approx -n_e e\underline{E} + \underline{J}_e \times \underline{B}c^{-1} - m_e \nu_{ei}\underline{J}e^{-1} - \nabla \cdot \underline{P}_e$, $n_e \approx n_i$. Ions are described as particles. A novel feature of these algorithms is the use of the electron fluid equation of motion to determine the electric field, which renders these numerical schemes remarkably simple and direct. The simulation plasma is either periodic, or bounded by particle reflecting conducting walls. Both fully nonlinear codes with spatial grids and linearized gridless codes have been implemented. We have implemented several variations of the basic algorithm which utilize the conservation of canonical momentum or which relax this constraint for more general situations. The numerical dispersion

* Work performed under the auspices of USERDA under contract No. W-7405-Eng-48.

and stability of the various algorithms are investigated analytically and verified by computer experiments. Stability generally requires that $(kV_A)_{eff}\Delta t \lesssim 1$ and $\omega_{ci}\Delta t \lesssim 1$, where the effective Alfvén frequency $(kV_A)_{eff}$ includes space-time grid effects. The goal is to simulate low-frequency $(\omega \sim \omega_{ci} << \omega_{pi})$ electrostatic and magnetostatic plasma collective behavior. Special emphasis is given to mirror physics applications, e.g. low-frequency microinstabilities and build-up to and stability of field-reversed configurations. Maxwell's equations are solved in the Darwin approximation (no transverse displacement current) with quasineutrality additionally imposed. The plasma is a nonrelativistic ensemble of particle ions and an electron fluid.

Quasineutrality:

$$\frac{\partial}{\partial t} e(n_i - n_e) + \nabla\cdot(\underline{J}_i + \underline{J}_e) = 0$$

Boundary conditions: $$(\underline{J}_i + \underline{J}_e)^{\ell}\cdot\hat{n}|_{surface} = 0$$

$$\therefore \quad (\underline{J}_i + \underline{J}_e)^{\ell} = 0$$

Therefore, quasineutrality + Darwin approximation $\frac{1}{c}\frac{\partial}{\partial t}\underline{E}^t \to 0$ reduces Ampère's law to

$$\nabla \times \underline{B} = \frac{4\pi}{c}(\underline{J}_i + \underline{J}_e)^t = \frac{4\pi}{c}(\underline{J}_i + \underline{J}_e)$$

To this add

$$\nabla \times \underline{E} = -\frac{1}{c}\frac{\partial}{\partial t}\underline{B} \qquad \text{(Faraday's law)}$$

and

$$n_e m_e\left(\frac{\partial}{\partial t} + \underline{V}_e\cdot\nabla\right)\underline{V}_e \approx 0 \approx n_e q_e\left(\underline{E} + \underline{V}_e \frac{\times}{c}\underline{B}\right) - n_e m_e(\underline{V}_i - \underline{V}_e)\nu_{ei} - \nabla\cdot\underline{\underline{P}}_e$$

We neglect the electron inertia, use $n_e \approx n_i$,
and solve the electron fluid equation for $\underline{E}$.

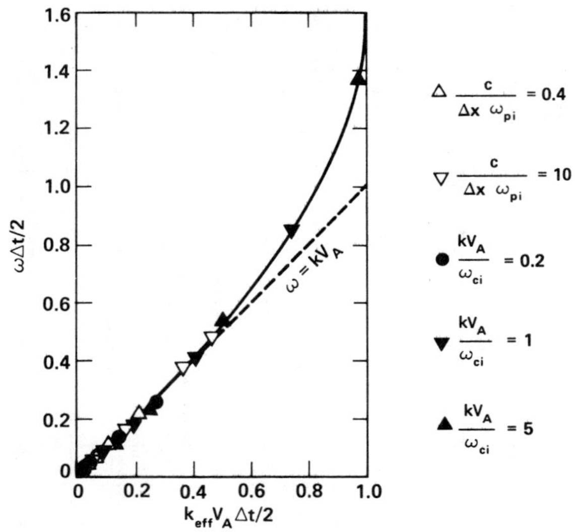

FIG.1. Compressional Alfvén waves.

Basic algorithm:

$$\underline{E} = -\frac{\nabla^2}{\omega_{pi}^2}\underline{A}^t c \times \underline{\omega}_{ci} - \frac{4\pi}{\omega_{pi}^2}\underline{J}_i \times \underline{\omega}_{ci} - (n_i q_i)^{-1}\,\nabla\cdot\underline{\underline{P}}_e + \frac{\nu_{ei}}{\omega_{pe}^2}\nabla^2\underline{A}^t c$$

(first group: $\underline{E} \times \underline{B}$; $\nabla\cdot\underline{\underline{P}}_e$ term: diamagnetic effects; last term: resistivity)

$$-\frac{1}{c}\frac{\partial \underline{A}^t}{\partial t} = \underline{E}^t \qquad \underline{E} = \underline{E}^t + \underline{E}^\ell \qquad \underline{B} = \nabla \times \underline{A}^t + \underline{B}_0 \qquad \underline{\underline{P}}_e = \underline{\underline{P}}_e(n_e, T_e), \quad n_e = n_i$$

$$d\underline{V}_i/dt = (q_i/m_i)\left(\underline{E} + \underline{V}_i \times \underline{B}c^{-1}\right)$$

$$d\underline{X}_i/dt = \underline{V}_i$$

$$\underline{X}_i,\ \underline{V}_i \rightarrow n_i,\ \underline{J}_i$$

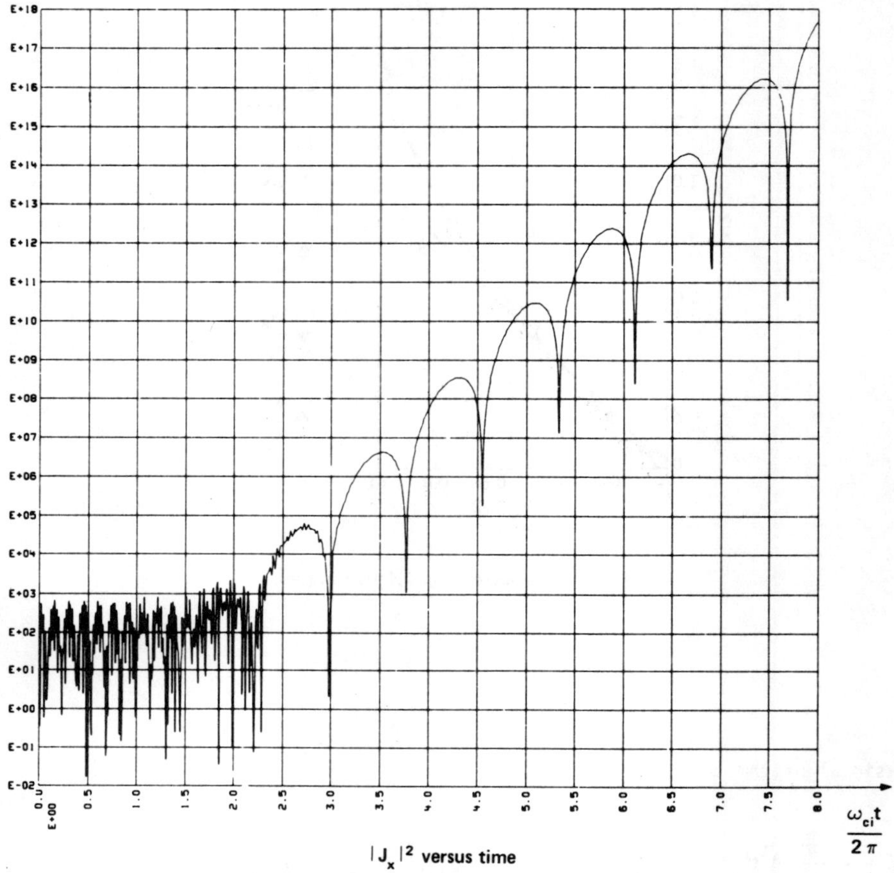

FIG. 2. Unstable drift-cyclotron-loss-cone mode.
$F(v) = \exp(-v^2) - \exp(-Rv^2)$;
$R = 10$; $k_x\rho_i = 7$; $\beta = 0.1$; $\rho_i/R_p = 0.1$.
Result shows initial high-frequency transient and then transitions to an instability with $\omega/\omega_{ci} = 0.65$ *and* $\gamma/\omega_{ci} = 0.45$.

Note that the vector potential is propagated ahead in time and that the electric field appears as part of the electron current, $\underline{J}_e = n\,\frac{\underline{E}\times\underline{B}}{B^2}$, and is obtained by solving for $\underline{J}_e = -\nabla^2\underline{A} - \underline{J}_i$. It is possible to construct algorithms that are perfectly time-centered and that require no iteration. In one dimension the

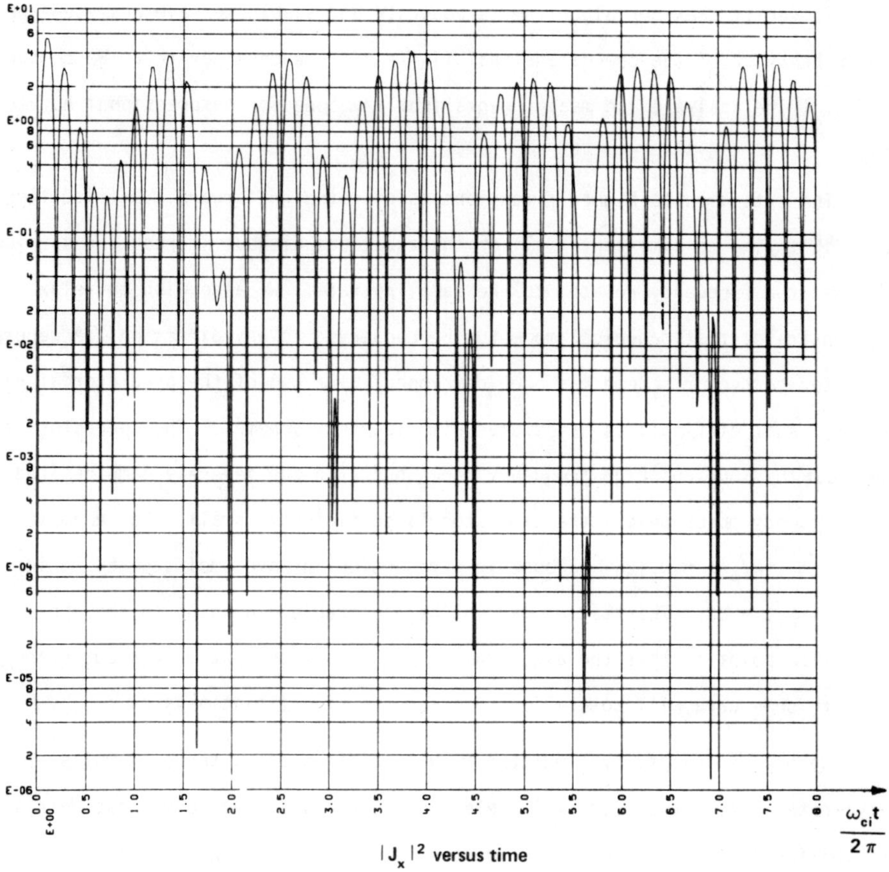

FIG.3. Stable drift-cyclotron-loss-cone mode.

$k_x\rho_i = 1$; *other parameters same as in Fig. 2.*

No growth is observed; the primary frequency is close to $\omega/\omega_{ci} = kV_A/\omega_{ci} = 3$.

algorithm is particularly straightforward. There is a longitudinal electric field E_x, a transverse electric field E_y, a vector potential A_y, and a magnetic field B_z. It is the geometric decoupling of transverse and longitudinal vector fields that makes one-dimensional versions of Darwin models especially simple. Another simplification possible in one dimension is the use of conservation of y momentum.

To verify that our algorithm is accurate and stable we have analysed the linear behavior of the one-dimensional finite difference algorithm. We compare in Fig. 1 the analysis and measurements from the code for compressional Alfvén waves.

The extension to higher dimensions is not straightforward. Difficulties arise in trying to time-center all of the equations without introducing spurious effects at a frequency related to the sampling rate. We also wish to relax the constraint that canonical momentum be conserved in any direction. A general predictor-corrector scheme has been developed. The use of the predictor alone has the usual difficulties associated with leapfrog schemes. The corrector damps out the spurious high-frequency modes but also can severely damp desired low-frequency modes unless the time step is sufficiently small. For example, with $\omega\Delta t = kV_A\Delta t = 0.125$ we observe a reduction in wave amplitude by 7% in a time $\omega_{ci}t = 50$. This may or may not be acceptable depending on the application. The point is that the algorithm is stable without extra spurious branches and with some care will not seriously alter the significant physics.

We have successfully simulated with linearized codes three examples of microinstability: two essentially electrostatic modes, the Dory-Guest-Harris instability[1,2,3] and the drift-cyclotron-loss-cone mode[4]; and the Alfvén-ion-cyclotron mode[5] which is electromagnetic. The simulation results agree well with linear analytical theory, and the generally stabilizing influence of the electromagnetic modifications of the dominantly electrostatic modes is demonstrated. We have also applied these simulation models to the study of field-reversed magnetic-mirror systems. Our simulations verify that electron return currents cancel an embedded, linearly rising, external current which is perpendicular to the vacuum magnetic field, only for times up to the Alfvén transit time of a plasma bounded by conducting walls [6]. After this, there is a growth of net current and concomitant magnetic field modification.

Examples of results from a linearized version of the code for a drift-cyclotron-loss-cone mode are shown in Figs 2, 3. For the case $k_x\rho_i = 7$, we see an instability with both growth rate and frequency agreeing with the theory.

For $k_x \rho_i = 1$, the mode is observed to be stable, again in agreement with the theory.

Similar agreement is obtained for the Dory-Guest-Harris instability the Alfvén ion cyclotron mode.

These algorithms are presently being extended to 2D nonlinear versions for inhomogeneous plasmas. They are expected to be important for a variety of mirror-confined plasma problems.

REFERENCES

[1] DORY, R.A., GUEST, G.E., HARRIS, E.G., Phys. Rev. Lett. **14** (1965) 131.
[2] CRAWFORD, F.W., TATARONIS, J.A., J. Appl. Phys. **36** (1965) 2930.
[3] CALLEN, J.D., GUEST, G.E., Phys. Fluids **14** (1971) 1588.
[4] TANG, W.M., PEARLSTEIN, L.D., BERK, H.L., Phys. Fluids **15** (1972) 1153.
[5] DAVIDSON, R.C., OGDEN, J.M., Phys. Fluids **18** (1975) 1045.
[6] BERK, H.L., PEARLSTEIN, L.D., Phys. Fluids **19** (1976) 1831.

IAEA-SMR-31/103

APPLICATIONS OF COMPUTATIONAL PLASMA PHYSICS IN PARTICLE BEAM FUSION*

J. R. FREEMAN
Sandia Laboratories,
Albuquerque, New Mexico,
United States of America

Abstract

APPLICATIONS OF COMPUTATIONAL PLASMA PHYSICS IN PARTICLE BEAM FUSION.

Inertial confinement fusion has relied heavily on advanced computer simulation techniques since its inception. The early computations at Lawrence Livermore Laboratory of laser-fusion target implosion and burn provided the vital spark for the rapid growth of this field. The more recent programmes in which electron and ion beams are being studied as alternatives to lasers have also been supported by large-scale plasma simulation activities. These activities range from particle-in-cell models for diode focusing studies, through combined Monte Carlo deposition and hydrodynamics calculations, to one- and two-dimensional target design computations using advanced multitemperature combined hydrodynamics and energy flow codes. These computational plasma physics applications are described.

1. INTRODUCTION: PARTICLE BEAM FUSION

Before describing computational methods and results, it is useful first to review briefly what inertial confinement and particle beam fusion [1–7] is about. Inertial confinement fusion requires that a high-power source of energy be focused on a small target. Possible sources include laser, electron and ion beams. The target contains D-T fuel which must be compressed and heated rapidly to ignition. The inertia of the imploding target must then confine the reacting D-T for a time long enough to consume a large fraction of the fuel.

Electron and ion beams are being studied as alternatives to lasers because of their higher efficiency and different deposition characteristics. The beams are produced by pulsed accelerators usually consisting of a Marx generator, a pulse-forming line and a vacuum diode. Electrical energy is transferred from the Marx generator to the pulse-forming line where a short-duration high-power pulse is formed and then applied to the diode. The pulse produces extremely high electric fields in the diode, so that a beam of electrons is emitted from the cathode. If a plasma source is present at the opposite electrode (the anode), a beam of ions will flow in the direction opposite to the electrons. Figure 1 shows the 8-TW PROTO-II accelerator [8] recently put into operation at Sandia. Its design parameters are a voltage of 1.5 MV, a current of 6 MA and a pulse length of 24 ns.

* Work supported by USERDA.

FIG.1. 8-TW PROTO-II electron beam accelerator.

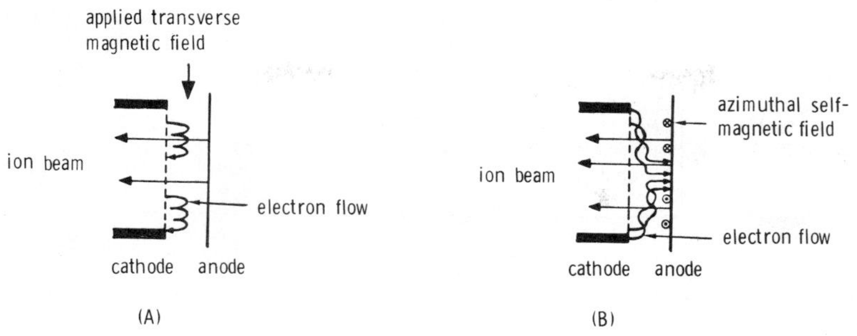

FIG.2. Versions of magnetically insulated diodes in which electron insulation is achieved by (A) externally applied magnetic fields and (B) beam self-fields.

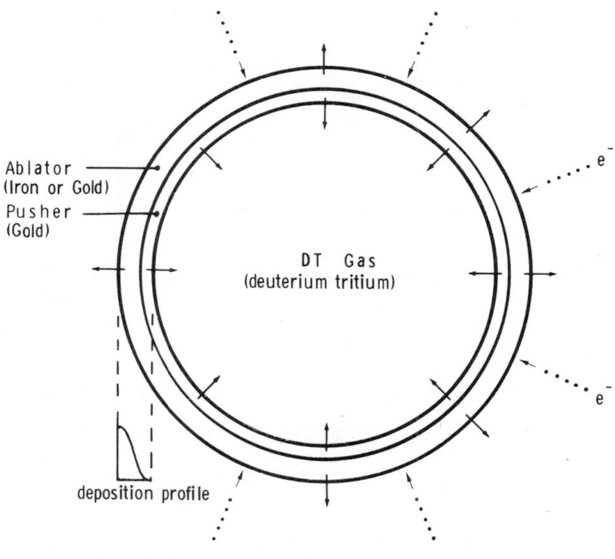

FIG.3. Schematic diagram of a simple gas-filled electron-beam fusion target.

To use electron beams for compressing targets, it is next necessary to focus the electrons to a small area at the anode. At high current levels, the self-magnetic field of the beam provides the necessary focusing forces. The ion flow in the diode can contribute to the focusing by increasing the magnetic field and by partially neutralizing the self-electrostatic fields and forces of the electrons. Current densities from 1 to 10 $MA \cdot cm^{-2}$ have been achieved experimentally by this process.

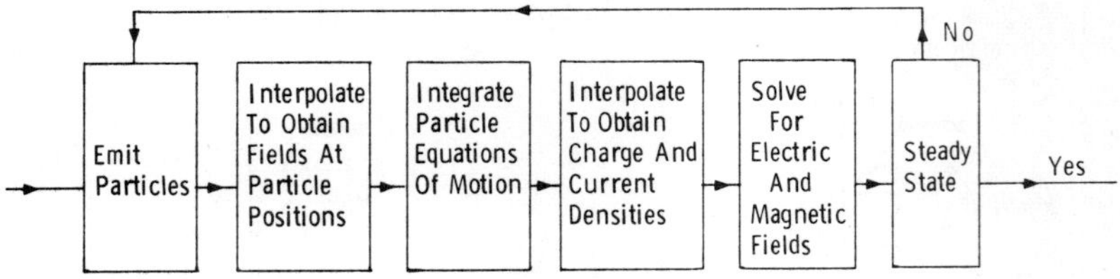

FIG.4. Diode code flow chart.

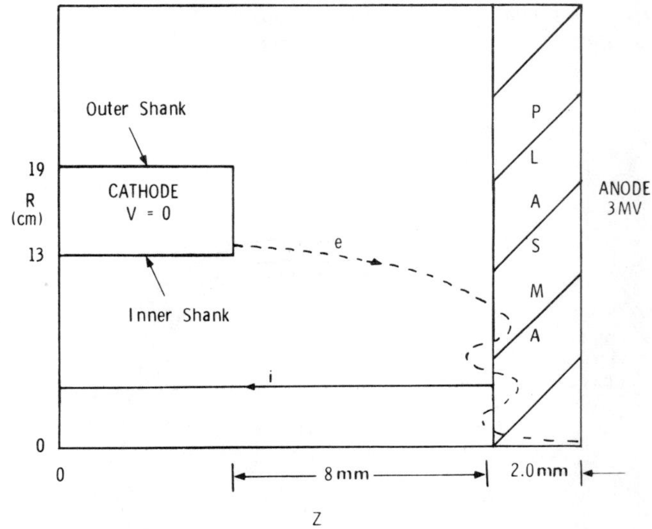

FIG.5. 3-MV hollow cathode diode configuration; the dashed line is a sample electron trajectory and the solid line is a sample ion trajectory. The diode is cylindrically symmetric about the z-axis.

It was suggested several years ago [9] that the ion flow itself in the diode be focused and used to implode targets. Target design calculations indicated that the more favourable ion deposition profile could lead to reduced peak power requirements and greatly relaxed machine impedance values. To use the ions efficiently, it is first necessary to suppress or reduce the electron flow by using magnetic insulation or using a design in which the electrons reflex back and forth through a thin anode. In addition, certain voltage and electrode configuration choices for self-pinched electron beam diodes provide a self-magnetic insulation and an enhanced ion flow, as shown in Fig.2. Proton currents of about 200 kA at 1.5 MeV have been reported by the Naval Research Laboratory. The ultimate choice between electrons and ions will be determined by such questions as efficiency, focusability and target design.

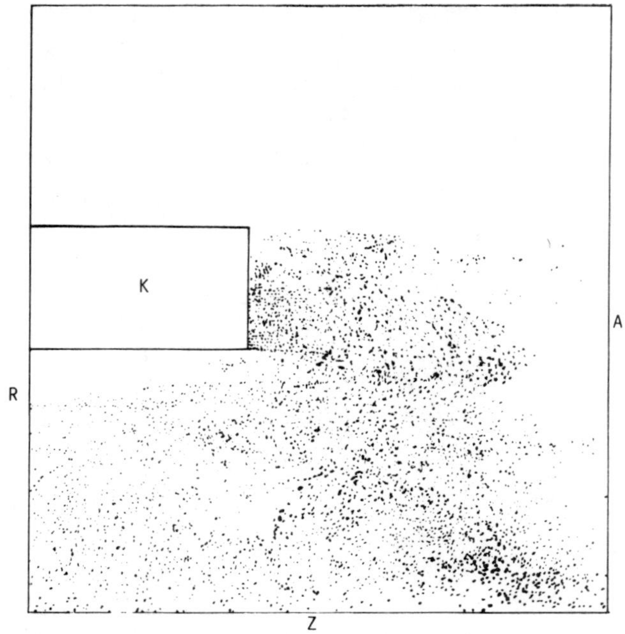

FIG. 6. Electron simulation particle map at a time near the steady state.

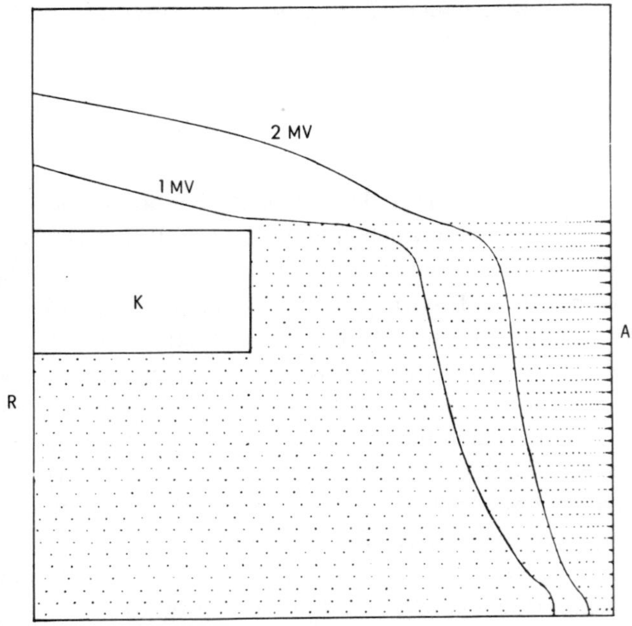

FIG. 7. Ion simulation particle map; 1-MV and 2-MV equipotentials are indicated.

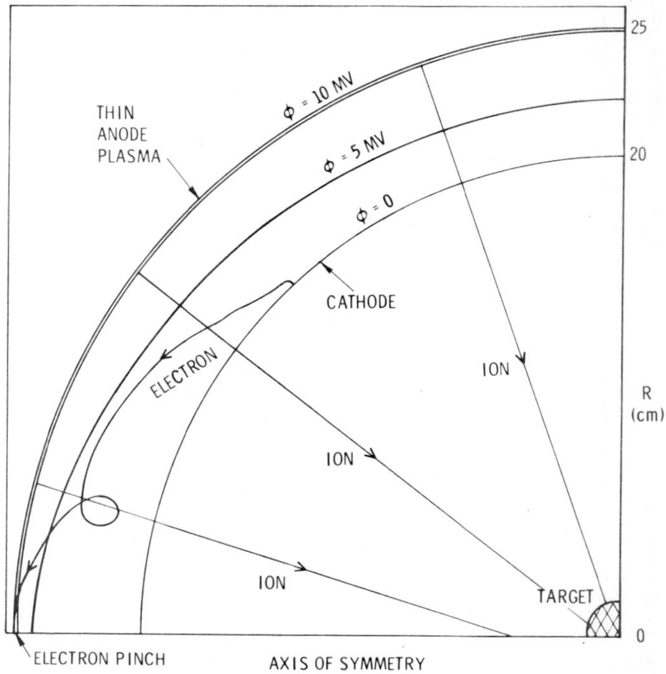

FIG.8. Code solution for a 10-MV hemispherical diode. The ions flow radially inward from anode to cathode; they are assumed to be charge neutralized inside the cathode.

The simplest target designs [10] for particle beam fusion, shown in Fig.3, consist of high-Z spheres (such as Au) which contain gaseous D-T fuel. The charged particles deposit their energy in the outer portion of the sphere, raising its temperature and pressure. This increase in pressure leads to the ablation of the outer portion; momentum conservation then causes the dense inner portion of the target to be imploded inwards. The magnetic and electric fields present in and near the target can contribute to the efficiency of the energy deposition. Calculations thus far have indicated that breakeven for electron beam targets requires somewhat more than 10^{14} W of power deposited. Ion beam targets [9] have requirements reduced by a factor of 5 to 10. Symmetry requirements and the stability of these targets are topics of present study.

Computational plasma physics has played a major role in the development of inertial confinement fusion concepts and in the analysis of present-day experiments. These applications, which can be broadly categorized as diode physics and target physics, are described in the next two sections.

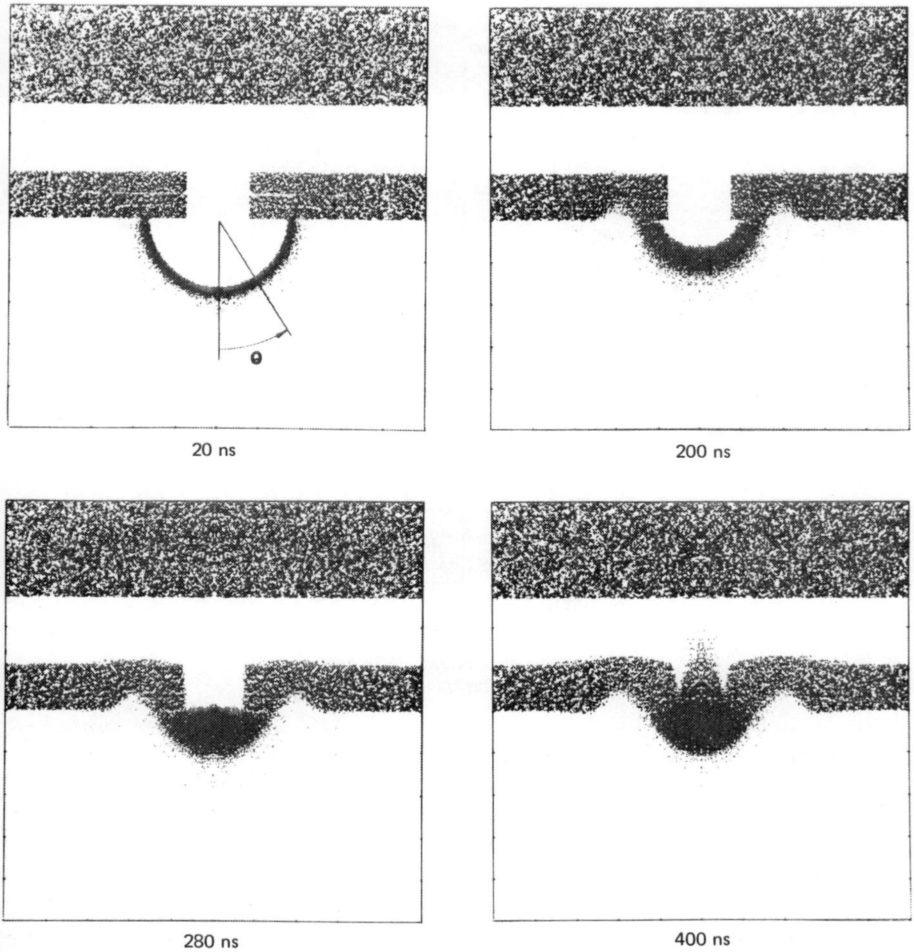

FIG.9. Hydrocode calculation of target implosion at several times showing the cross-sectional view of target density.

2. DIODE PHYSICS

Simulation of the electron and ion flow in diodes has been performed using two-dimensional particle-in-cell (PIC) type models [11–14]. The components of a diode code include a fast, accurate particle-pusher for both electrons and ions, a fast Poisson solver which can treat cylindrically symmetric regions with irregularly shaped electrodes, and an emission-law model. The charge and current densities typically present in high-power diodes require that the effects of self-magnetostatic and self-electrostatic fields be included, although displacement currents and inductive fields are neglected.

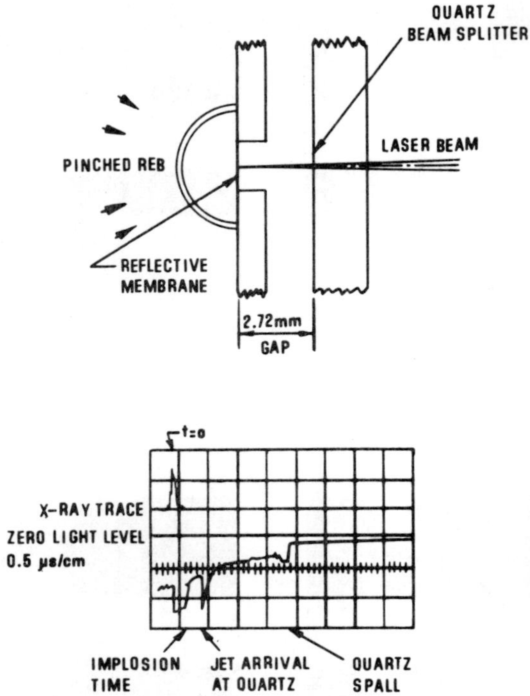

FIG.10. Experimental configuration to measure jet velocity and arrival time; the oscillogram records laser light reflected from both the membrane and the beam splitter.

Most diode code applications have studied quasisteady flow equilibria. The method is shown in Fig.4. A particular voltage is applied and particles are first emitted from assumed electrode plasmas based on locally one-dimensional Child's-law models. Next, the relativistic equations of motion for the particles are integrated for a small increment in time. The new particle positions and velocities are then interpolated to a finite difference mesh to determine new charge and current densities. The field solvers are finally used to obtain fields and forces for the next time step. Particles striking electrode surfaces are removed from the calculation, and new particles are emitted to continue the integration. The anode current and current density are monitored during the calculation. The final solution converges to an equilibrium in which fluctuations remain, but the macroscopic quantities are nearly constant.

Figure 5 shows the configuration for a diode studied recently by Quintenz and Poukey [13]. For the particular dimensions shown, the code predicts an electron current of 0.65 MA, and an ion current of 1.6 MA. Figure 6 shows the positions of the 10^4 simulation electrons at equilibrium, while Figure 7 shows a similar plot for the ions. The ion gyroradius is so large that little bending of the trajectories is observed.

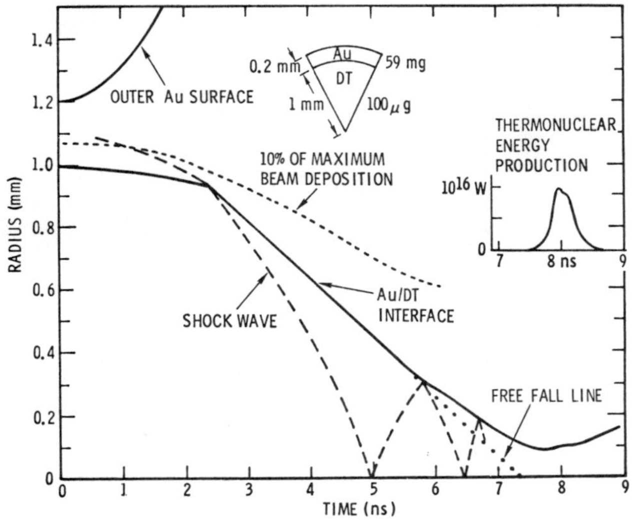

FIG.11. Computed radius versus time for various regions of a breakeven electron beam target.

The most recent versions of the codes include more complete physical models. Shank and corner emission models are included, and time-dependent effects are now being studied. The emission laws used are corrected for magnetic fields and returning particles. The pinch formation phase requires the use of a time-dependent emission model [14, 15] in which ions are emitted only in a local annular region of the anode where the electrons are striking. This model is based on the assumption that ions can only be emitted from an anode plasma and that no plasma exists until the anode experiences an electron flux.

The Poisson solver used in the code is based on a fast Fourier transform in the axial direction, followed by the usual fast one-dimensional tridiagonal method [16] in the radial direction. Capacity matrix methods [17] are used to allow irregularly shaped regions. The original two-step predictor-corrector particle pusher [11] has been replaced by a fourth-order Runge-Kutta method [13] to increase accuracy. The use of a high-accuracy method permits a larger time step to be taken with a significant reduction in run time.

Many variations of this PIC-type code have been constructed. A spherical version was written to study the behaviour of a spherical ion diode [18, 19]. This diode, shown in Fig.8, uses the self-pinched electron flow to enhance the ion current and a spherical geometry for ion focusing. Another version was written in a coaxial co-ordinate system to study laser-exciter diodes [20]. Similar methods have been used to study collective ion acceleration concepts, both in linear [21, 22] and electron ring accelerators [23]. The basic techniques which have been developed are sufficiently modularized to permit codes to be constructed which provide a quick analysis of many proposed experiments in particle acceleration.

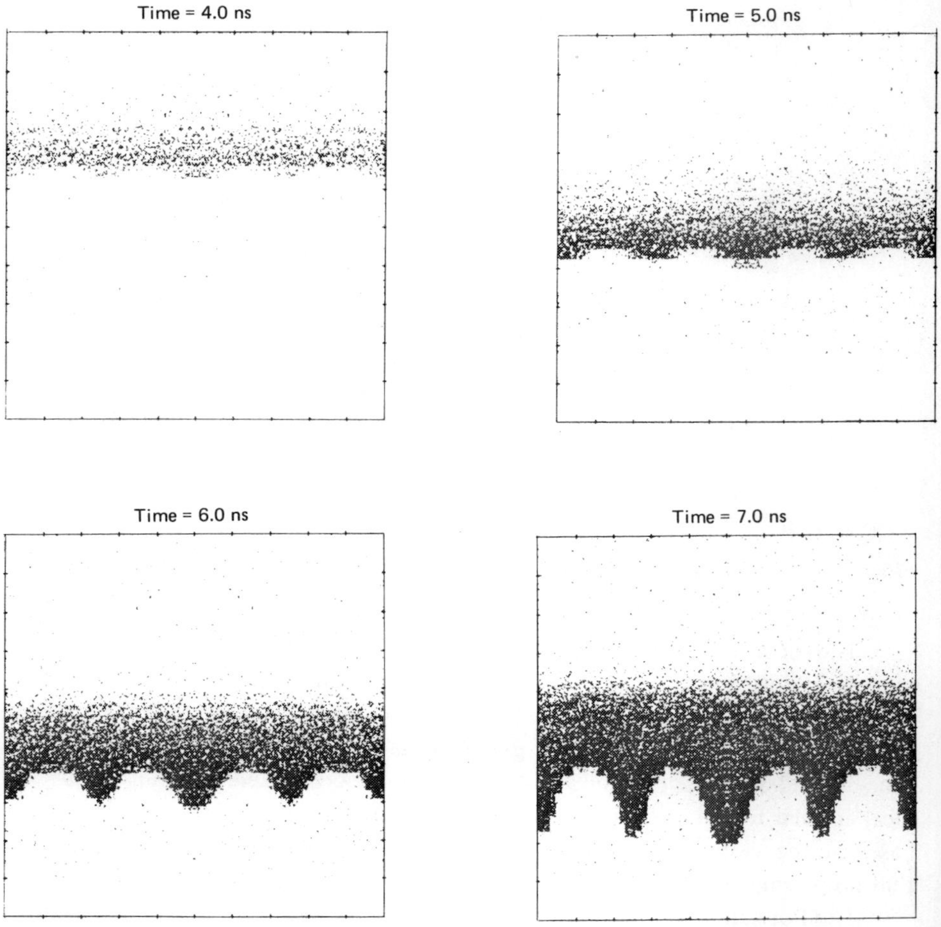

FIG.12. Sequence of frames (t = 4–7 ns) showing iron pusher (upper portion of frame) compressing deuterium fuel (lower portion of frame). Times are measured with respect to the time of initial pusher deceleration. Each frame is cylindrically symmetric about a vertical axis through the middle of the frame.

3. TARGET PHYSICS

Both target design calculations and the analysis of target response experiments rely on the use of one- and two-dimensional hydrodynamics codes. These codes are based on collision-dominated continuum models. At high beam powers the targets exhibit basically compressible fluid-like motion, while at low powers elastic solid properties can be important. The partial differential equations describing the conservation properties of the system are integrated using finite difference methods. One-dimensional computations usually employ Lagrangian

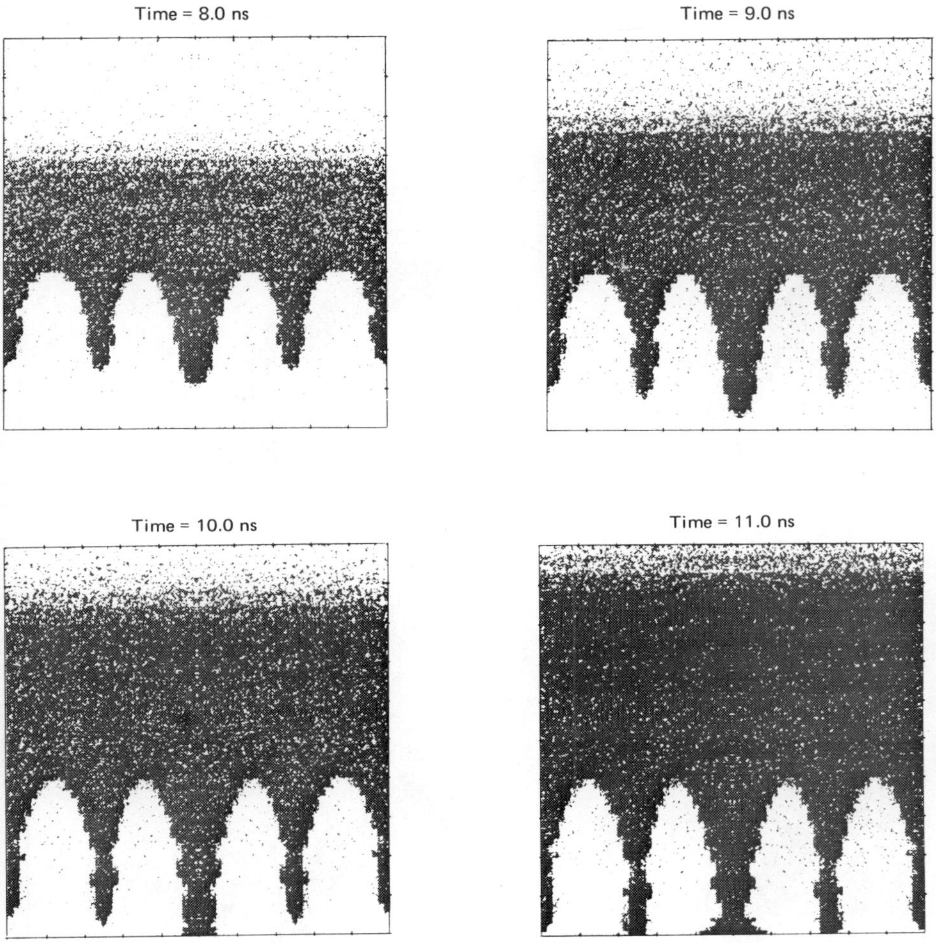

FIG.13. Continuation of sequence shown in Fig.12 for t = 8–11 ns.

meshes which follow the fluid motion. This enables the zones to follow and resolve the compressed regions, which are usually the regions of interest. For two-dimensional problems, Eulerian calculations which use a fixed mesh are often employed. These methods are helpful in problems where non-laminar fluid flow would distort a Lagrangian mesh.

An example of the simulation of a 2-D target response experiment [24] is shown in Fig.9. A 4-mm-diameter gold hemisphere mounted in a copper anode plate is driven by a 0.75-MeV electron beam. The momentum imbalance causes a dense gold jet to form and penetrate the central region. Figure 10 shows an

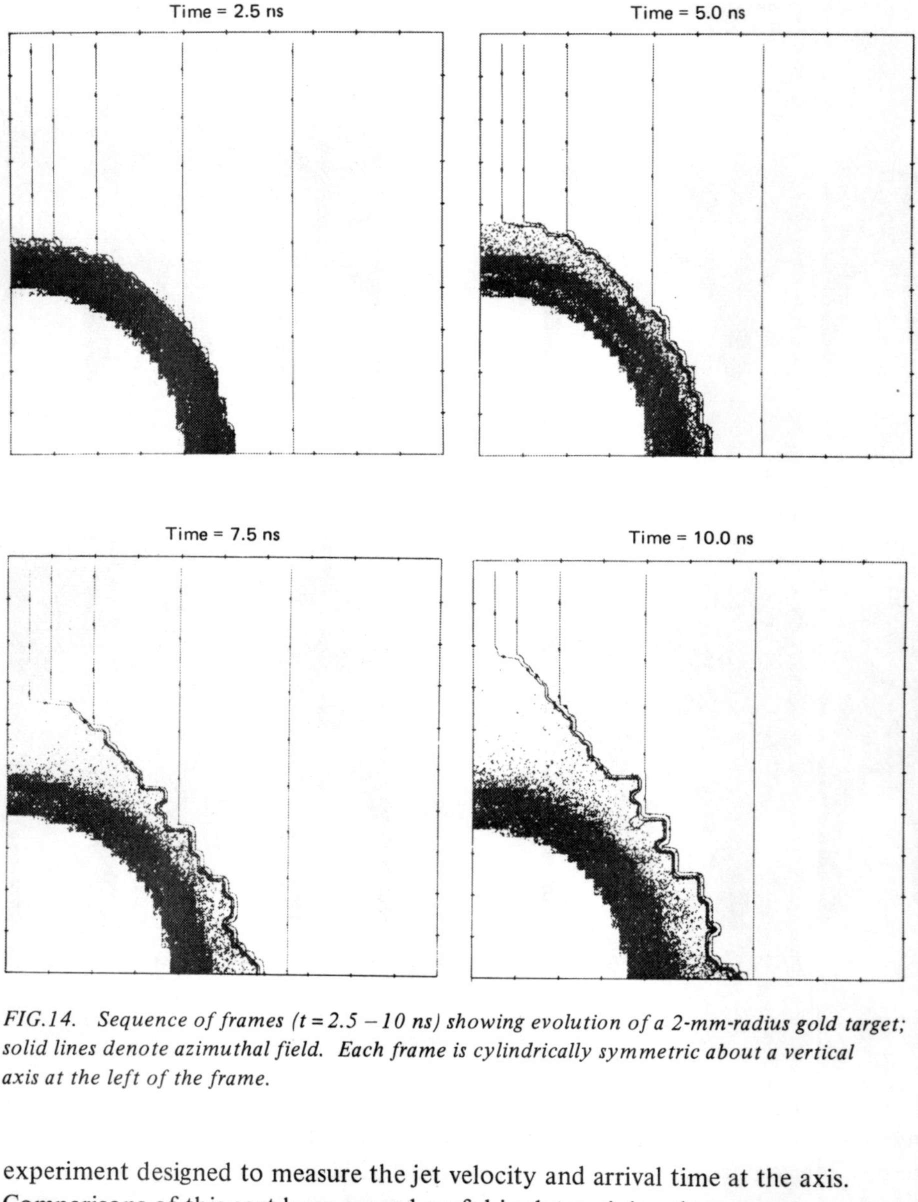

FIG.14. Sequence of frames (t = 2.5 – 10 ns) showing evolution of a 2-mm-radius gold target; solid lines denote azimuthal field. Each frame is cylindrically symmetric about a vertical axis at the left of the frame.

experiment designed to measure the jet velocity and arrival time at the axis. Comparisons of this sort have proved useful in determining the symmetry and efficiency of beam power deposition.

The preceding computation included a beam energy deposition model which was based on a single classical Monte Carlo deposition calculation. Such a calculation provides information on the rate at which a single electron deposits its energy as a function of $\rho \cdot dx$ (density times depth into target). It has been found, however, that the classical energy deposition picture must be modified

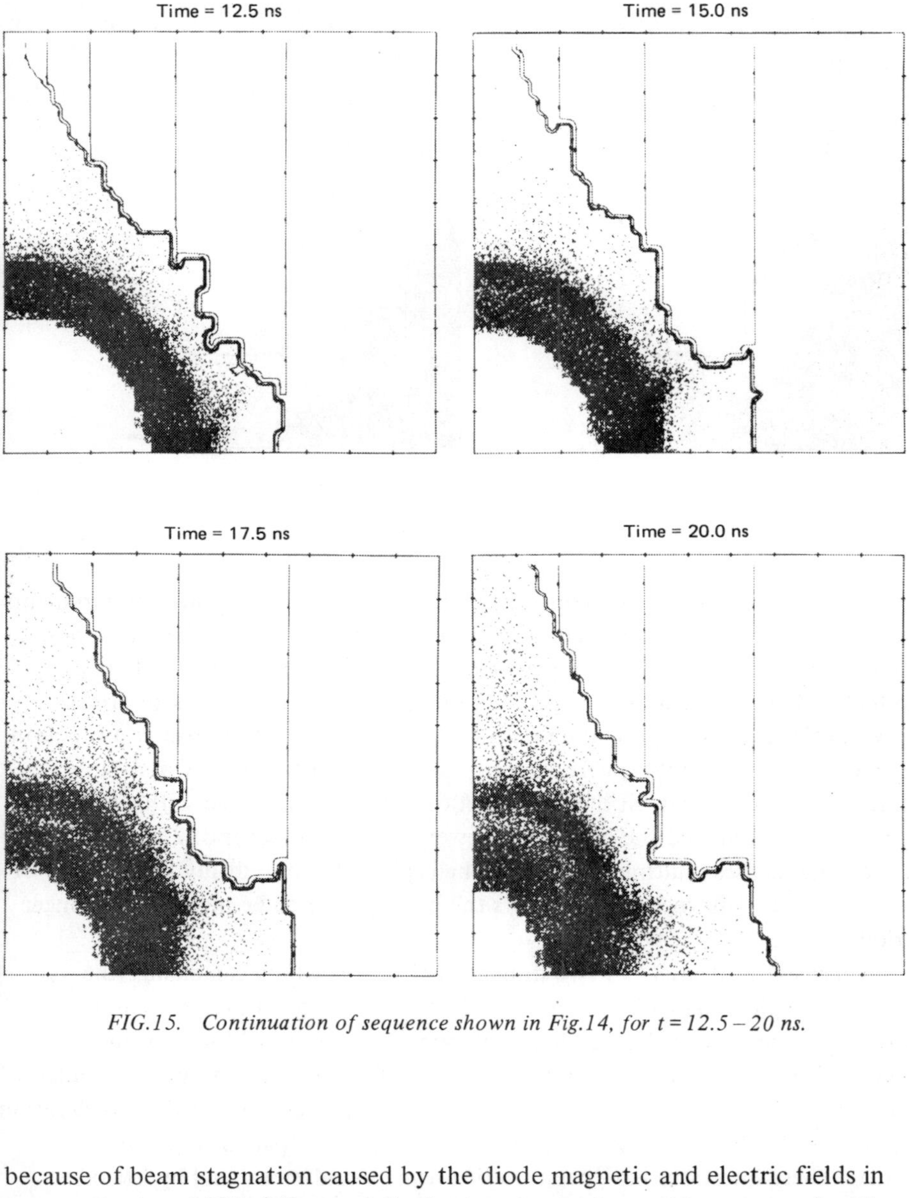

FIG.15. Continuation of sequence shown in Fig.14, for t = 12.5 – 20 ns.

because of beam stagnation caused by the diode magnetic and electric fields in or near the target [25, 26], especially for targets which are thin compared with the magnetic diffusion skin depth. This latter problem is now being studied with Monte Carlo codes which have been modified to include the effects of fields on electron deposition, and with a combined Monte Carlo and hydrodynamics code. The latter code alternates the integration of the conservation equations and the electron deposition calculations to attempt to provide a more self-consistent treatment of the problem.

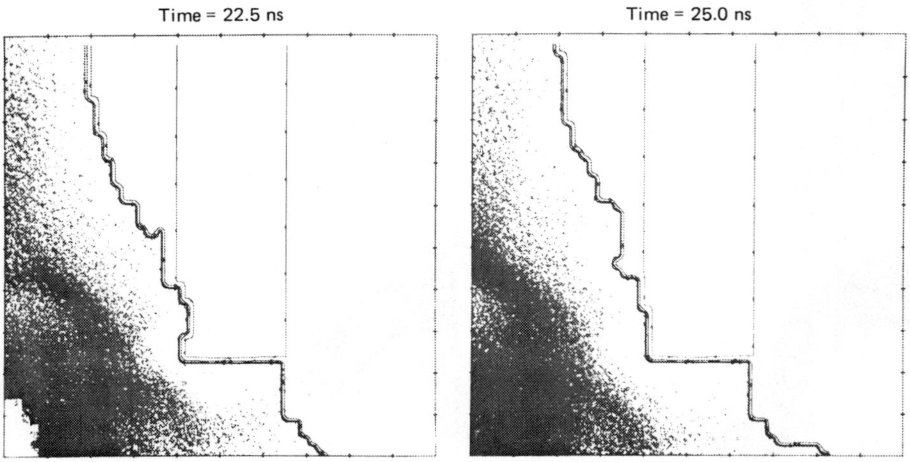

FIG.16. Continuation of sequence shown in Fig.14, for t = 22.5 – 25 ns.

Target design studies are concerned with determining the power requirements for breakeven and with the design of targets for use with present-day machines. The one-dimensional Lagrangian hydrocodes used for these studies include a model for thermonuclear energy production. A breakeven level design [10] is shown in Fig.11, which plots the radius of various regions of interest as a function of time. This simple gold sphere target was driven with 8×10^{14} W of 1-MeV electrons. Designs in which the outer part of the gold is replaced with a lower-Z layer [27] have reduced the breakeven power by about a factor of 2 to 3. Another class of targets uses multiple concentric shells [28–30] to multiply the velocity of the inner shell. This technique permits the beam power to be input over a longer period of time.

There exists a variety of two-dimensional effects which can change the relatively simple dynamics seen in the 1-D computations. Foremost among these is the question of the stability of these targets to spatial [30–33] or power [33] perturbations. The targets are potentially unstable to modes similar to the Rayleigh-Taylor instability both early in the power pulse during acceleration and later when the compressed fuel decelerates the inner part of the dense pusher. These questions have been studied using two-dimensional hydrocodes, and an example [33] is shown in Figs 12 and 13. These figures show the growth of a surface perturbation on a planar iron slab as it is decelerated by compressed deuterium fuel. The non-linear regime consists of narrow spikes coasting at a constant velocity. The intrusion of these spikes into the fuel volume can destroy the compression and heating of the fuel. Various stabilizing mechanisms are at present under study, and target designs which produce their output before spike intrusion would presumably be unaffected.

A final example of two-dimensional target calculations concerns the possible importance of magnetic pressures at high beam currents. At 10^{14} W, a 2-mm radius, uniform current density 1-MeV beam produces magnetic fields of up to 10^8 G. The pressures produced by these fields are sufficiently high to affect the symmetry of the implosion. This problem has been investigated using a 2-D MHD version of the hydrocode. Such a code is constructed by adding Lorentz force and Ohmic heating terms to the conservation equations and including an equation to describe the transport and diffusion of magnetic flux. Figures 14–16 show the results for a 2-mm-radius gold target and a 10^{14}-W, 1-MeV beam. Significant asymmetries are seen at $t = 22.5$ ns, just before the implosion time. Beams which are peaked on axis, as seen in present-day experiments, lead to much better symmetry.

4. SUMMARY

The development of particle beam fusion has been accompanied by the development of computer simulation models and techniques. These models have been applied to beam and plasma problems ranging from the low-density (10^{12} cm^{-3}) collisionless beam regime to the high-density (10^{23} cm^{-3}) collision-dominated conditions found in compressed fuel. Reasonably detailed two-dimensional time-dependent models have been constructed for many of the problems of physical interest. Present trends are in the direction of increasing self-consistency. One would ultimately like to combine the diode PIC class of model with the combined two-dimensional Monte Carlo deposition-hydrodynamics model, since the time scales for the beam focusing, deposition and hydrodynamics are often comparable. Moreover, the diffusion of the magnetic field into the target, which can alter the electron deposition, should also be treated self-consistently. Such a treatment is beyond present capabilities, but work in these directions is progressing.

ACKNOWLEDGEMENTS

Most of the work described here was developed in conjunction with the Sandia Particle Beam Fusion Program, managed by G. Yonas; it includes the contributions of J.W. Poukey, J.P. Quintenz, M.J. Clauser, M.M. Widner, M.A. Sweeney, L. Baker and S.L. Thompson. Many helpful discussions in the diode physics area with S.A. Goldstein and R.Lee, of the US Naval Research Laboratory, are also acknowledged.

REFERENCES

[1] YONAS, G., et al., Electron beam focusing and application to pulsed fusion, Nucl. Fusion **14** (1974) 731.

[2] YONAS, G., et al., 6th European Conf. on Controlled Fusion and Plasma Physics (Proc. Conf. Moscow, 1973) 483.

[3] RUDAKOV, L.I., SAMARSKY, A.A., ibid., 487.
[4] RUDAKOV, L.I., BABYKIN, M.V., in Proc. 7th European Conf. on Controlled Fusion and Plasma Physics,Lausanne, 1975.
[5] FREEMAN, J.R., et al., "Particle beam fusion research", Plasma Physics and Controlled Nuclear Fusion Research 1976 (Proc. 6th Int. Conf. Berchtesgaden, 1976) **1**, IAEA, Vienna (1977) 167.
[6] BOGOLYUBSKIJ, S.L., et al., "Demonstration of the possibility of using electron beams for heating thermonuclear targets", ibid., 177.
[7] SANDIA LABORATORIES, Sandia Technology – Particle Beam Fusion, Rep. SAND-76-0615 (1976).
[8] MARTIN, T.H., VANDEVENDER, J.P., JOHNSON, D.L., McDANIEL, D.H., AKER, M., Proc. Int. Top. Conf. Electron Beam Research and Technology, Albuquerque, 1976, Sandia Labs., Albuquerque (1976) 450.
[9] CLAUSER, M.J., Phys. Rev. Lett. **35** (1975) 848.
[10] CLAUSER, M.J., Phys. Rev. Lett. **34** (1975) 570.
[11] POUKEY, J.W., FREEMAN, J.R., YONAS, G., J. Vac. Sci. Technol. **10** (1973) 954.
[12] POUKEY, J.W., J. Vac. Sci. Technol. **12** (1975) 1214.
[13] QUINTENZ, J.P., POUKEY, J.W., J. Appl. Phys. **48** (1977) 2287.
[14] GOLDSTEIN, S.A., DAVIDSON, R.C., LEE, R., SIAMBIS, J.G., Proc.Int. Top. Conf. on Electron Beam Research and Technology, Albuquerque, 1976, 218.
[15] MILLER, S., ZINAMON, Z., Phys. Rev. Lett. **36** (1976) 1313.
[16] RICHTMYER, R.D., MORTON, K.W., Difference Methods for Initial-Value Problems, Interscience (1967) 198.
[17] HOCKNEY, R.W., Methods in Computational Physics **9**, Academic Press, New York (1970) 135.
[18] POUKEY, J.W., FREEMAN, J.R., CLAUSER, M.J., YONAS, G., Phys. Rev. Lett. **35** (1975) 1806.
[19] GOLDSTEIN, S.A., LEE, R., COOPERSTEIN, G., BLAUGRUND, A.E., Bull. Am. Phys. Soc. **20** (1975) 1252.
[20] FREEMAN, J.R., POUKEY, J.W., RAMIREZ, J.J., PRESTWICH, K.R., Bull. Am. Phys. Soc. **20** (1975) 1250.
[21] POUKEY, J.W., OLSON, C.L., Phys. Rev. **A11** (1975) 691.
[22] OLSON, C.L., POUKEY, J.W., VANDEVENDER, J.P., OWYOUNG, A., PEARLMAN, J.S., in Proc. 1977 Particle Accelerator Conference, Chicago, 1977.
[23] POUKEY, J.W., FREEMAN, J.R., REISER, M., Part. Accel. **6** (1975) 245.
[24] WIDNER, M.M., et al., J. Appl. Phys. **48** (1977) 1047.
[25] CLAUSER, M.J., et al.,Phys. Rev. Lett. **38** (1977) 398.
[26] WIDNER, M.M., POUKEY, J.W., HALBLEIB, J.A., Sr., Phys. Rev. Lett. **27** (1975) 483.
[27] SWEENEY, M.A., CLAUSER, M.J., Appl. Phys. Lett. **27** (1975) 483.
[28] KIRKPATRICK, R.C., et al.,Structured fusion target designs, Nucl. Fusion **15** (1975) 333.
[29] GULA, W.P., KIRKPATRICK, R.C., in Proc. First Int. Top. Conf. on Electron Beam Research and Technology, Albuquerque, 1976.
[30] LINDL, J.D., BANGERTER, R.O., ibid.
[31] BANGERTER, R.O., LINDL, J.D., MAX, C.E., MEAD, W.C., ibid.
[32] CLAUSER, M.J., SWEENEY, M.A., ibid.
[33] FREEMAN, J.R., CLAUSER, M.J., THOMPSON, S.L., Rayleigh-Taylor instabilities in inertial-confinement fusion targets, Nucl. Fusion **17** (1977) 223.

IAEA-SMR-31/104

NON-LINEAR PLASMA KINETICS
Solitons, cavitons and spikons

V.N. TSYTOVICH
Lebedev Institute of Physics,
Academy of Sciences of the USSR,
Moscow,
Union of Soviet Socialist Republics

Abstract

NON-LINEAR PLASMA KINETICS: SOLITONS, CAVITONS AND SPIKONS.
1. Introduction. 2. Modulation instability. 3. Soliton kinetics. 4. Caviton kinetics. 5. General kinetic theory of modulation interactions. 6. Fast particles and spikons. 7. General theoretical problems – concluding remarks.

1. INTRODUCTION

Although non-linear plasma theory is already a very well developed branch of physics with fundamental researches [1–5] and is used in many applications (laser-plasma interactions, astrophysics, accelerators, etc.), new and interesting problems have arisen in connection with high-power input in plasmas, for example by lasers or intense relativistic electron beams. The recent new explosion of interest in non-linear plasma physics is connected with the phenomenon of self-formation in plasmas of some non-linear self-consistent entities. By non-linear plasma kinetics we mean a kinetic description of these entities, the nature of which is non-linear. They are called solitons, cavitons, spikons, etc. and will be defined later on. The following problems need clarification:

(1) What is the physical mechanism responsible for creating non-linear entities in plasma?
(2) Can an appropriate description be found for a number of such entities, for example by using kinetic equations of some kind ?
(3) What kind of interactions can exist for these entities with each other and with plasma particles or plasma waves?

Several other problems will be discussed later, but one point should be emphasized here: the difference that can exist if the plasma is an open or a closed system. An example of an open system is a turbulent plasma with an input of energy in some domain of real space or k-space with energy transformation and with some sink of this energy. The appearance of non-linear entities can be considered as a self-ordering process. A self-ordering process cannot be excluded in an open system, but in a closed system the formation of non-linear entities should be accompanied by an increase of the entropy, for example by emission of some kind of plasma waves. Thus these entities should coexist with plasma waves and interact with them. The elementary entities of non-linear kinetics are some sort of clumps. Clumps have already been discussed by Dupree [6] and Kadomtsev and Pogutse [7] as random-phase short-lived correlations in phase space. Entities of the kind to be discussed are of a different nature, although some analogy exists. They have two essential features:

(1) The existence of a mechanism which forms them (as we shall see later, it is the modulational instability); and
(2) The important role of the phase correlations inside these entities.

It should be emphasized that concepts similar to clumps of some sort already exist in current plasma theories. We have in mind that the averaging usually used to describe the particle motion leads to a concept of quasiparticles, i.e. particles surrounded by clouds of particles of opposite charge sign. Such a concept appears in the derivation of the Landau-Balescu [8] collision integral by averaging an exact Klimontovich distribution function [9] and using the Bogolyubov hierarchy [10]. The average motion is described by quasiparticles with screened fields. It is obvious that the same particle cannot take part in both scattering of the other particle and the screening of this scattering. But the quasiparticles behave so that their interaction is screened and the screening is produced by the quasiparticles. This is the only physical result of the complicated mathematics usually employed to derive the collisional integral.

The more pronounced effects of an ensemble of particles with opposite charge signs which can be called quasiparticles or clumps, and which are some entities interacting with their surroundings in a compact way, are known in the weak non-linear theory of plasma. In the presence of a high level of oscillations one can proceed in a further averaging of the distribution function on a time scale larger than the period of oscillations. As was shown for the first time in Ref.[11], the equation found by this averaging describes the interaction of oscillations with 'dressed' charges, i.e. new quasiparticles described by the averaged distribution function. The scattering of oscillation on the new quasiparticles is described by the superposition of amplitudes of the two processes, namely the usual Thomson scattering and the non-linear or transition scattering. The latter process is due to the presence of the charge density which screens the charge of the quasiparticle. The induced oscillations of this charge density give the new process of transition scattering. It was proved that not only is the scattering of the quasiparticle different, but so is the bremsstrahlung process [12, 13]. The transition scattering [14] can also act to convert one type of wave into another and even convert gravitational waves into electromagnetic waves [15]. The most important physical result of a weak turbulent theory is a proof that this picture of 'dressed' quasiparticles is self-consistent and can be formulated in a rather general form [2].

These examples show the importance of the formation of some plasma clumps[1] in the description of a non-turbulent or weakly turbulent plasma. It can be expected that the effect of clumping will be larger when the energy input in plasma increases. Indeed, new phenomena can start for a high level of oscillations, known as modulation instability [16, 17]. The belief exists that this instability creates non-linear entities, such as solitons, cavitons and spikons. This phenomenon can also be described by some averaging of non-linear equations [18] and the equations found should be those which describe the kinetics of such entities. We shall not discuss here this straightforward way to obtain the equations of non-linear plasma kinetics but a much simpler, although not well grounded, way: (1) we consider the property of a single non-interacting non-linear entity; (2) we try to understand the mechanisms of its interactions with other entities or waves; and (3) we try to construct the kinetic equations describing the change in space and time of the occupation number for these entities.

2. MODULATION INSTABILITY

Modulation instability appears in the case when the energy level of oscillations is high and the dispersion of waves is weak. In the case of a weak dispersion the group velocity of the waves is small. Suppose there exists a density variation δn localized in a certain space domain. The group velocity of waves is changed there – say, decreased. The waves are accumulated in the region of density variation and can push the plasma away by the ponderomotive force. This can increase the δn, thus leading to an increase in the waves accumulated in this domain and to an increase of the ponderomotive force. This is the way the modulation instability can develop. Thus the plasma becomes locally inhomogeneous with possibly many density inhomogeneities present simultaneously. These inhomogeneities trap the waves and are the sort of clumps we are interested in here. The question is: under what conditions is the formation of density inhomogeneities energetically favourable? Let us consider two types of plasma wave dispersion:

---

[1] The word 'clump' is used here in a different sense from that in Refs [6, 7].

$$\text{(a)} \quad \omega^2 = \omega_0^2 + \alpha^2 k^2 \tag{1}$$

$$\text{(b)} \quad \omega^2 = \omega_0^2 - \beta^2/k^2 \tag{2}$$

The dispersion described by (1) is known for Langmuir waves,

$$\alpha = \sqrt{3}\, v_{Te}, \quad \omega_0 = \omega_{pe} = \sqrt{\frac{4\pi n_0 e^2}{m_e}}$$

as well as for lower hybrid waves, used in present plasma-heating experiments. The dispersion described by (2) is known for ion sound waves with k close to

$$k_d = \frac{1}{r_d} = \frac{\omega_{pe}}{v_{Te}}, \quad \beta = \omega_0 k_d, \quad \omega_0 \cong \omega_{pi}$$

for low-frequency waves with frequency close to $\omega_{Bi} = eB_0/m_i c$, etc.

Without loss of generality we can consider a plasma volume whose size is unity. Let us consider the two density distributions: a homogeneous one and a distribution when in one half of the volume the density is $n_0 - \delta n$ and in the other half the density is $n_0 + \delta n$ (see Fig. 1). We shall suppose that in the case of homogeneous density distribution there also exists a homogeneous and isotropic distribution of waves. It is necessary to calculate the change of thermal energy of the plasma and the change in the energy of the waves. To form the inhomogeneous distribution shown in Fig. 1 from the homogeneous distribution, it is necessary to compress the plasma and give an energy δW_p to the plasma:

$$\delta W_p = \frac{\gamma}{2} n_0 (T_e + T_i) \left(\frac{\delta n}{n_0}\right)^2 \tag{3}$$

where $\gamma = 1$ for isothermal compression and $\gamma = 5/3$ for adiabatic compression. To calculate the change in the energy of the waves, we suppose for simplicity that in the presence of the density variation δn the change in ω_0^2 can be described by

$$\delta\omega_0^2 = \omega_0^2 \frac{\delta n}{n_0} \tag{4}$$

The change in the frequency of the waves $\delta\omega^2$ can be found from

$$\int\limits_{k_\parallel^2(z) > 0} \sqrt{k_\parallel^2(z)}\, dz = k_\parallel ; \quad k_\parallel^2(z) = k_\parallel^2 + \frac{1}{\alpha^2}\left(\delta\omega^2 - \delta\omega_0^2(z)\right) \tag{5}$$

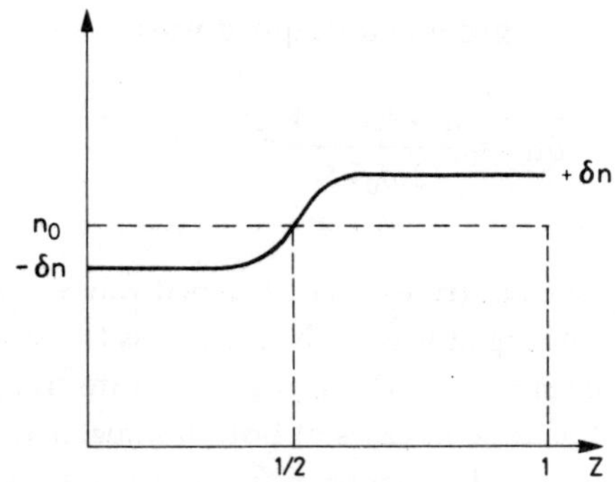

FIG.1. Sketch of density in homogeneity.

For the untrapped waves the integration domain is $0 < z < 1$, and Eq.(5) gives

$$\delta\omega^2 \equiv \delta\omega_f^2 = \frac{\omega_0^4}{4\alpha^2 k_\parallel^2}\left(\frac{\delta n}{n_0}\right)^2 \tag{6}$$

For trapped waves the integration in (5) is over $z < 1/2$. We find

$$\delta\omega^2 \equiv \delta\omega_{tr}^2 = -\,\omega_0^2\,\frac{\delta n}{n_0} + 3\alpha^2 k_\parallel^2 \tag{7}$$

The critical value k_{cr} of $k_\parallel$ dividing the trapped and untrapped waves can be found from the condition of marginal trapping (equalizing (6) and (7)):

$$k_{cr}^2 = \frac{\omega_0^2}{2\alpha^2}\,\frac{\delta n}{n_0} \tag{8}$$

For isotropic distribution of waves, only the average frequency change of waves of given k is of interest:

$$\overline{\delta\omega} \cong \frac{\overline{\delta\omega^2}}{2\omega_0} = \frac{1}{2\omega_0 k}\int_0^k \delta\omega^2(k_\parallel)\,dk_\parallel \tag{9}$$

The integration gives for trapped and untrapped waves, respectively,

$$\overline{\delta\omega_{tr}} = -\frac{k_{cr}^3\alpha^2}{2k\omega_0} \;;\quad \overline{\delta\omega_f} = \frac{\alpha^2 k_{cr}^3(k-k_{cr})}{2\omega_0 k^2} \tag{10}$$

In the case $k \gg k_{cr}$ the contribution of trapped waves is almost cancelled by the contribution of untrapped waves. Nevertheless the sum of these contributions is always negative and the sign comes from trapped waves. Since the number of waves is the same in cases of both homogeneous and inhomogeneous density distributions, the change in frequency is proportional to the change in the energy of the waves δW_w:

$$\delta W_w = -\left(\frac{\delta n}{n_0}\right)^2 \int \frac{\omega_0^2}{8\alpha^2 k^2}\, W_k dk \tag{11}$$

where $W = \int W_k dk$ is the energy density of the waves, and W_k is the spectral energy density of waves in the interval dk. Thus we found (cf. (3) and (11)) that the inhomogeneous density distribution becomes energetically favourable if

$$\int \frac{\omega_0^2}{4\alpha^2 k^2}\, W_k\, dk > \gamma n_0(T_e + T_i) \tag{12}$$

The numerical coefficient appearing in (12) can depend on the kind of $\delta n(z)$ we consider. The order-of-magnitude relation which one finds for Langmuir oscillations is

$$\int \frac{W_k dk}{k^2 r_d^2 n_0 (T_e + T_i)} > 1 \tag{13}$$

It should be mentioned that the way the conditions (12) and (13) are derived does not depend on the phase relations between waves (which could be regular or random waves).

Similar calculations can be performed for a dispersion determined by (2). For simplicity we consider only the case of wave propagation along the density gradient $k^2 = k_\parallel^2$ and suppose that $\beta^2/k^2\omega_0^2 \gg \delta n/n_0$. The change in the energy of the waves is

$$\delta W_w = -\left(\frac{\delta n}{n_0}\right)^2 \frac{3\omega_0^2}{4\beta^2} \int k^2 W_k dk \tag{14}$$

The criterion for modulation instability is then

$$\int \frac{3\omega_0^2}{\beta^2} k^2 W_k dk > \gamma n_0 (T_e + T_i) \tag{15}$$

To find the growth rate of modulation instability one can start with a simple approach, taking into account that the time development of the density inhomogeneity is much slower than that of one plasma oscillation. The δn can be considered adiabatically. Thus for dispersion (1) one has

$$\left(\omega - \omega_0 - \frac{\alpha^2 k^2}{2\omega_0}\right) \vec{E}_{\vec{k},\omega} = \tfrac{1}{2}\omega_0 \frac{\delta n}{n_0} \vec{E}_{\vec{k},\omega} \tag{16}$$

One then transforms this equation into the co-ordinate representation, taking into account that the E field is a longitudinal field $\vec{E}_{\vec{k},\omega} = (\vec{k}/k) E_{\vec{k},\omega}$; $\Delta = \nabla \cdot \nabla$,

$$\nabla \cdot \left(i \frac{\partial \vec{E}(\vec{r},t)}{\partial t} + \frac{\alpha^2}{2} \Delta \vec{E}(\vec{r},t) \right) = \frac{\omega_0}{2} \nabla \cdot \left(\frac{\delta n(\vec{r},t)}{n_0} \vec{E}(\vec{r},t) \right) \tag{17}$$

where $\vec{E}(\vec{r},t)$ is a complex amplitude of the strength of the electric field $\vec{E}(\vec{r},t)\exp(-i\omega_0 t)$. The second equation for $\delta n(\vec{r},t)$ can be derived using the continuity equation and the equation for ion motion:

$$\begin{gathered} \frac{\partial}{\partial t} \delta n(\vec{r},t) = - n_0 \nabla \cdot \vec{v}_i(\vec{r},t) \\ m_i \frac{\partial}{\partial t} \vec{v}_i(\vec{r},t) + \frac{T_e}{n_0} \nabla \delta n(\vec{r},t) = e\vec{E}_p(\vec{r},t) = \nabla \frac{|\vec{E}(\vec{r},t)|^2 e^2}{m_e \omega_0^2} \end{gathered} \tag{18}$$

where $\vec{E}_p(\vec{r},t)$ is the polarization ambipolar field which acts on ions. By neglecting the electron inertia one finds the polarization field. From (18) follows

$$\begin{gathered} \frac{\partial^2}{\partial t^2} \delta n(\vec{r},t) - v_s^2 \Delta \delta n(\vec{r},t) = \Delta \frac{|\vec{E}(\vec{r},t)|^2 \omega_{pe}^2}{4\pi m_i \omega_0^2} \\ v_s^2 = \frac{T_e}{m_i} \end{gathered} \tag{19}$$

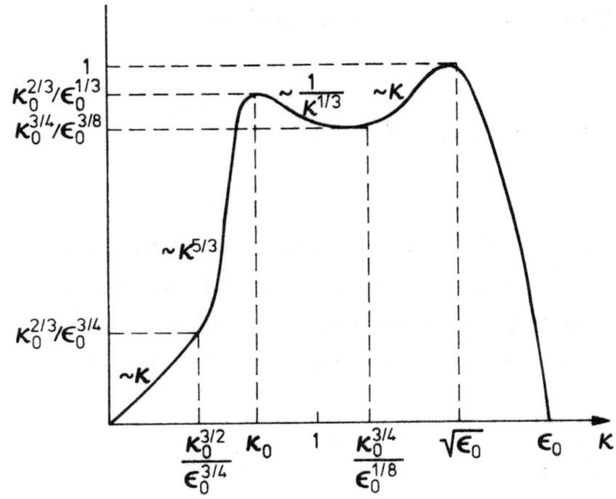

FIG. 2. Dependence of modulation instability growth rate on wavenumber.

The modulation instability can be found from the set of equations (17) and (19), linearizing them in the presence of a pump wave with amplitude E_0 and $\vec{k} = \vec{k}_0$. For the perturbations of fields and δn a dispersion equation is found:

$$1 = \frac{E_0^2}{4\pi n_0 T_e} \cdot \frac{k^2 v_s^2}{(\omega^2 - k^2 v_s^2)} \left\{ \frac{3k^2 v_{Te}^2}{\omega^2} + \frac{\omega_{pe}}{\omega} \, \psi(\vec{k},\vec{k}_0) \right\} \tag{20}$$

where ψ is a function of the order of unity for k of the order of k_0, diminishes as $(k_0/k)^3$ for $k \gg k_0$ and as $(k/k_0)^3$ for $k \ll k_0$ and is also essentially angular-dependent. It vanishes both for $\vec{k} \parallel \vec{k}_0$ and $\vec{k} \perp \vec{k}_0$. Shown in Fig.2 are the growth rates of Langmuir wave modulation instabilities Γ normalized to the maximum growth rate $\Gamma = \gamma/\gamma_{max}$ as a function of the wave number κ, normalized to the characteristic wavenumber $\kappa = k/k_*$:

$$\gamma_{max} = \omega_{pi} \sqrt{\frac{E_0^2}{4\pi n_0 T_e}} \; ; \quad k_* = \frac{\omega_{pe}}{v_{Te}} \sqrt{\frac{\mu}{12}}$$

$$\epsilon_0 = E_0 / \sqrt{4\pi n_0 T_e \mu} \; ; \; \mu = \frac{4m_e}{3m_i} \tag{21}$$

It must be emphasized that the modulation instability of the Langmuir field is automatically set up if the energy is continuously pumped into these waves and the collisional damping of the waves does not play an essential role. Indeed the non-linear Landau damping (induced scattering) in the absence of modulation instability leads to a redshift of Langmuir waves, thus diminishing the k-values up to the values when the threshold of modulation instability is reached (see Eq.(15)). Then the density in homogeneity will develop.

The diminishing size of these density inhomogeneities leads to an increase of the k-values. The numerical 1D simulation of Ref. [19] shows that, indeed, in the presence of a strong regular pump in the domain of small k's, the energy flows towards higher k's and the spectrum $W_k \sim 1/k^2$ is formed. The result of the development of the density inhomogeneities is different in 1D and 3D cases. In a 1D case, solitons can be formed.

3. SOLITON KINETICS

The formation of solitons in a 1D case is accompanied by ion-sound emission which is necessary for increasing the entropy of the plasma. We restrict ourselves here to the case of Langmuir solitons, although the results will be valid for any waves with the dispersion $\alpha^2 k^2$. It is useful to introduce the dimensionless variables:

$$\vec{\xi} = \frac{\vec{r}}{r_d}\sqrt{\frac{\mu}{3}}\ ; \quad \tau = \frac{\omega_{pe}\,\mu t}{2}\ ; \quad \nu = \frac{\delta n}{n_0 \mu}\ ; \quad \epsilon = \frac{E}{\sqrt{4\pi n_0 T_e \mu}}\ ; \quad \mu = \frac{4m_e}{3m_i} \tag{22}$$

Then Eqs (17) and (19) in the 3D case have the form:

$$\begin{aligned} &\frac{\partial}{\partial \vec{\xi}} \cdot \left[i\frac{\partial \vec{\epsilon}}{\partial \tau} + \left(\frac{\partial}{\partial \vec{\xi}} \cdot \frac{\partial}{\partial \vec{\xi}}\right)\vec{\epsilon}\right] = \frac{\partial}{\partial \vec{\xi}} \cdot (\nu \vec{\epsilon}) \\ &\frac{\partial^2 \nu}{\partial \tau^2} - \left(\frac{\partial}{\partial \vec{\xi}} \cdot \frac{\partial}{\partial \vec{\xi}}\right)\nu = \left(\frac{\partial}{\partial \vec{\xi}} \cdot \frac{\partial}{\partial \vec{\xi}}\right)|\vec{\epsilon}\,|^2 \end{aligned} \tag{23}$$

The velocities in the variables used are measured in units of ion-sound velocity. We define a soliton as an exact 1D non-linear solution of (23), whose density perturbation and amplitude of the electric field move with a constant velocity u and they both vanish for $\xi \to \pm\infty$. We find from (23)

$$\nu = -\frac{1}{1-u^2}\,|\epsilon|^2 = -\lambda|\epsilon|^2\ ; \ \lambda = \frac{1}{1-u^2} \tag{24}$$

Equation (23) is not valid for $u < \sqrt{T_i/T_e}$ (see below). In this case one finds $\lambda = T_e/(T_e + T_i)$. Since to trap the Langmuir waves it is necessary to have a density depletion, the solitons exist only for subsonic velocities. The soliton solution of (23) has the form [20]:

$$\epsilon = \frac{\epsilon_0 \exp\left[-i\Omega t + i\frac{u}{2}(\xi - u\tau)\right]}{\cos h\left[\epsilon_0\sqrt{\frac{\lambda}{2}}(\xi - u\tau)\right]} \; ; \quad \Omega = -\frac{u^2}{4} - \frac{\lambda}{2}\epsilon_0^2 \tag{25}$$

Certain integrals of Eq.(23) exist. These are the conservation of number of quanta N, the conservation of energy H and the conservation of momentum $\vec{P}$. One finds

$$H = \int \left[\left|\frac{\partial}{\partial\vec{\xi}}\cdot\vec{\epsilon}\right|^2 + \nu\,|\vec{\epsilon}\,|^2 + \frac{\nu^2 + \vec{v}^{\,2}}{2}\right] d\vec{\xi} \tag{26}$$

$$\vec{P} = \frac{1}{2}\int \left[i\epsilon^*\left(\frac{\partial}{\partial\vec{\xi}}\cdot\vec{\epsilon}\right) - \vec{\epsilon}\,i\left(\frac{\partial}{\partial\vec{\xi}}\cdot\vec{\epsilon}^{\,*}\right) + 2\nu\vec{v}\right] d\vec{\xi}$$

$$N = \int |\vec{\epsilon}\,|^2 d\vec{\xi} \; ; \quad \frac{\partial\nu}{\partial\tau} = -\frac{\partial}{\partial\vec{\xi}}\cdot\vec{v} \tag{27}$$

For a soliton with $\epsilon_0 \gg 1$ we find

$$H = -\frac{\epsilon_0^3\lambda\sqrt{2\lambda}}{3}(1 - 5u^2) \; ; \quad P = \frac{4}{3}u\epsilon_0^3\lambda\sqrt{2\lambda} \quad ; \quad N = 2\sqrt{\frac{2}{\lambda}}\,\epsilon_0 \tag{28}$$

It is easy to understand why for the 1D case the balance between the pressure and Langmuir field ponderomotive force can exist. Conservation of number of quanta means $|\epsilon|^2\xi_0 = \text{const}$, $|\epsilon|^2 \sim 1/\xi_0$, where ξ_0 is the characteristic size of the density depletion, the ponderomotive force $F_\epsilon \sim \partial/\partial\xi|\epsilon|^2 \sim 1/\xi_0^2$. The condition for wave trapping, $\delta n/n_0 \approx k^2 r_d^2$, means $\delta n \sim 1/\xi_0^2$ and the pressure force $F_T \sim (\partial/\partial\xi)T\delta n \sim 1/\xi_0^3$. Thus the size ξ_0 is diminishing up to the value when the pressure force prevents a further decrease in ξ_0. The balance established is stable, since any increase in ξ_0 from equilibria value will lead to the domination of the ponderomotive force which acts to diminish the ξ_0, and the decrease in ξ_0 from the equilibria value will increase the pressure force and thus increase the ξ_0.

The solitons are well localized in space, so one can imagine an ensemble of separated solitons. If the distance between them is essentially larger than their width they can be considered as a soliton gas. Such a picture was proposed by Kingsep, Rudakov and Sudan [21]. Let us analyse it, starting with a rough and simple model. The solitons could interact if they collide with each other. The numerical calculations of soliton interaction using Eq.(23) show that the solitons can merge with each other [22]. The merging process is possible only if the amplitudes of two solitons are large, $\epsilon_0 \gg 1$, and almost equal [22, 23]. Let us consider a model in which the ensemble of solitons consists of solitons which are the result of merging solitons with the amplitude $\epsilon_0^{(0)}$, i.e. $\epsilon_0^n = 2^n \epsilon_0^{(0)}$, where n is an integer number.

The last statement follows from the conservation of number of quanta during merging, confirmed by the calculations performed in Ref.[23]. Indeed, in merging two solitons with equal amplitudes ϵ_0, the new single soliton will have the amplitude $\epsilon_1 = 2\epsilon_0$. Using the conservation of energy $H_0 \sim -\epsilon_0^3$, one can find that the merging of solitons is possible only if ion-sound waves are emitted. Indeed $H_1 \sim -\epsilon_1^3 = -8\epsilon_0^3$, i.e. $H_1 = 8H_0$. The initial energy of the two solitons was $2H_0$. Thus the difference between the final and the initial soliton energy is $H_1 - 2H_0 = 6H_0$. This energy is negative since $H_0 < 0$. The same amount of positive energy, $-6H_0$, should be taken by ion-sound waves.

We now introduce the distribution function of solitons F_n which is the number of solitons with amplitude ϵ_0^n per unit length. The problem of non-linear kinetics of the soliton gas is the problem of finding a kinetic equation for F_n. We try to find it while not fixing the constant which determines the cross-section of merging. Let us count the rate $(\partial F_n/\partial t)_-$ due to the solitons which are leaving the state n. We should take into account both the merging of solitons and their splitting by sound waves, which is the reverse process of merging. Since the probability of soliton collision is proportional to F_n^2, and the merging leads to loss of the two solitons from the state n, we have $(\partial F_n/\partial t)_{-,merg} \sim -2F_n^2$. The sound waves with sufficient energy to split the soliton in the state n are emitted in the merging of solitons in the states n + k, where $k \geqslant 1$. The amount of energy emitted in sound waves due to the merging of solitons in the state n + k is $-6H_{n+k} = -6 \times 8^k H_n$. Only half of it, $-3 \times 8^k H_n$, is available for splitting since the sound waves have both positive and negative density variation, but the solitons have only negative energy variation. This assumption was verified in the numerical calculations of Ref.[22]. The splitting of solitons with the amplitude ϵ_0^n leads to the formation of two solitons with the amplitude $\epsilon_0^n/2$. The energy needed for splitting is proportional to $-(\epsilon_0^n)^3 + 2(\epsilon_0^n/2)^3$, i.e. it is $(-3/4)H_n$.

Thus the number m of splittings produced by sound waves emitted in the state n + k is $-3 \times 8^k H_n = m((-3/4)H_n)$, i.e. $m = 4 \times 8^k$, and one finds the rate at which the solitons escape from the state n:

$$\left(\frac{\partial F_n}{\partial t}\right)_- \sim - 2F_n^2 - \sum_{k=1}^{\infty} 4 \times 8^k F_{n+k}^2 \tag{29}$$

Similar calculations also give the rate at which the solitons reach the state

$$\left(\frac{\partial F_n}{\partial t}\right)_+ \sim F_{n-1}^2 + \sum_{k=1}^{\infty} 8^{k+1} F_{n+k+1}^2 \tag{30}$$

The equation

$$\frac{dF_n}{dt} = \left(\frac{\partial F_n}{\partial t}\right)_- + \left(\frac{\partial F_n}{\partial t}\right)_+ \tag{31}$$

is an example of an equation describing non-linear kinetics. Let us search for the stationary distribution of the type

$$F_n \sim \frac{1}{(\epsilon_0^n)^\gamma} = \frac{1}{2^{n\gamma}(\epsilon_0^{(0)})^\gamma}$$

For $\gamma > \frac{3}{2}$ it is possible to sum up the series in (29) and (30) and find that γ should satisfy the equation

$$y^3 - 2y^2 - 12y - 8 = 0 \;\; ; \;\; y = 8 \times 2^{-2\gamma} < 1 \tag{32}$$

The approximate solution of (32) with an accuracy of three figures is [41]

$$\gamma \cong \frac{\ell n\, 90/7}{2\, \ell n\, 2} \cong 1.84 \tag{33}$$

This is obviously rough, but at least it gives a picture of the diffusion of solitons in the amplitudes ϵ_0 toward higher amplitudes. The larger the amplitude of the soliton the less its width and the larger the value of the effective k-number. It is simple to see that the space Fourier-components of the field of solitons has the same value up to $k_{max} \sim \epsilon_0^n$ for solitons of any amplitude. Thus to calculate the total energy of solitons for a given k, one needs to count the number of solitons with $n > n_k$ where $\epsilon_0^{n_k} \sim k$, i.e.

$$W_k \sim \sum_{n_k} F_n \sim \frac{1}{k^\gamma}$$

The spectrum $\gamma = 1.84$ is close to the result of the numerical simulation [19], $\gamma = 2$, although the model used was rough. The energy is diffused to larger k's until it reaches the tail of the thermal particles and will be absorbed by fast particles. But even before the interaction with resonant particles – linear Landau damping – non-linear Landau damping can influence the diffusion of the energy to larger k's [24–26]. Non-linear Landau damping is scattering by electrons moving through the soliton. This process should diminish the frequency and lead to a decrease in the group velocity of the waves in the soliton, thus stopping the soliton (see [32]). The number of quanta in the process of non-linear Landau damping is conserved, and thus the amplitude of the soliton is approximately conserved. Two consequences are possible due to soliton stopping:

(1) In the absence of a constant pump of energy in the waves, the stopping of solitons will decrease the rate of collisions. The mean distance between solitons is enlarged by merging. When the mean free path for soliton stopping becomes of the order of the mean distance between the solitons, the merging stops. Thus one should expect a rapid decrease of the spectra W_k for some $k > k_N$, where k_N is determined by the equalization of the mean free path for stopping with the mean distance between the solitons.

(2) In the presence of a constant pump, the spectra should be changed essentially for $k > k_N$ also. But instead of having the spectra W_k reduced, one should expect a higher spectral energy density than in the absence of non-linear Landau damping. Indeed, for $k > k_N$ the solitons will be accumulated until the mean distance between solitons is equal to the mean free path which is in inverse proportion to ϵ_0^n [24, 25]. Thus

$$F_n \sim \epsilon_0^n \quad ; \qquad W_k \sim \sum_{n_k}^{n_{max}} \epsilon_0^n \sim \epsilon_0^{max} - \epsilon_0^{n_k} \sim k_{max} - k$$

The n_{max} is determined by linear Landau damping on fast particles. The final result of soliton diffusion is the absorption on fast particles and the creation of the fast particle tail. Thus the model described gives an example of a mechanism working as the energy sink of Langmuir waves in fast particles.

4. CAVITON KINETICS

The 3D development of modulational instability is not so well understood, since the balance between the pressure force and the Langmuir wave ponderomotive force is not possible. This statement is, strictly speaking, valid only for a cavity the sizes of which in all three dimensions are of the same order of magnitude, i.e. $\xi_1 \approx \xi_2 \approx \xi_3 \approx \xi_0$. The conservation of number of quanta gives $|\epsilon|^2 \xi_0^3 = \text{const}$ and $F_\epsilon \sim (\partial/\partial\xi)|\epsilon|^2 \sim 1/\xi_0^4$ while $F_T \sim T(\partial/\partial\xi)\delta n \sim 1/\xi_0^3$. Thus the smaller ξ_0 the larger the ponderomotive force, and the pressure force cannot stop the self-contraction. The non-stationary development of the modulational contraction leads to a possibility of contraction of density depletion up to a size of the order of a Debye radius. In this connection the problem of so-called Langmuir collapse has been raised [27]. During this contraction some self-similar motions can exist for finite time intervals. One of the set of self-similar motions can be changed to another.

Let us start with definitions. We define a caviton as a self-similar non-stationary process of the field self-contraction. We say that the collapse is the process of modulational field self-contraction which develops finally up to the size of Debye radius. The cavitons can exist on time scales much less than the time of the collapse, and the collapse (if it exists) can be formed by a set of self-similar solutions, i.e. by a set of cavitons which replace one another. The problem is whether collapse is possible or not. In this connection the questions arise:

(1) Whether during self-contraction the characteristic sizes of the cavity in all three directions should be of the same order of magnitude;

(2) Whether the number of the quanta N in the cavity must be constant or, in other words, whether the cavity can lose the quanta during contraction;

(3) Whether one self-similar solution (e.g. subsonic) can be simply converted to another self-similar solution (e.g. supersonic) or some additional irreversible processes (such as ion-sound emission) can occur in the intermediate stage;

(4) Whether the development of a hierarchy of modulational disturbances similar to that of eddies in an incompressible liquid is possible.

The complete answers to these questions do not at present exist, although there are some indications of the kind of answers one can expect to have in the future.

Let us first take up the question of spherical subsonic motions. In Refs [27, 28] it was shown that the characteristic size ξ_0^2 of any subsonic cavity with initially negative H tends to zero in a finite time interval. The real problems are:

(1) Whether the motion becomes supersonic in a finite time interval before ξ_0^2 tends to zero (further analyses have shown that this is indeed the case);

(2) Whether the spherical symmetry is broken in a finite time interval during the self-contraction (further investigation has shown that the subsonic cavitons are usually not spherically symmetric).

Thus one has not proceeded much further from the simple argument that in the 3D case the balance between the ponderomotive force and the pressure force is not possible for a cavity with sizes of the same order of magnitude in all three dimensions.

Note that near the threshold the modulation instability behaves similarly to the gravitation instability. This analogy helps our intuitive understanding of what can happen in the 3D case. Obviously the gravitational contraction cannot go at the same rate in all three directions. Even a small difference in the initial contraction along one of the axes will lead to a structure similar to a pancake which is nothing but the formation of galaxies. Although rotation helps in the formation of such structures, pancake structures should be formed even in the absence of rotation. The further development of the gravitational instability can lead to galaxy arm formation.[2] In the 3D development of modulation instability the formation of a pancake structure is also quite probable. Indeed, a balance between the pressure force and the ponderomotive force tends to be established, but this balance is possible in one direction only. The pancake structure should then continue to contract due to forces acting on its edges. Numerical simulations have verified this picture. The contracting entity with a balance established in one direction can be called a quasi-soliton since its parameters vary slowly compared to the time of the balance established in the direction perpendicular to its plane. It was first shown in Ref.[26] that the spherical contraction develops to a spherical quasi-soliton. The use of conservation of number of quanta N and the energy H leads [26] to the law of the change of the distance of the spherical layers $R(\tau)$ from the centre $R(\tau) \to \tau/\sqrt{5}$, i.e. the velocity reaches $1/\sqrt{5}$ of the velocity of sound, and $\epsilon_0 \to 1/\tau^2$, i.e. the thickness of the layer diminishes faster than the radius. This is a good example of all the motion being subsonic and not converted to a supersonic motion.

---

[2] The theory of galaxy arms as solitons was developed in Refs [29 – 31].

Spherical quasi-solitons, however, pose the following problems:

(1) Azimuthal modulation instability can develop during the process of contraction and thus the spherical symmetry will be lost;

(2) It is not certain whether or not sound emission can take essential energy from the quasi-soliton during contraction.

If the ion-sound emission takes place it is not possible to use the conservation of energy H as in Ref.[26]. The statement in [33] is that to take the sound emission into account one should use the conservation of the Lagrangian L instead of H. The analyses [34, 35] show that for some initial R(0) the contraction is converted to an expansion and the energy and momentum are taken by the ion-sound waves emitted. This could be an example of an irreversible process preventing the collapse discussed above, but the problem needs a further, more detailed, analysis.

Let us consider a more realistic model of a pancake type subsonic self-similar motion, i.e. a subsonic caviton. Let us suppose that the number of quanta in the caviton is conserved,

$$\int |\vec{\epsilon}\,|^2\, d\vec{\xi} \cong \pi R^2(\tau)\xi_0(\tau)|\epsilon_0(\tau)|^2 = \text{const} \tag{34}$$

where $\xi_0(\tau)$ is the thickness of the pancake shape of the caviton and $R(\tau)$ its radius. The subsonic motions are described by the equation

$$\frac{\partial}{\partial \xi}\cdot\left(i\frac{\partial \vec{\epsilon}}{\partial \tau} + \left(\frac{\partial}{\partial \vec{\xi}}\cdot\frac{\partial}{\partial \vec{\xi}}\right)\vec{\epsilon}\right) = -\frac{\partial}{\partial \vec{\xi}}\cdot(\vec{\epsilon}\,|\vec{\epsilon}\,|^2) \tag{35}$$

Neglecting in the first approximation the term with the time derivative, we find an estimation

$$\frac{\partial}{\partial \vec{\xi}}\cdot\frac{\partial}{\partial \vec{\xi}}\,\epsilon \cong \frac{\epsilon_0}{\xi_0^2} \quad ; \quad \epsilon_0^2 \sim \frac{1}{\xi_0^2}$$

and, using (34), $R^2(\tau) \sim \xi_0(\tau)$. Thus, in time, the pancake structure is more and more pronounced. Supposing the phase of the field is changing essentially (about 2π) only in the distance R, we find in the next approximation $R\xi_0 \sim \tau$ or $\xi_0^3 \sim \tau^2$, $\xi_0 \sim \tau^{2/3}$, $R \sim \tau^{1/3}$. This gives a self-similar solution (caviton) [26]:

$$\epsilon = \frac{1}{\tau^{2/3}} \phi\left(\frac{\xi}{\tau^{2/3}}, \frac{\vec{R}}{\tau^{1/3}}\right) \tag{36}$$

where ϕ is a function of ξ and $\vec{R}$ determined by the initial conditions. The τ in (36) is time counted in the past from $\tau = 0$, which is the time at which the singularity can occur. But really the singularity in (36) cannot occur since Eq.(36) is valid only for subsonic motions. Equation (36) shows that the characteristic velocity is increasing in time. Indeed

$$u \cong \frac{\xi}{\tau} \sim \frac{\tau^{2/3}}{\tau} \sim \left(\frac{1}{\tau}\right)^{1/3}$$

The motion becomes supersonic for $\tau < 1$. Thus (36) presents a self-similar solution valid only for a finite time interval, i.e. a subsonic caviton. The density variations reached at $\tau \cong 1$ are still very small:

$$\frac{\delta n}{n_0} \cong \frac{m_e}{m_i}$$

A different self-similar solution of (36) can even be found by supposing that all the terms in (36) are of the same order of magnitude:

$$\epsilon = \frac{1}{\tau^{1/4}} \phi\left(\frac{\xi}{\tau^{1/2}}, \frac{\vec{R}}{\tau^{1/4}}\right) \tag{37}$$

This caviton does not conserve the number of quanta $N \sim \sqrt{\tau} \to 0$, i.e. the caviton emits quanta during contraction. The numerical calculations performed to date give an indication of the appearance of the cavitons (36) but not (37). The essential point is that both cavitons (36) and (37) convert themselves to a supersonic motion at $\tau < 1$. Strictly speaking, we know only the tendency since the intermediate stage τ of the order of unity is not investigated in detail. Let us now consider a supersonic caviton $u \gg 1$. The numerical investigation (see e.g. [26]) shows that in this case the pancake structure is not very pronounced, i.e. R is larger than ξ_0, but their ratio is not enlarged in time ($R/\xi_0 \sim 3$). So we shall consider for simplicity the case when all the dimensions of the caviton are of the same order of magnitude. Since the motion is supersonic, the $\partial^2\nu/\partial\tau^2$ term in the equation of density variation dominates, and, by order of magnitude, $\nu \cong \tau^2 |\epsilon|^2/\xi^2$. Neglecting the term $i(\partial/\partial\tau)$ in the equation for the field, we have

$|\epsilon|^2 \sim 1/\tau^2$. Supposing $N \sim |\epsilon|^2 \xi_0^3 = \text{const}$, we find $\xi \sim \tau^{2/3}$. This gives a self-similar solution:

$$\epsilon = \frac{1}{\tau}\,\phi_\epsilon\left(\frac{\vec{\xi}}{\tau^{2/3}}\right) \; ; \; \nu = \frac{1}{\tau^{4/3}}\,\phi_\nu\left(\frac{\vec{\xi}}{\tau^{2/3}}\right) \tag{38}$$

The neglected terms are small for $\tau \ll 1$ if the phases ψ of all the waves tend to be equal, $\psi \to \tau^{1/3} \to 0$. This phenomenon of self-phasing is the most characteristic for the caviton described by (38). It is similar to the phenomenon for the explosive instability in the three-wave interactions. The problems connected with the caviton (38) are the following:

(1) Whether or not the emission of ion-sound waves plays an essential role during the contraction;

(2) Whether or not the supersonic caviton (38) is stable in the frame of the equation used and whether or not the additional small terms can change its behaviour essentially;

(3) What is the domain of validity of the equation used to find the caviton (38) and whether or not the additional terms not taken into account can stop or change the behaviour of self-contraction.

We shall consider the last problem in the next section. The first problem was considered in Refs [25, 30]. Note that one can expect the process of emission to be essential since the motion is supersonic. In [25], perturbation theory was used to calculate the total energy emitted in ion-sound waves during the total time of contraction described by (38). There are two possible cases when the total energy emitted is large: (a) when it is of the order of H, i.e. the initial energy of the caviton; and (b) when it is of the order of $N\omega_{pe}$, i.e. the total electric field energy of the caviton. When the energy emitted is of the order of H, the process of emission can increase the rate of self-contraction. Indeed, the greater the amount of energy emitted in S-waves that is positive, the larger is the absolute value of H, since H is negative. When the total energy emitted is of the order of $N\omega_{pe}$ the solution (38) will be a contradictory one, since in deriving it we start with the conservation of the number of quanta. The perturbation theory of Ref.[25] shows that the total energy emitted is of the order of $N\omega_{pe}$.

The second problem was taken up in [26] by including in the equations a small term due to Landau damping. The authors observed in the computations a sudden change in the behaviour of the solution, which becomes subsonic. The other small terms seem to do the same thing.

One can also ask whether another self-similar solution which does not conserve the N can exist.

Searching for a solution of the type

$$\epsilon = |\epsilon|\, e^{i\psi} \quad ; \quad |\epsilon| = \tau^{\mu}\phi\left(\frac{\vec{\xi}}{\tau^{\nu}}\right) \quad ; \quad \psi = \tau^{\sigma}\phi\left(\frac{\vec{\xi}}{\tau^{\nu}}\right)$$

in the case when all the terms of the equation for the field are of the same order of magnitude (not neglecting the $i(\partial\epsilon/\partial\tau)$ term) one finds a caviton [34]:

$$|\epsilon| = \frac{1}{\tau}\,\phi_{\epsilon}\left(\frac{\vec{\xi}}{\sqrt{\tau}}\right) \; ; \; \psi = \phi_{\psi}\left(\frac{\vec{\xi}}{\sqrt{\tau}}\right) \tag{39}$$

which does not conserve the number of quanta $N \sim 1/\sqrt{\tau}$. One concludes then that in the case when a source of quanta does not exist in the centre of the cavity, Eq.(39) can present only the expansion process. It is uncertain whether or not the caviton (38) describing self-contraction due to some instability, small additional non-linearity, or sound emission will convert itself to the expansion caviton (39). In the expansion the velocity $\xi/\tau \sim 1/\sqrt{\tau}$ is decreased (since τ increases) and the process (39) can proceed only to the stage when motion becomes subsonic. The question is open as to whether such a simultaneous substitution of a supersonic by a subsonic caviton with oscillation around sound velocity can indeed happen. The numerical analysis [26] indicates that the motion becomes near-sonic, i.e. it proceeds with velocities close to but less than the sound velocity. One can find a near-sonic caviton by assuming $\xi \approx \tau$, which gives $\nu \cong -|\epsilon|^2$; $|\epsilon|^2 R^2 \xi = \text{const}$; $|\epsilon|^2 \sim 1/\xi^2$, i.e. $R^2 \sim \xi \sim \tau$;

$$|\epsilon| = \frac{1}{\tau^2}\,\phi\left(\frac{\xi}{\tau}, \frac{\vec{R}}{\sqrt{\tau}}\right) \tag{39'}$$

The questions that need an answer for a more detailed understanding of the pancake caviton (39′) are:

(1) What kind of mechanism keeps the motion near-sonic?

(2) Is the ion-sound emission important or not?

(3) What is the role of ion-non-linearities, i.e. steepening of the density profiles?

(4) Since the motion is slightly subsonic, the sound waves emitted by one caviton can reach another, and the question arises whether the splitting and merging of cavitons can occur in the same way as described for the 1D solitons.

5. GENERAL KINETIC THEORY OF MODULATION INTERACTIONS

It is possible to construct a general kinetic theory of modulation interactions [36, 37] and find the domain of validity of Eqs (17) and (19) as well as all necessary corrections to these equations. We shall show that the standard theory [1, 2] of non-linear plasma responses found by perturbation in the strength of the electric field leads to general equations which describe the kinetic effects in modulation interactions. The difference between the non-linear equations for wave-wave interactions and the equations which we find is that the $\Delta\omega$ – the difference of the frequencies of the two interacting fields E_{k_1} and E_{k_2} – should not be fixed arbitrarily to be equal to the difference of the frequencies of the waves $\Delta\omega = \omega_{k_1} - \omega_{k_2}$ but should be found as a result of the solution of the equations. The non-linear frequency shift is larger than the frequencies of the waves if the threshold of modulation instability is reached. Thus $\Delta\omega$ should be determined by the process itself. The use of two first terms in the expansion in fields is possible since the non-linear responses are regular functions of $\Delta\omega$, even diminishing as $\Delta\omega$ increases. The starting equation cubic in E is a standard one [2]:

$$ik\epsilon_k E_1 = 4\pi \int S_{1,2} E_1 E_2 d_{1,2} + 4\pi \int \Sigma_{1,2,3} E_1 E_2 E_3 d_{1,2,3} \tag{40}$$

where

$$E_1 = E_{k_1},\ E_2 = E_{k_2},\ E_3 = E_{k_3};\ k = \{\vec{k}, \omega\}$$

$$dk = d\vec{k}\, d\omega\ ;\ d_{1,2} = \delta(k - k_1 - k_2)\, dk_1 dk_2$$

$$d_{1,2,3} = \delta(k - k_1 - k_2 - k_3)\, dk_1 dk_2 dk_3\ ;\ \epsilon_1 = \epsilon_k$$

$$n = \frac{\vec{k}}{k},\ \lambda = \frac{1}{\omega - \vec{k}\vec{v}}$$

and

$$S_{1,2} = -\frac{e^3}{2} \int \lambda \left\{ \left(\vec{n}_1 \cdot \frac{\partial}{\partial \vec{p}}\right) \lambda_2 \left(\vec{n}_2 \cdot \frac{\partial}{\partial \vec{p}}\right) + \left(\vec{n}_2 \cdot \frac{\partial}{\partial \vec{p}}\right) \right.$$

$$\left. \times\ \lambda_1 \left(\vec{n}_1 \cdot \frac{\partial}{\partial \vec{p}}\right)\right\} \frac{\phi\, d\vec{p}}{(2\pi)^3} \tag{41}$$

$$\Sigma_{1,2,3} = -\frac{e^4}{2i}\int \lambda\left(\vec{n}_1 \cdot \frac{\partial}{\partial \vec{p}}\right)\lambda_{2+3}\left\{\left(\vec{n}_2 \cdot \frac{\partial}{\partial \vec{p}}\right)\lambda_3\left(\vec{n}_3 \cdot \frac{\partial}{\partial \vec{p}}\right)\right.$$

$$\left. + \left(\vec{n}_3 \cdot \frac{\partial}{\partial \vec{p}}\right)\lambda_2\left(\vec{n}_2 \cdot \frac{\partial}{\partial \vec{p}}\right)\right\}\phi \frac{d\vec{p}}{(2\pi)^3} \quad (42)$$

We suppose that $\Delta\omega$ is arbitrary but $|\Delta\omega| \ll \omega_{pe}$, and introduce the positive part of the field E^+ with the frequency close to ω_{pe} and the negative part of the field with the frequency close to $-\omega_{pe}$. Then we write Eq.(40) for E^+ on the left-hand side. The δ-function in the first term on the right-hand side of this equation gives $\omega_{pe} \cong \omega_1 + \omega_2$, i.e. only one of the fields could be the Langmuir field. The second one should be either a low-frequency field ($\omega_{pe} \cong \omega_{pe} + \omega_{low\ freq.}$) or the double plasma frequency field ($\omega_{pe} = -\omega_{pe} + 2\omega_{pe}$). We shall call these the virtual fields. For the second term on the right-hand side of (40), all the three fields can be Langmuir fields since the conservation law $\omega_{pe} \cong \omega_1 + \omega_2 + \omega_3$ can be satisfied by the two possible substitutions $\omega_{pe} = \omega_{pe} + \omega_{pe} - \omega_{pe}$ and $\omega_{pe} = -\omega_{pe} + \omega_{pe} + \omega_{pe}$ (note that $\Sigma_{1,2,3}$ is symmetric on the change in indices $2 \rightleftarrows 3$). Since the amplitude of the virtual fields is small compared to the Langmuir field, it is possible to use Eq.(40) with only the first term on the right-hand side to find the virtual fields through the Langmuir field. This is possible since $\omega_{low\ freq.} \approx \omega_{pe} - \omega_{pe}$, $2\omega_{pe} = \omega_{pe} + \omega_{pe}$.

This is how we find an equation which contains only the Langmuir fields:

$$\epsilon E^+ = \int \Sigma^{eff}_{1,2,3} E_1^+ E_2^+ E_3^- d_{1,2,3} \quad , \quad \Sigma^{eff}_{1,2,3} = \widetilde{\Sigma}_{1,2,3} + \widetilde{\Sigma}'_{1,2,3} \; ; \; g_1 = \frac{4\pi}{i\epsilon_1 k_1} \quad (43)$$

$$\widetilde{\Sigma}_{1,2,3} = \frac{8\pi}{ik}\left(\Sigma_{1,2,3} + 2S_{1,2+3}\, g_{2+3}\, S_{2,3}\right) \quad (44)$$

$$\widetilde{\Sigma}'_{1,2,3} = \frac{4\pi}{ik}\left(\Sigma_{3,1,2} + 2S_{3,1+2}\, g_{1+2}\, S_{1,2}\right) \quad (45)$$

The $\widetilde{\Sigma}$ describes the processes going on through the low-frequency virtual field, and $\widetilde{\Sigma}'$ describes the processes through the double plasma frequency wave. We can proceed further by using an expansion of the non-linear responses in the natural small parameter

$$\alpha = \frac{\Delta\omega}{\omega_{pe}} \quad \text{and} \quad \beta = \frac{k^2 v_{Te}^2}{\omega_{pe}^2}$$

Then $\widetilde{\Sigma}'$ is less than $\widetilde{\Sigma}$ by a factor β and we find

$$\Sigma^{\text{eff}} \cong \widetilde{\Sigma}_0 = -\frac{1}{4\pi n_0 T^{\text{eff}}_{2+3}} \frac{\epsilon^{(i)}_{2+3}}{\epsilon_{2+3}} \frac{(\vec{k}\cdot\vec{k}_1)(\vec{k}_2\cdot\vec{k}_3)}{k k_1 k_2 k_3} \tag{46}$$

where T^{eff}_{2+3} is the effective temperature

$$\frac{1}{T^{\text{eff}}_{2+3}} = \frac{(\epsilon^{(e)}_{2+3} - 1)(\vec{k}_2 + \vec{k}_3)^2}{\omega^2_{pe}\; m_e}\; ; \; \omega_2 + \omega_3 = \Delta\omega_2 + \Delta\omega_3 \tag{47}$$

and $\epsilon^{(i)}$ and $\epsilon^{(e)}$ are the ion and electron part of $\epsilon = \epsilon^{(e)} + \epsilon^{(i)} - 1$.

Note that the ratio α/β is arbitrary in (46). In the weak non-linear theory or weak turbulence $\alpha/\beta \approx 1$. The equations derived from (43) and (46) are more general than those from (17) and (19). First of all, they are valid for any non-Maxwellian distributions of electrons and ions and take into account the kinetic effects such as non-linear and linear Landau damping. The equations (17) and (19) can be derived from (43) and (46) if the following assumptions are made:

(1) The imaginary parts of $\epsilon^{(i)}_{2+3}$ and $\epsilon^{(e)}_{2+3}$ are neglected;
(2) The distribution functions are Maxwellian and $T^{\text{eff}}_{2+3} \cong T_e$;
(3) The real parts of $\epsilon^{(i)}_{2+3}$ and $\epsilon^{(e)}_{2+3}$ are essentially larger than unity and

$$|\vec{k}_2 + \vec{k}_3|\, v_{Ti} \ll \Delta\omega_2 + \Delta\omega_3 \ll |\vec{k}_2 + \vec{k}_3|\, v_{Te}$$

Then defining a density variation $\delta n(\vec{r},t)$ and the amplitude of the electric field $\vec{E}(\vec{r},t)$ by the relations

$$\frac{\delta n(\vec{r},t)}{n_0} = \int \frac{\delta n_k}{n_0}\, dk \exp[i(\vec{k}\vec{r} - \omega t)]$$

$$\vec{E}(\vec{r},t) = \int \frac{\vec{k}}{k} E^+_k\, dk \exp[i(\vec{k}\vec{r} - \omega t + \omega_{pe} t)] \tag{48}$$

$$\frac{\delta n_{k_1}}{n_0} = \frac{\epsilon^{(i)}_{k_1}}{\epsilon_{k_1}} \int \frac{(\vec{k}_2\cdot\vec{k}_3)}{k_2 k_3} \frac{E^+_{k_2} E^-_{k_3}}{4\pi n_0 T_e} \delta(k_1 - k_2 - k_3)\, dk_2\, dk_3 \tag{49}$$

we find that Eqs (43) and (49) are converted to Eqs (17) and (19). The advantages of this approach are:

(1) A general and systematic method exists in deriving the equations for the Langmuir field valid for any relation between the dispersion corrections and the non-linear frequency shift;

(2) It is simple to have a generalization for arbitrary particle distributions taking into account various kinetic effects;

(3) It is possible to find the necessary corrections to the simplest set of Eqs (17) and (19) derived in the first approximation in the parameters α and β;

(4) The relations of the modulation interactions to the other non-linear processes can be easily understood.

Let us start with the last problem. The two known non-linear processes, the decay process and non-linear Landau damping, are indeed included in (46). The imaginary parts of $\epsilon^{(i)}_{2+3}$ and $\epsilon^{(e)}_{2+3}$ describe non-linear Landau damping. This process will lead to the soliton-breaking already discussed above. The processes of sound emission discussed are no more than the decay processes and are described by the zeros of ϵ_{2+3} in (46).

Let us take up the problem of corrections to Eqs (17) and (19) [36]. The presence of the $\epsilon^{(i)}/\epsilon$ term in (46) means a severe cancellation of the contributions of the two terms in (44) appearing correspondingly from the cubic non-linear plasma response and from the iteration of a quadratic non-linear plasma response. Because of this cancellation, the terms of further expansion in the parameters α and β can be of the same order of magnitude[3] as that taken into account in (46). The corrections are due to: (a) the interactions through the virtual field with frequencies close to the double plasma frequency, (b) the further expansion in α and β of the interaction through the low-frequency virtual field, and (c) the terms coming from the partial loss of local quasineutrality. The result of calculating the sum of these in the 1D case is $\delta\Sigma^{\mathrm{eff}} = 0$ but in the 3D case $\delta\Sigma^{\mathrm{eff}} \neq 0$. This means that the corrections are not essential for 1D solitons and decrease in time for the pancake quasisolitons in the 3D case. For supersonic cavitons the pancake structure is not pronounced and one finds [36, 38] that the corrections are essential even for

$$\frac{\delta n}{n_0} > \left(\frac{m_e}{m_i}\right)^{1/3} \; ; \; r < r_d \left(\frac{m_i}{m_e}\right)^{1/6}$$

[3] In each of the two terms (44).

This means that the supersonic caviton cannot lead to a collapse. Even before reaching

$$\frac{\delta n}{n_0} \approx \left(\frac{m_e}{m_i}\right)^{1/3}$$

the supersonic caviton can become unstable due to the corrections $\delta\Sigma^{eff}$ which are still small. If indeed the supersonic caviton is then converted to a near-sonic one,the corrections $\delta\Sigma^{eff}$ become less important since the near-sonic caviton has a pancake structure. But also, independent of this argument, the estimates show that the corrections $\delta\Sigma^{eff}$ are small during the total time of self-contraction of a near-sonic caviton. Thus the caviton (40) can lead to a collapse if indeed the problems with the caviton (40) already listed above are solved.

There also exists an essentially kinetic effect which stops the collapse [39] (and see [42]). This is an $\ell \to s$ non-linear Landau damping [2]. Suppose that the near-sonic contraction proceeds up to $r \cong r_d$. The result of that will be fast-particle creation accompanied by an accumulation of density fluctuations. Owing to the Landau damping on fast particles, the cavitons become empty of Langmuir waves much faster than the density depletion of the caviton can be dissipated. Then the transition radiation of fast particles on the accumulated density inhomogeneities gives the emission and reabsorption of the Langmuir waves with $\omega \cong \omega_{pe}$, $k \cong 0$. Thus the empty cavitons transform the energy of the Langmuir waves of the non-empty cavitons to the fast particles and stop the collapse. The damping rate can be estimated from the probability of the $\ell \to s$ scattering given in Ref.[2]:

$$\gamma = \frac{1}{27}\int \frac{\delta n_{k_1}^2}{n_0^2} \frac{\gamma_{k_1}^L \, dk_1}{k_1^4 r_d^4} \tag{50}$$

and differs from the result in [39,40] only by a numerical factor 1/3. The $\gamma_{k_1}^L$ is the rate of linear Landau damping on fast particles $v > v_{Te}/k_1 r_d$, $k_1 r_d \ll 1$. The estimation shows that the very small density inhomogeneity can stabilize the collapse. The stabilization occurs even for

$$\frac{\overline{\delta n^2}}{n_0} \cong \frac{n'}{n_0}\sqrt{\frac{m_e}{m_i}\frac{W^\ell}{n_0 T_e}} \tag{51}$$

if the energy density in the fast particles $n'T'$ is of the order of the thermal plasma energy density $n_0 T_e$ (see the discussion in §6).

6. FAST PARTICLES AND SPIKONS

The result of the previous discussion can be summarized as follows. The accumulation of the energy in the domain of the small k's leads to the development of modulation instability which drives the energy back to the high k-values until it reaches the tail of the particle distribution and is damped by linear Landau damping. The fast-particle creation can be considered as a dangerous effect for laser fusion and electron beam/plasma interaction. This situation leads to a tendency to exclude from the experiments the conditions when the modulation instability is settled. The main question is whether indeed the modulation instability leads to the transform of the energy into the fast particles only or whether there are conditions when the main body of the particle can be heated. There are some indications from the numerical calculations that at the last stage of development of the modulation instability some new non-linear motions of the soliton type can be formed which have high velocity and are rapidly damped. They are called spikons. Perhaps they should be connected with the fast particles present and have velocities higher than that of sound.

We now discuss whether it is possible to have supersonic solitons and whether or not the main-body particles can be heated. In the absence of fast particles there are only two possibilities for supersonic solitons:

(1) The antisoliton (or the hole soliton) for which instead of $|\epsilon| \to 0$ at $\xi \to \pm\infty$ one has $|\epsilon| \to \epsilon_* = \text{const}$ for $\xi \to \pm\infty$:

$$\nu = -\frac{\epsilon_*^2 - |\epsilon|^2}{u^2 - 1} \; ; \; |\epsilon|^2 = \epsilon_*^2 - \frac{\epsilon_0^2}{\left\{\cosh\left[\dfrac{(\xi - u\tau)}{\sqrt{2(u^2-1)}}\right]\right\}^2} \tag{52}$$

The antisoliton moves with velocity higher than that of sound. The processes of formation and interaction of antisolitons were not investigated, but apparently the antisolitons can be damped on the fast particles only.

(2) The soliton described by higher non-linearity [36, 37]: This is the special case when $1/T^{eff}$, which determines the coupling constant in the cubic non-linear term, is essentially less than T. It can happen only for non-thermal particle distribution [36].

One finds:

$$\epsilon = \frac{\epsilon_0 \exp[-i\Omega\tau + i(u/2)(\xi - u\tau)]}{\sqrt{\cosh\left[\dfrac{\chi\epsilon_0^3(\xi - u\tau)}{\sqrt{3}(u^2 - v_s^2)^{3/2}}\right]}} \tag{53}$$

$$\Omega = -\frac{u^2}{4} - \frac{\epsilon_0^4 \chi}{3(u^2 - v_s^2)^{3/2}} \;;\; \chi = \frac{\mu(T^{eff})^3}{2T(\widetilde{T}^{eff})^2}$$

$$\frac{1}{T^{eff}} = -\frac{1}{m_e}\int \frac{1}{v}\frac{\partial\phi}{\partial v}\frac{dp}{2\pi} \;;\; \frac{1}{(\widetilde{T}^{eff})^2} = \frac{1}{m_e^2}\int \frac{1}{v}\frac{\partial}{\partial v}\frac{1}{v}\frac{\partial\phi}{\partial v}\frac{dp}{2\pi} \tag{54}$$

Such solitons can exist for supersonic velocities $u > v_s$ but u should be not very different from v_s (they can exist only for large ϵ_0). The non-linear Landau damping diminishes the ϵ_0. The existence of such a soliton in plasma would be rather exceptional.

The most important and most likely possibility of having a supersonic soliton appears in the presence of fast particles, i.e. in the case when fast particles change the dispersion of Langmuir waves. In what follows we describe the proposal in Ref.[41] to explain the spikon phenomenon. We should mention that in the absence of thermal particle heating, the energy of the external source is converted to the fast particles. We shall consider the case when there is a real change in the energy of the plasma by the external source. Since the energy is released in the fast particles, we have $n'T' \gg n_0 T_e$. Here we consider the simplest case when the fast particles have thermal distribution with temperature $T' \gg T_e$ and density $n' \ll n_0$. For $\omega_{pe}/v_{Te} \gg k \gg \omega_{pe}/v'_{Te}$ the damping of the Langmuir waves on fast particles will still be small but their dispersion can be determined by the fast particles:

$$\omega_{\vec{k}}^{\ell} \cong \omega_{pe} - \frac{\beta^2}{k^2} \;;\; \beta^2 = \frac{\omega_{pe}^2}{2v_{Te}^2}\frac{T_e n'}{T' n_0} \tag{55}$$

The conditions necessary to have approximately the dispersion (55) are:

$$\frac{T_e}{T'} \ll k^2 r_d^2 \ll \sqrt{\frac{T_e n'}{T' n_0}} \tag{55'}$$

We find a non-linear equation which should now be used instead of (23):

$$\frac{\partial^2}{\partial\xi^2}\left(i\frac{\partial}{\partial\tau} - \nu\right)\epsilon = \epsilon \;;\; \frac{\partial^2\nu}{\partial\tau^2} - \frac{\partial^2\nu}{\partial\xi^2} = \frac{\partial^2}{\partial\xi^2}|\epsilon|^2 \tag{56}$$

We have introduced the following dimensionless variables:

$$\nu = \frac{\delta n}{2n_0}\,\sigma^{1/3} \;;\; \epsilon = \frac{E}{\sqrt{8\pi\, n_0 T_e}}\,\sigma^{1/3}$$

$$\tau = \frac{t\omega_{pe}}{\sigma^{1/3}} \;;\; \xi = \frac{x\omega_{pe}}{v_s\sigma^{1/3}} \;;\; \sigma = \frac{2m_i n_0 T'}{m_e n' T_e} \tag{57}$$

One finds from (56) the conservation of number of quanta N, energy H and momentum P. The expressions for N and P are the same as usual (see (26), (27)) and

$$H = \int \left\{ \frac{\nu^2 + v^2}{2} + \nu|\epsilon|^2 - \left|\frac{\partial g}{\partial \xi}\right|^2 \right\} d\xi \;;\; g = i\frac{\partial \epsilon}{\partial \tau} - \nu\epsilon \tag{58}$$

One can derive from (56) the modulational instability of a given pump of amplitude ϵ_0 and frequency

$$\omega_0 = -\frac{1}{k_0^2} - |\epsilon_0|^2$$

We find

$$1 = \frac{k^2|\epsilon_0|^2}{\omega^2 - k^2}\left\{\left[\frac{2kk_0 - k^2}{(k-k_0)^2 k_0^2} - \omega\right]^{-1} + \left[\frac{-2kk_0 - k^2}{(k+k_0)^2 k_0^2} + \omega\right]^{-1}\right\} \tag{59}$$

The maximum growth rate appears for $\omega \gg k$, $k \gg k_0$, $\omega \gg k/k_0^3$ (ω, k, k_0 are dimensionless):

$$\gamma \cong \frac{k}{k_0^{3/2}}\left(\frac{|\epsilon_0|^2}{2}\right)^{1/4} \tag{60}$$

Since $d^2\omega/dk^2 < 0$, the sign of the non-linear term should be positive in order to have a soliton type of solution. This means that the soliton, which we call a spikon, should be supersonic: $\nu = |\epsilon|^2/(u^2 - 1)$; $u > 1$. Putting

$$\epsilon = \epsilon_0 y e^{-i\Omega\tau}\ ; \quad \Omega = \frac{3}{2(u^2-1)}\,\epsilon_0^2$$

$$\widetilde{\xi} = (\xi - u\tau)\left(\frac{2(u^2-1)}{3\epsilon_0^2}\right)^{1/2} \tag{61}$$

we find

$$\left(\frac{dy}{d\xi}\right)^2 = \frac{y^2(1-y^2)}{(1-2y^2)^2} \tag{62}$$

Obviously $y = 0$ and $y = 1$ are the turning points; $y = 1$ at $\widetilde{\xi} = 0$ and $y = 0$ at $\widetilde{\xi} \to \pm\infty$. In between there exists $y^2 = 1/2$ where $dy/d\xi \to \pm\infty$. Thus the spikon has steep gradients and looks more like a shock wave. Really, the gradients of the electric field are not infinite if one takes into account the dispersion of the waves due to the cold plasma or the linear Landau damping on fast particles. But the presence of the steep gradients, as in the case of shocks, leads to the dissipation of the spikon on thermal particles. Equation (62) can be simply integrated. One of the most essential features is that the width of the spikon,

$$\xi_0 = \sqrt{\frac{3}{2}}\,\frac{\epsilon_0}{\sqrt{u^2-1}}$$

is proportional to its amplitude ϵ_0 instead of being inversely proportional to the amplitude in the case of solitons described previously. This leads to three essential conclusions: (a) the spikons are stable for 3D perturbations, (2) collapse is impossible, and (3) the rate of thermal particle heating is larger than that of fast particle heating [41]. Indeed, collapse is a phenomenon which is due to a simple thing: the smaller the entity the larger the field, and the larger the field the larger the force pushing the plasma from the entity. When $\xi_0 \sim \epsilon_0$ this cannot happen. The same is true for perturbation of the spikon in the direction perpendicular to its motion. Then due to $dy/d\xi \sim (\xi - \xi_1)^{-1/2}$ near to the $y^2(\xi_1) = 1/2$ point, we have $E_k \sim 1/k^{3/2}$, $W_k \sim 1/k^3$, and the quasilinear equation leads to an estimate of the heating rate of thermal particles:

$$\gamma^T \cong \omega_{pe}\left(\frac{r_d}{x_0}\right)^2\ ; \quad x_0 = \xi_0 v_s \sigma^{1/3}/\omega_{pe} \tag{63}$$

The rate of Landau damping on the fast particles is

$$\gamma_{T'} \cong \omega_{pe} \frac{n'}{n_0} \left(\frac{T_e}{T'}\right)^{3/2} \left(\frac{x_0}{r_d}\right)^3 \tag{64}$$

The condition for the rate of heating of thermal particles to be larger than that of fast particles is expressed by the left-hand side of the inequalities:

$$\left(\frac{T' n_0}{T_e n'}\right)^{2/5} \left(\frac{T'}{T_e}\right)^{1/5} \gg \frac{x_0^2}{r_d^2} \gg \left(\frac{T' n_0}{T_e n'}\right)^{1/2} \tag{65}$$

while the right-hand side of (65) is the inequality (55). The left-hand side of (65) is indeed larger than the right-hand side if $n'T' \gg n_0 T_e$, which was the starting point of the discussion.

Thus one can expect the following sequence of events in the case of a high-power input in plasma:

(a) The energy is stored in the Langmuir plasma waves;
(b) The energy is accumulated in the condensation ($k \cong 0$);
(c) The modulational instability develops and converts the energy towards higher k's;
(d) The energy is converted to the fast particles which are heated;
(e) The new modulational instability develops which creates the spikons;
(f) The spikons heat the thermal particles;
(g) Since the thermal particles are heated faster, the condition $n'T' \gg n_0 T_e$ should be violated and the fast particle start to receive the energy.

On average, the equipartition between the energy density in fast and thermal particles should be established as $\overline{n'T'} \cong \overline{n_0 T_e}$. For this picture to work it is necessary to preserve the fast particles in the system. Other applications of the theory are mentioned in Ref.[41]: anomalous resistivity and the possibility of thermal ion heating by ion-sound turbulence; runaway electrons in tokamaks, and the conversion of their energy to thermal particles; the heating of thermal particles in beam-plasma interactions.

7. GENERAL THEORETICAL PROBLEMS: CONCLUDING REMARKS

7.1. General theoretical description

(1) The general kinetic description of modulation interactions can be given in any power of the field strength and the validity of the expansion in the field strength is $E^2/(4\pi n T) \ll 1$.

(2) This description of Langmuir waves takes the parameter $(\Delta\omega\,\omega_{pe}/k^2 v_{Te}^2)$ as an arbitrary value and thus makes it possible to consider the so-called strong non-linear case when the non-linear frequency shifts are larger than the linear shift.

(3) The general theory shows that all the non-linear processes, namely modulation interactions, $\ell \rightleftarrows \ell$ scattering, $\ell \rightleftarrows s$ scattering and $\ell \rightleftarrows \ell + s$ decay processes, are strongly tied together and need to be considered simultaneously.

(4) The kinetic description generalizes the theorem of renormalization of ω_{pe} known in the frame of the hydrodynamic description but also shows that in the general case the renormalization is not possible if the thermal motion of particles is taken into account.

7.2. New non-linear entities

(1) The various new non-linear entities, solitons, cavitons, antisolitons, spikons, can exist but all of them need strong phase correlations or phasing of waves of self-contractions.

(2) The interaction of these entities with each other and with the free untrapped waves can determine the non-linear plasma kinetics.

(3) The major problem of these entities is the problem of their stability and the influence of additional even weak non-linearities. The instability forced by the neighbourhood entity can be regarded as a mechanism of interaction between entities.

7.3. Statistics and turbulence

The problem of a statistical description of turbulence is very complicated. We can compare it with the known problem of hydrodynamic turbulence, which has been discussed for more than twenty years without essential progress. The equations of hydrodynamic incompressible motion have the quadratic non-linearity:

$$v_{ik} = \int S^{ij\ell}_{1,2}\, v_{j_1} v_{\ell_2}\, d_{1,2}\,; \quad S^{ij\ell}_{1,2} = \frac{1}{\Delta\omega}\, k_{2\ell}\left(\delta_{ij} - \frac{k_i k_j}{k^2}\right) \tag{66}$$

while the equation for Langmuir waves has cubic non-linearity:

$$\epsilon E_k = \int \Sigma^{\mathrm{eff}}_{1,2,3}\, E_1^+ E_2^+ E_3^-\, d_{1,2,3} \tag{67}$$

Hydrodynamic and strong Langmuir turbulence have many difficulties in common and one can expect the development of the problems of the theory of strong

Langmuir turbulence to be discussed for many years to come. One point seems to be certain: the best way to start a theory is to use the procedure in Eq.(67) for statistical averaging [37].

7.4. Astrophysical applications

There are many astrophysical applications (see [42]). A brief list follows:

(1) The problem of cosmic ray acceleration of the Fermi type by non-linear entities.

(2) The problem of formation of collisionless shock waves, where the modulation instability can act as a mechanism of the dissipation in the shock front.

(3) The problem of fast-particle creation in the front of the shock and behind it, and the X-ray emission of the fast particles behind the shock (for solar flares see Ref.[43]).

(4) The influence of the magnetic field on modulational instabilities.

(5) Stabilization of electron beams by modulational effects and application to the problem of type-III solar radio bursts [44 – 46].

(6) Generation of magnetic fields by modulation instabilities [42].

(7) The problem of solitons in relativistic plasmas and its application to the problem of pulsar emission [42, 47, 48].

In conclusion, I should like to say that although some of the necessary fundamental experiments in this field have already begun [49, 50], it is obvious that to understand this field, which is of great importance to the whole of modern plasma physics, new fundamental experiments are needed.

REFERENCES

[1] KADOMTSEV, B.B., Plasma Turbulence, Academic Press, New York (1965).
[2] TSYTOVICH, V.N., Nonlinear Effects in Plasma, Plenum Press, New York (1970); Introduction to the Theory of Plasma Turbulence, Pergamon Press, Oxford (1972).
[3] KAPLAN, S.A., TSYTOVICH, V.N., Plasma Astrophysics, Pergamon Press, Oxford (1973).
[4] SAGDEEV, R.Z., GALEEV, A.A., Nonlinear Plasma Theory, Benjamin, New York (1969).
[5] NACKACH, R., WILLHELMSSON, H., Fiz. Plazmy (1977).
[6] DUPREE, T.H., Phys. Fluids **15** (1972) 334.
[7] KADOMTSEV, B.B., POGUTSE, O.P., Phys. Rev. Lett. **25** (1970) 1155.
[8] LANDAU, L.D., Zh. Ehksp. Teor. Fiz. 7 (1936) 203; BALESCU, R., Phys. Fluids **3** (1960) 52.

[9] KLIMONTOVICH, Yu.L., The Statistical Theory of Nonequilibrium Processes in a Plasma, Pergamon Press, Oxford (1967).
[10] BOGOLYUBOV, N.N., "Problems of a dynamical theory in statistical physics", Studies in Statistical Mechanics **1**, North-Holland, Amsterdam (1962).
[11] GAILITIS, A.A., TSYTOVICH, V.N., Zh. Ehksp. Teor. Fiz. **46** (1964) 1726.
[12] (a) TSYTOVICH, V.N., Proc. P.N. Lebedev Physics Inst. **66** (1973);
(b) AKOPIAN, A.V., TSYTOVICH, V.N., Fiz. Plazmy **1** (1975) 673.
[13] AKOPIAN, A.V., TSYTOVICH, V.N., Zh. Ehksp. Teor. Fiz. **71** (1976) 166.
[14] GINZBURG, V.L., TSYTOVICH, V.N., Zh. Ehksp. Teor. Fiz. **65** (1973) 1818.
[15] GINZBURG, V.L., TSYTOVICH, V.N., Izv. Vyssh. Uchebn. Zaved., Radiofiz. **18** (1974) 173.
[16] VEDENOV, A.A., RUDAKOV, L.I., Sov. Phys.-Dokl. **159** (1964) 769.
[17] GAILITIS, A.A., Candidate Thesis, P.N. Lebedev Phys. Inst. Moscow (1964); Latv. PSR Zinat. Akad. Vestis, Fiz. Teh. Zinat. Ser. **4** (1965) 13.
[18] KHAKOMOV, F.Kh., TSYTOVICH, V.N., Zh. Ehksp. Teor. Fiz. **64** (1973) 1261.
[19] KRUER, W., ESTABROOK, K., THOMSON, T., Lawrence Livermore Lab. Rep. UCRL-74947 (1973).
[20] RUDAKOV, L.I., Dokl. Akad. Nauk USSR **207** (1972) 391.
[21] KINGSEP, A.S., RUDAKOV, L.I., SUDAN, R.N., Phys. Rev. Lett. **31** (1973) 1482.
[22] DEGTIAREV, L.M., MACKHAN'KOV, V.G., RUDAKOV, L.I., Zh. Ehksp. Teor. Fiz. **67** (1974) 533.
[23] LEBEDEV, A.A., TSYTOVICH, V.N., Phys. Scripta **11** (1975) 266.
[24] GOREV, V.V., KINGSEP, A.S., Zh. Ehksp. Teor. Fiz. **66** (1974) 2048.
[25] VERIAEV, A.A., TSYTOVICH, V.N., Izv. Vyssh. Uchebn. Zaved., Radiofiz. **20** (1977).
[26] GOREV, V.V., KINGSEP, A.S., RUDAKOV, L.I., Izv. Vyssh. Uchebn. Zaved., Radiofiz. **19** (1976) 691.
[27] ZAKHAROV, V.E., Zh. Ehksp. Teor. Fiz. **62** (1972) 1745.
[28] NISHIKAWA, K., LEE, I.C., LIU, C.S., Comments Plasma Phys. and Contr. Fusion 2 (1975) 63.
[29] NORMAN, K., TER HAAR, D., Plasma Physics, Nonlinear Theory and Experiment (Proc. 37th Nobel Symp. Stockholm, 1976), Plenum Press, New York (1977).
[30] TSYTOVICH, V.N., P.N. Lebedev Inst. Rep. 178 (1976).
[31] KAPLAN, S.A., KHODATAEV, K.V., TSYTOVICH, V.N., Sov. Astron. Lett. (1977).
[32] PEREIRA, N.R., SUDAN, R.N., DANAVIT, J., Lab. of Plasma Studies, Cornell Univ. Rep. PPS 191 (1976).
[33] OSTROVSKY, L.A., Izv. Vyssh. Uchebn. Zaved., Radiofiz. **17** (1974) 454.
[34] LITVAK, A.G., FRAIMAN, G.M., Zh. Ehksp. Teor. Fiz. **69** (1975) 4.
[35] LITVAK, A.G., FRAIMAN, G.M., VUAKOBOVSKY, A.D., Pis'ma Zh. Ehksp. Teor. Fiz. **19** (1974) 23.
[36] TSYTOVICH, V.N., in Proc. 30th Nobel Symposium on Physics of Hot Plasma in the Magnetosphere, 1975, Plenum Press, New York (1976).
[37] KHAKOMOV, F.Kh., TSYTOVICH, V.N., Zh. Ehksp. Teor. Fiz. **67** (1976) 312.
[38] TSYTOVICH, V.N., Proc. 12th Int. Conf. Phenomena of Ionized Gases, Eindhoven (1975) Part 2, North-Holland, Amsterdam (1976) 141.
[39] GALEEV, A.A., et al., Pis'ma Zh. Ehksp. Teor. Fiz. **21** (1975) 539.
[40] GALEEV, A.A., et al., Fiz. Plazmy **1** (1975) 10.
[41] RUDAKOV, L.I., TSYTOVICH, V.N., P.N. Lebedev Inst. Rep. (1977).
[42] TSYTOVICH, V.N., Conf. on Plasma Astrophysics, Irkutsk (1976), P.N. Lebedev Inst. Rep. 178 (1976).

[43] HOYNG, P., Rep. High-Altitude Observatory, Boulder (1976).
[44] BARDWELL, S., GOLDMAN, M., Astrophys. J. **209** (1976) 912.
[45] SMITH, D.F., Astron. Astrophys. **7** (1970) 147.
[46] PAPADOPOULOS, K., Phys. Fluids **18** (1975) 1769.
[47] KARPMAN, V.I., et al., Phys. Scr. **11** (1975) 271.
[48] KHAKIMOVA, M., TSYTOVICH, V.T., Astrophysics **12** (1976) 605.
[49] STENZEL, R.L., et al., Phys. Rev. Lett. **32** (1974) 6341.
[50] IKEZI, H., et al., J. Phys. Soc. Jpn. **37** (1974) 766.

THIRD INTERNATIONAL ('KIEV') CONFERENCE ON PLASMA THEORY

Directors

B.B. KADOMTSEV
V.N. TSYTOVICH
Union of Soviet Socialist Republics

THIRD INTERNATIONAL ('KIEV') CONFERENCE ON PLASMA THEORY, TRIESTE, 1977

The Third International ('Kiev') Conference on Plasma Theory took place at the ICTP, Trieste, from 5 to 9 April 1977. The two previous International Conferences on Plasma Theory, organized by the Academies of Sciences of the Ukrainian SSR and the USSR were held in 1971 and 1974 at the Institute for Theoretical Physics of the Academy of Sciences of the Ukrainian SSR in Kiev. The main purpose of the International Kiev Conferences on Plasma Theory is the discussion of plasma physics problems basic to current work on such subjects as controlled thermonuclear fusion, magnetohydrodynamic conversion of thermal energy into electric energy, the creation of new radiowave sources, and the conquest of space.

During the five-day Conference the participants presented more than eighty papers reflecting the state of practically all the main directions of research in plasma theory. The following topics were discussed in detail: general problems of statistics and kinetics; waves and instabilities (Alfvén waves, drift and lower-hybrid waves in a magnetic field, dissipative instability); non-linear wave interactions (parametric excitation, strong Langmuir turbulence, solitons, collapse); CTR theory (high-beta plasmas, equilibrium, transport). The complete programme of the Conference is given in Appendix B.

B.B. KADOMTSEV

Statistics and kinetics

SOME STATISTICAL ASPECTS OF PARTIALLY IONIZED SYSTEMS

G. ECKER
Institut für Theoretische Physik,
Lehrstuhl I,
Ruhr-Universität Bochum,
Federal Republic of Germany

Abstract

SOME STATISTICAL ASPECTS OF PARTIALLY IONIZED SYSTEMS.
1. Introduction. 2. The equilibrium state. 3. States with non-equilibrium radiation. 3. Stationary general non-equilibrium states. 5. Concluding remark.

1. INTRODUCTION

The statistics of the partially ionized system is even more complicated and involved than that of the fully ionized system, and we can therefore only present in the limited space here some typical aspects which we consider important.

The statistical problem discussed is the distribution on the various particle states of a partially ionized system. There are, of course, other interesting statistical questions, for instance the properties of the microfield distribution (see e.g. Ref. [1]) but space does not permit all these phenomena to be studied here.

2. THE EQUILIBRIUM STATE

The least complicated case – although it is by no means simple – is a partially ionized system in the state of equilibrium. It has been said that this state might not be relevant to real plasma systems, but it is. In particular, the local thermodynamic equilibrium (LTE) is of prominent importance, for instance, in arc theory and in astrophysics.

To concentrate on the essentials, we shall consider the equilibrium of the partially ionized hydrogen system, which consists of electrons, protons and neutral hydrogen atoms only.

To calculate the composition, the Helmholtz free energy can be expressed through the partition function Q in the Γ-space in the semi-classical approximation

$$Q = \mathrm{tr}(\exp(-\hat{H}/kT)) \simeq \frac{\hbar^{-\ell}}{N_-!N_+!N_o!} \int \exp(-\epsilon/kT)\, d\Omega \tag{1}$$

which is valid in the range

$$\frac{N}{V}\left(\frac{h^2}{2\pi\, m_+\, kT}\right)^{3/2} \ll 1 \tag{2}$$

N_α denotes the particle numbers, V the volume, T the temperature, ϵ the energy, m_α the particle mass, k the Boltzmann constant, and h Planck's constant.

Separating – naively – the energy contributions of the three kinds of particle, we have

$$Q = \frac{\varphi_-(T)^{N_-} \cdot \varphi_+(T)^{N_-} \cdot \varphi_o(T)^{N_o}}{N_-!N_+!N_o!} \tag{3a}$$

$$\varphi_{\mp} = \rho_{\mp} \; ; \; \varphi_o = \rho_o \cdot \sigma_o$$

$$\rho = 2V\left(\frac{2\pi\, m_-\, kT}{h^2}\right)^{3/2} \tag{3b}$$

$$\sigma = \sum_i g(i) \exp[-\epsilon(i)/kT]$$

where φ indicates the partition function in the μ-space composed in general of the free-motion contribution ρ and an internal contribution σ; $\epsilon(i)$ is the internal energy and g(i) the weight factor of the ith internal state. The well-known minimum requirement for the free energy using the Stirling formula results in the *Saha equation*:

$$\frac{n_+\, n_-}{n_o} = \frac{2}{\sigma_o}\left(\frac{2\pi\, m_-\, kT}{h^2}\right)^{3/2} \cdot \exp(-\chi/kT)$$

$$\chi = \epsilon_+ - \epsilon_o(0) \tag{4}$$

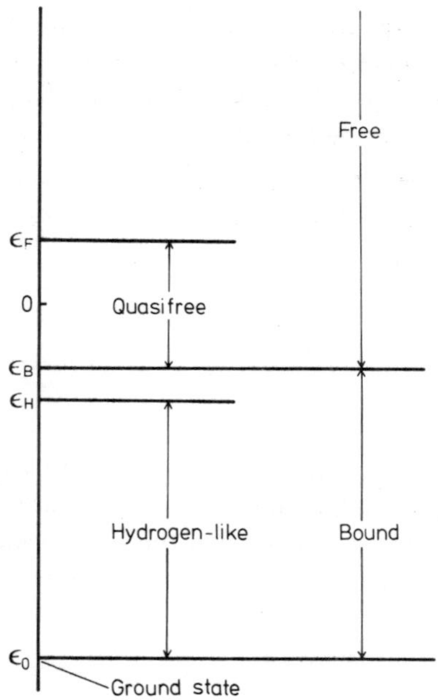

FIG.1. Schematic drawing of the particle states of the proton-electron pair.

The failure of this approach is already reflected in the fact that the partition function of the atoms σ_0 is obviously divergent. The reason must be seen in the neglect of the internal interaction of the three particle components. This contribution cannot be split up and therefore separation is altogether impossible. We have to go back to the Γ-space and keep in mind that the states of the whole system are all more or less affected by this interaction.

The discussion of this problem in the present paper is not devoted to the mathematical aspects of quantum-mechanical developments since they can be found in the literature (e.g. Refs [2–4]). Rather, it aims at a solution which also provides physical insight and is adapted to the concept of 'free' and 'bound' particles essential to all applications.

There are bound states of electron-ion pairs and free states of electron-ion pairs, but there are also 'quasifree states'. The bound states of low quantum number are little affected by the electrostatic interaction with a third partner since the average value of the binding energy is large compared to the interaction energy e^2/r_0, r_0 being the average particle distance and e the elementary charge. The free particle states of high or average kinetic energy ($\epsilon \geqslant kT$) are also little affected since we are well below the critical density, so that $e^2/r_0 \ll kT$ holds. However, the bound states of high quantum number and the free states of low kinetic energy – the quasifree states – will have no similarity to the hydrogen-like bound states or the free-particle wave functions.

In a quantum-mechanical treatment of this problem [5] we derived the magnitude of the limits for the bound states, the hydrogen-like bound states, the quasifree states and the free states. The lengthy calculation, which we do not present here, results in the following data for these limits (see also Fig. 1):

$$\begin{aligned} \epsilon_o &= -Ry \quad ; \quad \epsilon_H = \epsilon_B = -O(e^2/r_o) \\ \epsilon_F &= +O(e^2/r_o) \end{aligned} \tag{5}$$

where ϵ_0 is the ground state; ϵ_H the upper limit for hydrogen-like states; ϵ_B the upper limit for bound states; and ϵ_F the lower limit for free states.

Applying the above *pair concept* and the subdivision in *bound, free* and *quasifree* states, we can formulate the partition function in the form [6, 7]:

$$\left.\begin{aligned} Q &= \rho^N(\sigma_o + \sigma_1 + \sigma_2)^N \\ \sigma_\mu &= \int_{\epsilon_\mu}^{\bar{\epsilon}_\mu} g_\mu(\epsilon)\, e^{-\epsilon}\, d\epsilon \quad , \qquad \mu = 0,1,2 \\ g_o &= \sum_i Ry\, \delta(\epsilon - \epsilon_i) \cdot \frac{1}{\epsilon_i} \quad ; \quad g_2 = \left(\frac{V\, 4\pi\, (2m)}{h^3}\right)^{3/2} \cdot \epsilon^{1/2} \end{aligned}\right\} \tag{6}$$

where the limits to be introduced for ϵ_μ, $\bar{\epsilon}_\mu$ can be taken from

$$\begin{aligned} \epsilon_o &= -Ry \quad ; \quad \bar{\epsilon}_o = \epsilon_1 = -e^2/r_o \\ \epsilon_2 &= \bar{\epsilon}_1 = e^2/r_o \quad ; \quad \bar{\epsilon}_2 = \infty \end{aligned} \tag{7}$$

N denotes here the total number of electrons, and with that the total number of ions, in the system.

The contribution of the bound states is no longer a problem here since the divergence is automatically removed due to the upper limit $\bar{\epsilon}_0$ for the bound states.

The effect of the interaction on the contribution of the free particles to the partition function is well known. It can be formulated by simply replacing in the weight function g_2, the quantity ϵ by $(\epsilon - e^2/D)$. Principally, the calculation of this term is complicated due to a divergence that occurs for small separations of electrons and ions. It follows from the quantum-mechanical

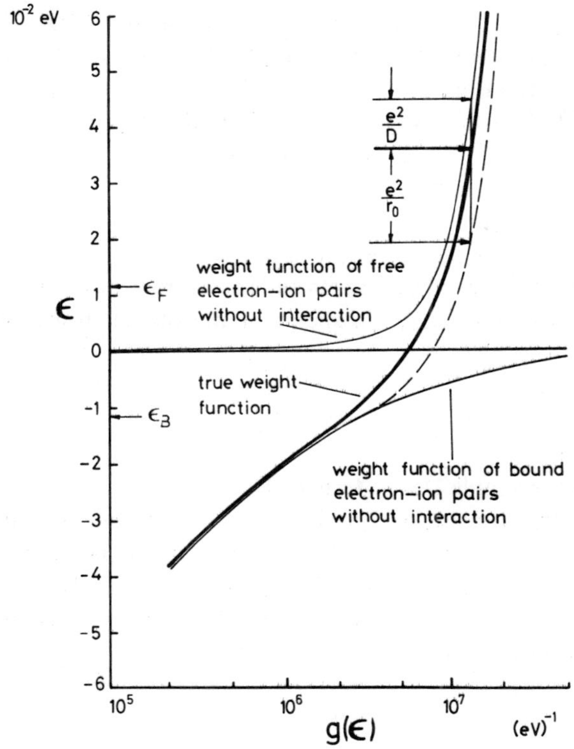

FIG. 2. The thin line gives the weight function $g(\epsilon)$ as calculated for an isolated atom and a free electron-ion pair without interaction. The heavy line is the correct $g(\epsilon)$ curve. The broken line shows $g(\epsilon)$ as derived under the false assumption of a constant interaction effect for all free and quasifree states identical with that experienced by the low-energy quasifree states.

treatment [5] that this divergence disappears since, due to quantum-mechanical effects, distances of the order

$$r^2 < O(\lambda\, r_w) \quad ; \quad r_w = e^2/kT \quad ; \quad \lambda = h/(2\pi m\, kT)^{1/2} \tag{8}$$

have to be excluded.

A problem is presented by the quasifree states where the relation between the weight factor and the energy is unknown and in the greater part cannot be calculated by perturbation methods. As we may see from Fig. 2, it is, however, not difficult to interpolate between the weight function of the free states and that of the bound states and thus account for the contribution of the quasifree states to the partition function.

With this knowledge of the partition function, the minimization of the free energy using a discontinuous free-bound transition [6, 7] yields the Saha equation with an important correction.

There is an effective lowering of the ionization energy

$$\Delta \chi_{saha} \simeq 1.4\ e^2/D \tag{9}$$

where D denotes the Debye length of the system.

In contrast to other investigations in the literature, it is found to be neither identical with the effective ionization energy of an isolated atom in the plasma, which is given by

$$\Delta \chi_{atom} \simeq e^2/r_0 + e^2/D \tag{10}$$

nor identical with the simple Debye term e^2/D.

The reason is that for the low quasifree states the free-bound interaction dominates through the term e^2/r_0. With increasing energy, however, the free-bound interaction becomes negligible in comparison with the Debye term of the free-free interaction. Since all quasifree and free states contribute to the Saha equation, it is evident that

$$e^2/D < \Delta \chi_{saha} < e^2/r_0 \tag{11}$$

holds. This situation is demonstrated in Fig. 2. Here we can see why the term e^2/r_0 is not the proper correction term. The correction $0.4\ e^2/D$ to the usual Debye term as given in Eq. (9) is stronger than the correction found from classical cluster expansions. The reasons are the free-bound definition, the accounting for the free-bound interaction, quantum-mechanical effects, and strong correlations below the true free limit. It is of course clear that, owing to the uncertainty of the quasifree states, accuracies of order 1% as claimed in the literature cannot be substantiated.

We further point out that the plasma composition is also affected through the limitation of the single-particle partition function σ_0. For temperatures low enough this effect is, of course, negligible. As it becomes important with increasing temperature, its limitation is governed by e^2/r_0 and not by e^2/D.

3. STATES WITH NON-EQUILIBRIUM RADIATION

Under this heading we consider systems which deviate from equilibrium owing to deviations of the radiation from equilibrium. In contrast, all particle components are assumed to be in a Maxwell distribution – possibly with different temperatures (T_-, T_+, T_0). Within these limits, various situations are possible, depending on the parameters of the system and on the characteristics of the non-equilibrium radiation distribution.

It is not difficult to describe the corresponding stationary states if one emphasizes the assumption of detailed balancing, on one hand, and negligible cumulative effects on the other. Balancing the collision and photoprocesses connecting the ground state with the ionized state results in the formula:

$$\frac{n_{\nu+1}\, n_-}{n_\nu} = \frac{n_- C_\nu^+ + P_\nu^+}{n_- C_{\nu+1}^- + P_{\nu+1}^-} \tag{12}$$

Here it is instructive to consider systems more general than the hydrogen system of the previous section and the index ν denotes the degree of ionization accordingly. The terms C_ν^+ and $C_{\nu+1}^-$ describe, respectively, collisional ionization and three-body recombination, whereas P_ν^+ and $P_{\nu+1}^-$ denote, respectively, photo-ionization and radiative recombination from and into the state ν. n_- denotes the electron density.

The problem of this formula is the calculation of the terms in the denominator. It now becomes clear why the assumption of equilibrium for the particle components is so crucial for this class of system. Knowing the electrons are Maxwell-distributed with a temperature $T_-(f^M_{T_-})$, we can now remove these terms in the denominator by introducing the detailed balances for the equilibrium state, looking at collisional and photo effects separately. In this way, we express the three-body recombination and radiative recombination term through the collisional ionization and photo-ionization terms of the equilibrium, arriving at

$$\frac{n_{\nu+1}\, n_-}{n_\nu} = \left(\frac{n_{\nu+1}\, n_-}{n_\nu}\right)^{Sa}_{T_-} \cdot \frac{n_- C_\nu^+(f^M_{T_-}) + P_\nu^+(I_\omega)}{n_- C_\nu^+(f^M_{T_-}) + P_\nu^+(B_\omega^{T_-})} \tag{13}$$

where I_ω is the radiation distribution and $B_\omega^{T_-}$ denotes the Planck distribution at temperature T_-. The index Sa refers to the Saha value. The relation (13) is the generalization of the Saha formula to systems with the radiation not in equilibrium with the electron Maxwell distribution.

Further progress can only be made if the evaluation of the collisional and photo-ionization processes C_ν^+, P_ν^+ is advanced. This is not the place to go into these relatively straightforward details. We therefore refer to a paper by Elwert [8], where these coefficients are evaluated using a simplified form of the collision ionization cross-section and, for the photo-ionization coefficient, a diluted Planck distribution and a cross-section known from quantum-mechanical calculations. One thus arrives at the general formula:

$$\frac{n_{\nu+1}\, n_-}{n_\nu} = \left(\frac{n_{\nu+1}\, n_-}{n_\nu}\right)^{Sa}_{T_-} \frac{1 + \kappa \dfrac{B}{An_-} \dfrac{x_-^{1/2}}{x_*} e^{x - x_*} G(x_-, x_*)}{1 + \dfrac{B}{An_-} \dfrac{1}{x_-^{1/2}} G(x_-, x_-)} \tag{14}$$

with the abbreviations

$$x_- = \chi_\nu / kT_- \quad ; \quad x_* = \chi_\nu / kT_* \quad ; \quad I_\omega = \kappa B_\omega^{T_*}$$

$$G(x_-, x_*) \simeq (1 - \frac{1}{x_*}) / (1 - \frac{2}{x})$$

$$A = q_\nu (8\chi_\nu / \pi m)^{1/2} \tag{15}$$

$$B = q_T\, c(\zeta_\nu \chi_\nu^2 / \bar{n}_\nu \chi_H^2)/a^3 \pi^2 3^{1/2}$$

$$q_T = (8\pi/3)(e^2/mc^2)^2$$

where q_T is the Thompson cross-section, c the velocity of light, a the Bohr radius, ζ_ν the number and $\bar{n}_\nu$ the principal quantum number of equivalent electrons for the photo process, κ the dilution factor and T_* the characteristic temperature of the Planck distribution.

Let us introduce a critical density through the relation

$$n_c(T_-, T_* / T_-) = \frac{B}{A} \frac{x_-^{1/2}}{x_*} e^{x_- - x_*} G(x_-, x_*) \tag{16}$$

Then we see from Eq. (14) that collisional or photo processes dominate the ionization or the recombination mechanism, depending on whether the $>$ or the $<$ sign holds, in the relations:

$$n_- \gtrless \kappa\, n_c(T_-, T_*/T_-) \quad ; \quad n_- \gtrless n_c(T_-, 1) \tag{17}$$

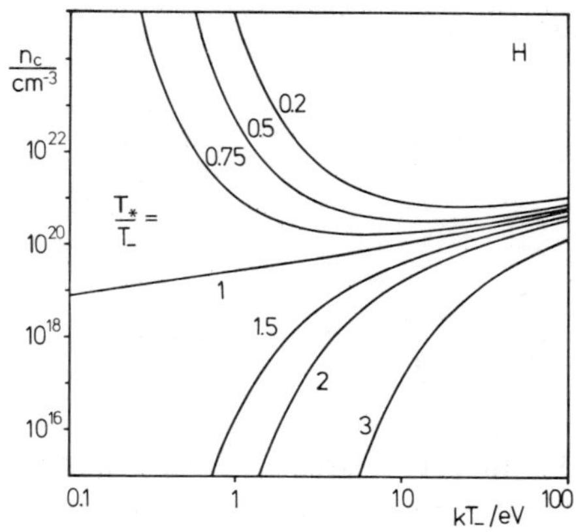

FIG. 3. Critical density n_c which is suitable to separate the regions of dominance of collisional and radiative effects in stationary systems with a diluted Planck radiation. T_- is the electron temperature and T_* the temperature of the diluted Planck distribution of the radiation.

The critical electron density n_c as a function of the electron temperature T_- and the ratio of radiation temperature and electron temperature (T_*/T_-) are shown in Fig. 3.

Equation (14),owing to the parameters n_-, T_-, T_* and the dilution factor κ, still describes a manifoldness of systems. Two special ones have drawn particular attention:

(a) **The Corona case** assumes that the radiation density is so small that photo-ionization is negligible in comparison with electron collision ionization, and the electron density is still so small that three-body recombination is negligible in comparison with radiative recombination.

(b) **The Eddington case** assumes the electron density so small that ionization is dominated by photo-ionization and three-body recombination by radiative recombination.

With these specifications it is not difficult to find the Corona formula and the Eddington formula from the general formalism (14).

In the Corona case we have

$$n_c(T_-, 1) \gg n_- \gg \kappa\, n_c(T_-, T_*/T_-) \tag{18}$$

and with that the relation

$$\frac{n_{\nu+1}\, n_-}{n_\nu} = \left(\frac{n_{\nu+1}\, n_-}{n_\nu}\right)^{Sa}_{T_-} \cdot \frac{A\, n_-\, x^{1/2}}{B\, G(x, x)}$$
$$= \left(\frac{n_{\nu+1}\, n_-}{n_\nu}\right)^{Sa}_{T_-} \cdot \frac{n_-}{n_c(T_-, 1)} \qquad (19)$$

We see from (18) and (19) that the degree of ionization in the Corona range is necessarily smaller than that of the degree of ionization in the equilibrium state Sa. We also see that in the Corona the degree of ionization n_+/n_0 is independent of the electron density.

In the Eddington case we have

$$n_c(T_-, 1) \gg n_- \quad ; \quad \kappa\, n_c(T_-, T_*/T_-) \gg n_- \qquad (20)$$

and find the relation

$$\frac{n_{\nu+1}\, n_-}{n_\nu} = \left(\frac{n_{\nu+1}\, n_-}{n_\nu}\right)^{Sa}_{T_-} \cdot \kappa\, \frac{x_-}{x_*}\, e^{x_- - x_*}\, \frac{G(x_-, x_*)}{G(x_-, x_-)} \qquad (21)$$

Obviously, this relation applies to plasmas of low electron density and high radiation density. Observe that, in spite of the dominant influence of the radiation, the formula still depends on the electron temperature T_- via the radiative recombination mechanism. For $T_- = T_*$ we find the Saha formula of the equilibrium reduced with the dilution factor.

In closing, we note that two other cases – complementary to the Corona and Eddington formulas – could be considered:

(c) The collision processes dominate the photo processes.

(d) Photo-ionization dominates electron collision ionization and three-body recombination dominates radiative recombination.

Case (c):

$$n_- \gg n_c(T_-, 1) \quad ; \quad n_- \gg \kappa\, n_c(T_-, T_*/T_-) \qquad (22)$$

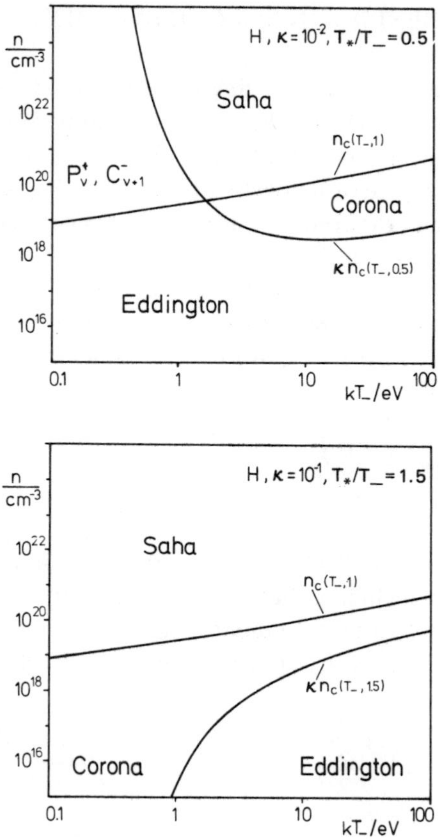

FIG.4. Examples of the validity ranges of the Saha, Corona and Eddington formulas in hydrogen (H) for different dilution factors κ, and ratios of the radiation temperature T_* and electron temperature T_- (n denotes electron density).

is not interesting since, trivially, it will reproduce the Saha formalism of equilibrium because the only non-equilibrium effect – the radiation – is assumed to be negligible.

Case (d):

$$\kappa\, n_c (T_- , T_*/T_-) \gg n_- \gg n_c (T_- , 1) \qquad (23)$$

is hardly of practical interest.

Figure 4 shows examples for the applicability of various models.

4. STATIONARY GENERAL NON-EQUILIBRIUM STATES

The systems treated in the preceding section allowed a simple approach for two reasons: we assumed the particle components to be Maxwellian-distributed and neglected the influence of excited states. In this section we consider the more general type where there are deviations from the Maxwellian distribution due to inhomogeneity in the configuration space, field effects and contributions of the excited states of the atom.

First, we consider systems homogeneous in the configuration space but allow for the presence of an electric field. In these circumstances, the distribution function of the electrons is governed by the kinetic equation:

$$\frac{e}{m}\,\underline{E}\,\frac{\partial f}{\partial \underline{v}} = \sum_{s} \left(\frac{\delta f}{\delta t}\right)_s^c \tag{24}$$

where the term $(\delta f/\delta t)_s^c$ denotes the effect on the electron distribution function through collisions, also including the internal levels of the atom. These terms naturally depend on the occupation number n_k of the kth state and therefore we have to calculate this quantity from a corresponding set of equations:

$$\begin{aligned} -n_k &\left\{ n_- C_k^+ + P_k^+ + \sum_{k\neq i} (n_- C_k^i + P_k^i) + \sum_{i<k} E_k^i \right\} \\ &+ n_-^2\, n_{\nu+1}\, C_{k,\nu+1}^- \\ &+ n_-\, n_{\nu+1}\, P_{k,\nu+1}^- + \sum_{k\neq i} n_i (n_- C_i^k + P_i^k) + \sum_{i>k} n_i E_i^k = 0 \end{aligned} \tag{25}$$

In these equations C_k^+ denotes collisional ionization, C_k^i collisional excitation $(i > k)$ or de-excitation $(i < k)$, P_k^+ photo-ionization, P_k^i photo-absorption $(i > k)$ or induced emission $(i < k)$; $C_{k,\nu+1}^-$ describes three-body recombination, $P_{k,\nu+1}^-$ radiative recombination and E_k^i spontaneous emissions. It is clear that the coefficients C_k^+, C_k^i, C_i^k and $C_{k,\nu+1}^-$ depend on the electron distribution f, and with that couple the set of equations (25) with Eq. (24). Moreover, the coefficients P_k^+, P_k^i, P_i^k and $P_{k,\nu+1}^-$ depend on the radiation distribution $I(\omega)$, which is either a given quantity or must be calculated consistently. The latter would then require the radiation transport equation as an additional equation.

In Eq. (24) we introduce in the velocity space spherical polar co-ordinates with the axes along the field using $\mu = \cos\theta$, θ being the polar angle. At the same time, we separate the collision term into the two contributions of the electron-electron interaction and the electron interaction with the heavy particles. This is sensible since for the electron-electron interaction an expansion is advised which is different from that used for the electron heavy-particle interaction. This transformation and separation results in

$$\frac{e}{m} E \mu \frac{\partial f}{\partial v} + (1 - \mu^2) \frac{e}{m} \frac{E}{v} \frac{\partial f}{\partial \mu} = \left(\frac{\delta f}{\delta t}\right)_{--} + \sum_{\alpha} \left(\frac{\delta f}{\delta t}\right)_{\alpha} \tag{26}$$

with

$$\left(\frac{\delta f}{\delta t}\right)_{\alpha} = \int \Bigg[f_{\alpha}(v_j') f(v', \mu') u' \sigma_{\alpha}(u') \left|\frac{d\underline{v}_{\alpha}'}{d\underline{v}_{\alpha}}\right| \left|\frac{d\underline{v}'}{d\underline{v}}\right| - f_{\alpha}(v_{\alpha}) f(v, \mu) u \sigma(u) \Bigg] d^2\Omega \tag{27}$$

The sum over α covers all heavy-particle components and all corresponding internal states. Moreover, Eq. (27) describes elastic and inelastic collisions.

To achieve further progress one expands f in Legendre polynomials:

$$f = \sum_{\nu} P_{\nu}(\mu) f^{\nu}(v) \tag{28}$$

and applies the Lorentz approximation $m \ll m_{\alpha}$. Further, we restrict ourselves to moderate ionizations for which the electron-electron collision term can be reduced to its isotropic part. With this we find for the zero and first-order term of (28) the equations

$$\frac{e}{m} \frac{E}{3v^2} \frac{d}{dv} \left\{ v^2 f^1 \right\} = \left(\frac{\delta f^0}{\delta t}\right)_{--} + \sum_{\alpha} \left(\frac{\delta f^0}{\delta t}\right)_{\alpha} \tag{29a}$$

$$\frac{e}{m} E \frac{df^0}{dv} = \sum_{\alpha} \left(\frac{\delta f^1}{\delta t}\right)_{\alpha} \tag{29b}$$

The evaluation of the elastic-heavy-particle collision contributions yields in the Lorentz approximation

$$\left(\frac{\delta f^1}{\delta t}\right)_{el} = - \nu_1 (v)\, f^1 (v) \tag{30a}$$

$$\left(\frac{\delta f^0}{\delta t}\right)_{el} = \frac{1}{v^2} \frac{d}{dv} \left\{ \frac{m}{M} v^3 \nu_1 (v) \left(f^0 + \frac{kT_0}{mv} \frac{df^0}{dv} \right) \right\} \tag{30b}$$

with the momentum transfer collision frequency

$$\nu_1 (v) = \int f_0\, d\underline{v}_0 \int (1 - \cos \theta)\, \sigma (v,\theta)\, d^2\Omega \tag{31}$$

Since under the conditions specified ($kT_- \ll \epsilon_1$), the elastic effect on f^1 dominates the corresponding inelastic influence, it is sufficient to keep the inelastic effect only in Eq. (29b). With this, Eqs (29) and (30) result in

$$\frac{1}{v^2} \frac{d}{dv} \left\{ \frac{m}{M} v^3 \nu_1 (v) f^0 + \left(\frac{1}{3} \frac{e^2}{m^2} \frac{E^2}{\nu_1 (v)} + \nu_1 (v) \frac{kT_0}{M} \right) v^2 \frac{df^0}{dv} \right\} + \left(\frac{\delta f^0}{\delta t}\right)_{--} + \sum_\alpha \left(\frac{\delta f^0}{\delta t}\right)_{in,\alpha} = 0 \tag{32}$$

which couples f^0 with (25) via $(\delta f^0/\delta t)_{in,\alpha}$.

In the case where inelastic collisions and electron-electron interaction are negligible, the solution is straightforward and yields the *Davydov* distribution:

$$f^0 = C \exp \left\{ - \int_0^v \frac{mv'\, dv'}{kT_0 + \frac{M}{3} \frac{e^2}{m^2} \frac{E^2}{\nu_1^2 (v')}} \right\} \tag{33}$$

which for constant ν_1 is a *Maxwell* distribution with an electron temperature:

$$kT_- = kT_0 + \frac{1}{3} \frac{M}{m} \frac{e^2 E^2}{m\, \nu_1^2} \tag{34}$$

and for a constant mean free path with $\nu_1(v) \sim v$ yields the *Druyvesteyn* distribution:

$$f^0 = C \exp\left\{-\left(\varepsilon/\left[eE\lambda(4M/3m)^{1/2}\right]\right)^2\right\} \tag{35}$$

which with increasing velocity drops much faster than the Maxwell distribution.

The effect of the electron interaction can be formulated via the Fokker-Planck formalism and gives

$$\left(\frac{\delta f}{\delta t}\right)_{--} = \frac{1}{v^2}\frac{d}{dv}\left\{v^2 \nu_-\left[a(v)f^0(v) + b(v)\, v\frac{df^0}{dv}\right]\right\} \tag{36}$$

with $\nu_- = n_- v q_-$ (q_- = Coulomb cross-section) and the abbreviations

$$a(v) = \frac{4\pi}{n_-}\int_0^v (v')^2 f^0(v')dv' \tag{37}$$

and

$$b(v) = \frac{4\pi}{3n_-}\left\{\frac{1}{v^2}\int_0^v (v')^4 f^0(v')\,dv' + v\int_v^\infty v' f^0(v')\,dv'\right\} \tag{38}$$

Including (36) in Eq. (33) results in the solution

$$f^0 = C \exp\left\{-\int_0^v \frac{\left[mv' + a(v')\, M\frac{\nu_-}{\nu_1}(v')\right]dv'}{\left[kT_0 + \frac{M}{3}\frac{e^2}{m^2}\frac{E^2}{\nu_1^2(v')} + v'b(v')\cdot M\frac{\nu_-}{\nu_1}(v')\right]}\right\} \tag{39}$$

instead of the Davydov distribution. The 'Maxwellizing influence' of the electron interaction is demonstrated in Fig. 5.

In the evaluation of Eq.(32) with (36) presented in Fig. 5 a fundamental simplification was introduced since we used n_- as an independent parameter. Actually, it would be necessary to calculate n_- from the set of equations (32) and (36) consistently with the set of equations (25). This would then involve the other internal states of the atom and the corresponding inelastic processes.

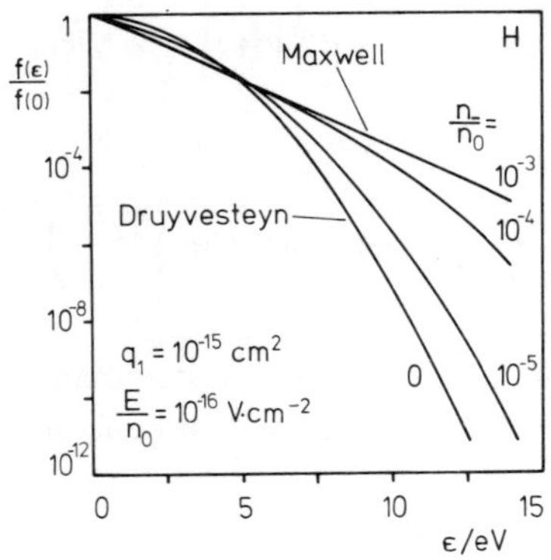

FIG.5. The transition from a Druyvesteyn distribution to a Maxwellian with increasing ratio of the electron density n_- and ground-state density n_0. A constant collision cross-section q_1 is assumed. E denotes the electric field in V·cm, and f the distribution function of the electron energy ε.

The excited states and the inelastic phenomena are the main feature still missing in our treatment. It requires the simultaneous solution of the set of equations (25), (32) and (36), where we recall that the coefficients in (25) are also dependent on the distribution function f found from (32) and (36).

In the formulation of the inelastic contributions, we proceed in a similar way as for elastic collisions, neglecting within the Lorentz approximation the term m/M altogether, since there is now an effective energy exchange via the inelastic collisions. In that approximation, we have then for the zero-order ($\nu = 0$) and first-order ($\nu = 1$) inelastic contribution

$$\left(\frac{\delta f^{\nu}}{\delta t}\right)_{in} = \sum_{i,j} \int \left[\left| \frac{d^3 \underline{v}_{ij}}{d^3 \underline{v}} \right| f^{\nu}(v_{ij})\, v_{ij}\, \sigma_{ij}(v_{ij}) \left\{\cos\theta\right\}^{\nu} - f^{\nu}(v)\, v\, \sigma_{ij}(v) \right] f_i(\underline{v}_i)\, d^3\underline{v}_i\, d^2\Omega \qquad (40)$$

Here the indices i, j refer to the various internal states; $f_i(v_i)$ is the distribution function of the excited states (i); v_{ij} denotes the electron velocity before the collision, and v that after the collision. The energy law relates them through

$$v_{ij}^2 = v^2 + \frac{2}{m}(\epsilon_j - \epsilon_i) \qquad (41)$$

We want to rewrite this double sum, in which each pair i, j appears twice, as a sum in which each transition occurs only in one combination ($i < j$). This requires a straightforward rearrangement and an application of the detailed balancing. The result is:

$$\left(\frac{\delta f^{\nu}}{\delta t}\right)_{in} = \sum_{i<j} n_i \left\{ \overbrace{\frac{v_{ij}^2}{v} Q^{\nu}_{ij}(v_{ij}) f^{\nu}(v_{ij})}^{(GI)} - \overbrace{v\, Q^{0}_{ij}(v) f^{\nu}(v)}^{(LI)} \right\}$$

$$+ \frac{g_i}{g_j} n_j \left\{ \underbrace{v\, Q^{\nu}_{ij}(v) f^{\nu}(v_{ji})}_{(GII)} - \underbrace{\frac{v_{ij}^2}{v} Q^{0}_{ij}(v_{ij}) f^{\nu}(v)}_{(LII)} \right\} \tag{42}$$

The symbols are: G is gain; L is loss; I, II are collisions of the first and second kind; $\nu = 1;2$. The quantities Q^{ν}_{ij} are defined through

$$Q^{\nu}_{ij} = \int \sigma_{ij} \left\{\cos\theta\right\}^{\nu} d^2\Omega \tag{43}$$

g_i is the weight factor of the level i.

As one would expect, each term ($i < j$) of this sum due to the reduction now consists of four terms, and it is not difficult to interpret these. To make it easier, we have indicated in the formula the type of collision (first or second kind, I or II) and the type of process (gain or loss, G or L), where we emphasize that these latter expressions refer to the electron distribution function and not to the levels. The first term (GI) describes gain through transitions from v_j to $v < v_j$; the second term (LI) describes loss due to transitions from v to $v_{ij} < v$. The third term (GII) represents gain due to collisions increasing the velocity from v_{ji} to v. Finally, the fourth term (LII) describes collisions that change the velocity from v to $v_{ij} > v$.

A look at Eqs (25,(32),(36) and (42) shows that this problem certainly needs simplification. An extreme reduction can be reached if the equations describing the distribution function are simply replaced by the assumption of a Maxwellian distribution. Another crude approximation allows for only one effective internal state to describe the structure of the atom. More sophisticated is the approach to neglect in the electron balance gains of the first kind and losses of the second kind in comparison with the complementary processes. This

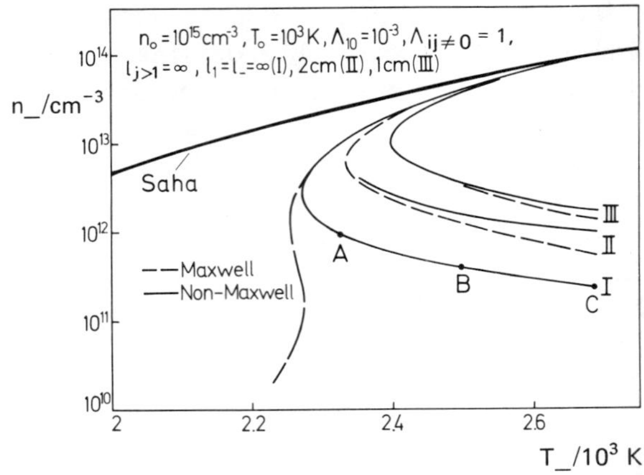

FIG. 6. Demonstration of the deviations from the equilibrium density due to radiation escape, diffusive effects and deviations from the Maxwell distribution. n_0 *is the ground-state density,* T_0 *the heavy-particle temperature,* $T\_$ *the electron temperature,* $n\_$ *the electron density,* Λ_{10} *the resonance radiation escape factor,* Λ_{ij} *the corresponding escape factor for all other transitions,* l_i *the diffusion lengths,* l_1 *and* $l\_$ *the diffusion lengths specifically for resonance states and electrons, respectively.*

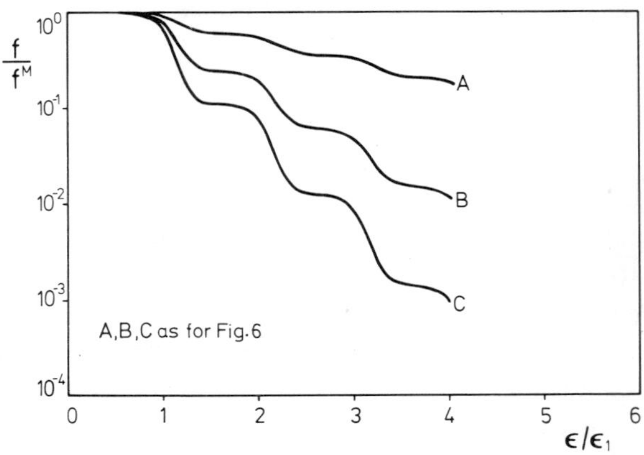

FIG. 7. Deviations of the electron distribution f from the Maxwell distribution f^M *for three typical states A, B and C, indicated in Fig. 6.* ϵ_1 *is the energy of the resonance level.*

assumption is justified if the electron distribution function decreases fast enough with increasing energy and if the average electron energy kT_ is small in comparison with the resonance level energy ϵ_1.

There is much information about approximate solutions of this kind in the literature (e.g. [9–16]). We present here results that have been calculated recently for a caesium plasma, accounting analytically by a recursion method for deviations from the Maxwell distribution up to energies well above the ionization energy.

We have included in the analysis 41 atomic energy levels, 25 of which are assumed to be in a non-equilibrium state. We have also allowed for diffusion of excitation states and ambipolar diffusion, total radiation escape for higher levels, and partial radiation escape for the resonance level [17]. In this model, the deviation of the electron distribution is dominated by the imbalance of the resonance level and governed by gains of the second kind and losses of the first kind. The results of this evaluation, which starts from the basis described above but requires a lot of additional analysis that can be found in the References, are shown in Figs 6 and 7, which are quite instructive.

Figure 6 presents various curves of the electron density n_- as a function of the electron temperature T_- for a given neutral ground-state density n_0 and heavy-particle temperature T_0, a constant resonant radiation loss factor Λ_{10}, total radiation loss for all non-resonance transitions and no diffusion loss for all excited states except the resonance state. The three curves, I, II and III, refer to different diffusion lengths of electrons l_- and resonance state l_1. The deviation from the equilibrium value is strong everywhere. The deviations of the solutions with the assumption of Maxwell distribution from those with a consistent solution of the electron distribution function is noticeable in all cases but very strong for small values of electron and resonance state diffusion. The n_- (T_-) functions are multivalued above certain critical values of the electron temperature and the plasma loss parameters. This ambiguity can be explained by the cumulative ionization mechanism, as all the features of this figure are open to physical interpretation. However, since space is limited and the problem relatively simple, we can leave this task to the reader.

Figure 7 shows the deviations from the Maxwellian electron distribution for the points A, B and C indicated in Fig. 6. The deviations are obviously quite strong and show a stepwise character. This feature can be understood from the fact that the deviations are dominated by gain of collisions of the second kind and loss of the first kind which both couple to the lower energies. Moreover, it is essential that only the resonance level contributes decisively to the deviation from the Maxwell distribution.

5. CONCLUDING REMARK

From the preceding section, which presents as an example one of the advanced descriptions of a plasma in a stationary non-equilibrium state, it should be obvious that a great distance has still to be covered in this field. The description of the diffusion processes in the configuration space through diffusion lengths is rather crude. Field effects which would cause effects similar to those demonstrated in the distributions of Fig. 5 are neglected. The assumption of moderate ionization degree and low plasma temperatures $(kT_- \ll \epsilon_1)$ is restricting. Its relaxation would complicate the analysis extremely.

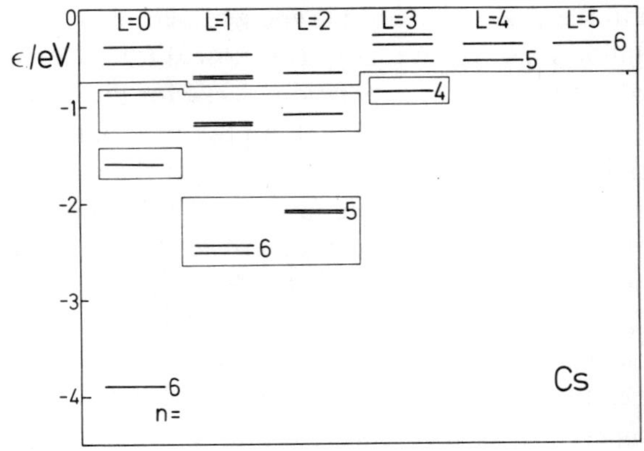

FIG.8. Level scheme of caesium. Grouping of the levels into an effective scheme is indicated according to Ref. [17].

Finally, and most important, the assumptions about the radiation are not too realistic and, in general, for a proper account of the radiation, it would be necessary to add the transport equation of radiation to the system of equations (25), (32), (36) and (42).

There is little hope for a general solution including all these effects if some kind of a basic simplification is not found. One such possibility may be the replacement of the true scheme of the internal states by a simpler effective one, as has been tried in a recent publication [17], where it is claimed that for most applications it is possible to replace the complicated scheme of the caesium atom by a small number of effective levels, as indicated in Fig. 8.

If this simplification or a similar one is not possible, then I am afraid we shall probably have no choice but to try for specific results on a computer with programs which, as our experience shows, will be frighteningly extensive and expensive.

REFERENCES

[1] ECKER, G., in Proc. 2nd Int. Conf. on Plasma Theory, Kiev, 1974.
[2] EBELING, W., Physica **43** (1969) 293.
[3] EBELING, W., SÄNDIG, R., Ann. Phys. **28** (1973) 289.
[4] EBELING, W., Physica **73** (1974) 573.
[5] ECKER, G., KRÖLL, W., Z. Naturforsch. A. **21** (1966) 2012.

[6] ECKER, G., KRÖLL, W., Phys. Fluids **6** (1963) 62.
[7] ECKER, G., KRÖLL, W., Z. Naturforsch. A. **21** (1966) 2023.
[8] ELWERT, G., Z. Naturforsch. A. **7** (1952) 432, 703.
[9] BATES, D.R., KINGSTON, A.E., McWHIRTER, R.W.P., Proc. R. Soc. (London) A. **267** (1962) 297; **270** (1962) 155.
[10] NORCROSS, D.W., STONE, P.W., Quant. Spectrosc. Radiat. Transfer **8** (1968) 655.
[11] DRAWIN, H.W., Z. Naturforsch. A. **25** (1970) 145.
[12] SAYER, B., JEANNET, J.C., BERLANDE, J., J. Phys. (Paris) **33** (1972) 993.
[13] SHAW, J.F., MITCHNER, M., KRUGER, C.H., "Necessary conditions for the validity of the two-temperature plasma model", Electricity from MHD 1968 (Proc. Symp. Warsaw, 1968) **1**, IAEA, Vienna (1968) 53.
[14] BAKSHT, F.G., MOIZHES, B.Ya., NEMCHINSKIJ, V.A., Sov. Phys. – Tech. Phys. **13** (1969) 1401.
[15] SHAW, D.T., J. Appl. Phys. **42** (1971) 490.
[16] CLAASSEN, H.A., Z. Naturforsch. A. **30** (1975) 451.
[17] HELRICH, C.S., Kernforschungsanlage Jülich Rep. JÜL.-1108-TP (1974).

IAEA-SMR-32/2

A THEORY FOR CORRELATIONS IN STRONGLY TURBULENT PLASMAS

J.H. MISGUICH
Association Euratom-CEA,
Département de la physique du plasma
et de la fusion contrôlée,
Centre d'études nucléaires de Fontenay-aux-Roses,
Fontenay-aux-Roses, France

R. BALESCU
Association Euratom-Etat belge,
Faculté des sciences,
Université libre de Bruxelles,
Brussels, Belgium

Abstract

A THEORY FOR CORRELATIONS IN STRONGLY TURBULENT PLASMAS.

A kinetic theory for binary correlations in turbulent plasmas is presented which is an extension of the Dupree-Weinstock method. It is shown that the effects associated with "clumps" can be obtained as a lowest approximation of this general theory.

INTRODUCTION

Let us begin with a remark on 'collisionless' plasma turbulence. Turbulence in a high-temperature low-density plasma appears to be very different from fluid turbulence:

(a) In neutral fluid, indeed, the energy is fed into the system in the form of macroscopic eddies. These eddies are transformed non-linearly into smaller ones and, for wavelengths of the order of the mean free path, the energy is dissipated by viscosity, thus by particle collisions.

(b) In hot fusion plasmas the mean free path is extremely long. The collective motions are due to the long-range interactions and fields, but the spatial range of the waves remains much smaller than the mean free path. The particle collisions thus play a minor role.

An essential problem of plasma turbulence is to understand the *occurrence of a new type of dissipative mechanism of evolution in the absence of the binary collisions* which explain irreversibility in neutral fluids or cold plasmas. In a so-called collisionless plasma the collective excitations of the turbulence appear spontaneously through the interactions (wave-wave and wave-particle interactions). The problem is to describe the *generation* mechanism for particle correlations in the absence of true particle collisions.

About clumps, Kadomtsev and Pogutse on one hand [1] and Dupree on the other [2–4], during their stay at the ICTP, Trieste, in 1970, suggested that particle correlations in a turbulent plasma have a peculiar structure leading to a kind of 'granulation' of the fluid in phase space. It is said that the turbulent plasma consists of 'clumps' treated as true macroparticles. These objects could explain the anomalous transport processes observed in plasmas.

To understand the formation of clumps, it is necessary to explain how the correlations of particles which are close together in phase space can be enhanced, and how the lifetime of these correlations can be significantly dilated as a result of the turbulence. Following a very important comment of Engelmann and Morrone [5], these clump effects can only be explained if one can demonstrate the *existence of a mechanism by which the correlations are continuously generated from the average one-particle distribution function* (DF) *through the plasma interactions.*

In a collisional fluid this mechanism is provided by the particle collisions. In a kinetic regime there exists a general Bogolyubov relation which expresses binary correlations as a functional of the one-particle DF.

To understand the formation of clumps in a hot plasma it is thus necessary to demonstrate the existence of a similar relation which holds even in the absence of true collisions.

1. KINETIC THEORY FOR THE ONE-PARTICLE DISTRIBUTION FUNCTION

At the last Kiev Conference [6, 7] we presented a generalization of the Dupree-Weinstock method [8, 9]. We start from the Klimontovich equation for the exact (microscopic) DF

$$\frac{\partial}{\partial t} f_1(\vec{x}, \vec{v}, t) = L_1 f_1(\vec{x}, \vec{v}, t) \tag{1.1}$$

with

$$f_1(\vec{x}, \vec{v}, t) = \sum_{i=1}^{N} \delta\,[\vec{x} - \vec{x}_i(t)]\,\delta\,[\vec{v} - \vec{v}_i(t)] \tag{1.2}$$

$$L_1 = -\vec{v}\cdot\nabla - \frac{q}{m}\vec{E}(\vec{x},t)\cdot\frac{\partial}{\partial\vec{v}} \tag{1.3}$$

The electric field is to be determined afterwards in terms of f_1 [10].

We take the ensemble average of this equation:

$$\partial_t \bar{f} = \bar{L}\,\bar{f} + \langle L'f'\rangle \tag{1.4}$$

and solve formally the fluctuating equation for f'. One obtains in this way the *exact master equation* of Weinstock:

$$\partial_t \bar{f}_1(t) = \bar{L}_1(t)\bar{f}_1(t) + \int_{t_0}^{t} dt' M_{11}(t,t')\,\bar{f}_1(t') + s_1(t,t_0) \tag{1.5}$$

This equation is non-Markovian, and depends on the initial fluctuation $f'(t_0)$ through the source term s_1. We have proved that one can also write this equation as [7]

$$\partial_t \bar{f}_1(t) = [\bar{L}_1(t) + G_{11}(t)]\,\bar{f}_1(t) + \hat{S}_1(t,t_0) \tag{1.6}$$

in terms of the turbulent collision operator G_{11}. Contrary to s_1, the source term $\hat{S}_1$ does not vanish for $f'(t_0) = 0$. This source term is proved to vanish for a general class of initial fluctuations given by

$$f_1'(t_0) = \mathbb{C}_1(t_0)\,\bar{f}_1(t_0) \tag{1.7}$$

The remarkable property of (1.7) is that this functional relation, giving the fluctuations in term of $\bar{f}_1$, is conserved by exact motion. The creation operator is such that the turbulent collision operator is given by

$$G_{11}(t) = \langle L_1'(t)\,\mathbb{C}_1(t)\rangle \tag{1.8}$$

In simple cases, the latter reduces to a generalized diffusion operator in velocity space.

A *kinetic regime* of a developed plasma turbulence is defined as being independent of the precise value of the initial fluctuation. This regime is thus obtained when (1.7) holds: $\hat{S}_1 = 0$. We thus obtain the *general kinetic equation* [7]:

$$\partial_t \bar{f}_1(t) = [\bar{L}_1(t) + G_{11}(t)]\,\bar{f}_1(t) \tag{1.9}$$

where the turbulent collision operator takes non-Markovian effects into account:

$$G_{11}(t) = \int_0^{\infty} d\tau \; \langle L_1'(t)\, \Lambda_1(t, t-\tau)\, L_1'(t-\tau)\rangle\, \overline{V}_1(t-\tau, t) \qquad (1.10)$$

by means of the one-particle turbulent propagator $\overline{V}_1$. This propagator takes into account the effect of the turbulence on particle trajectories, and yields to the usual effects of resonance broadening and frequency shift.

2. ICHIMARU'S THEORY FOR BINARY CORRELATIONS

The theory developed by Ichimaru [11] for strongly correlated plasma starts from the BBGKY hierarchy:

$$\partial_t \overline{f}_1 = \overline{L}_1 \overline{f}_1 + \int d2\, \theta_{12} \overline{g}_{12} \qquad (2.1)$$

$$\partial_t \overline{g}_{12} = (\overline{L}_1 + \overline{L}_2)\overline{g}_{12} + (1 + P_{12}) \int d3\, \theta_{13} [\underbrace{\overline{f}_1 \overline{g}_{23}}_{\text{BLG}} + \underbrace{\overline{g}_{123}}_{\text{TRIPLE}}]$$

$$+ \vec{\mathscr{V}}_{12} \cdot \left(\frac{\partial}{\partial \vec{v}_1} - \frac{\partial}{\partial \vec{v}_2}\right) [\underbrace{\overline{f}_1 \overline{f}_2}_{\text{L}} + \underbrace{\overline{g}_{12}}_{\text{B}}] \qquad (2.2)$$

including the 'discrete' terms L and B. Let us recall that the usual kinetic equations are obtained by only considering in (2.2) the free flow term along with:

term L	for the Landau equation
terms L and B	for the Boltzmann equation
terms L and BLG	for the Balescu-Lenard-Guernsey equation

and by neglecting triple correlations (TC).

The characteristic feature of Ichimaru's approach is the *crucial role of the TC*. For plasma turbulence he introduced, without too strong a justification, the

Kirkwood superposition approximation relating TCs to the lower binary correlations. Assuming the existence of a velocity-independent non-equilibrium radial distribution function, he used this spatial part of binary correlations to *renormalize the microscopic field* in the hierarchy ($\overline{L}, \theta, \vec{v} \Rightarrow \overline{L}^*, \theta^*, \vec{v}^*$). This represents an effect of strong turbulence in modifying the interactions by a form factor. He obtains [11]

$$\partial_t \overline{g}_{12} = [\overline{L}_1^* + \overline{L}_2^*]\, \overline{g}_{12} + (1 + P_{12}) \int d3\ \theta_{13}^* \overline{f}_1\, \overline{g}_{23} + H_{12} \tag{2.3}$$

This equation has the same form as the equation leading to the BLG equation in the collisional regime. The source term,

$$H_{12} = (1 + P_{12}) \left[\theta_{12}^* \overline{f}_1 \overline{f}_2 + \int d3\ \theta_{13}^*\ \overline{g}_{12} \overline{g}_{23} / \overline{f}_2\right] \tag{2.4}$$

however, contains a non-linear contribution from TCs.

In this approximate description, the turbulence thus introduces two effects: (a) modification of the interactions, and (b) non-linear modification of the source term. The latter remains significant even in the limit of rare collisions ($L \to 0$), provided the level of turbulence is strong enough.

We shall see how this renormalization and non-linear source term can be obtained in a general kinetic formulation.

3. KINETIC THEORY FOR BINARY CORRELATIONS

The generalized Dupree-Weinstock method can be extended to the study of two-particle functions [12]. We start from the two-particle Klimontovich equation:

$$\partial_t\, f_1 f_2 = [L_1 + L_2]\, f_1 f_2 \tag{3.1}$$

and define the average two-particle function:

$$\overline{m}_{12} \equiv \langle f_1 f_2 \rangle = \overline{f}_1 \overline{f}_2 + \overline{c}_{12} \tag{3.2}$$

where the correlation $\overline{c}_{12}$ includes the usual correlation $\overline{g}_{12}$ along with the 'self-correlation':

$$\overline{c}_{12} = \overline{g}_{12} + \overline{g}_{12}^{SELF} \tag{3.3}$$

$$\bar{g}_{12}^{SELF} = \delta(\vec{x}_1 - \vec{x}_2)\ \delta(\vec{v}_1 - \vec{v}_2)\ \bar{f}_1 \qquad (3.4)$$

The fluctuating two-particle function is given by the difference:

$$f'_{12} = f_1 f_2 - \bar{m}_{12} \qquad (3.5)$$

By using similar methods, we derive an exact equation for the average two-particle function [12]:

$$\partial_t \bar{m}_{12}(t) = [\bar{L}_1(t) + \bar{L}_2(t) + \mathcal{G}_{12}(t)]\ \bar{m}_{12}(t) + \hat{S}_{12}(t, t_0) \qquad (3.6)$$

in terms of a new turbulent collision operator:

$$\mathcal{G}_{12} = G_{11} + G_{22} + G_{12} + G_{21} + \mathcal{L}^{+}_{12} \qquad (3.7)$$

and of a source term $\hat{S}_{12}$ which depends on the initial fluctuation $f'_{12}(t_0)$ through the difference

$$f'_{12}(t_0) - \mathbb{C}_{12}(t_0)\ \bar{m}_{12}(t_0)$$

This source term is thus vanishing when the fluctuation satisfies

$$f'_{12}(t_0) = \mathbb{C}_{12}(t_0)\ \bar{m}_{12}(t_0) \qquad (3.8)$$

In the kinetic regime, $\hat{S}_{12} = 0$ and we obtain the *general kinetic equation* [12] *for the average two-particle function:*

$$\partial_t \bar{m}_{12}(t) = [\bar{L}_1 + \bar{L}_2 + \mathcal{G}_{12}(t)]\ \bar{m}_{12}(t) \qquad (3.9)$$

The corresponding turbulent propagator is denoted by $\bar{V}_{12}$, which takes into account the trajectory correlations between two electrons, due to the turbulent fields: $\bar{V}_{12} \neq \bar{V}_1 \bar{V}_2$. This property is due to the interaction Liouvillian $\mathcal{L}'_{12}$ in the following decomposition:

$$\bar{L}_1 + \bar{L}_2 + \mathcal{G}_{12} = \underbrace{\bar{L}_1 + G_{11} + \bar{L}_2 + G_{22}}_{\mathcal{L}^0_1 + \mathcal{L}^0_2 \equiv \mathcal{L}^0_{12}} + \underbrace{G_{12} + G_{21} + \mathcal{L}^{+}_{12}}_{\mathcal{L}'_{12}} \qquad (3.10)$$

From the kinetic equations for $\bar{m}_{12}$ and $\bar{f}_1$ or $\bar{f}_2$ we easily derive the following *kinetic equation for binary correlation* [12]:

$$\partial_t \bar{c}_{12}(t) = [\bar{L}_1 + \bar{L}_2 + \underline{G_{11} + G_{22} + \{G_{12} + G_{21} + \mathcal{L}^{+}_{12}\}}]\ \bar{c}_{12}$$
$$+ \{G_{12} + G_{21} + \underline{\mathcal{L}^{+}_{12}}\}\ \bar{f}_1 \bar{f}_2 \qquad (3.11)$$

The essential feature of this result is the occurrence of the *source term*. This term is reponsible for the generation of correlations from the one-particle DF.

By using Poisson's equation and comparing with the BBGKY hierarchy, one can prove that the underlined terms in (3.11) correspond to the contribution of TCs (along with the discrete Boltzmann term). The first part of the source term $(G_{12}+G_{21})\,\overline{f}_1\,\overline{f}_2$ can be proved to correspond to the BLG term (along with the discrete Landau term) and this term thus corresponds to the usual source term of the kinetic theory of plasmas. The contribution of TCs in the source term $(\mathscr{L}^{+}_{12}\,\overline{f}_1\,\overline{f}_2)$ is, however, essential in *strong* turbulence where discrete terms are negligible [11]. Let us remember that this last feature has also been found in Ichimaru's description, where the 'most non-linear' part of TCs acts as a source for binary correlations.

In the kinetic regime, binary correlations do not depend on initial correlations, and the solution of Eq.(3.11) is given by:

$$\overline{c}_{12}(t) = \int_0^\infty d\tau\,\overline{V}_{12}(t,t-\tau)\,[G_{12}+G_{21}+\mathscr{L}^{+}_{12}]\,\overline{V}_1\,\overline{V}_2(t-\tau,t)\,\overline{f}_1(t)\overline{f}_2(t)$$

$$\equiv \chi_{12}(t)\,\overline{f}_1(t)\,\overline{f}_2(t) \qquad (3.12)$$

We thus obtain a functional relation between correlations and one-particle DF, which is similar to the Bogolyubov relation that holds in the kinetic stage of collisional systems. *This result describes the creation mechanism of binary correlations through the internal interactions of the turbulent plasma.* This is the result we were looking for.

Here $\overline{V}_{12}$ is the binary propagator associated with $\overline{m}_{12}$, which is different from the factorized propagator associated with $\overline{f}_1\,\overline{f}_2$: *this effect of trajectory correlations is the basic mechanism for the existence of clumps.*

When we substitute this solution (3.12) in the first hierarchy equation, we obtain a closed non-linear equation for $\overline{f}_1$ [12]:

$$\partial_t\overline{f}_1(t) = \overline{L}_1(t)\,\overline{f}_1(t) + \int d2\,\theta_{12}\,\chi_{12}(t)\,\overline{f}_1(t)\overline{f}_2(t)$$

This equation has the same structure as the BLG equation, but for strong turbulence it has a very different physical content. It can be proved that this general kinetic equation reduces to the BLG equation in the limit of vanishing TCs. Generalized BLG equations of this form were obtained previously by Hatori [13], Dupree [14] and Pelletier and Pomot [15].

4. 'ANOMALOUS' CORRELATIONS AND CLUMPS

Let us see how these general results can be evaluated explicitly in a rather simple approximation called QL 2. This approximation represents the first step beyond the renormalized quasilinear (RQL) description. It takes into account, to the lowest order, trajectory correlations; thus the property $\overline{V}_{12} \neq \overline{V}_1 \overline{V}_2$. The binary propagator can be calculated as [16]

$$\overline{V}_{12}(t, t-\tau) = \mathscr{H}(t, t-\tau)\, \overline{V}_1(t, t-\tau)\, \overline{V}_2(t, t-\tau) \tag{4.1}$$

in terms of a trajectory correlator $\mathscr{H}$ and the renormalized one-particle turbulent propagators. When acting on a function of space, this operator $\overline{V}_{12}$ introduces

$$\overline{V}_{12}(t, t-\tau) \exp(i\vec{k}\cdot\vec{r}) = \left\langle \exp\,[i\vec{k}\cdot\vec{r}\,(t-\tau)] \right\rangle \xrightarrow{\text{QL2}} \exp\,[i\vec{k}\cdot\langle \vec{r}(t-\tau)\rangle]$$
$$\times \exp[-\tfrac{1}{2}\,\vec{k}\vec{k} : \langle \tilde{\vec{r}}\,(t-\tau)\, \tilde{\vec{r}}\,(t-\tau)\rangle] \tag{4.2}$$

a cumulant expansion in terms of fluctuating particle trajectories. The QL 2 approximation is responsible for the truncation after the second cumulant. All this is very similar to what happens in the usual RQL approximation. Our formalism makes the connection between the propagator calculations and the particle trajectory calculations.

The present QL 2 approximation goes, however, beyond the RQL or Dupree-Weinstock description, because $\langle \tilde{\vec{r}}\,\tilde{\vec{r}}\rangle$ is modified here by trajectory correlations:

$$\langle \tilde{\vec{r}}\,\tilde{\vec{r}}\rangle = \langle \tilde{\vec{x}}_1 \tilde{\vec{x}}_1\rangle + \langle \tilde{\vec{x}}_2\, \tilde{\vec{x}}_2\rangle \; - \; [\langle \tilde{\vec{x}}_1\, \tilde{\vec{x}}_2\rangle + \langle \tilde{\vec{x}}_2\, \tilde{\vec{x}}_1\rangle] \tag{4.3}$$

Because of the trajectory correlator $\mathscr{H}$, the bracket is different from zero. It can even be proved that this bracket is positive for neighbouring particles in phase space, so that *trajectory correlations actually reduce the relative diffusion* $\langle \tilde{\vec{r}}^{\,2}\rangle$ of these particles. This is the physical mechanism responsible for 'clump' effects.

The binary correlations can be evaluated in this QL 2 approximation, which takes into account trajectory correlations beside the usual Dupree-Weinstock renormalization. The form of the result is similar to Dupree's result [17]:

$$\overline{c}_{12}(t) = \int d\vec{k}\, e^{i\vec{k}\cdot\vec{r}}\, \tau_{\vec{k}}\,(\vec{r}, \vec{v}_1, \vec{v}_2)\, \overleftrightarrow{d}_{\vec{k}}(\vec{v}_1, \vec{v}_2) : \frac{\partial}{\partial \vec{v}_1} \frac{\partial}{\partial \vec{v}_2}\, \overline{f}_1(t)\, \overline{f}_2(t) \tag{4.4}$$

Here $\vec{\vec{d}}_{\vec{k}}$ is the Fourier transform of a diffusion coefficient which depends on $\vec{r}$, $\vec{v}_1$ and $\vec{v}_2$, while $\tau_{\vec{k}}$ represents the *correlation lifetime.* Using the previous result for the trajectory correlator, this time can be written:

$$\tau_{\vec{k}}(\vec{r}, \vec{v}_1, \vec{v}_2) = \int_0^{\infty} d\tau \exp(-i\vec{k}\cdot\vec{r}) \exp[i\vec{k}\cdot\langle\vec{r}(t-\tau)\rangle]$$

$$\times \exp[-\tfrac{1}{2}\vec{k}\vec{k}:\langle\tilde{\vec{r}}(t-\tau)\,\tilde{\vec{r}}(t-\tau)\rangle] \tag{4.5}$$

Its calculation is thus reduced to the solution of a relative diffusion problem, thus to the study of the average relative distance, and the dispersion due to the fluctuating fields. By evaluating these quantities we find that in the RQL approximation (which does *not* contain clump effects)

$$\langle\tilde{r}^2(t-\tau)\rangle^{RQL} = \tfrac{1}{3}\lambda_0^2\left(\frac{\tau}{\tau_0}\right)^3 \tag{4.6}$$

where λ_0 is the lengthscale of the turbulent spectrum and τ_0 the corresponding time (λ_0/τ_0 is the trapping velocity). The τ^3 behaviour is well known and is related to the usual Dupree damping which produces resonance broadening. In this approximation, the correlation lifetime is of the order of τ_0:

$$\tau_{\vec{k}}^{RQL} \approx \tau_0 \tag{4.7}$$

On the other hand, if we now take the trajectory correlations into account, we find that for small values of the initial distance r and the initial velocity difference g,

$$\langle\tilde{r}^2(t-\tau)\rangle^{QL2} = \frac{1}{6}(r^2 + \frac{1}{10}\tau^2 g^2)\left(\frac{\tau}{\tau_0}\right)^3 \tag{4.8}$$

This term goes to zero for weak r and g, and we find *an important reduction of relative diffusion.* This explains why particles tend to move together and to form a so-called 'clump'.

Following Dupree [2] the *clump lifetime* can be defined as the time necessary for two particles, initially separated by r and g, to diffuse up to a dispersion equal to λ_0:

$$\langle\tilde{r}^2(t+\tau_{cl})\rangle = \lambda_0^2\ , \quad \Rightarrow \tau_{cl}(r,g) \tag{4.9}$$

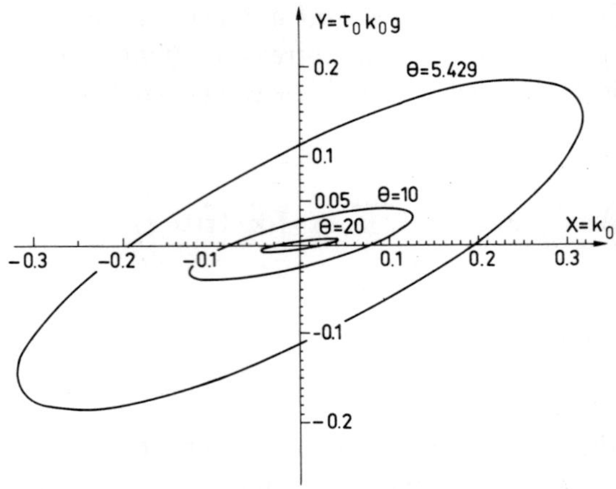

FIG.1. Phase-space contours of 1D clump lifetime.

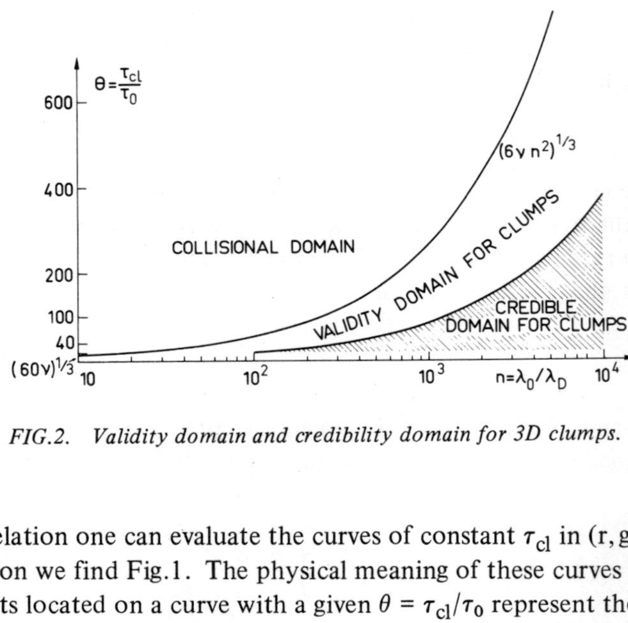

FIG.2. Validity domain and credibility domain for 3D clumps.

From this relation one can evaluate the curves of constant τ_{cl} in (r, g) space. In one dimension we find Fig.1. The physical meaning of these curves is as follows. All the points located on a curve with a given $\theta = \tau_{cl}/\tau_0$ represent the initial relative distance (in phase space) of two particles which will diffuse over a distance λ_0 in a time $\theta\,\tau_0$. Thus, all the particles located in phase space within an ellipse parametrized by θ are considered as belonging to the same 'clump', with lifetime $\tau_{cl}(r, g) = \tau_0\theta(r, g)$. The clump is said to be destroyed after that time. As diffusion is decreased by the effects of trajectory correlations, the particles are found to 'move together' for a long time. This is why Dupree calls such a set of particles a 'macroparticle'. We think that the essence of a clump is the enhanced correlation between particles; this is not quite equivalent to a discrete macroparticle description.

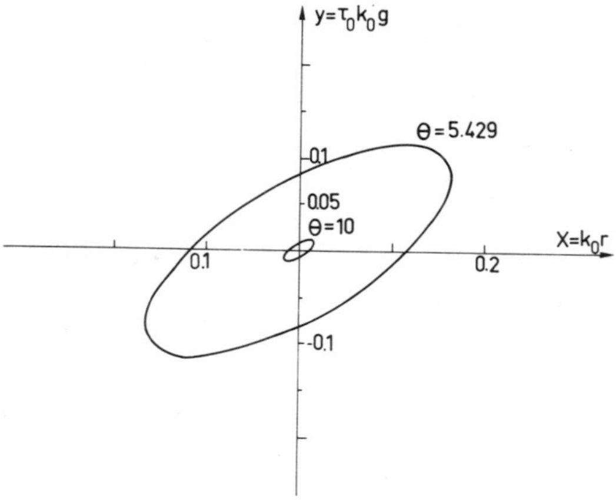

FIG.3. Phase-space contours of 1D clump lifetime, from Dupree's formula.

There exists of course a *validity domain* for this result (Fig.2). We have an upper and a lower bound for the distance of the particles in phase space (the particles have to be close to each other, but not too near in order to avoid collisional repulsions). This gives an upper and a lower bound for the value of the clump lifetime, as function of the turbulent lengthscale $n \equiv \lambda_0/\lambda_D$.

The *dimension of a clump* in phase space, for a given clump lifetime, is given by the extension of the corresponding ellipse. Exactly like a Debye sphere, a clump cannot be localized in the physical space: around each particle one can draw a Debye sphere and a clump for each θ. In physical space a clump θ has a maximum spatial extension $L(\theta)$, corresponding to the maximum r of the ellipse, but the clump is composed only of the particles with a velocity nearly equal to the velocity of the central particle.

The essential point in our result is that the *spatial dimension of clumps is much larger than predicted by Dupree*. His result in one dimension is given in Fig.3. The ratio between the spatial dimension of clumps in both theories is represented in Fig.4; this factor can go up to 400 in the validity domain. For instance, when the turbulent lengthscale is about 10^3 Debye lengths, the characteristic dimension of clumps is about several tens of Debye lengths.

In velocity space, however, the dimension of clumps remains very small, of the order of λ_D/τ_0, which is much smaller than the trapping velocity. This justifies in part the 'singular clump' model considered by Dupree by assuming that all particles in the clump have the same velocity:

$$r \lesssim \lambda_0 = n\lambda_D \xrightarrow{n=10^3} r \sim \text{several } 10\ \lambda_D$$

$$g \sim \frac{\lambda_D}{\tau_0} = \frac{1}{n}\frac{\lambda_0}{\tau_0} \ll \Delta\, v_{\text{trapping}} \quad (n \gg 1)$$

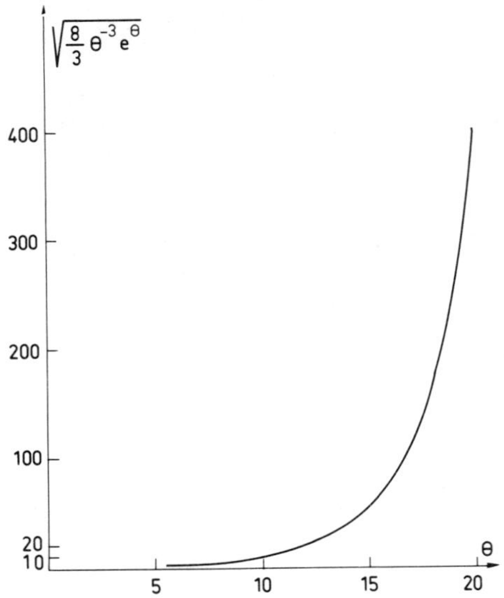

FIG.4. Enhancement of the 1D clump dimension with respect to Dupree's formula.

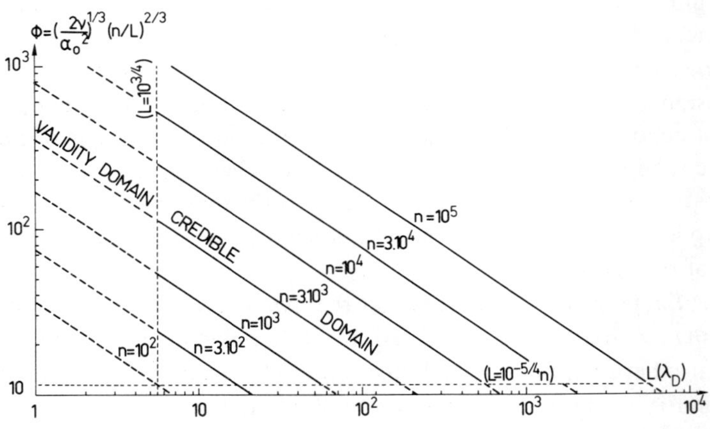

FIG.5. Enhancement factor for binary correlation, for various turbulent lengthscales $n = \lambda_0/\lambda_D$.

We have seen that *trajectory correlations reduce the relative diffusion* for neighbouring particles. This means that the characteristic diffusion time τ_{cl} is much larger than τ_0 (and θ larger than one). As a consequence, the correlation lifetime is found to be enhanced:

$$\tau_{\vec{k}}^{QL2} \sim \tau_{cl} \gg \tau_{\vec{k}}^{RQL} \sim \tau_0 \qquad (4.10)$$

We thus find an *anomalous enhancement of binary correlations* for particles with small r and g, and this is the main point about clumps. It corresponds to long-range correlations ($\gg \lambda_D$). The enhancement factor is given in Fig.5 for several values of the turbulent lengthscale $\lambda_0 = n\lambda_D$: it can be as high as 10^2 up to distances of $10^2\ \lambda_D$.

A physical interpretation of this important effect arises from the fact that clumps also modify the dielectric function of the plasma [17]. As the screening mechanism is modified, we interpret the clump as *a generalization of the concept of Debye polarization cloud in a turbulent plasma.* This could explain the occurrence of a long-range 'turbulent effective potential' from the point of view of statistical mechanics, in agreement with the result of Tegeback and Stenflo [18]. They have shown indeed that exponential shielding occurs only at distances of the order of the turbulent lengthscale.

Further, as clumps are a result of trajectory correlations, we can say very roughly that neighbouring particles move together essentially because they feel the same long-range fluctuations: clumps could be thought of as a residue of all trapping phenomena in the turbulent waves.

In summary, we have derived a kinetic theory for turbulent correlations which describes the generation mechanism for correlations even without the usual binary collisions.

In a lowest approximation, the effect of trajectory correlations can be taken into account, and binary correlations take a form similar to the Dupree result. The characteristic time associated with spatial relative diffusion is defined as the clump lifetime. It has been shown that the effect of trajectory correlations reduces the relative diffusion and thus enhances the correlation lifetime; thus also the amplitude and the range of the correlations. The appearance of these long-range correlations is the main point about clumps. It is interpreted as a modified screening mechanism which appears in turbulent plasmas and is associated with a long-range effective potential.

REFERENCES

[1] KADOMTSEV, B.B., POGUTSE, O.P., Phys. Fluids **14** (1971) 2470.

[2] DUPREE, T.H., Phys. Fluids **15** (1972) 334.

[3] HUI, B.H., DUPREE, T.H., Phys. Fluids **18** (1975) 235.

[4] DUPREE, T.H., WAGNER, C.E., MANHEIMER, W.M., Phys. Fluids **18** (1975) 1167.

[5] ENGELMANN, F., MORRONE, T., Comments Plasma Phys. Contr. Fusion **1** (1972) 75.

[6] MISGUICH, J.H., BALESCU, R., in Proc. 2nd Int. Conf. Plasma Theory, Kiev, 1974.

[7] MISGUICH, J.H., BALESCU, R., Physica C **79** (1975) 373.

[8] DUPREE, T.H., Phys. Fluids **9** (1966) 1773.

[9] WEINSTOCK, J., Phys. Fluids **12** (1969) 1045; **15** (1972) 454.

[10] MISGUICH, J.H., BALESCU, R., Assoc. Euratom CEA Rep. EUR-CEA-FC Rep. 860 (1976).

[11] ICHIMARU, S., Basic Principles of Plasma Physics, Benjamin, New York (1973).
[12] MISGUICH, J.H., BALESCU, R., J. Plasma Phys. **19** (1978) 147.
[13] HATORI, T., J. Phys. Soc. Jpn. **27** (1969) 203.
[14] DUPREE, T.H., Phys. Rev. Lett. **25** (1970) 789.
[15] PELLETIER, G., POMOT, C., J. Plasma Phys. **14** (1975) 153.
[16] MISGUICH, J.H., BALESCU, R., Plasma Phys. **19** (1977) 611.
[17] MISGUICH, J.H., BALESCU, R., Assoc. Euratom-CEA Rep. EUR-CEA-FC 859 (1976) submitted for publication to Plasma Phys.
[18] TEGEBACK, R., STENFLO, L., Plasma Phys. **17** (1975) 991.

GROWING SECOND-ORDER WAVE IN THE LANDAU-VLASOV PROBLEM*

R.W.B. BEST
Association Euratom-FOM,
FOM-Instituut voor Plasmafysica,
Rijnhuizen, Jutphaas,
The Netherlands

Abstract

GROWING SECOND-ORDER WAVE IN THE LANDAU-VLASOV PROBLEM.

A short derivation is given of the relation between the Van Kampen mode and the dielectric function for a Vlasov plasma. A numerical example is given of a second-order growing low-frequency wave.

The one-dimensional motion of electrons with charge e and mass m, in a neutralizing homogeneous background of fixed ions with constant charge density −Ne, is governed by the equations of Vlasov and Poisson:

$$\frac{\partial f}{\partial t} + v\frac{\partial f}{\partial x} + \frac{e}{m}E\frac{\partial f}{\partial v} = 0, \quad \frac{\partial E}{\partial x} = \frac{e}{\varepsilon_o}\left(\int f dv - N\right) \tag{1}$$

in which f(x, v, t) is the electron distribution function and E(x, t) the electric field. The weakly non-linear solution consists of the following expansion to second order in the field amplitude:

$$E = \int E_1 e^{i(kx-\omega t)} dk d\omega + \int E_2 e^{i(k''x-\omega''t)} dk dk' d\omega d\omega'$$

$$f = F + \int \frac{\varepsilon_o}{e} ikE_1 K e^{i(kx-\omega t)} dk d\omega$$

$$+ \int \left(\frac{\varepsilon_o}{e} ik''E_2 K'' - \frac{\varepsilon_o}{2m} E_1 E_1' L\right) e^{i(k''x-\omega''t)} dk dk' d\omega d\omega' \tag{2}$$

* Work performed as part of the research programme of the association agreement of Euratom and the Stichting voor Fundamenteel Onderzoek der Materie, with financial support from the Nederlandse Organisatie voor Zuiver-Wetenschappelijk Onderzoek and Euratom.

The indices 1 and 2 indicate the order:

$$k'' = k + k', \quad \omega'' = \omega + \omega', \quad E_1 = E_1(k,\omega)$$

$$E_2 = E_2(k,k',\omega,\omega'), \quad F = F(v), \quad K = K(k,\omega,v)$$

$$E_1' = E_1(k',\omega'), \quad K'' = K(k'',\omega'',v), \quad L = L(k,k',\omega,\omega',v)$$

Substitution of (2) into (1) converts the problem into the following set of equations (symmetrizing second-order terms in k, k′ and in ω, ω'):

$$(kv-\omega)K = \frac{e^2}{\varepsilon_o mk}\frac{dF}{dv}, \quad \int K dv = 1, \quad \int F dv = N$$

$$(k''v-\omega'')L = \frac{\partial}{\partial v}(kK+k'K'), \quad \int L dv = 0 \tag{3}$$

The initial-value problem can be solved without explicit expressions for K and L which can be found in Ref. [1]. Putting t = 0 in Eq. (2) yields to first and second order:

$$ik\int E_1 K d\omega = (e/\varepsilon_o) g_1$$

$$ik''\int E_2 K'' d\omega d\omega' = (e/2m)\int E_1 E_1' L d\omega d\omega' \tag{4}$$

in which $g_1(k, v)$ represents the initial condition:

$$f(x,v,o) = F + \int g_1 e^{ikx} dk \tag{5}$$

The properties of K (see Eq. (3)) imply that ($\Omega_\pm = \Omega \pm i0$ defining the integration contour)

$$\int \frac{\Omega-\omega}{\Omega_\pm - kv} K dv = \int \left(1 + \frac{kv-\omega}{\Omega_\pm - kv}\right) K dv$$

$$= 1 + \frac{e^2}{\varepsilon_o mk}\int \frac{1}{\Omega_\pm - kv}\frac{dF}{dv} dv \equiv \varepsilon^\pm(k,\Omega) \tag{6}$$

so the relation between k and $\epsilon^{\pm}$ is

$$\int \frac{K dv}{\Omega_{\pm} - kv} = \frac{\varepsilon^{\pm}}{\Omega_{\pm} - \omega} \tag{7}$$

Dividing the first of Eqs (4) by $\Omega_{\pm} - kv$ and integrating over v gives the solution:

$$E_1^{\pm} = - ieg_1^{\pm}/\varepsilon_o \varepsilon^{\pm} k \tag{8}$$

in which $E_1^{\pm}$ and $g_1^{\pm}$ are defined by

$$E_1^{\pm}(k,\Omega) = \frac{\pm i}{2\pi} \int \frac{E_1 d\omega}{\Omega_{\pm} - \omega}, \quad g_1^{\pm}(k,\Omega) = \frac{\pm i}{2\pi} \int \frac{g_1 dv}{\Omega_{\pm} - kv} \tag{9}$$

$E_1(k, \Omega) = E_1^{+} + E_1^{-}$; E_1^{+} contributes only to E(x, t) for $t > 0$ and E_1^{-} for $t < 0$. Further, Eqs (3) imply that:

$$\int \frac{\Omega - \omega''}{\Omega_{\pm} - k''v} L dv = \int \left(1 + \frac{k''v - \omega''}{\Omega_{\pm} - k''v}\right) L dv$$

$$= \int \frac{1}{\Omega_{\pm} - k''v} \frac{\partial}{\partial v}(kK + k'K) dv = k'' \frac{\partial}{\partial \Omega} \int \frac{kK + k'K'}{\Omega_{\pm} - k''v} dv$$

$$= k'' \frac{\partial}{\partial \Omega} \left(k \frac{\varepsilon^{\pm}(k, \Omega k/k'')}{\Omega_{\pm} - \omega k''/k} + k' \frac{\varepsilon^{\pm}(k', \Omega k'/k'')}{\Omega_{\pm} - \omega' k''/k'} \right) \tag{10}$$

in which $\epsilon^{\pm}(k, \Omega k/k'')$ is given by Eq. (6) with $\Omega_{\pm}$ replaced by $\Omega_{\pm} k/k''$, not $(\Omega k/k'')_{\pm}$. The relation to solve the second equation (4) is:

$$\int \frac{L dv}{\Omega_{\pm} - k''v} = \frac{k''}{\Omega_{\pm} - \omega''} \frac{\partial}{\partial \Omega} \left(\quad \right) \equiv k'' M^{\pm} \tag{11}$$

in which the brackets contain the same expression as in (10) and $M^{\pm}$ agrees with Eq.(16) of Ref. [1]. The above derivation of $M^{\pm}$ is shorter and more direct than in Ref. [1]. In the further calculation, only the Landau poles in E_1 are retained:

$$E_1^{\pm}(k,\omega) = \frac{R_1(\beta\pm i\gamma)}{\omega-\beta\mp i\gamma} + \frac{R_1(-\beta\pm i\gamma)}{\omega+\beta\mp i\gamma} \tag{12}$$

to perform the ω and ω' integration in

$$E_2^{\pm} = \pm \frac{e}{4\pi m} \frac{1}{\varepsilon^{\pm}(k'',\Omega)} \int E_1 E_1' M^{\pm} d\omega d\omega' \tag{13}$$

where $E_2^{\pm}$ is defined by

$$E_2^{\pm}(k,k',\Omega) = \frac{\pm i}{2\pi} \int \frac{E_2 d\omega d\omega'}{\Omega_{\pm}-\omega''} \tag{14}$$

Note that

$$\int E_2 e^{-i\omega'' t} d\omega d\omega' = \int (E_2^{+}+E_2^{-}) e^{-i\Omega t} d\Omega \tag{15}$$

in which E_2^+ contributes only for $t > 0$ and E_2^- for $t < 0$. The long expressions for $E_2^{\pm}$ resulting from (13) are given by Eqs (20) and (21) of Ref. [1] and are not reproduced here. It turns out that E_2^+ has a double pole at $\Omega = (\beta + i\gamma)k''/k$ if $k/k' < -1$, and many other poles which are listed in Ref. [1] but neglected in this paper. The first term of the Laurent expansion of E_2^+ around $(\beta + i\gamma)k''/k$ reads:

$$-\pi \frac{e}{m} \frac{k''^3}{k} \frac{k-k'}{k+k''} \frac{R_1(\beta+i\gamma)}{[\Omega-(\beta+i\gamma)k''/k]^2}$$

$$\times \left[\left(\frac{R_1'(\beta'+i\gamma')}{k'(\beta+i\gamma)-k(\beta'+i\gamma')} + \frac{R_1'(\beta'-i\gamma')}{k'(\beta+i\gamma)-k(\beta'-i\gamma')} \right) + (-\beta') \right] \tag{16}$$

in which R_1' is the residue of E_1' at $\beta' \pm i\gamma'$, the Landau poles of E_1', and $(-\beta')$ means the preceding expression between brackets with β' replaced by $-\beta'$. The amplitude of the growing wave (16) is now evaluated for a Maxwellian electron plasma which is initially disturbed as follows:

$$g_1 = v_1(k)F'(v) \tag{17}$$

This disturbance corresponds approximately to $f(x, v, 0) = F(v + \overline{v_1}(x))$. Then it follows from (8), (9) and (12) that:

$$R_1(\beta \pm i\gamma) = \frac{-ie}{\varepsilon_o k} \left. \frac{g_1^{\pm}}{\partial\varepsilon^{\pm}/\partial\Omega} \right|_{\Omega = \beta \pm i\gamma}$$

$$= \pm \frac{mv_1}{2\pi e} \left. \frac{\varepsilon^{\pm} - 1}{\partial\varepsilon^{\pm}/\partial\Omega} \right|_{\Omega = \beta \pm i\gamma}$$

$$= \mp \frac{mv_1}{2\pi e} \left. \left(\frac{\partial\varepsilon^{\pm}}{\partial\Omega}\right)^{-1} \right|_{\Omega = \beta \pm i\gamma} \tag{18}$$

To lowest order in kv_t/ω_p (ω_p is the plasma frequency and v_t the thermal velocity),

$$\left. \frac{\partial\varepsilon^{\pm}}{\partial\Omega} \right|_{\Omega = \beta \pm i\gamma} = \frac{2}{\beta} \mp 2i \frac{\gamma}{k^2 v_t^2} \tag{19}$$

so that

$$R_1(\beta \pm i\gamma) = \frac{mv_1\beta}{4\pi e} \left(\mp 1 - i \frac{\gamma}{\beta} \frac{\omega_p^2}{k^2 v_t^2}\right) \tag{20}$$

With (20) the expression between square brackets in (16) proves to be, to lowest order in kv_t/ω_p,

$$\frac{mv_1'\beta'}{4\pi e} \frac{4i\gamma'}{k'(k^2 - k'^2)v_t^2}$$

and the growing second-order wave has the form:

$$\frac{i}{2} \frac{e}{m} \frac{k''^2 \hat{E}_1 \hat{E}_1'}{kk'(k+k'')v_t^2} \gamma' t \, e^{\gamma t k''/k} \, e^{i(k''x - \beta t k''/k)} \tag{21}$$

in which $\hat{E}_1$ and $\hat{E}_1'$ are the initial amplitudes of the first-order waves. The relation with the residues is

$$R_1(\beta+i\gamma) = (i/2\pi)\hat{E}_1\delta(k-k_o)$$

As an example take

$$k\lambda_D = 0.35, \qquad \gamma/\omega_p = -0.06$$

in which λ_D is the Debye length. As a function of k′ the expression $\gamma' k''^2/k'(k+k'')$ has a sharp maximum; the half-width is between $k'\lambda_D = -0.32$ and -0.24. Take

$$k'\lambda_D = -0.3, \; \gamma'/\omega_p = -0.02$$

The second-order wave has a low wavenumber, frequency and damping:

$$k''\lambda_D = 0.05, \quad \beta k''/k\omega_p = 0.15, \quad \gamma k''/k\omega_p = -0.01$$

The ratio of the amplitudes of the second-order wave and the k′-wave, which is the least damped first-order wave, is

$$0.08 \frac{\omega_b^2}{\omega_p^2} |\gamma' t| \, e^{0.6|\gamma' t|}$$

in which $\omega_b^2 = ek\hat{E}_1/m$ is the bounce frequency. Owing to the exponential factor this ratio can exceed unity for some t smaller than ω_b^{-1} provided that ω_b/ω_p is very small ($<10^{-3}$).

Discussion

It has been shown that two Landau-damped waves of the form

$$\hat{E}_1 e^{\gamma t} e^{i(kx-\beta t)}, \quad \hat{E}_1 e^{\gamma' t} e^{i(k'x-\beta' t)}$$

give rise to (apart from other well-known non-linear waves) a second-order wave of the form (21) provided that $k/k' < -1$. A heuristic argument for the existence of this wave is the following. The perturbation of the distribution f corresponding to the first Landau wave is of the form

$$\frac{e^{ik(x-vt)}}{\beta+i\gamma-kv}\,\frac{dF}{dv}$$

Integration over v shows that this perturbation in f yields a perturbation in the density of the form $e^{\gamma t}e^{i(kx-\beta t)}$. Then in the case of simultaneous excitation of two Landau waves, a second-order perturbation in f of the form

$$\frac{e^{ik'(x-vt)}}{\beta'+i\gamma'-k'v}\,\frac{\partial}{\partial v}\,\frac{e^{ik(x-vt)}}{\beta+i\gamma-kv}$$

can be expected. Integration over v shows that the pole at $v = (\beta + i\gamma)/k$ contributes terms of the form $te^{\gamma tk''/k}e^{i(k''x-\beta tk''/k)}$ provided that k and k″ have the same sign. Other features of the second-order wave, such as the condition $kk' < 0$, cannot be derived from this simple argument.

The form (21) for the growing wave has been derived assuming the special initial condition (17) to make the calculation simpler. A different form of the initial condition changes, in general, the relation (20) between $R(\beta+i\gamma)$ and $R(\beta-i\gamma)$, corresponding to Landau waves for $t > 0$ and $t < 0$. Therefore the form (21), and consequently the asymptotic behaviour of a Vlasov plasma, depends on details of the initial disturbance, even when smoothness is assumed.

ACKNOWLEDGEMENT

Discussions with Prof. Dr. F. Engelmann are gratefully acknowledged.

REFERENCE

[1] BEST, R.W.B., Physica **74** (1974) 183.

Waves and instabilities

STABLE MHD EQUILIBRIA*

D. LORTZ, J. NÜHRENBERG
Max-Planck-Institut für Plasmaphysik,
Garching,
Federal Republic of Germany

Abstract

STABLE MHD EQUILIBRIA.
1. Introduction. 2. Stability criteria: Hamada co-ordinates; the necessary stability criterion; sufficient stability criteria. 3. Equilibrium and stability of a toroidal configuration near its magnetic axis: equilibrium formulae; stagnation point discussion and beta estimate; applications (a) the $\ell = 2$ stellarator, (b) $\iota = 0$ equilibria. 4. Conclusions.

1. INTRODUCTION

Various methods at different stages of development and with different main virtues are in use for investigating the magnetohydrodynamic stability of magnetohydrostatic equilibria. The method with the best record of achievement is numerical mode analysis of equilibria with one ignorable co-ordinate. This method has been developed in two ways: numerical evaluation of the full MHD spectrum for axisymmetric equilibria (e.g.[1] [1–3]) and numerical solution of the linearized MHD equations as initial value problem for axisymmetric (e.g. [4]) and helical [5] equilibria to obtain the growth rates of the most unstable gross modes. For three-dimensional configurations the equilibrium itself becomes an unsolved problem [6] and several numerical codes have been developed for solving the non-linear MHD or artificially modified MHD equations to study equilibria and their stability properties (e.g. [7–9]).

In this paper we consider a third method which consists in development of stability criteria and their use as tools for the construction of stable magnetohydrostatic equilibria. Several specific advantages and drawbacks are associated with this approach. If a sufficient stability criterion is used, complete stability is proved, which will be difficult with numerical codes. There exists a consistent combination of equilibrium and stability analysis that simplifies even three-dimensional configurations to the degree that studies in the space of configurations become

---

* Work performed under the terms of the agreement on association between Max-Planck-Institut für Plasmaphysik and Euratom.

[1] No attempt is made to give a complete or weighted list of references; they are meant as illustrative examples.

possible. If a necessary stability criterion is used, no firm conclusion can be drawn with respect to stability. On the other hand, the violation of a necessary criterion may be tolerable. Existing sufficient stability criteria probably tend to be too pessimistic in many cases so that their usefulness may be doubtful because the β-values obtainable are too low.

2. STABILITY CRITERIA

The sufficient stability criteria which will be described can only be formulated in Hamada [10] co-ordinates. We therefore introduce the Hamada formalism as far as is necessary and then describe the available stability criteria.

2.1. Hamada co-ordinates

The magnetohydrostatic equilibrium equations written in Hamada co-ordinates are:

$$
\begin{aligned}
\vec{B} &= \dot{\chi}\,\vec{r}_{,\theta} + \dot{\phi}\,\vec{r}_{,\zeta} \\
\vec{j} &= \dot{I}\,\vec{r}_{,\theta} + \dot{J}\,\vec{r}_{,\zeta} \\
\dot{p} &= \dot{I}\dot{\phi} - \dot{J}\dot{\chi} \\
\vec{r}_{,V}\cdot(\vec{r}_{,\theta}\times\vec{r}_{,\zeta}) &= 1 \\
(\dot{\phi} g_{\zeta\zeta} + \dot{\chi} g_{\theta\zeta})_{,\theta} - (\dot{\phi} g_{\theta\zeta} + \dot{\chi} g_{\theta\theta})_{,\zeta} &= 0 \\
(\dot{\phi} g_{V\zeta} + \dot{\chi} g_{V\theta})_{,\zeta} - (\dot{\phi} g_{\zeta\zeta} + \dot{\chi} g_{\theta\zeta})_{,V} &= \dot{I} \\
(\dot{\phi} g_{\theta\zeta} + \dot{\chi} g_{\theta\theta})_{,V} - (\dot{\phi} g_{V\zeta} + \dot{\chi} g_{V\theta})_{,\theta} &= \dot{J} \qquad (1)
\end{aligned}
$$

where $\vec{r}(V, \theta, \zeta)$ is the position vector; V is the volume inside a flux surface (or, for $\iota \equiv 0$, inside a $\int dl/B$ = const surface); θ, ζ are angular-type co-ordinates increasing by unity the short and the long way round the torus, respectively; χ, ϕ and I, J are the fluxes and currents; ι is the rotational transform; p is the pressure; and $\dot{}$ = d/dV. We shall also need the relation of the Hamada co-ordinates to the geometrical co-ordinates of toroidal configurations in the neighbourhood of the magnetic axis. As geometrical co-ordinate system we use Mercier's [11] co-ordinates given by

$$ds^2 = d\rho^2 + \rho^2 d\varphi^2 - 2\tau\rho^2 d\varphi\, d\ell + [(1 - \varkappa\rho\cos\varphi)^2 + \tau^2\rho^2]\, d\ell^2$$
$$\sqrt{g} = \rho(1 - \varkappa\rho\cos\varphi) \tag{2}$$

where ℓ is the arc length along the magnetic axis with curvature κ and torsion τ, and ρ, φ are polar co-ordinates in the plane perpendicular to the magnetic axis at position ℓ. With the position vector given by

$$\vec{r}(V, \Theta, \ell) = \vec{r}_0(\ell) + \vec{r}_1(\Theta, \ell)V^{\frac{1}{2}} + O(V) \tag{3}$$

with

$$\ell(V, \Theta, \zeta) = \ell_0(\zeta) + \ell_1(\Theta, \zeta)V^{\frac{1}{2}} + O(V)$$

where $\vec{r}_0(\ell)$ describes the magnetic axis, the following relations hold [12]:

$$\left.\begin{aligned}
\ell_{0,\zeta} &= c_0/\dot{\phi}_0\,, \quad \dot{\phi}_0 = q_0^{-1}\,, \quad q_0 = \oint \frac{d\ell}{c_0} \\
\zeta &= q_0^{-1}\int_0^\ell \frac{d\ell'}{c_0} + O(V^{\frac{1}{2}}) \\
\vec{r}_1 &= \xi(\Theta, \ell)\,\vec{n} + \eta(\Theta, \ell)\,\vec{b} \\
\xi &= \xi_c \cos 2\pi\Theta + \xi_s \sin 2\pi\Theta \\
\eta &= \eta_c \cos 2\pi\Theta + \eta_s \sin 2\pi\Theta \\
\ell_1 &= \ell_c \cos 2\pi\Theta + \ell_s \sin 2\pi\Theta
\end{aligned}\right\} \tag{4}$$

$$\xi_c = (\pi\ell_{0,\zeta})^{-\frac{1}{2}}(e^{\frac{1}{2}} \sin K \sin\alpha + e^{-\frac{1}{2}} \cos K \cos\alpha)$$
$$\xi_s = (\pi\ell_{0,\zeta})^{-\frac{1}{2}}(e^{\frac{1}{2}} \cos K \sin\alpha - e^{-\frac{1}{2}} \sin K \cos\alpha)$$
$$\eta_c = (\pi\ell_{0,\zeta})^{-\frac{1}{2}}(e^{\frac{1}{2}} \sin K \cos\alpha - e^{-\frac{1}{2}} \cos K \sin\alpha)$$
$$\eta_s = (\pi\ell_{0,\zeta})^{-\frac{1}{2}}(e^{\frac{1}{2}} \cos K \cos\alpha + e^{-\frac{1}{2}} \sin K \sin\alpha)$$
$$K(L) - K(0) = 2\pi m + \alpha(L) - \alpha(0)$$
$$K'(\ell) = \frac{1}{e + \frac{1}{e}}\left(\frac{\dot{J}_0}{\dot{\phi}_0} + 2\tau + 2\alpha'\right) - \frac{2\pi\iota_0}{c_0 q_0}$$

$$2\pi\Theta = \arctan[e^{-1}\tan(\varphi + \alpha)] - K(\ell) + O(V^{\frac{1}{2}}) \tag{5}$$

$$
\begin{aligned}
l_{c,\zeta} - (l_{0,\zeta\zeta}/l_{0,\zeta})\, l_c + 2\pi \iota_0 l_s &= 2\varkappa\, l_{0,\zeta}\, \xi_c \\
l_{s,\zeta} - (l_{0,\zeta\zeta}/l_{0,\zeta})\, l_s - 2\pi \iota_0 l_c &= 2\varkappa\, l_{0,\zeta}\, \xi_s
\end{aligned}
\tag{6}
$$

where c_0 is the field on the magnetic axis; $\vec{n}, \vec{b}$ are the normal and binormal of the magnetic axis; e is the half-axis ratio of the elliptical (in second order in the distance from the magnetic axis) plasma cross-section; α (for $e > 1$) is the angle between the binormal of the magnetic axis and the major half-axis; and m is the number of full turns of the normal over the length L of the magnetic axis.

2.2. The necessary stability criterion

In Hamada co-ordinates, Mercier's necessary stability criterion [13] reads [14]:

$$
\frac{1}{4}\Sigma^2 - \dot{p}\left[W_1 \dot{\chi}\ddot{\chi} + W_2 (\dot{\chi}\dot{\phi})^{\cdot} + W_3 \dot{\phi}\ddot{\phi}\right] - \dot{p}^2 (W_1 W_3 - W_2^2) > 0 \tag{7}
$$

where $\Sigma = \dot{\chi}\ddot{\phi} - \ddot{\chi}\dot{\phi}$

$$
W_1 = \left\langle \frac{g_{\Theta\Theta}}{g^{VV}} \right\rangle, \quad W_2 = \left\langle \frac{g_{\Theta\zeta}}{g^{VV}} \right\rangle, \quad W_3 = \left\langle \frac{g_{\zeta\zeta}}{g^{VV}} \right\rangle
$$

$$
\langle \cdots \rangle = \int \cdots \frac{dl}{B} \Big/ \int \frac{dl}{B}
$$

Note that this is a form of the criterion without cancellation of diverging terms as one approaches the magnetic axis, where the criterion reduces to (for $\dot{p} < 0$)

$$
\dot{\phi}_0 \ddot{\phi}_0 + \dot{p}_0 \left\langle \frac{g_{\Theta\Theta}}{g^{VV}} \right\rangle_0 > 0 \tag{8}
$$

For irrational ι the mean values are replaced by $\int \ldots d\theta\, d\zeta$. Evaluation of W_1 on the magnetic axis in terms of the quantities introduced in § 2.1 then yields [12]

$$
W_{10} = \oint \frac{d\zeta}{l_{0,\zeta}^2} \left\{ 1 + \frac{1}{2}\pi l_{0,\zeta} \left[l_c^2 + l_s^2 - \frac{(l_s^2 - l_c^2)(\eta_s^2 - \eta_c^2 + \xi_s^2 - \xi_c^2) + 4 l_c l_s (\eta_c \eta_s + \xi_c \xi_s)}{\frac{2}{\pi l_{0,\zeta}} + \eta_c^2 + \eta_s^2 + \xi_c^2 + \xi_s^2} \right] \right\} \tag{9}
$$

For equilibria with one ignorable co-ordinate (e.g. in axial and helical symmetry) Mercier's criterion has been evaluated over the whole plasma cross-section. In all hitherto known instances (e.g. [15–17]) it becomes marginal at first on the magnetic axis, and existing mode analyses ([1, 2, 4]) show no unstable internal modes (i.e. modes vanishing at and outside the plasma boundary) if it is satisfied on the magnetic axis[2]. This leads us to two conjectures:

(a) For the specific low-β large-aspect-ratio limit of a conceptual sequence of configurations obtained by keeping the profiles fixed and increasing the aspect ratio by moving the $p = 0$ surface towards the magnetic axis, Mercier's criterion on the magnetic axis is sufficient for the stability of internal modes.

(b) There exist profile conditions for which it is possible to reverse the above sequence (and obtain finite-aspect ratios and β-values) and keep the internal modes stable.

As long as these conjectures are uncertain, known sufficient-stability criteria, which are, in general, more restrictive than Mercier's criterion, have to be used to prove stability. Since the necessary stability criterion is capable, at most, of predicting stability with respect to internal modes, the restrictions imposed by stability against external modes are, in general, more severe (e.g. [4, 5]).

2.3. Sufficient stability criteria

Two cases have to be distinguished: non-vanishing and vanishing current density j_0 on the magnetic axis.

A. $j_0 \neq 0$

In this case the sufficient stability criterion [18] is also restricted to internal modes. Stability is proved by constructing one-dimensional energy estimates of the energy variation and holds if the following three inequalities are satisfied:

[2] Meanwhile a new necessary stability criterion [32] has been found, which, off the magnetic axis, is more stringent than Mercier's criterion.

$$\left.\begin{aligned}
&\ddot{J}\dot{\chi} - \ddot{I}\dot{\phi} + \frac{\pi^2}{4V_p^2}\dot{\phi}^2 \min_V M_2 > 0 \\
&\frac{V_p}{\dot{\phi}^2}\left[M_3 V\dot{\phi}^2 - \dot{J}^2\frac{V}{M_1} + V(\dot{J}\ddot{\chi} - \ddot{I}\dot{\phi})\right] + \lambda^2 \min_V M_2 > 0 \,, \quad \lambda \approx 1.2 \\
&-c^{-2}\left[(\nu\dot{J} + \dot{I})^2 + 2(\nu\dot{J} + \dot{I})(M_2\ddot{\phi} - \nu M_1\ddot{\chi})\right] + \dot{I}\ddot{\phi} - \dot{J}\ddot{\chi} \\
&+ M_3(\nu\dot{\phi} + \dot{\chi})^2 - \left[c^{-2}(\nu\dot{\phi} + \dot{\chi})(\nu M_1\dot{I} - M_2\dot{J})\right] > 0
\end{aligned}\right\} \quad (10)$$

for all real ν, where V_p is the total plasma volume and

$$M_1 = \min_{\Theta,\zeta} \frac{\Gamma_1}{|\nabla\Theta|^2} \,, \quad M_2 = \min_{\Theta,\zeta} \frac{\Gamma_2}{|\nabla\zeta|^2} \,, \quad M_3 = (2\pi)^2 \min_{\Theta,\zeta} \frac{\Gamma_3}{|\nabla V|^2}$$

$$\Gamma_i > 0 \,, \quad \sum_i \Gamma_i = 1 \,, \quad c^2 = \nu^2 M_1 + M_2$$

In the neighbourhood of the magnetic axis the following sufficient criterion is obtained with $\Gamma_1, \Gamma_2 \to 0$:

$$\dot{\phi}_0 \ddot{\phi}_0 + \dot{p}_0 |\nabla\zeta|^2_0 > 0 \qquad (11)$$

where

$$|\nabla\zeta|^2_0 = \frac{1}{\ell_{0,\zeta}^2} + \pi^2\left[(\eta_s \ell_c - \eta_c \ell_s)^2 + (\xi_s \ell_c - \xi_c \ell_s)^2\right] \qquad (12)$$

A comparison of the necessary and the sufficient criterion on the magnetic axis has been made for the case of axial symmetry [14–16] and reveals that the gap between the two criteria is of a quantitative and not qualitative nature, so that critical values of the rotational transform on axis as well as the β-values are not seriously limited if the sufficient criterion is taken into account only on the magnetic axis.

An improved form of the stability criterion (10) (minimization of the one-dimensional estimates occurring in the δW estimate [18]) was applied over the whole plasma cross-section of axially symmetric equilibria [16]. The results are that stable equilibria with β-values of about 1% and rather large aspect ratio (A ~ 15) and stable equilibria with moderate aspect ratio (A ~ 5) but rather low β-values (~ 0.2%) can be constructed. These results show that the one-dimensional estimate of the energy variation used in [18] is probably too crude.

B. $j_0 = 0$

The first criterion obtained by Solov'ev [19] for this case was later improved to the following stability criterion [20]. Stability holds if a single-valued function Λ exists for which

$$\vec{B}\cdot\nabla\Lambda - \Lambda^2|\nabla V|^2 - |\nabla V|^2 A > 0 \tag{13}$$

with

$$|\nabla V|^2 A = 2|\nabla V|^{-4}(\vec{j}\times\nabla V)(\vec{B}\cdot\nabla)\nabla V$$

$$= \frac{\vec{\dot{j}}^{\,2}}{|\nabla V|^2} + \dot{I}\ddot{\Phi} - \dot{J}\ddot{\chi} - (\vec{B}\cdot\nabla)(\dot{I}g^{V\zeta} - \dot{J}g^{V\Theta})\frac{1}{|\nabla V|^2}$$

throughout the plasma. Since this guarantees that δW_{fluid} is positive without boundary conditions on the perturbations, this criterion provides for complete stability. In the neighbourhood of the magnetic axis, Λ can be determined and the following stability criterion is obtained [12]:

$$\dot{\Phi}_0\ddot{\Phi}_0 + \dot{p}_0\oint|\nabla\zeta|^2_0\,d\zeta > 0 \tag{14}$$

which is less restrictive than Eq.(11) although it implies full stability. This is, of course, a consequence of the vanishing current density. Toroidal configurations without internal conductors are necessarily three-dimensional for $j_0 = 0$. A type of equilibrium with one ignorable co-ordinate to which this stability criterion can be applied is helical symmetry. Similar to the result in axial symmetry, it is found [21] that taking into account the stability criterion on the magnetic axis alone does not limit the β-value severely; an evaluation of the criterion over the whole cross-section of helical equilibria is still lacking.

3. EQUILIBRIUM AND STABILITY OF A TOROIDAL CONFIGURATION NEAR ITS MAGNETIC AXIS

Several lines of thought for β-enhancement in toroidal configurations are currently being investigated (e.g. [22]). Here we describe the search for stable three-dimensional configurations with the help of necessary and sufficient stability criteria on the magnetic axis and an appropriate approximation of the equilibrium. This is a rather crude approach because only the most primitive features of equilibrium and stability are incorporated in the theory. Nevertheless we think that this method is a useful tool, mainly for two reasons. It is the only method currently available for a systematic search in the total space of configurations and it is capable of singling out many configurations which will probably have β-values too low to be acceptable. Having already listed the stability criteria to be used, we now describe the equilibrium calculation necessary to obtain the magnetic well $\ddot{\phi}_0/\dot{\phi}_0$ on the magnetic axis and then the stagnation point discussion of the third order (in the distance from the magnetic axis) flux surfaces leading to β-estimates and, finally, applications of this theory.

3.1. Equilibrium formulae

The first consistent equilibrium calculation up to third order in the distance from the magnetic axis for a general three-dimensional configuration was done by Mercier [11]. Here we present a more recent concise version of this calculation [12]. Let V be the volume inside a magnetic surface; then

$$V = c_0 F \oint \frac{dl}{c_0} + O(\rho^4)$$

where F is the cross-sectional area of the magnetic surfaces normal to the magnetic axis. We choose the following third-order representation for F:

$$F = \pi \Big[e(\bar{x}\cos\alpha - \bar{y}\sin\alpha)^2 + \frac{1}{e}(\bar{x}\sin\alpha + \bar{y}\cos\alpha)^2 - 4\rho^3 \frac{\pi}{L}(\delta\cos u\sin^2 u + \Delta\cos^2 u\sin u)\Big] \qquad (15)$$

with

$$u = \varphi + \alpha$$

$$\bar{x} = \rho\cos\varphi + \frac{2\pi}{L}\frac{S}{e}\rho^2\left(e\cos^2 u + \frac{1}{e}\sin^2 u\right)$$

$$\bar{y} = \rho\sin\varphi + \frac{2\pi}{L}\frac{s}{e}\rho^2\left(e\cos^2 u + \frac{1}{e}\sin^2 u\right)$$

The meaning of e and α has been explained above. The non-dimensional third-order quantities δ and Δ describe symmetric and antisymmetric triangular deformations of the surface, where the symmetry holds with respect to u = 0. The quantities S and s describe shifts of the magnetic surfaces with respect to the magnetic axis, S and s being positive for shifts antiparallel to the normal and binormal respectively. Let us also introduce the shifts in the direction of the larger (b) and smaller (a) half-axis of the elliptical cross-section by

$$\bar{S}_a = S\cos\alpha - s\sin\alpha$$

$$\bar{S}_b = S\sin\alpha + s\cos\alpha$$

and the following transformed quantities:

$$S_a^* = c_o^{-\frac{1}{2}} e^{-\frac{1}{2}} \bar{S}_a \ , \quad S_b^* = c_o^{-\frac{1}{2}} e^{-\frac{3}{2}} \bar{S}_b$$

$$\delta^* = c_o^{-\frac{1}{2}} e^{\frac{1}{2}} \delta \ , \quad \Delta^* = c_o^{-\frac{1}{2}} e^{-\frac{1}{2}} \Delta$$

The equilibrium equations between these four quantities then read

$$\begin{aligned} \left(e+\frac{3}{e}\right)\left[S_a^{*\prime} + K_o' S_b^*\right] - \frac{1}{e}\delta^{*\prime} - e K_o' \Delta^* &= R_1^* \\ \left(3e+\frac{1}{e}\right)\left[S_b^{*\prime} - K_o' S_a^*\right] - e\Delta^{*\prime} + \frac{1}{e} K_o' \delta^* &= R_2^* \end{aligned} \qquad (16)$$

where

$$K_o' = \left(\frac{j_o}{c_o} + 2\tau + 2\alpha'\right)\Big/\left(e + \frac{1}{e}\right)$$

$$R_1^* = \frac{L\varkappa}{8\pi c_o^{\frac{1}{2}} e^{\frac{1}{2}}} \left\{ 2\sin\alpha\left[-K_o' + \alpha'\left(3e - \frac{2}{e}\right) + 4\tau e\right]\right.$$

$$\left. + \cos\alpha\left[\frac{c_o'}{c_o}\left(e - \frac{2}{e}\right) + 2\frac{\varkappa'}{\varkappa}\left(e + \frac{2}{e}\right) - \frac{e'}{e}\left(3e + \frac{2}{e}\right)\right]\right\} + b_s^*$$

$$
\begin{aligned}
R_2^* = \frac{L\varkappa e^{\frac{1}{2}}}{8\pi c_0^{\frac{1}{2}}} & \left\{ 2\cos\alpha \left[K_0' + \alpha'\left(2e - \frac{3}{e}\right) - 4\tau/e \right] \right. \\
& \left. + \sin\alpha \left[\frac{c_0'}{c_0}\left(\frac{1}{e} - 2e\right) + 2\frac{\varkappa'}{\varkappa}\left(2e + \frac{1}{e}\right) + \frac{e'}{e}\left(2e + \frac{3}{e}\right) \right] \right\} - b_c^*
\end{aligned}
\tag{17}
$$

Here,

$$
\begin{aligned}
b_s^* &= \dot{p}_0 L q_0 \left(b_r \cos\alpha + b_i \sin\alpha \right) \\
b_c^* &= \dot{p}_0 L q_0 \left(b_i \cos\alpha - b_r \sin\alpha \right)
\end{aligned}
$$

where the complex quantity $b = b_r + ib_i$ solves the equation

$$
b' + i(K_0' - \alpha')b = -\exp(i\alpha)\, c_0^{-\frac{3}{2}} \varkappa \left(e^{-\frac{1}{2}}\cos\alpha - i e^{\frac{1}{2}} \sin\alpha \right) \tag{18}
$$

Note that the quantity b characterizes the influence of the finite pressure gradient (i.e. plasma-β) on the relation between the shifts and triangularities.

Using the above quantities, the magnetic well on the magnetic axis is given by the following formula:

$$
\begin{aligned}
\frac{\ddot{\Phi}_0}{\dot{\Phi}_0} = \frac{1}{2\pi q_0^2} \oint \frac{dl}{c_0^2} & \left\{ \frac{\varkappa^2}{2}\left[e + \frac{1}{e} - \left(e - \frac{1}{e}\right)\cos 2\alpha \right] \right. \\
& - \left(\frac{j_0}{c_0} \right)^2 \frac{1}{e + \frac{1}{e}} - \frac{\left(e - \frac{1}{e}\right)^2}{e + \frac{1}{e}} (\tau + \alpha')^2 \\
& \left. - \frac{3}{4}\left(\frac{c_0'}{c_0}\right)^2 \left(e + \frac{1}{e}\right) - \frac{1}{4}\left(\frac{e'}{e}\right)^2 \left(e + \frac{1}{e}\right) + \frac{c_0'}{c_0}\frac{e'}{e}\left(e - \frac{1}{e}\right) \right\} \\
& - \frac{\dot{p}_0}{q_0} \oint \frac{dl}{c_0^3} \\
& - \frac{2}{L q_0^2} \oint \frac{\varkappa}{c_0^2} \left[\frac{4}{e} S - (\Delta \sin\alpha + \delta \cos\alpha) \right] dl
\end{aligned}
\tag{19}
$$

Since only two of the four quantities S, s, δ, Δ can be prescribed, the well can only be evaluated with the help of the third-order equilibrium equations.

In the applications we shall restrict our discussion to cases without longitudinal current on the magnetic axis or c_0 = const if there is a current on the magnetic axis, because for these cases the explicit diamagnetic deepening of the magnetic well (the second last term in Eq. (19)) cancels with the first term in Eqs (9) and (12) inserted into the stability criteria (8), (11) and (14).

3.2. Stagnation point discussion and beta estimate

Defining the plasma-β by

$$\beta = \frac{\int_0^{V_p} p\,dV}{\int_0^{V_p} \frac{1}{2}\vec{B}^2 d^3\tau} \tag{20}$$

where V_p is the plasma volume, we see that stagnation points in the third-order flux surfaces, which limit the attainable volume, lead to the following β-estimate:

$$\beta = -\frac{\dot{p}_o V_s}{\dot{\phi}_o \oint c_o dl} + O\left[\left(\frac{4\pi V_s}{L^3}\right)^2\right] \tag{21}$$

where V_s is the volume limited by the third-order stagnation points. Equation (15) yields the following expression for V:

$$V = \frac{L^2}{2\pi} q_o \left[\frac{1}{2}x^2 + \frac{1}{2}y^2\right.$$

$$\left. + S_a^* x^3 + (S_b^* - \Delta^*)x^2 y + (S_a^* - \delta^*)xy^2 + S_b^* y^3\right] \tag{22}$$

where

$$x = \frac{2\pi}{L} c_o^{\frac{1}{2}} e^{\frac{1}{2}} \varrho \cos u \;, \quad y = \frac{2\pi}{L} c_o^{\frac{1}{2}} e^{-\frac{1}{2}} \varrho \sin u$$

so that V_s may be estimated by

$$V_s \leq \frac{1}{27} \frac{L^2}{2\pi} q_o \min_{\ell}\left[\frac{1}{2S_a^{*2}}, \frac{1}{2S_b^{*2}},\right.$$

$$\left.\frac{4}{(2S_a^* + 2S_b^* - \Delta^* - \delta^*)^2}, \frac{4}{(2S_a^* - 2S_b^* + \Delta^* - \delta^*)^2}\right] \tag{23}$$

This estimate would only fail if $S_a^* = S_b^* = \Delta^* = \delta^* \equiv 0$, which would impose $R_1^* = R_2^* \equiv 0$. Toroidal equilibria of this type have not yet been found and would obviously be hard to construct[3]. Considering Eqs (16), (21), (22) together, we see that the following procedure has to be observed in order to obtain a correct β-estimate. Either S_a^*, S_b^* or δ^*, Δ^* and $\dot{p}_0$ may be prescribed; then δ^*, Δ^* (or S_a^*, S_b^*) are calculated according to Eqs (16); the stagnation points of the third-order flux surfaces, Eq. (22), are calculated for each value of ℓ, which leads to a maximum admissible volume $V_s(\ell)$; the minimum value of the volumes $V_s(\ell)$ (an upper estimate of this quantity is provided by Eq. (23)) has to be inserted in Eq. (21); to eliminate the arbitrary choices for S_a^*, S_b^* and $\dot{p}_0$, the β-value has to be optimized with respect to these quantities. Only this optimized β-value may be considered as a quantity which characterizes a basic property of the configuration which is studied. If one wants to obtain a β-estimate taking into account a stability criterion, an additional inequality resulting from one of the Eqs (8), (11), (14) for the quantities S_a^*, S_b^* and $\dot{p}_0$ has to be satisfied. Thus we see that stability criteria on the magnetic axis have to be looked at in two ways. Taken per se, they indicate stability or instability for given equilibrium functions in the neighbourhood of the magnetic axis. In order to calculate stable β-values, they are properly used as side conditions in the equilibrium β-estimates.

3.3. Applications

(a) The $\ell = 2$ stellarator

We consider an $\ell = 2$ stellarator with circular magnetic axis of radius R, vanishing current density on the magnetic axis, constant e, constant $d\alpha/d\ell = n/(2R)$, where n is the number of field periods, and two shifts of the magnetic surfaces with respect to the magnetic axis: a shift S_0 in the direction of the normal to the magnetic axis and a shift S_2 rotating with 2α as one proceeds along the magnetic axis. This is the simplest three-dimensional configuration of a type for which the $q = \int d\ell/B$ = const surfaces are generated by means of the rotational transform alone. In the equilibrium equations this is reflected by the singularity appearing for ι_0, $(K_0' - \alpha') \to 0$ in Eqs (6) and (18). This type of stellarator has been considered several times, particularly with respect to the stability criteria on the magnetic axis [11, 23–26]. With respect to the β-values that can be obtained in a stable manner, the situation still remained unclear, which led us to a systematic study in accordance with the procedure in § 3.2. The result [27] was that, irrespective of the five parameters (e, n, $\dot{p}_0$, S_0, S_2) characterizing this type of stellarator, there are universal bounds for the β-estimates of 0.66% and 0.22% for the necessary (Eq. (8)) and the sufficient (Eq. (14)) stability criterion, respectively.

[3] Meanwhile toroidal equilibria of this type have been constructed [33].

(b) $\iota = 0$ equilibria

Again we consider vanishing current density on the magnetic axis. The q = const surfaces are then solely determined by external shaping fields. A condition for the existence of such equilibria, namely that q be stationary on the magnetic axis, was in its general form first discussed by Shafranov [28] and may easily be obtained from Eq. (6):

$$0 = \oint \varkappa \xi_c d\zeta = \pi^{-\frac{1}{2}} q_0^{-\frac{3}{2}} \oint \varkappa c_0^{-\frac{3}{2}} \left(e^{\frac{1}{2}} \sin\alpha \sin K + e^{-\frac{1}{2}} \cos\alpha \cos K\right) dl$$

$$0 = \oint \varkappa \xi_s d\zeta = \pi^{-\frac{1}{2}} q_0^{-\frac{3}{2}} \oint \varkappa c_0^{-\frac{3}{2}} \left(e^{\frac{1}{2}} \sin\alpha \cos K - e^{-\frac{1}{2}} \cos\alpha \sin K\right) dl \tag{24}$$

We restrict our further discusssion to a plane magnetic axis and reflectional symmetry across a plane perpendicular to the magnetic axis, $\alpha = K \equiv 0$, so that the condition

$$\oint \varkappa c_0^{-\frac{3}{2}} e^{-\frac{1}{2}} dl = 0 \quad , \quad k' = \varkappa c_0^{-\frac{3}{2}} e^{-\frac{1}{2}}$$

is left. In a consistent calculation up to third order in the distance from the magnetic axis, an additional condition has to be met in order that

$$q = q(0) + \dot{q}(0) V + O(V^{\frac{3}{2}})$$

be satisfied. This condition reads

$$\oint dl \left\{ c_0 k'^2 (e^2-1) + \frac{3}{4} \frac{c_0'^2}{c_0^4 e} (e^2-1) + \frac{1}{4} \frac{e'^2}{c_0^2 e^3} (e^2-1) \right.$$

$$- \frac{c_0'}{c_0^3} \frac{e'}{e^2} (e^2+1) + 3 c_0 e e' k k' - c_0' k k' e^2$$

$$\left. + \frac{4\pi}{L} k S_\alpha^{*\prime} (e^2+1) + 4\pi q_0 \dot{p}_0 e k^2 \right\} = 0 \tag{25}$$

while

$$\oint d\ell \left\{ c_0 k'^2 (e^2+1) + \frac{3}{4} \frac{c_0'^2}{c_0^4 e} (e^2+1) + \frac{1}{4} \frac{e'^2}{c_0^2 e^3} (e^2+1) \right.$$

$$- \frac{c_0'}{c_0^3} \frac{e'}{e^2} (e^2-1) + 3 c_0 e e' k k' - c_0' k k' e^2$$

$$\left. + \frac{4\pi}{L} k S_a^{*\prime} (e^2-1) - 4\pi q_0 \dot{p}_0 e k^2 \right\} < 0 \tag{26}$$

is obtained for the sufficient criterion (14). Considering the structure of these equations, we see that Eq. (25) imposes a severe additional restriction on the functions k, c_0, e,S_a^* compatible with Eq. (26). Another interesting result [29] is that the sufficient criterion Eq. (26) is identical with the necessary criterion Eq. (8). This comes about because, for $\iota = 0$, Eq. (8) has to be evaluated on each field line and

$$\max_{\Theta} \frac{g_{\Theta\Theta}}{g^{VV}} = |\nabla \xi|^2$$

for this type of equilibria. This appears to be the first incidence of identical necessary and sufficient criteria.

For the case c_0 = const, β-estimates have been obtained several times [29, 30]. Combining Eqs (21), (23),(25) and (26) and using $\oint \kappa \, d\ell = 2\pi$, we obtain as upper bound for the β-estimate

$$\beta < \frac{1}{11664} e_{max} A^2$$

where

$$A^2 = \frac{L^3}{4\pi V_s}$$

This estimate is still much too optimistic, as a numerical optimization [29] of the β-value shows ($\beta \sim 2\%$ for A ~ 50; $\beta \sim 11\%$ for A ~ 1000). It seems that inclusion of a variable c_0 does not change these results very much if one is not willing to consider extreme variations of e and c_0 [31].

4. CONCLUSIONS

The situation with regard to sufficient stability criteria and their application to the total space of toroidal equilibria needs to be improved. If it is decided to pursue such efforts, it would be desirable to achieve progress in several directions:

(a) For cases with non-vanishing current density on the magnetic axis, it should be possible to obtain better sufficient stability criteria for the plasma region off the magnetic axis.

(b) For cases with vanishing current density on the magnetic axis, examples should be had in which the sufficient stability criterion is evaluated over the whole cross-section in order to study its behaviour off the magnetic axis. For helical equilibria this would be feasible.

(c) A complete classification of the total space of toroidal configurations is still lacking even on the relatively primitive theoretical level discussed in §3. This classification should, in particular, answer the question whether or not it is possible to obtain sizeable stable β-values in geometrically reasonable toroidal configurations. So far, there appear to be no compelling arguments that render this goal impossible. The analysis of §3 shows that configurations with rotational transform, but which satisfy the condition that q be stationary on the magnetic axis for vanishing rotational transform, are possible candidates.

REFERENCES

[1] GRIMM, R.C., GREENE, J.M., JOHNSON, J.L., Methods in Computational Physics (ALDER, B., et al., Eds) **16**, Academic Press, New York (1976) 253.

[2] KERNER, W., Nucl. Fusion **16** (1976) 643.

[3] BERGER, D., et al., "Numerical computation of MHD spectrum of non-circular small-aspect-ratio tokamaks", Plasma Physics and Controlled Nuclear Fusion Research 1976 (Proc. 6th Int. Conf. Berchtesgaden, 1976) **2**, IAEA, Vienna (1977) 411.

[4] WESSON, J.A., SYKES, A., "Toroidal calculations of tokamak stability", Plasma Physics and Controlled Nuclear Fusion Research 1974 (Proc. 5th Int. Conf. Tokyo, 1974) **1**, IAEA, Vienna (1975) 449.

[5] HERRNEGGER, F., SCHNEIDER, W., Nucl. Fusion **16** (1976) 925.

[6] GRAD, H., Phys. Fluids **10** (1967) 137.

[7] BRACKBILL, J.U., BARNES, D.C., Bull. Am. Phys. Soc. II **21** (1976) 1074.

[8] BETANCOURT, O., GARABEDIAN, P., Proc. Natl. Acad. Sci. USA **73** (1976) 984.

[9] CHODURA, R., SCHLÜTER, A., 2nd Europ. Conf. on Computational Physics, Garching, 1976 (BISKAMP, D., Ed.), paper C 2.

[10] HAMADA, S., Nucl. Fusion **2** (1962) 23.

[11] MERCIER, C., Nucl. Fusion **4** (1964) 213.

[12] LORTZ, D., NÜHRENBERG, J., Z. Naturforsch. A. **31** (1976) 1277.

[13] MERCIER, C., "Critère de stabilité d'un système toroïdal hydromagnétique en pression scalaire", Plasma Physics and Controlled Nuclear Fusion Research (Proc. Int. Conf. Salzburg, 1961), Nucl. Fusion 1962 Suppl. **2**, IAEA, Vienna (1962) 801.
[14] LORTZ, D., NÜHRENBERG, J., Nucl. Fusion **13** (1973) 821.
[15] LORTZ, D., NÜHRENBERG, J., JET Workshop on Theory, Culham, 1974.
[16] LORTZ, D., NÜHRENBERG, J., "MHD-stable axisymmetric equilibria", Plasma Physics and Controlled Nuclear Fusion Research 1974 (Proc. 5th Int. Conf. Tokyo, 1974) **1**, IAEA, Vienna (1975) 439.
[17] HERRNEGGER, F., NÜHRENBERG, J., Nucl. Fusion **15** (1975) 1025.
[18] LORTZ, D., Nucl. Fusion **13** (1973) 817.
[19] SOLOV'EV, L.S., Soviet Phys.-JETP **26** (1968) 1167.
[20] LORTZ, D., REBHAN, E., SPIES, G., Nucl. Fusion **11** (1971) 583.
[21] BRAUN, W., et al., "Garching toroidal high-beta stellarator experiments", Plasma Physics and Controlled Nuclear Fusion Research 1974 (Proc. 5th Int. Conf. Tokyo, 1974) **3**, IAEA, Vienna (1975) 25.
[22] CALLEN, J.D., et al., "Tokamak plasma magnetics", Plasma Physics and Controlled Nuclear Fusion Research 1976 (Proc. 6th Int. Conf. Berchtesgaden, 1976) **2**, IAEA, Vienna (1977) 369.
[23] SHAFRANOV, V.D., YURCHENKO, E.I., Nucl. Fusion **8** (1968) 329.
[24] SOLOV'EV, L.S., SHAFRANOV, V.D., YURCHENKO, E.I., "Plasma stability in closed systems", Plasma Physics and Controlled Nuclear Fusion Research (English translations of Russian papers from Int. Conf. Novosibirsk, 1968) Nucl. Fusion Suppl., IAEA, Vienna (1969) 25.
[25] SHAFRANOV, V.D., YURCHENKO, E.I., Nucl. Fusion **9** (1969) 285.
[26] MERCIER, C., LUC, H., Lectures in Plasma Physics, CEC Rep. EUR 5127e (1974) 74.
[27] LORTZ, D., NÜHRENBERG, J., Nucl. Fusion **17** (1977) 125.
[28] SHAFRANOV, V.D., Sov. At. Energy **22** (1967) 449.
[29] HERRNEGGER, F., et al., "Experimental and theoretical study of the FLR stabilization in the high-beta stellarator", Plasma Physics and Controlled Nuclear Fusion Research 1976 (Proc. 6th Int. Conf. Berchtesgaden, 1976) **2**, IAEA, Vienna (1977) 183.
[30] SHAFRANOV, V.D., YURCHENKO, E.I., Sov. At. Energy **29** (1970) 801.
[31] FURTH, H.P., ROSENBLUTH, M.N., Phys. Fluids **7** (1964) 764.
[32] CONNOR, J.W., HASTIE, R.J., TAYLOR, J.B., Phys. Rev. Lett. **40** (1978) 396.
[33] LORTZ, D., NÜHRENBERG, J., submitted for publication to Nucl. Fusion.

IAEA-SMR-32/5

DISSIPATIVE MHD STABILITY*

H. TASSO
Max-Planck-Institut für Plasmaphysik,
Garching,
Federal Republic of Germany

Abstract

DISSIPATIVE MHD STABILITY.

A short survey of the literature on dissipative magnetohydrodynamic instabilities is given as introduction. A mathematical technique allowing 'energy principles' is developed and applied to toroidal equilibria and to cylindrical tokamak-like equilibria, taking into account resistivity and the full macroscopic tensor, i.e. finite Larmor radius (FLR) effects and viscosity. This allows general statements to be made about MHD stability in the presence of viscosity and FLR, and permits, without much computation, a qualitative and comparative study of resistive perturbations under the influence of FLR and viscosity. Applications to tokamak observations are also sketched. Finally, it is proved that the stability of dissipative time-dependent force-free fields can be analysed by a simple functional containing only the perturbed vector potential. This proof is valid even if all non-ideal effects of the two-fluid theory are considered. The conclusion contains a discussion of the open problems, and suggestions are given for their solution.

INTRODUCTION

Dissipative instabilities in hydrodynamics have been considered since the beginning of this century in several well-known works, described in, for example, the book by Lin [1]. Dungey [2] is apparently the initiator of an instability mechanism due to resistivity which seems to play in MHD the role of viscosity in hydrodynamics. Several papers dealing with particular resistive instabilities appeared between 1958 and 1963. They are referred to in the paper of Furth, Killeen and Rosenbluth [3] dealing with the sheet-pinch finite-resistivity stability. These authors discussed qualitatively the influence of other non-ideal effects such as viscosity and thermal conductivity on the eigenvalues (see also Coppi's paper [4]). Several later papers based on [3] tried to introduce more sophistication in geometry and physics by using scaling and expansion techniques. Most of the contributions came from Coppi, Furth, Frieman, Greene, Johnson, Rosenbluth and Rutherford, and many references can be found in, for example,

---

* Work performed under the terms of the agreement on association between the Max-Planck-Institut für Plasmaphysik and Euratom.

the book by Čap [5]. Some references are also given by Glasser, Greene and Johnson [6], which is a sort of culmination point of expansion techniques and scalings involving geometry and physics.

Besides this progress in the physics of resistive instabilities, some mathematical progress has been achieved by Barston [7], who was the first to prove for the sheet pinch a necessary and sufficient condition similar to an energy principle. He also gave exact estimates for the growth rates. In fact, this energy principle was already used by Furth [8] but without proof (as far as the reviewer is aware). An extension of Barston's work to two-dimensional plasmas was made by Tasso [9].

It was noticed [3, 8] early that gyroviscosity and viscosity could play an important role in the tiny resistive sheath of the modes. Generally, non-ideal effects and realistic geometry should be taken into account. This is too much for an eigenmode analysis, but other methods such as energy principles, if they exist, could at least give qualitative answers to such problems. A recent paper by Tasso [10] shows how to obtain a quasi-energy principle for realistic geometries taking into account resistivity, gyroviscosity and viscosity.

Sections 1–5 of this paper closely follow Ref. [10]. They describe the equations and geometries for which energy principles can be given as well as the results and applications which can be expected. In addition, a sufficient condition [11] for the stability of general dissipative force free fields is given, and its relation to Taylor's invariant [12] is discussed.

1. STABILITY EQUATION ALLOWING 'ENERGY PRINCIPLE'

Let us consider the following equation:

$$N\ddot{\underline{\xi}} + (F+M)\dot{\underline{\xi}} + Q\underline{\xi} = 0 \tag{1}$$

where $\underline{\xi}$ is a complex multidimensional representation vector in a functional space, N and M are hermitic and positive operators, and Q is a hermitic and F an antihermitic operator.

It can be seen that this equation contains several limiting cases. If F = M = 0, it is the ideal MHD case [13, 14] for static equilibria. If M = 0, F ≠ 0,it is the case of linearized conservative systems such as the linearized Vlasov equation [15]. If F = 0, M ≠ 0,it is the case of a resistive plasma in one-dimensional [7] and two-dimensional [9] geometries.

(a) Sufficient condition for stability

Let us first recall the definition of the scalar product for two vectors $\underline{\xi}$ and $\underline{\eta}$ denoted by (ξ, η):

$$(\xi,\eta) = \int d\tau\ \underline{\xi}^{*}\cdot\underline{\eta} \tag{2}$$

the Hilbert space to which $\underline{\xi}$ and $\underline{\eta}$ belong being restricted to functions fulfilling specific boundary conditions suggested by physics.

Let us now consider the scalar product of $\underline{\dot{\zeta}}$ with the left-hand side of Eq. (1) and add to the expression its complex conjugate. The $(\dot{\zeta}, F\dot{\zeta})$ terms cancel because of the antihermiticity of F, and one obtains

$$\frac{1}{2}\left[(\dot{\zeta},N\dot{\zeta}) + (\zeta,Q\zeta)\right]^{\bullet} = -(\dot{\zeta},M\dot{\zeta}) \tag{3}$$

If Q is positive, i.e. $(\zeta,Q\zeta) > 0$ for all $\underline{\zeta}$, the system is stable owing to the positivity of N and M.

(b) Necessary and sufficient condition for stability

If $M \not\equiv 0$ and if $(\zeta,Q\zeta) < 0$ for any $\underline{\zeta} = \underline{\eta}$, the system is unstable. Together with the previous result this leads to a necessary and sufficient condition.

Proof: The proof is done by demonstrating incompatibility of stable $\underline{\zeta}$ and negative values of $(\zeta,Q\zeta)$. Indeed, it is then possible to choose $\underline{\zeta} = \underline{\eta}$ at a particular time with $(\zeta,Q\zeta) < 0$, and then, integrating Eq. (3), we obtain $(\zeta,Q\zeta)$ at later times:

$$(\zeta,Q\zeta) = -2\int_{t_0}^{t}(\dot{\zeta},M\dot{\zeta})dt' - (\dot{\zeta},N\dot{\zeta}) + (\eta,Q\eta) \tag{4}$$

From Eq. (4) it follows that $(\zeta,Q\zeta)$ remains negative and at least finite for all later $t > t_0$. This excludes the possibility that $\zeta \to 0$ as $t \to \infty$. An oscillation of $\underline{\zeta}$ around a finite value at $t \to \infty$ is also in contradiction to Eq. (4), the integral becoming infinite because $\underline{\dot{\zeta}}$ vanishes only on a countable set. The last possibility for a stable $\underline{\zeta}$ would be to tend to a constant in time, but this is in contradiction to Eq. (1) itself since $Q\underline{\zeta}$ cannot vanish because of Eq. (4).

It is appealing to conjecture that the growth will be exponential because any power growth is incompatible with Eq. (4). A rigorous proof of exponential growth cannot be done in the same way [7, 16] as for $F \equiv 0$, in which case overstability is forbidden. We conclude this section by saying that for Eq. (1) with $M \not\equiv 0$

$$(\zeta,Q\zeta) \geq 0 \tag{5}$$

is necessary and sufficient for stability.

2. SIMPLEST MODEL AND TIME SCALES

Let us consider a second-order differential equation with constant coefficients which is a particular case of Eq. (1):

$$\frac{1}{2}\ddot{y} + (a + i\,b)\dot{y} + c\,y = 0 \tag{6}$$

with $a > 0$. The solution is $y = e^{\omega t}$, with

$$\omega = -(a+ib) \pm \sqrt{(a+ib)^2 - 2c} \tag{7a}$$

and $c > 0$ is necessary and sufficient for stability.

a) If $a^2+b^2 \ll |c|$

$$\omega \simeq -(a+ib) \pm i\sqrt{2|c|}\left(1 - \frac{(a+ib)^2}{4|c|}\right) \tag{7b}$$

In the unstable case the growth rate is given by $\sqrt{|c|}$ as expected.

b) If $a^2+b^2 \gg |c|$

$$\omega \simeq -(a+ib) \pm (a+ib)\left(1 - \frac{c}{(a+ib)^2}\right)$$

The unstable case gives a growth rate:

$$\mathrm{Re}(\omega) = \frac{-a\,c}{a^2+b^2} \tag{7c}$$

c) If $a^2 \ll b^2 \approx |c|$

$$\omega \approx -(a+ib) \pm i\sqrt{b^2+2c}\left(1 - \frac{iab}{b^2+2c}\right) \quad \text{for } b^2+2c \approx |c|$$

$$\text{and } \omega \approx -(a+ib) \pm \sqrt{i2a\,b} \quad \text{for } b^2+2c = 0$$

$$\text{so that } \mathrm{Re}(\omega) \approx a\left(\frac{b}{\sqrt{|c|}} - 1\right) \quad \text{for } b^2+2c = |c| \tag{7d}$$

$$\mathrm{Re}(\omega) \approx \sqrt{ab} \quad \text{for } b^2+2c = 0 \tag{7e}$$

d) If $a^2 \approx b^2 \approx |c|$

$$\mathrm{Re}(\omega) \approx \sqrt{|c|} \tag{7f}$$

This particular example shows that although the sign of c governs stability independently of the values of a and b, the growth rates are strongly dependent on the relative magnitudes of a, b and c.

3. THREE-DIMENSIONAL PLASMAS WITH FULL-PRESSURE TENSOR

The macroscopic equations are of the following form:

$$\begin{aligned}
&\rho \frac{d\underline{v}}{dt} = \underline{j} \times \underline{B} - \nabla p - \nabla\cdot\underline{\underline{\Pi}} \\
&\underline{E} + \underline{v} \times \underline{B} = 0 \\
&\dot{\rho} + \nabla\cdot\rho\underline{v} = 0 \\
&p = f(\rho) \\
&\nabla \times \underline{B} = j \\
&\nabla \cdot \underline{B} = 0 \\
&\nabla \times \underline{E} = - \dot{\underline{B}}
\end{aligned} \tag{8}$$

The pressure tensor [17] Π is given by

$$\begin{aligned}
-\Pi_{xx} &= \alpha(\Gamma_{xx} + \Gamma_{yy}) + \beta\,\Gamma_{xy} + \frac{\beta^2}{4\alpha}(\Gamma_{xx} - \Gamma_{yy}) \\
-\Pi_{yy} &= \alpha(\Gamma_{xx} + \Gamma_{yy}) - \beta\,\Gamma_{xy} - \frac{\beta^2}{4\alpha}(\Gamma_{xx} - \Gamma_{yy}) \\
-\Pi_{zz} &= 2\,\alpha\,\Gamma_{zz} \\
-\Pi_{xy} &= -\Pi_{yx} = \frac{\beta}{2}(\Gamma_{yy} - \Gamma_{xx}) + \frac{\beta^2}{2\alpha}\Gamma_{xy} \\
-\Pi_{xz} &= -\Pi_{zx} = 2\,\beta\,\Gamma_{yz} + 2\,\Gamma_{xz}\frac{\beta^2}{\alpha} \\
-\Pi_{yz} &= -\Pi_{2z} = 2\frac{\beta^2}{\alpha}\Gamma_{yz} - 2\,\beta\,\Gamma_{xz}
\end{aligned} \tag{9}$$

$$\Gamma_{ij} = \frac{1}{2}\left(\frac{\partial v_i}{\partial x_j} + \frac{\partial v_j}{\partial x_i}\right) - \frac{1}{3}\frac{\partial v_n}{\partial x_n}\delta_{ij}$$

$$\beta = p/\omega_{ci}$$

$$\alpha = \frac{2}{3}\,\beta\,\omega_{ci}\,\tau_{ii}$$

x, y, z are a local system of Cartesian co-ordinates, z being along the magnetic field; ω_{ci} is the ion cyclotron frequency and τ_{ii} the ion-ion collision time.

After linearizing the system (8) around a static equilibrium and expressing all physical quantities in the perturbed velocity, we obtain (see Ref. [18])

$$\rho_o \frac{\partial^2 \underline{v}}{\partial t^2} + \frac{\partial}{\partial t} \nabla \cdot \underline{\underline{\Pi}} + Q\, \underline{v} = 0 \tag{10}$$

where Q is the MHD stability operator. To find out what the properties of the operator $\nabla \cdot \underline{\underline{\Pi}}$ are, let us consider $\underline{\underline{\Pi}}$ in a general co-ordinate system x^1, x^2, x^3; then

$$\begin{aligned} \int \left[\underline{v}' \cdot (\nabla \cdot \underline{\underline{\Pi}})\right] d\tau &= \int v'_r\, \Pi^{rn}_{\ \ ,n} \sqrt{g}\, dx^1\, dx^2\, dx^3 \\ &= - \int \Pi^{rn}\, v'_{r,n} \sqrt{g}\, dx^1\, dx^2\, dx^3 \\ &+ \int \frac{\partial}{\partial x^n} (\sqrt{g}\, v'_r\, \Pi^{rn}) dx^1\, dx^2\, dx^3 \end{aligned} \tag{11}$$

where ,n is the covariant derivative with respect to x^n. The last integral vanishes because of the boundary conditions on $\underline{v}$, and $\Pi^{rn}\, v'_{r,n}$ can be evaluated in the local co-ordinates system.

It turns out [18] that the terms in α and β^2/α are symmetric and positive definite, and the terms in β are antisymmetric. The pure β terms are due to the finite Larmor radius (FLR), the α terms to the magnetic-free viscosity, and the β^2/α terms to the magnetic viscosity.

Equation (10) is of the same type as Eq. (1), the FLR effects correspond to the operator F, and the viscosity to the operator M. The stability is decided by the MHD operator but the growth rates can be affected by FLR effects and viscosity. Such viscous destabilization has already been proposed by Greene and Coppi [19] but not in the general form done here. Let us make estimates of the reduction in MHD growth rates.

We know that the most dangerous MHD modes are nearly divergence-free. This means that

$$\Gamma_{xx} + \Gamma_{yy} = \frac{\partial v_x}{\partial x} + \frac{\partial v_y}{\partial y} = - \frac{\partial v_z}{\partial z}$$

but because of $v_\perp = \frac{E \times B}{B^2}$ and $E \approx \nabla \phi$,

$$\text{div } v_\perp \approx \frac{v_\perp}{r} \frac{r}{R}$$

where r is the radial extent of the mode, and R the large radius of the torus. On the other hand, one has $\Gamma_{xy} \approx v_\perp/r$, so that the ion-ion term dominates the FLR term if

$$\frac{\alpha}{\beta} > \left(\frac{R}{r}\right)^2$$

For an ion temperature of 1 keV, a magnetic field of 30 kg and a density of 10^{14} one has $\omega_{ci}\, \tau_{ii} \approx 5 \times 10^4$.

Very roughly, we can associate the coefficient 'a' of Eq. (6) with the ion-ion term in Eq. (10) by the relation

$$a \approx \frac{\alpha}{n_i MR^2} \approx \frac{kT\,\tau_{ii}}{MR^2} \approx 3 \times 10^7$$

if one takes $R \approx 100$ cm.

A full MHD growth rate $\gamma_{MHD} \approx \sqrt{|c|}$ is usually also of this order. If for particular reasons (geometry, near-marginality condition) an MHD growth rate is not full, then $a^2 \gg |c|$ and Eq. (7c) lead to the reduced growth rate:

$$\mathrm{Re}(\omega) \approx \frac{|c|}{a}$$

(a) Tokamak case

The m = 1, n = 1 mode observed [20–22] in tokamaks near the magnetic axis has a rather small growth rate [23] (resistive kink or ideal internal kink) of the order of 10^5 to 10^6 s^{-1} compared with typical MHD growth, but still too high to explain the experiment. Viscosity leads to a reduced $\gamma \approx |c|/a \approx 10^3$ to 10^4, which agrees with observation.

(b) High-β, $\ell = 1$ stellarator

The m = 2 mode has a typical MHD growth rate and for the previous plasma parameters we should obtain $a^2 \approx b^2 \approx |c|$, so that we would have the case (7e) with a small reduction in growth rate.

This mode has rarely been observed [24] in the present experiments, and this is because $\omega_{ci}\, \tau_{ii}$ is two to three orders of magnitude lower ($T_i \approx 100$ eV) than calculated before, so that $a^2 \ll b^2 \approx |c|$, which is the case of Eq. (7d). The FLR stabilization dominates in this case.

The derivation of the pressure tensor as given by Eqs (9) is done for $r_L/L \ll 1$ and $\Lambda_{MFP} \ll L$, with r_L = Larmor radius, L = inhomogeneity length, Λ_{MFP} = mean free path. For incompressible motion, the validity of the pressure tensor can be extended to the domain $\Lambda_{MFP} \gg L_\perp$ but $\Lambda_{MFP} \ll L_\parallel$.

4. TWO-DIMENSIONAL PLASMAS WITH RESISTIVITY AND THE FULL-PRESSURE TENSOR

It is well known that resistivity leads [3] to new modes and one can expect a much more difficult behaviour. At present only the two-dimensional case for straight plasmas can be solved, as we shall see.

The equilibrium is characterized [9] by

$$\begin{aligned}
\underline{j}_o &= \underline{e}_z\, J_o(\psi) \\
\underline{E}_o &= \underline{ct} = \underline{e}_z\, \eta_o(\psi)\, J_o(\psi) \\
\underline{B}_o &= \underline{e}_z \times \nabla\psi\ (\underline{e}_z \cdot \underline{B}_o)\ \underline{e}_z \\
\nabla^2\psi &= J_o(\psi) = -\frac{dP_o}{d\psi}
\end{aligned} \qquad (12)$$

where z is the co-ordinate along the straight plasma, ψ is the meridional magnetic flux, and η_0 is the resistivity.

The meridional currents are assumed to be zero in order to have a static plasma in equilibrium, which is important for this kind of formulation but not necessarily for the physical results.

After linearizing the equation of motion we obtain

$$\rho_o\, \ddot{\underline{\xi}} + \nabla p_1 + \nabla \cdot \underline{\underline{\Pi}}(\underline{\xi}) - \underline{j}_1 \times \underline{B}_o - \underline{j}_o \times \underline{B}_1 = 0 \qquad (13)$$

$$\nabla \cdot \underline{\xi} = 0 \qquad (14)$$

$$\dot{\underline{A}}_1 + \eta_o \nabla \times \nabla \times \underline{A}_1 + \eta_1 \underline{j}_o - \dot{\underline{\xi}} \times \underline{B}_o = 0 \qquad (15)$$

$$\underline{B}_1 = \nabla \times \underline{A}_1 \qquad (16)$$

$$\eta_1 = -\,\xi \cdot \nabla \eta_o \qquad (17)$$

Apart from the pressure tensor term, these equations are the same as in Ref. [9]. Equation (17) is valid as long as the heat conductivity is small enough. This will be discussed later.

Restriction to two-dimensional perturbations leads to

$$\underline{\xi} = \underline{e}_z \times \nabla U = -\nabla \times \underline{e}_z U , \quad \underline{B}_1 = -\underline{e}_z \times \nabla A = \nabla \times \underline{e}_z A$$

$$\underline{j}_1 = -\underline{e}_z \nabla^2 A , \qquad \nabla \times \underline{j}_1 = -\nabla(\nabla^2 A) \times \underline{e}_z$$

where U and A are two scalars: the stream function and the z-component of the vector potential.

Taking the curl of Eq. (13), we obtain

$$- \nabla \cdot \rho_o \nabla \ddot{U} + e_z \cdot \nabla \times \nabla \cdot \underline{\underline{\Pi}}(\nabla \times \underline{e}_z \dot{U}) - \underline{B}_o \cdot \nabla (\nabla^2 A) \tag{18}$$

$$- \nabla \times j_o \cdot \nabla A = 0$$

$$\dot{A} + \underline{B}_o \cdot \nabla \dot{U} - \eta_o \nabla^2 A + J_o(\underline{e}_z \times \nabla \eta_o \cdot \nabla U) = 0 \tag{19}$$

If $\nabla^2 A$ is taken from Eq. (19) ($\eta_0 \neq 0$) and inserted in Eq. (18), we obtain the following system of equations in matrix operatorial form:

$$\begin{pmatrix} -\nabla \cdot \rho_o \nabla & 0 \\ 0 & 0 \end{pmatrix} \begin{pmatrix} \ddot{U} \\ \ddot{A} \end{pmatrix}$$

$$+ \begin{pmatrix} \dfrac{-(\underline{B}_o \cdot \nabla)^2}{\eta_o} & \dfrac{-(\underline{B}_o \cdot \nabla)}{\eta_o} \\ \dfrac{\underline{B}_o \cdot \nabla}{\eta_o} & \dfrac{1}{\eta_o} \end{pmatrix} \begin{pmatrix} \dot{U} \\ \dot{A} \end{pmatrix}$$

$$+ \begin{pmatrix} e_z \nabla \times \nabla \cdot \underline{\underline{\Pi}} (\nabla \times \underline{e}_z \ldots) & 0 \\ 0 & 0 \end{pmatrix} \begin{pmatrix} \dot{U} \\ \dot{A} \end{pmatrix}$$

$$+ \begin{pmatrix} -J_o(\psi)\underline{B}_o \cdot \nabla e_z \times \dfrac{\nabla \eta_o}{\eta_o} \cdot \nabla & -\nabla \times \underline{j}_o \cdot \nabla \\ \nabla \times \underline{j}_o \cdot \nabla & -\nabla^2 \end{pmatrix} \begin{pmatrix} U \\ A \end{pmatrix} = 0 \tag{20}$$

Apart from the $\nabla \cdot \underline{\underline{\Pi}}$ term, Eq. (20) is identical with Eq. (17) of Ref. [9]. Let us investigate the operator $e_z \cdot \nabla \times \nabla \cdot \underline{\underline{\Pi}} (\nabla \times e_z \ldots)$. We know from the previous section and from Ref. [18] that $\nabla \cdot \underline{\underline{\Pi}} (\underline{\xi})$ contains α terms which are positive definite operators and pure β terms which are antisymmetric, but if $\underline{\xi}$ is a curl and if $\nabla \cdot \underline{\underline{\Pi}} (\underline{\xi})$ is replaced by $\nabla \times \nabla \cdot \underline{\underline{\Pi}}(\underline{\xi})$ the symmetry properties will not be changed, as the following equation shows:

$$\int d\tau \; v \; \underline{e}_z \cdot \nabla \times \nabla \cdot \Pi \quad (\nabla \times \underline{e}_z \; u)$$

$$= \int \nabla \times \underline{e}_z v \cdot \nabla \cdot \underline{\underline{\Pi}} \quad (\nabla \times \underline{e}_z \; u) \; d\tau$$

The symmetry properties of $\nabla \cdot \underline{\underline{\Pi}}$ are thus the same as $e_z \cdot \nabla \times \nabla \cdot \underline{\underline{\Pi}}(\nabla \times e_z \ldots)$. Then Eq. (20) has the same character as Eq. (1) and the stability condition is given by the last matrix operator of Eq. (20), which is the same as in Ref. [9]. This leads to the energy principle derived in [9] for zero pressure tensor and extended here to FLR and ion-ion collisions:

$$\delta W = \int d\tau \left(-\frac{dJ_o}{d\psi}\right) (\underline{e}_z \times \nabla\psi \cdot \nabla U)^2$$
$$+ 2\int d\tau \left(-\frac{dJ_o}{d\psi}\right) A \; (\underline{e}_z \times \nabla\psi \cdot \nabla U)$$
$$+ \int d\tau \; |\nabla A|^2 \qquad (21)$$

A simple application to tokamaks is the instability of skin currents. If one localizes a test function with ∇U finite at $dJ_0/d\psi > 0$ and $A \approx 0$, then $\delta w < 0$. This mode is similar to the rippling mode in one-dimensional geometry [3].

Another application would be the stability of configurations with stagnation points (such as doublets or for islands in tokamaks). This necessitates numerical calculations which are not easy to do because of the stagnation point, but Eq. (21) at least allows the problem to be correctly formulated.

5. ALL PERTURBATION WITH RESISTIVITY, VISCOSITY AND FLR IN THE TOKAMAK SCALING

The energy principle of Section 4 can be extended to three-dimensional perturbation if one goes to helical co-ordinates for the representation of the equilibrium and perturbations and in the approximation of the tokamak scaling $Kr \approx B_\theta / B_z \approx \epsilon$.

The proof and calculations are given in Ref. [25]. The necessary and sufficient criterion is given by:

$$\delta W = \int \left(-\frac{dj_o}{dr}\right) \left(\underline{u} \times \underline{e}_r \cdot \nabla G\right) \left(\underline{u} \times \nabla F_o \cdot \nabla G\right) d\tau$$

$$+ 2 \int \left(-\frac{dj_o}{dr}\right) \left(\underline{u} \times \underline{e}_r \cdot \nabla G\right) F \, d\tau$$

$$- \int F \, L \, F \, d\tau \qquad (22)$$

where

$$\underline{B}_o = f_o(r) \, \underline{u} + \underline{u} \times \nabla F_o(r)$$
$$\underline{B}_j = f(r,u,t) \, \underline{u} + \underline{u} \times \nabla F(r,u,t)$$
$$\underline{\xi} = g(r,u,t) \, \underline{u} + \underline{u} \times \nabla G(r,u,t)$$

$$\underline{u} = \frac{l \, \underline{e}_z + hr \, \underline{e}_\theta}{l^2 + h^2 r^2} , \qquad hr \simeq \varepsilon$$

$$u = l \, \theta - hz$$

A test function can always be found to make the system unstable as soon as $dj_0/dr \neq 0$. This test function can be taken as:

> $F \approx 0$ and G concentrated on the side of the resonance ($\nabla F_0 = 0$) at which the first integrand is negative,which is always possible. An estimate of the growth rate can also be made as in Section 3.

This test function characterizes the rippling mode. A similar test function was used by Furth [8] in the sheet-pinch geometry. The tearing mode test function is not localized but can, in principle, be found in the same way as in Ref. [8] and can be affected by cylindrical geometry (particularly the mode m = 1, n = 1). The FLR effects and viscosity do not stabilize the resistive modes but can appreciably reduce the growth rates (see estimate in Section 3).

Stabilization of the tearing mode alone seems possible [26]. To obtain it, one has to assume a non-fluctuating resistivity and shape the current density in a step-like form [26]. If the plasma has to be stable to all resistive modes (rippling included), then the current density as demonstrated by expression (22) has to be constant up to the boundary. Kadomtsev [27] came to this last result using Taylor's invariant [12], which is discussed in the next section.

6. STABILITY OF FORCE-FREE FIELDS

Resistive force-free fields have to be time-dependent and have to be restricted to the class [28] $j = \lambda \underline{B}$ with $\lambda = ct$ (see also Ref. [11]) and $\underline{B} = \underline{B}_0 \exp(-\eta\lambda^2 t)$.

The linearized equations of motion around such solutions are:

$$\rho_0 \ddot{\underline{\xi}} = \underline{j}_1 \times \underline{B}_0 - \lambda \underline{B}_1 \times \underline{B}_0 \tag{23}$$

$$-\dot{\underline{A}}_1 + \dot{\underline{\xi}} \times \underline{B}_0 = \eta \underline{j}_1 + \eta_1 \underline{j}_0 \tag{24}$$

$$\underline{B}_1 = \nabla \times \underline{A}_1 \,, \quad \underline{j}_1 = \nabla \times \nabla \times \underline{A}_1 \tag{25}$$

The investigation is restricted to the case $\lambda = ct$, $\eta = ct$ consistent with $\eta_1 = 0$ and the gauge is chosen such that $\underline{E} = -\dot{A}$. The scalar product of Eq. (23) with $\dot{\underline{\xi}}$ yields

$$\rho_0 \ddot{\underline{\xi}} = -(j_1 - \lambda \underline{B}_1) \cdot (\dot{\underline{A}}_1 + \eta j_1)$$

Integrating over the plasma volume limited by a perfectly conducting wall, we obtain

$$\frac{1}{2}\frac{\partial}{\partial t}\left[(\rho_0 \dot{\xi}, \dot{\xi}) + (\nabla \times \underline{A}_1, \nabla \times \underline{A}_1) - (\lambda A_1, \nabla \times \underline{A}_1)\right]$$

$$= -\eta \left[(j_1, j_1) - (\lambda j_1, B_1\right] \tag{26}$$

where

$$(\alpha, \zeta) = \int_V d\tau \, \underline{\alpha} \cdot \underline{\zeta}$$

Let δW_R be defined as

$$\delta W_R = (\nabla \times \underline{A}_1, \nabla \times \underline{A}_1) - \lambda(A_1, \nabla \times \underline{A}_1) \tag{27}$$

The variation of δW_R leads to the following Euler eigenvalue equation:

$$\nabla \times \nabla \times \underline{A}_1 - \lambda \nabla \times \underline{A}_1 = \alpha \underline{A}_1 \tag{28}$$

The variation of the right-hand side of Eq. (26) leads to

$$\nabla \times \nabla \times \underline{B}_1 - \lambda \nabla \times \underline{B}_1 = \beta \underline{B}_1 \tag{29}$$

The curl of Eq. (28) is identical with Eq. (29). This means that any solution of Eq. (29) verifying $\underline{n} \cdot \underline{B}_1 = 0$ at the boundary is also a solution of Eq. (28) with $\underline{n} \times \underline{A}_1 = 0$ at the boundary.

It follows that $\delta W_R > 0$ implies the negativeness of the right-hand side of Eq. (26). This means that $\delta W_R > 0$ is sufficient for stability with respect to MHD + resistive modes. This condition is found necessary and sufficient if one ignores resistivity and uses instead Taylor's hypothesis of a global invariant [12, 29]. This result is somewhat to be expected if one considers Woltjer's [30] proof that λ = ct force-free fields represent the state of minimal energy in a closed system.

The important practical question for fusion plasmas is to know how much one can deviate [31] from λ = ct force-free fields without appreciably affecting the gross stability properties and without appreciably diminishing the confinement time. This question might require an understanding of the non-linear problem, which would exceed the scope of this paper.

7. DISCUSSION AND CONCLUSION

The method pursued in this paper allows statements about stability without going to the solution of eigenmodes. This is only possible if the representation variable ζ in which the linearized equations of motion are of the same type as Eq. (1) can be found. This depends, of course, on the physical equations used.

Effects such as thermal conductivity affect the symmetry properties of the operators of Eqs (1) and (20). In fact, Ohm's law in hot plasmas is not known; it can be affected by trapped particles in toroidal geometry and generally by turbulence. Even cylindrical geometry presents difficulties: energy principle (22) would not have been possible without making a tokamak expansion. Answers to these questions can be found as follows:

(1) One can restrict the investigation to a class of resistive modes (essentially the 'tearing' modes) as done in, for example, Refs [6] and [26], which generally leads to optimistic results. But then it remains for us to understand the meaning of the restriction in the stable case and to know how the growth rates are affected by the ignored physical terms (FLR, viscosity, etc.) in the unstable case.

(2) This dilemma may require general stability conditions to be found as in, for example, Ref. [10] and patient searching for the representation ζ if it exists, for which the linearized physical equations become of the type of Eq. (1). This is demonstrated here for some non-ideal effects and geometries.

(3) The hardest way, but the nearest to real plasmas, is to develop methods dealing with the stability of equations more general than Eq. (1), particularly those for which Q is not symmetric.

REFERENCES

[1] LIN, C.C., The Theory of Hydrodynamic Stability, Cambridge Univ. Press (1955).
[2] DUNGEY, J.W., Cosmic Electrodynamics, Cambridge Univ. Press (1958).
[3] FURTH, H.P., KILLEEN, J., ROSENBLUTH, M.N., Phys. Fluids **6** (1963) 459.
[4] COPPI, B., in Propagation and Instabilities, Stanford Univ. Press (1963) 70.
[5] CAP, F.F., Handbook on Plasma Instabilities, Academic Press, New York (1976).
[6] GLASSER, A.H., GREENE, J.M., JOHNSON, J.L., Phys. Fluids **18** (1975) 875.
[7] BARSTON, E.M., Phys. Fluids **12** (1969) 2162.
[8] FURTH, H.P., in Propagation and Instabilities, Stanford Univ. Press (1963) 87.
[9] TASSO, H., Plasma Phys. **17** (1975) 1131.
[10] TASSO, H., "Energy principles for non-ideal MHD", Plasma Physics and Controlled Nuclear Fusion Research 1976 (Proc. 6th Int. Conf. Berchtesgaden, 1976) **3**, IAEA, Vienna (1977) 371.
[11] TASSO, H., On the Stability of Force-Free Fields, Max-Planck Inst. Plasmaphysik, Garching, Rep. IPP 6/151 (1976).
[12] TAYLOR, J.B., Phys. Rev. Lett. **33** (1974) 1139.
[13] BERNSTEIN, I.B., et al., Proc. R. Soc. London A **244** (1958) 17.
[14] HAIN, K., LÜST, R., SCHLÜTER, A., Z. Naturforsch. A **12** (1957) 833.
[15] LOW, F.E., Proc. R. Soc. London A **248** (1958) 283.
[16] PFIRSCH, D., TASSO, H., Nucl. Fusion **11** (1971) 259.
[17] CHAPMAN, S., COWLING, T., The Mathematical Theory of Non-uniform Gases, Cambridge Univ. Press (1953).
[18] TASSO, H., SCHRAM, P.P.J.M., Nucl. Fusion **6** (1966) 284.
[19] GREENE, J., COPPI, B., Phys. Fluids **8** (1965) 1745.

[20] VON GOELER, S., et al., Phys. Rev. Lett. **33** (1974) 1201.
[21] LAUNOIS, P., Proc. 7th Europ. Conf. Controlled Fusion and Plasma Physics, Lausanne, 1975, p.1.
[22] KARGER, F., et al., "On the origin of the disruptive instability in the Pulsator I tokamak", Plasma Physics and Controlled Nuclear Fusion Research 1976 (Proc. 6th Int. Conf. Berchtesgaden, 1976) **1**, IAEA, Vienna (1977) 267.
[23] COPPI, B., et al., Proc. Annual Meeting on Theoretical Aspects of Controlled Thermonuclear Research, Madison, 1976, paper 1 A-4.
[24] FÜNFER, E., et al., Proc. 7th Europ. Conf. on Controlled Fusion and Plasma Physics, Lausanne, 1975, p.151.
[25] TASSO, H., Plasma Phys. **19** (1977) 177.
[26] GLASSER, A.H., FURTH, H.P., Rutherford, P.H., Phys. Rev. Lett. **38** (1977) 234.
[27] KADOMTSEV, B.B., "Reconnection of field lines and disruptive instability in tokamaks", Plasma Physics and Controlled Nuclear Fusion Research 1976 (Proc. 6th Int. Conf. Berchtesgaden, 1976) **1**, IAEA, Vienna (1977) 555.
[28] JETTE, A.D., J. Math. Anal. Appl. **29** (1970) 109.
[29] JUKES, J.D., "Relaxation of toroidal discharges", Plasma Physics and Controlled Nuclear Fusion Research 1976 (Proc. 6th Int. Conf. Berchtesgaden, 1976) **1**, IAEA, Vienna (1977) 479.
[30] WOLTJER, L., Proc. Natl. Acad. Sci. **44** (1958) 489.
[31] TAYLOR, J.B., in Proc. 3rd Topical Conf. Pulsed High-Beta Plasmas, Culham, 1975, Pergamon, Oxford (1976) paper B1.6, p. 5g.

LINEAR AND NON-LINEAR CALCULATIONS OF THE TEARING MODE*

D. SCHNACK, J. KILLEEN
Lawrence Livermore Laboratory,
Livermore, California,

and

Department of Applied Science,
University of California,
Davis/Livermore, California,
United States of America

Abstract

LINEAR AND NON-LINEAR CALCULATIONS OF THE TEARING MODE.

The results are presented of the application of a two-dimensional, non-linear MHD computer code to the tearing mode. For the case of the sheet pinch, linear growth rates are obtained and non-linear behaviour investigated. In addition, the linear and non-linear behaviour of the mode is studied for the case when two singular surfaces are in close proximity. For this double tearing mode, it is found that the linear growth rate is greatly enhanced over that of the single tearing mode. The mode is followed into the non-linear regime and is found to result in magnetic field configurations which could lead to enhanced diffusion. It is found that approximately 10% of the energy released goes into kinetic energy.

1. INTRODUCTION

The existence of sheared or reversed magnetic fields in many recent controlled fusion devices has led to renewed interest in the tearing mode, one of a class of finite-resistivity instabilities which were first studied in detail by Furth, Killeen, and Rosenbluth [1]. In that work (henceforth referred to as FKR) a dispersion relation is derived for a sheet pinch, and growth rates are obtained as functions of wavelength and magnetic Reynolds number $S = t_R/t_H$. This analytical work has been extended to tokamak geometries [2,3], and more recently there have been attempts to understand the non-linear aspects of this mode [4,5,6].

* Work performed under the auspices of USERDA, Contract No. W-7405-Eng-48.

In addition, the tearing mode has been studied for several years in connection with the problem of solar flares [7,8,9], where it has been advanced as a candidate for the onset of the flare event. Also, there is currently considerable interest in the double-tearing mode [10,11], which can arise when two singular ($\underline{k}\cdot\underline{B} = 0$) surfaces are in close proximity. The resulting interaction can lead to enhanced cross-field diffusion and field annihilation.

Concurrently, the rise in speed and memory size of computers has made feasible the solution of more and more complex problems by numerical, rather than analytical, means. For example, ideal MHD instabilities are now being studied as initial value problems in more than one space dimension [12,13]. In this way the equations need only be modified for physical, not analytical, reasons, and non-linearities and complex geometries can be treated.

This paper describes the application of this approach to the study of the tearing mode. The problem here is more complex than for ideal MHD, as we now have phenomena occurring on two often widely separated time scales: the hydromagnetic time and the resistive diffusion time. Thus, to make the problem computationally realizable, implicit methods should be used. This was first done by Killeen [14], who obtained growth rates by solving the linearized, Fourier-decomposed equations in one space dimension. This work was extended for cylindrical geometries by Dibiase [15], who included the effects of compressibility, viscosity and thermal conductivity. Recently, non-linear multi-dimensional studies have appeared [16]. However, these are tailored to special geometries and solve reduced sets of equations. In the present work we apply a more primitive model: the resistive MHD equations in two space dimensions. No assumptions about the ordering of terms are made, and the equations are written in orthogonal curvilinear coordinates so that geometrical effects can be accounted for in the metric.

In section 2 of this paper we describe the mathematical model and display the equations to be solved. Section 3 is a brief discussion of some computational techniques employed by the numerical model. In section 4 we present results obtained by applying the code to a sheet pinch. We find linear growth rates for the tearing mode in good agreement with previous work [1,14], and follow the mode into the non-linear regime, where saturation is observed. The remainder of the paper deals with the double tearing instability. In section 5 we solve the linearized resistive MHD equations, as derived in [14], with an appropriate model for the zero-order magnetic field. We obtain linear growth rates as a function of the magnetic Reynolds number and of the separation distance between the singular surfaces. The eigenfunctions are also found. We show that the growth rates can be greatly enhanced over those of the ordinary tearing mode for the same magnetic shear. In section 6 we display the non-linear behavior of this mode. We show that the interaction between the adjacent singular surfaces leads to enhanced vortex flow and magnetic field distortion in the region between the surfaces.

2. MATHEMATICAL MODEL

The equations appropriate for the description of low-frequency phenomena in a conducting fluid are the MHD equations, which usually appear in the form:

$$\frac{\partial \underline{B}}{\partial t} = \nabla x(\underline{v} x \underline{B} - \frac{\eta c^2}{4\pi} \nabla x \underline{B}) \tag{1}$$

$$\rho \frac{\partial \underline{v}}{\partial t} + \rho \underline{v} \cdot \nabla \underline{v} = -\nabla \cdot \underline{\underline{P}} + \frac{1}{4\pi} (\nabla x \underline{B}) \ x \ \underline{B} \tag{2}$$

$$\frac{\partial \rho}{\partial t} = - \nabla \cdot (\rho \underline{v}) \tag{3}$$

$$\frac{\partial(\rho\varepsilon)}{\partial t} = -p\nabla\cdot\underline{v} - \nabla\cdot(\rho\varepsilon\underline{v} - \kappa\nabla T) + \eta\left(\frac{c}{4\pi}\right)^2 (\nabla\times\underline{B})^2 - \nu\underline{\underline{V}}:\nabla\underline{v} \tag{4}$$

along with the equation of state $p = \rho T$. Here, $\varepsilon = kT/m_i(\gamma-1)$ is the internal energy per unit mass, η is the resistivity, κ is the thermal conductivity, ν is the viscosity, and $\underline{\underline{P}}$ is the stress tensor defined by

$$\underline{\underline{P}} = p\,\underline{\underline{I}} - \nu\,\underline{\underline{V}} \tag{5}$$

where $\underline{\underline{V}}$, the viscous stress tensor, is given by

$$\underline{\underline{V}} = \frac{2}{3}\nabla\cdot\underline{v}\,\underline{\underline{I}} - (\nabla\underline{v} + \nabla\underline{v}^{\dagger}) \tag{6}$$

We use the notation $(\;)^{\dagger}$ to designate the transpose. While this formulation assumes a collisional plasma of equal ion and electron temperatures, the generalization to the anisotropic case is well known, and in this paper we shall deal only with equation (1) thru (4).

For computational purposes, it is preferable to recast the equations in conservation form [17]. In terms of the dimensionless quantities

$$\underline{x}' = \underline{x}/a\,, \quad t' = t/t_H\,, \quad \underline{B}' = \underline{B}/B_o$$
$$\underline{v}' = \underline{v}/v_A\,, \quad \rho' = \rho/\rho_o\,, \quad p' = p/p_o$$
$$T' = T/T_o\,, \quad u' = u/u_o\,, \quad \eta' = \eta/\eta_o$$
$$\kappa' = \kappa/\kappa_o\,, \quad \nu' = \nu/\nu_o$$

equations (1-4) can be combined to yield [18]

$$\frac{\partial\underline{B}'}{\partial t'} = \nabla\times\left(\underline{v}'\times\underline{B}' - \frac{\eta'}{S}\nabla\times\underline{B}'\right) \tag{7}$$

$$\frac{\partial(\rho\underline{v}')}{\partial t'} = -\nabla\cdot\left(\rho'\underline{v}'\underline{v}' + \tfrac{1}{2}(p'+B'^2)\underline{\underline{I}} - \underline{B}'\underline{B}' - \frac{\nu'}{R}\underline{\underline{V}}'\right) \tag{8}$$

$$\frac{\partial\rho'}{\partial t'} = -\nabla\cdot(\rho' v') \tag{9}$$

$$\frac{\partial u'}{\partial t'} = -\nabla\cdot\Big((u' + p')\underline{v}' + \frac{2\eta'}{S}(\underline{B}'\cdot\nabla\underline{B}' - \nabla\underline{B}'\cdot\underline{B}') + (B'^2\underline{\underline{I}} - 2\underline{B}'\underline{B}')\cdot\underline{v}' - \frac{2\nu'}{R}\underline{\underline{V}}'\cdot\underline{v}' - \frac{\kappa'}{K}\nabla T'\Big) \tag{10}$$

which express the conservation of magnetic flux, momentum, mass and energy. The quantity $u' = \rho v'^2 + B'^2 + p'/(\gamma-1)$ is the total energy density of the fluid. In the normalization above, subscripts $(\ )_o$ refer to characteristic values, "a" is a characteristic length, $v_A = B_o/\sqrt{4\pi\rho_o}$ is the Alfvén velocity, and $t_H = a/v_A$ is the hydromagnetic (or Alfvén) transit time. The normalization of the thermodynamic properties is chosen such that

$$u_o = p_o = B_o^2/8\pi = \tfrac{1}{2}\rho_o v_A^2 \tag{11}$$

The nondimensional numbers R, S and K appearing in (7) thru (10) are the hydrodynamic, magnetic, and thermal Reynolds numbers, defined as the ratios of the corresponding diffusion times to the hydromagnetic transit time, i.e.

$$R \equiv t_{visc}/t_H = \rho_o a v_A/\nu_o$$

$$S \equiv t_R/t_H = 4\pi a v_A/c^2\eta_o$$

$$K \equiv t_{THERM}/t_H = \rho_o a v_A/\kappa_o$$

where t_{visc}, t_R, and t_{THERM} are the viscous, resistive and thermal diffusion times, respectively.

Equation (7) can also be written as

$$\frac{\partial\underline{B}'}{\partial t'} = \nabla\cdot\Big(\underline{B}'\underline{v}' - \underline{v}'\underline{B}' + \frac{\eta'}{S}(\nabla\underline{B}' - \nabla\underline{B}'^{\dagger})\Big) \tag{12}$$

which emphasizes the flux-divergence form.

Equations (7), (8), (9) and (10), along with the equation of state and the definition of u' , define a set of 8 equations in 8 unknowns:

three components of the magnetic flux density $\underline{B}'$, three components of the momentum density $\rho'\underline{v}'$, the mass density, and the total energy density. In orthogonal curvilinear coordinates the two-dimensional form of the equations is

$$\frac{\partial B_1}{\partial t} = \frac{1}{h_2h_3}\frac{\partial}{\partial x_2}\left\{h_3(v_1B_2-v_2B_1) + \frac{1}{S}\frac{h_3\eta}{h_1h_2}\left(\frac{\partial}{\partial x_2}(h_1B_1) - \frac{\partial}{\partial x_1}(h_2B_2)\right)\right\} \tag{13}$$

$$\frac{\partial B_2}{\partial t} = \frac{1}{h_1h_3}\frac{\partial}{\partial x_1}\left\{h_3(v_2B_1-v_1B_2) + \frac{1}{S}\frac{h_3\eta}{h_1h_2}\left(\frac{\partial}{\partial x_1}(h_2B_2) - \frac{\partial}{\partial x_2}(h_1B_1)\right)\right\} \tag{14}$$

$$\begin{aligned}\frac{\partial B_3}{\partial t} = {} & \frac{1}{h_1h_2}\frac{\partial}{\partial x_1}\left\{h_2(v_3B_1-v_1B_3) + \frac{1}{S}\frac{h_2\eta}{h_1h_3}\frac{\partial}{\partial x_1}(h_3B_3)\right\} \\ & + \frac{1}{h_1h_2}\frac{\partial}{\partial x_2}\left\{h_1(v_3B_2-v_2B_3) + \frac{1}{S}\frac{h_1\eta}{h_2h_3}\frac{\partial}{\partial x_2}(h_3B_3)\right\}\end{aligned} \tag{15}$$

$$\begin{aligned}\frac{\partial(\rho v_1)}{\partial t} = {} & -\frac{1}{h_1h_2h_3}\left\{\frac{\partial}{\partial x_1}\left(h_2h_3(\rho v_1^2-B_1^2)\right) + \frac{\partial}{\partial x_2}\left(h_1h_3(\rho v_1v_2-B_1B_2)\right)\right. \\ & + h_3\frac{\partial h_1}{\partial x_2}(\rho v_1v_2-B_1B_2) - h_3\frac{\partial h_2}{\partial x_1}(\rho v_2^2-B_2^2) \\ & \left. - h_2\frac{\partial h_3}{\partial x_1}(\rho v_3^2-B_3^2)\right\} - \frac{1}{2}\frac{1}{h_1}\frac{\partial}{\partial x_1}(p+B^2)\end{aligned} \tag{16}$$

$$\begin{aligned}\frac{\partial(\rho v_2)}{\partial t} = {} & -\frac{1}{h_1h_2h_3}\left\{\frac{\partial}{\partial x_1}\left[h_2h_3(\rho v_1v_2-B_1B_2)\right] + \frac{\partial}{\partial x_2}\left[h_1h_3(\rho v_2^2-B_2^2)\right]\right. \\ & + h_3\frac{\partial h_2}{\partial x_1}(\rho v_1v_2-B_1B_2) - h_3\frac{\partial h_1}{\partial x_2}(\rho v_1^2-B_1^2) \\ & \left. - h_1\frac{\partial h_3}{\partial x_2}(\rho v_3^2-B_3^2)\right\} - \frac{1}{2}\frac{1}{h_2}\frac{\partial}{\partial x_2}(p+B^2)\end{aligned} \tag{17}$$

$$\frac{\partial(\rho v_3)}{\partial t} = -\frac{1}{h_1h_2h_3}\left\{\frac{\partial}{\partial x_1}\left[h_2h_3(\rho v_1v_3-B_1B_3)\right] + \frac{\partial}{\partial x_2}\left[h_1h_3(\rho v_2v_3-B_2B_3)\right]\right.$$

$$\left. + h_2\frac{\partial h_3}{\partial x_1}(\rho v_1v_3-B_1B_3)+h_1\frac{\partial h_3}{\partial x_2}(\rho v_2v_3-B_2B_3)\right\} \tag{18}$$

$$\frac{\partial\rho}{\partial t} = -\frac{1}{h_1h_2h_3}\left\{\frac{\partial}{\partial x_1}(h_2h_3\rho v_1)+\frac{\partial}{\partial x_2}(h_1h_3\rho v_2)\right\} \tag{19}$$

and

$$\frac{\partial u}{\partial t} = -\frac{1}{h_1h_2h_3}\frac{\partial}{\partial x_1}\left[h_2h_3\left[(u+p)v_1-(B_1^2-B_2^2-B_3^2)v_1-2B_1(B_2v_2+B_3V_3)\right.\right.$$

$$\left.\left.-\frac{2\eta}{S}\left\{\frac{1}{h_1h_3}B_3\frac{\partial}{\partial x_1}(h_3B_3)+\frac{1}{h_1h_2}B_2\left[\frac{\partial}{\partial x_1}(h_2B_2)-\frac{\partial}{\partial x_2}(h_1B_1)\right]\right\}\right]\right]$$

$$-\frac{1}{h_1h_2h_3}\frac{\partial}{\partial x_2}\left[h_1h_3\left[(u+p)v_2+(B_1^2-B_2^2+B_3^2)v_2-2B_2(B_1v_1+B_3v_3)\right.\right.$$

$$\left.\left.+\frac{2\eta}{S}\left\{\frac{1}{h_1h_2}B_1\left[\frac{\partial}{\partial x_1}(h_2B_2)-\frac{\partial}{\partial x_2}(h_1B_1)\right]-\frac{1}{h_2h_3}B_3\frac{\partial}{\partial x_2}(h_3B_3)\right\}\right]\right] \tag{20}$$

We have assumed that an element of arc length can be written $ds^2 = h_1^2dx_1^2+h_2^2dx_2^2+h_3^2dx_3^2$, that there is no x_3 dependence, and have dropped the primed notation. In addition, we have set $\nu = \kappa = 0$ to conform with results to be presented in later sections.

3. COMPUTATIONAL TECHNIQUES

The set of equations (13) thru (20), together with the appropriate boundary and initial conditions, define an initial value problem which we solve by finite difference techniques on an Eulerian mesh. The spatial differencing is fully conservative on a variably zoned grid with densities defined at grid points and fluxes at half grid points.

First derivatives are differenced as

$$\left(\frac{\partial F}{\partial x_1}\right)_i = \frac{F_{i+1/2}-F_{i-1/2}}{x_{1i+1/2}-x_{1i-1/2}} = \frac{F_{i+1}-F_{i-1}}{\Delta_+ x_1+\Delta_- x_1} \tag{21}$$

where $F_{i\pm1/2}=(F_{i\pm1}+F_i)/2$, and Δ_+ and Δ_- are the forward and backward difference operators at x_i. Second derivatives become

$$\frac{\partial}{\partial x_1}\left(a\frac{\partial F}{\partial x_1}\right)_{ij} = \frac{2}{\Delta_+x_1\Delta_-x_1(\Delta_+x_1+\Delta_-x_1)}\Big[\Delta_-x_1\ a_{i+1/2,j}\left(F_{i+1,j}-F_{ij}\right)$$

$$-\ \Delta_+x_1\ a_{i-1/2,j}\left(F_{ij}-F_{i-1,j}\right) \tag{22}$$

while the two types of mixed derivatives are differenced as

$$\frac{\partial}{\partial x_1}\left(a\frac{\partial F}{\partial x_2}\right)_{ij} = C\Big[a_{i+1/2,j}\left(F_{i+1,j+1}+F_{i,j+1}-F_{i+1,j-1}-F_{i,j-1}\right)$$

$$-a_{i-1/2,j}\left(F_{i,j+1}+F_{i-1,j+1}-F_{i,j-1}-F_{i-1,j-1}\right)\Big] \tag{23}$$

and

$$\frac{\partial}{\partial x_2}\left(a\frac{\partial F}{\partial x_1}\right)_{ij} = C\Big[a_{i,j+1/2}\left(F_{i+1,j+1}+F_{i+1,j}-F_{i-1,j+1}-F_{i-1,j}\right)$$

$$-a_{i,j-1/2}\left(F_{i+1,j}+F_{i+1,j-1}-F_{i-1,j}-F_{i-1,j-1}\right)\Big] \tag{24}$$

where $C = (\Delta_+x_1 + \Delta_-x_1)^{-1}\ (\Delta_+x_2 + \Delta_-x_2)^{-1}$. The temporal differencing is Alternating Direction Implicit [19]. The mixed derivatives (23) and (24) are always placed at the 'old' time step. This differs from the method used by Killeen and Marx [20] and Lindemuth [21], but it is conservative at each half step.

The resulting difference equations are non-linear, and are solved by iterating the solution several times over a time step. Since the momenta are taken as the dependent variables, all velocities appearing in the difference equations are considered known, and their values at the appropriate time level from the most recent iteration are used. The boundary points are advanced by an implicit algorithm which assures total conservation on the difference mesh.

4. THE TEARING MODE IN A SHEET PINCH

The computer code MHDG, which has been described briefly in the previous sections, has been used to study the evolution of the tearing mode in a sheet pinch. The purpose of the study is twofold: 1) to serve as a check on the code by comparison with the analytic results of FKR; and 2) to display the non-linear behavior of the mode.

In our model of the sheet pinch, the zero-order magnetic field is in the x-direction and reverses at the singular surface y=0. The exact form of this field is

$$B_x = \tanh \frac{\pi}{2} y \qquad (25)$$

which results in a current sheet given by

$$J_z = - \frac{2}{\pi} \operatorname{sech}^2 \frac{\pi}{2} y \qquad (26)$$

The pressure, which is determined from the zero-order force balance condition

$$p + B^2 = \beta + 1 \qquad (27)$$

is also peaked at the singular surface, and since we assume that the plasma is initially of uniform density, there is a corresponding peak in temperature. In (27), $\beta = 8\pi p_\infty / B_\infty^2$ is the usual plasma beta, with p_∞ and B_∞ referring to values far from the current sheet.

We place conducting walls at $y = \pm y_w$ and assume a periodic structure in the x-direction. The scale length in the y-direction is chosen to be the half-width of the current sheet, while that in the x-direction is taken as the wavelength of the imposed perturbation. The resulting cartesian scale factors are $h_1 = \alpha/2\pi$, $h_2 = h_3 = 1$, where $\alpha = ka$. In addition, we have set $\nu = \kappa = 0$ to avoid unnecessary complications.

The perturbation quantities are obtained from a linear code, RIPPLE3, which solves the linearized equations described in FKR as an initial value problem [14]. These perturbation quantities, along with the equilibrium described above, constitute the initial conditions for the problem.

Before continuing, it should be noted that an objection could be raised to using the MHD model in a region where the magnetic field vanishes (in this case near the singular surface) due to the large gyroradius of the particles there. This problem has been studied by Furth [22] and Drake and Lee [23], who conclude that, in the collisional regime where our equations are valid, kinetic and fluid theories yield comparable results for this mode.

Several cases were run with initial perturbations and time step of small enough magnitude to assure that the mode was in the linear regime. Other parameters of significance were $y_w = 5$, $\alpha = .5$, $\beta = 1$. Two difrerent models for resistivity were used: 1) an analytic model, $\eta = \cosh^2 \frac{\pi}{2}y$; and 2) the Spitzer formula, $\eta = T^{-3/2}$. We define the growth rate to be

$$p = \frac{1}{\Delta\phi}\frac{d\Delta\phi}{dt} \tag{28}$$

where $\Delta\phi$ is the amplitude of the reconnected flux at the singular surface; i.e. with

$$\phi(x) = \int_{-y_w}^{0} B_x(0,y')dy' - \int_0^x B_y(x',0)dx' \tag{29}$$

as the flux at the singular surface as a function of x, we have $\Delta\phi = \phi_{max} - \phi_{mi}$

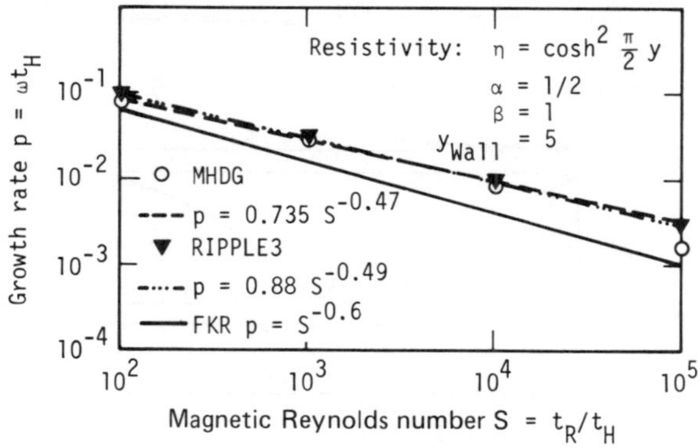

FIG.1. Linear growth rate versus magnetic Reynolds number for the sheet pinch.

This growth rate was monitored as a function of time and was found to be constant to 4 signficant figures for 50 time steps. This value, as a function of S, the magnetic Reynolds number, was then compared with results obtained from the RIPPLE3 code and those predicted by the linear analysis of FKR, and are presented in figure 1. In studying these results it should be noted that, in the present work, time is measured in units of the Alfvén transit time, while, in FKR, time is measured in units of the resistive diffusion time. Thus the usual FKR formula $p \equiv \omega t_R = S^{2/5}$ must be converted to $p \equiv \omega t_H = S^{-3/5}$. This normalization more clearly displays the decrease in growth rate with decreasing resistivity (increasing S).

We find that the linear code, the non-linear code, and FKR agree quite well over a wide range of magnetic Reynolds numbers, as seen in Fig. 1 for the analytic resistivity. Similar results were obtained for the Spitzer model. Note that the three sets of results are in better agreement for low S than for high S. It is felt that this is due to the computational difficulties encountered as the width of the singular layer shrinks with increasing S. Typical initial profiles are shown in figures 2 and 3 for $S = 10^2$ and $S = 10^5$. We see that the momentum eigenfunctions ($\rho v_1 \equiv \rho v_x$, $\rho v_2 \equiv \rho v_y$) become quite localized near $y = 0$ for large S, requiring finer and finer resolution of this region.

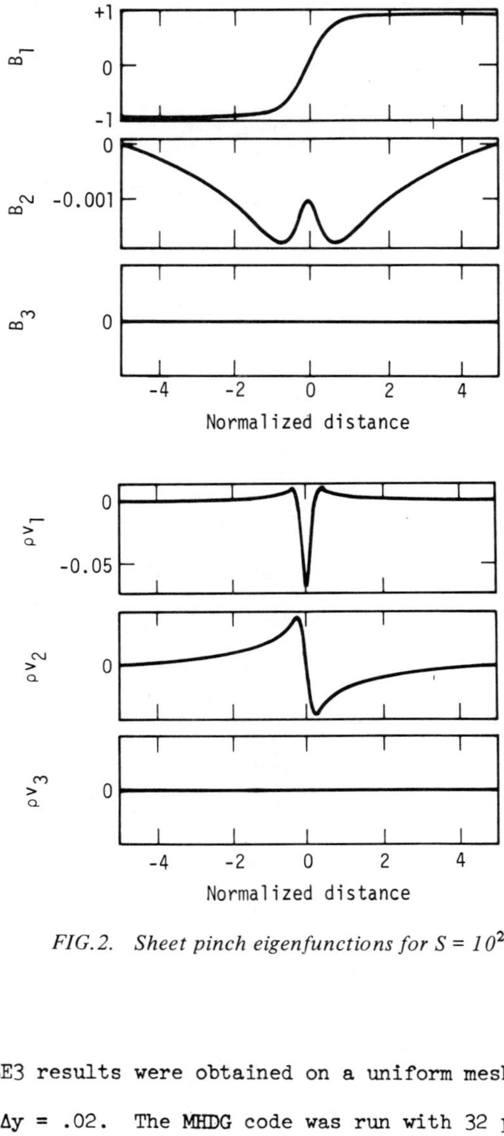

FIG. 2. Sheet pinch eigenfunctions for $S = 10^2$.

All the RIPPLE3 results were obtained on a uniform mesh of 501 points in y, giving $\Delta y = .02$. The MHDG code was run with 32 points in x, and 101 points in y. The x-mesh was uniform, and the y-mesh was variable, with $\Delta y = .02$ near the singular surface and increasing away from it. While this resolution is suitable for lower values of S, it may be too coarse to accurately resolve the singular region for high S.

Next, the size of the perturbation and timestep was increased and the mode was allowed to run into the non-linear regime. The

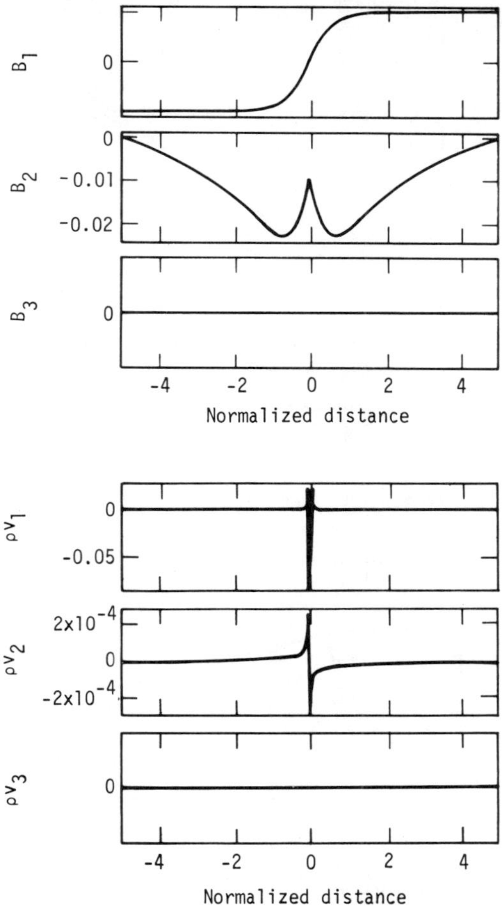

FIG. 3. Sheet pinch eigenfunctions for $S = 10^5$.

reconnected flux, $\Delta\phi$, as a function of time for the case $S = 10^2$ is shown in figure 4. This mode is found to grow exponentially for some time and then begins to saturate. The growth rate as a function of time is shown in figure 5. Figure 6 shows the magnetic flux surfaces for the $S = 10^2$ case after 10 Alfvén transit times. This places it well into the non-linear region. The magnetic island is evident. In figure 7 we show the corresponding velocity field. Note the tight vortex motion which has been set up in the vicinity of the singular surface.

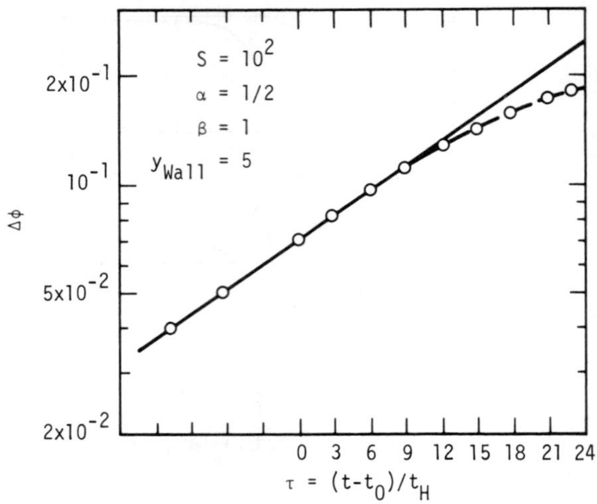

FIG.4. Reconnected flux versus time for the sheet pinch.

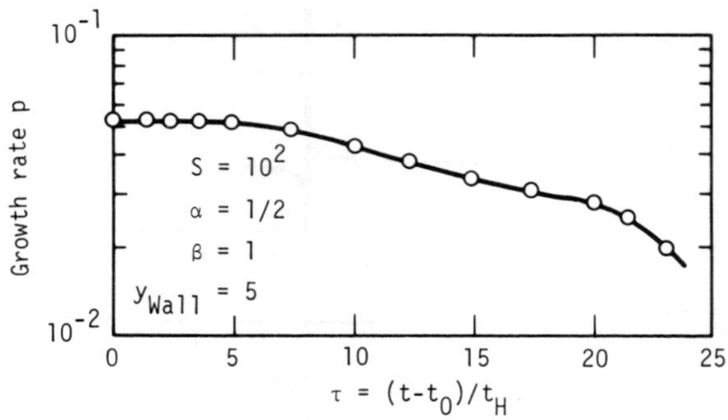

FIG.5. Growth rate versus time for the sheet pinch.

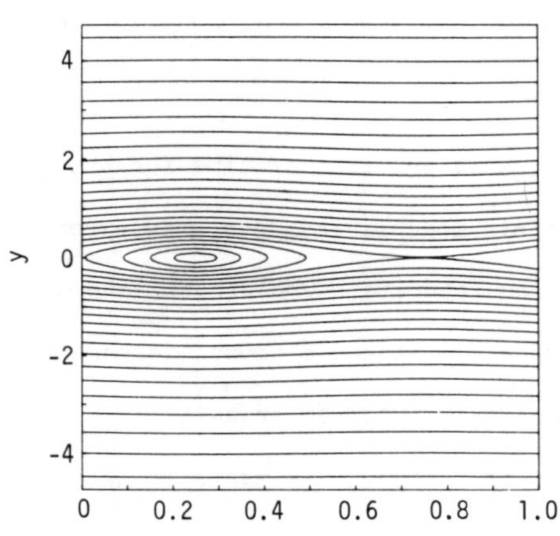

FIG.6. Magnetic flux surfaces, $S = 10^2$, $t = 10t_H$.

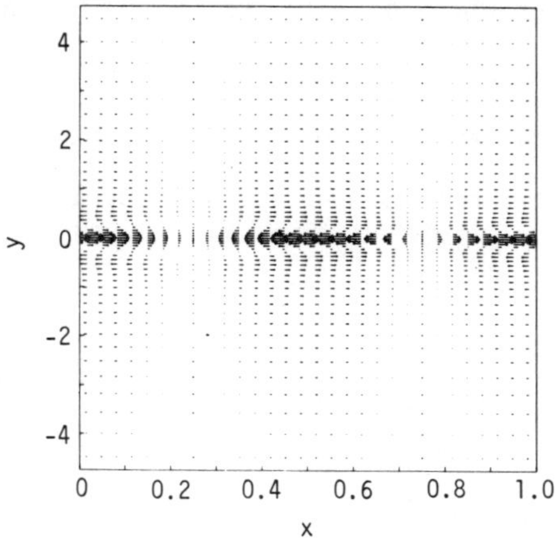

FIG. 7. Velocity field, $S = 10^2$, $t = 10t_H$.

5. LINEAR GROWTH OF THE DOUBLE TEARING MODE

The linearized resistive MHD equations in Cartesian coordinates as derived in FKR can be written in the form

$$\frac{\partial\psi}{\partial t} = -FW + \eta\left[\frac{\partial^2\psi}{\partial y^2} - \alpha^2\psi\right] \tag{30}$$

$$\frac{\rho}{\alpha^2 S^2}\frac{\partial}{\partial t}\left[\frac{\partial^2 W}{\partial y^2} - \alpha^2 W\right] = F\left[\frac{\partial^2\psi}{\partial y^2} - \alpha^2\psi\right] - F''\psi \tag{31}$$

We have taken the usual FKR normalization and have not Fourier-analysed in time. The quantities ψ, W and F are the non-dimensional first-order y-component of magnetic field, first-order y-component of velocity, and zero-order x-component of magnetic field, respectively. The parameter $\alpha = ka$ is a nondimensional wave number and S is the magnetic Reynolds number. A particular choice of the function F determines the background configuration from which unstable modes may arise.

Killeen [14] has numerically solved these equations as an initial value problem with equilibria appropriate to a single tearing mode (i.e. one reversal of the background field), and has obtained growth rates $p = (\partial\psi/\partial t)/\psi$ in good agreement with the analytic work presented in FKR. In the present work we choose a form of F relevant to the double tearing mode, i.e. two reversals of the background field. Our particular choice is

$$F = 1 - \left|1 + F(0)\right| \operatorname{sech}\left[\frac{y}{y_s} \operatorname{sech}^{-1}\left(\frac{1}{1 + F(0)}\right)\right] \tag{32}$$

which has the property that $F = 0$ at $y = \pm y_s$, $F = -F(0)$ at $y = 0$, and $F \to 1$ as $y \to \pm\infty$. We place conducting walls at $y = \pm y_w$ and solve equations (30,31) as an initial value problem, i.e. we specify an initial perturbation and advance the solution forward in time until exponentially growing eigenmodes evolve.

For the results to be presented here, we have chosen F(0) such as to make $F'(y_s)$ (the shear) the same for all runs. The particular value chosen was $F'(y_s) = \pi/2$ to allow for easy comparison with the results presented in section 5. The conducting walls were placed at $y_w = \pm 5$ and we have set $\alpha = .5$. Growth rates were then obtained as functions of S and y_s, the separation of the singular surfaces. We find that for small separation ($y_s = .25$), $p = .26\ S^{-.254}$, which is close to ideal MHD time scales, while for larger separation ($y_s = .7$), $p = .97\ S^{-.47}$. These are to be compared with the single tearing mode results $p = .75\ S^{-.542}$ as obtained numerically, and $p = S^{-.6}$ as derived in FKR. We see that the proximity of two singular surfaces can greatly enhance the growth rate of the tearing mode.

In figure 8 we plot growth rate as a function of separation distance for various values of magnetic Reynolds number. We see that the growth rates peak at smaller values of y_s for larger values of S. This is to be expected in light of the reduced size of the singular layer for higher

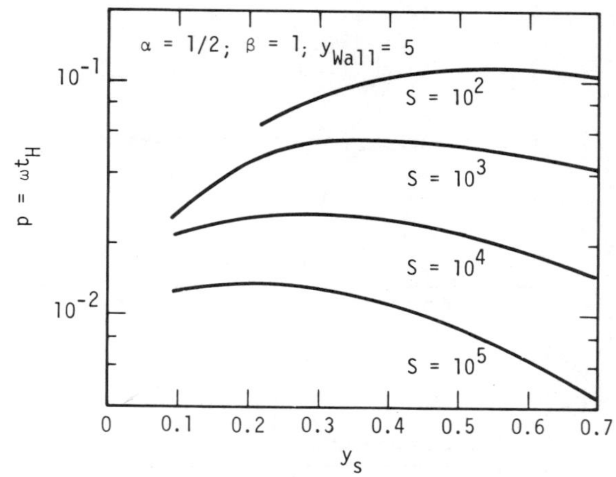

FIG.8. Linear growth rate p versus y_s, the separation of the singular surfaces, for the double tearing mode.

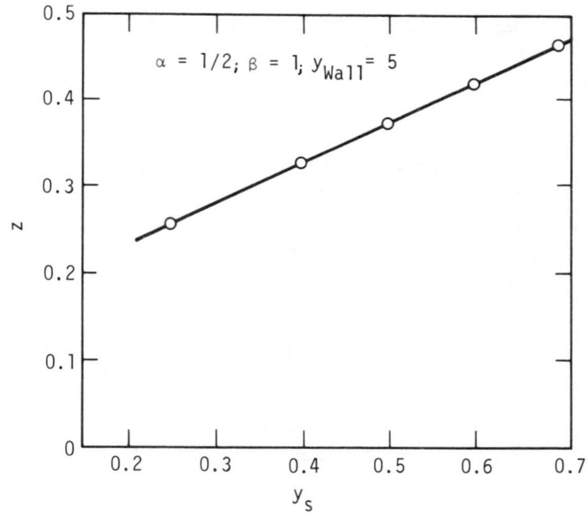

FIG.9. The exponent z in $p \sim S^{-z}$ versus y_s.

S modes. In figure 9 we plot the exponent z in $p \sim S^{-z}$ as a function of y_s. Recalling that $z \to 0$ implies that $\omega \sim 1/t_H$ (i.e. growth on the MHD time scale) we again see the enhanced growth rate for smaller separation distance. The linear behavior of the function $z(y_s)$ as displayed in this figure cannot be continued for indefinitely large y_s, since for $y_s \to \infty$ we expect to regain the single layer results, z = .542.

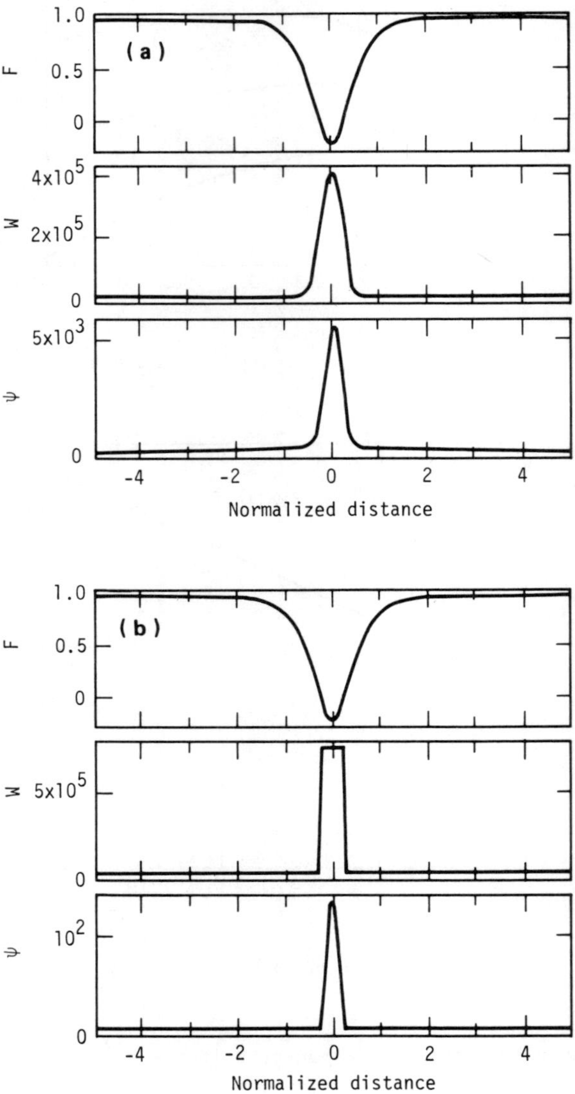

(a)
F
1.0
0.5
0
W
4x10^5
2x10^5
0
ψ
5x10^3
0
-4
-2
0
2
4
Normalized distance
(b)
F
1.0
0.5
0
W
5x10^5
0
ψ
10^2
0
-4
-2
0
2
4
Normalized distance

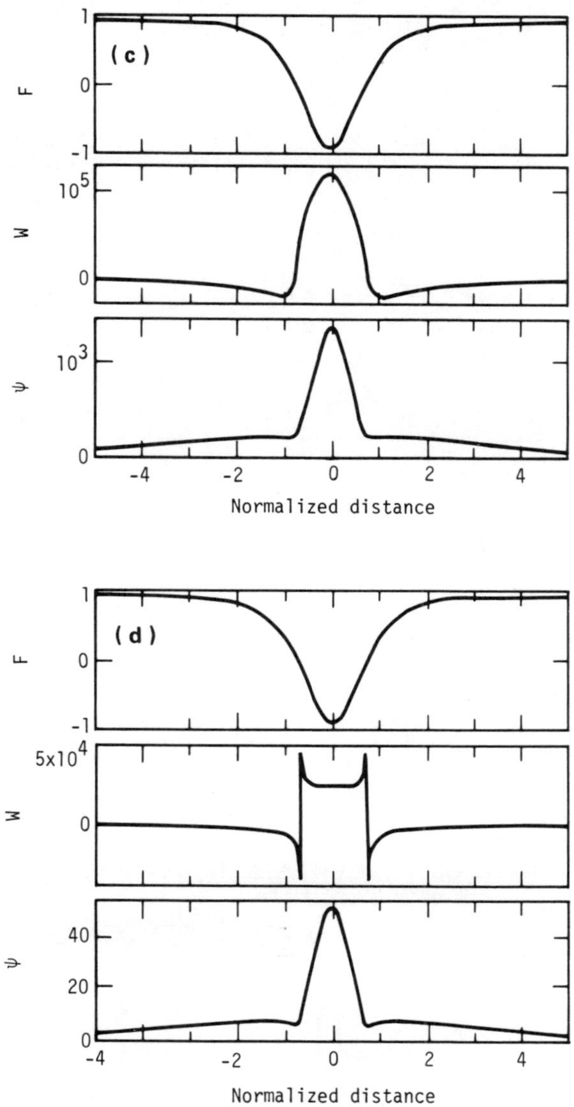

FIG.10. Linear eigenfunction of the double tearing mode for:

(a) $S = 10^2$, $y_s = 0.25$ *(c)* $S = 10^2$, $y_s = 0.7$

(b) $S = 10^5$, $y_s = 0.25$ *(d)* $S = 10^5$, $y_s = 0.7$

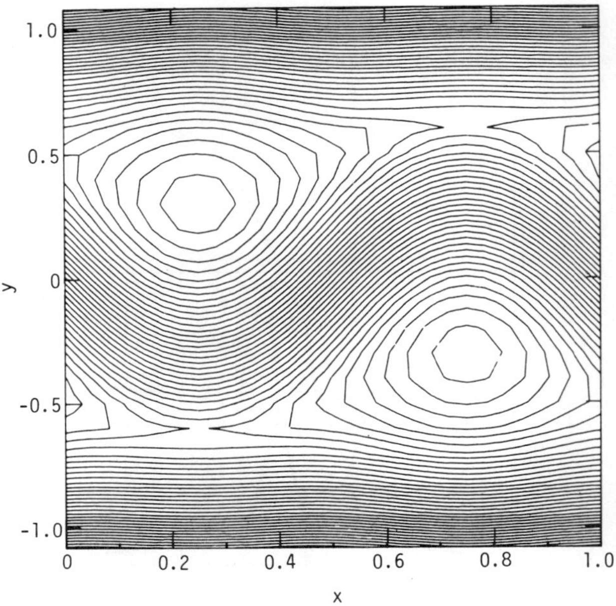

FIG.11. Flux surfaces in region between singular surfaces for the double tearing mode.

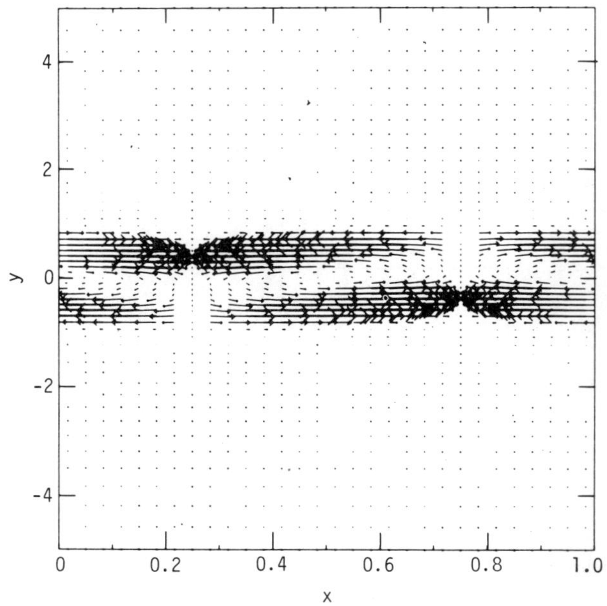

FIG.12. Velocity field corresponding to Fig.11. Note that the singular surfaces are at $y = \pm 0.5$ and the conducting walls are at $y = \pm 5$.

In figure 10 we display the linear eigenfunctions for various values of S and y_s. Note that the y-components of both the velocity and the magnetic field are large over the entire region between the singular surfaces. Thus the perturbation is not limited to the singular layer about each surface, but is more evenly distributed. This increased y-directed flow convects more magnetic field into the reconnection region of each singular surface, thus enhancing the growth rate.

6. NONLINEAR BEHAVIOR OF THE DOUBLE TEARING MODE

We use the zero-order equilibrium and the eigenfunctions obtained from the linear code as initial conditions and follow the behavior of the mode into the non-linear regime. We choose S = 100, y_s = .5 due to its large growth rate. We find that the magnetic islands on the adjacent singular surfaces grow toward one another, thus creating an extended region of y-directed magnetic field between the two surfaces. This could account for enhanced diffusion in the direction normal to the zero-order field (i.e. in the y-direction). This behavior of the flux surfaces is shown in figure 11.

The corresponding velocity field is shown in figure 12. We note the extended vortex motion, the region between the surfaces. As mentioned in section 5, this enhanced cross-field flow causes stronger convection of magnetic field into the reconnection regions and could account for an increased rate of field annihilation.

In figure 13 we plot the y-component of the magnetic field at a point half-way between the singular surfaces as a function of time. Note the initial exponential growth (at a rate in agreement with the linear results) followed by non-linear saturation. Figure 14 shows the kinetic energy in the entire plasma as a function of time, and we note that it is still growing approximately exponentially, well after $B_y(0)$

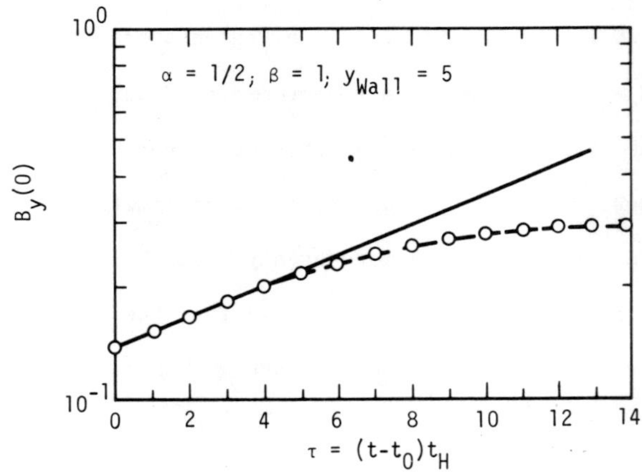

FIG.13. $B_y(0)$ versus time.

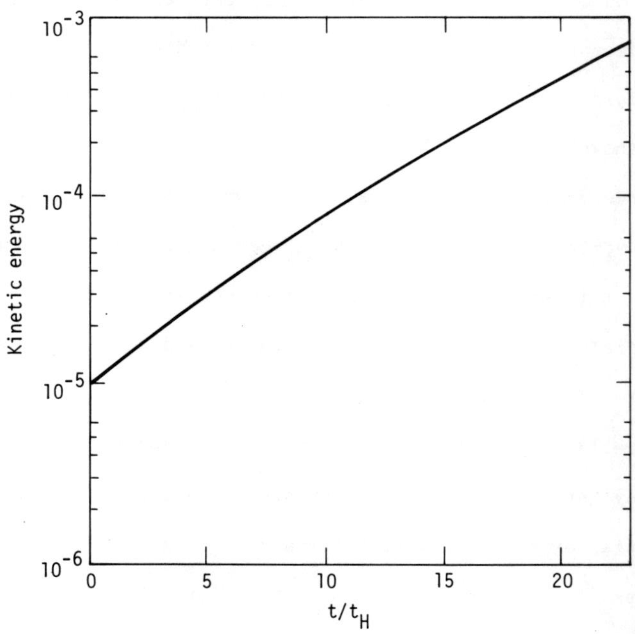

FIG.14. Kinetic energy versus time.

has saturated. This could be due to the fact that reconnection is still taking place near each singular surface, releasing magnetic energy into thermal and kinetic energy long after the field at the mid-plane has reached its maximum value. We find that after 14 Alfvén transit times (the approximate time at which $B_y(0)$ is completely quenched), 4.17% of the initial magnetic energy has been dissipated. Of that, 10.1% has gone into kinetic energy, and 89.9% into thermal energy.

7. CONCLUSIONS

We have obtained good agreement between the results of our non-linear code and previous linear numerical and analytic results for the case of the sheet pinch. We have followed the tearing mode for long times and have observed non-linear saturation. For the case of the double tearing mode, we have found that the linear growth rate can be greatly enhanced when two singular surfaces are in close proximity. This may be due to strong convection of field into the reconnection region of each singular surface. The non-linear interaction of the two surfaces leads to the evolution of magnetic field mainly in a direction normal to the original equilibrium field. This could account for enhanced cross-field diffusion. We also find that ~10% of the energy released goes into directed plasma motion.

ACKNOWLEDGEMENTS

The authors wish to thank C. Finan for help in coding parts of the computer program, and A. Mirin and K. D. Marx for many helpful discussions.

REFERENCES

[1] FURTH, H.P., KILLEEN, J., ROSENBLUTH, M.N., Phys. Fluids **6** (1963) 459.
[2] FURTH, H.P., RUTHERFORD, P.H., SELBERG, H., Phys. Fluids **16** (1973) 1054.
[3] GLASSER, A., GREENE, J., JOHNSON, J., Phys. Fluids **19** (1976) 567.

[4] VAN HOVEN, G., CROSS, M.A., Phys. Rev. **A7** (1973) 1347.
[5] RUTHERFORD, P.H., Phys. Fluids **16** (1973) 1903.
[6] DRAKE, J.F., LEE, Y.C., Phys. Fluids **20** (1977) 1341.
[7] PETSCHEK, H.E., in Physics of Solar Flares (HESS, W.N., Ed.), NASA SP-50 (1964).
[8] COPPI, B., FRIEDLAND, A., Astrophys. J. **169** (1971) 379.
[9] SPICER, D., US Naval Res. Lab. Rep. NRL 8036, to appear in Sol. Phys.
[10] STIX, T.H., Phys. Rev. Lett. **36** (1976) 521.
[11] RECHESTER, A.B., STIX, T.H., Phys. Rev. Lett. **36** (1976) 587.
[12] BATEMAN, G., SCHNEIDER, W., GROSSMANN, W., Nucl.Fusion **14** (1974) 669.
[13] SYKES, A., WESSON, J.A., in Controlled Fusion and Plasma Physics (Proc. 7th Europ. Conf. Lausanne, 1975) **1** (1975) 111.
[14] KILLEEN, J., in Physics of Hot Plasma, (RYE, B.J., TAYLOR, J.C., Eds) Plenum Press, New York (1970) 202.
[15] DIBIASE, J., KILLEEN, J., J. Comput. Phys. **24** (1977) 158;
DIBIASE, J., PhD Thesis, Univ. of California, Davis, LLL Rep. UCRL-51591 (1974).
[16] WADDELL, B.V., et al., Nucl. Fusion **16** (1976) 528.
[17] POTTER, D., Computational Physics, Wiley, London (1973).
[18] JEFFREY, A., Magnetohydronamics, Oliver and Boyd, London (1966).
[19] RICHTMYER, R., MORTON, K.W., Difference Methods for Initial Value Problems, 2nd edn, Interscience, New York (1967).
[20] KILLEEN, J., MARX, K.D., in Methods in Computational Physics (ALDER, B., et al., Eds) **9**, Academic Press, New York (1970).
[21] LINDEMUTH, I., KILLEEN, J., J. Comput. Phys. **13** (1973) 181.
[22] FURTH, H., "The mirror 'instability' for finite particle gyro-radius", Plasma Physics and Controlled Nuclear Fusion Research (Proc. Int. Conf. Salzburg, 1961) Nucl. Fusion Suppl. Part 1, IAEA, Vienna (1962) 169.
[23] DRAKE, J.F., LEE, Y.C., UCLA Rep. PPG-282 (1976).

ADIABATIC AND NON-ADIABATIC ELECTRON OSCILLATIONS IN A STATIC ELECTRIC FIELD

C. WAHLBERG
Institute of Technology,
Uppsala University, Uppsala,
Sweden

Abstract

ADIABATIC AND NON-ADIABATIC ELECTRON OSCILLATIONS IN A STATIC ELECTRIC FIELD.

The influence of a static electric field on the oscillations of a one-dimensional stream of electrons is investigated. In the weak-field limit the oscillations are adiabatic and mode coupling is negligible but becomes significant if the field is stronger. The latter effect is believed to be of importance for the stability of, e.g., potential double layers.

1. Introduction. Recent developments in plasma physics have shown that static electric fields are involved in quite a few plasma phenomena where the classical resistivity is essentially zero. These phenomena include e.g. parallel fields in magnetic mirrors [1], plasma beam-curved magnetic field interaction experiments [2], potential double layers [3] and turbulent resistivity [4]. As regards the further properties of such plasmas very little is known. Specifically, the difficult problem of stability of general BGK-equilibria, of which the one-dimensional double layers form a subclass, is as yet unsolved. The aim of the following analysis is to demonstrate a few basic features of the interaction between a stream of electrons and a static electric field, which might have consequences for the stability problem mentioned.

2. Basic equations. Neglecting the thermal spread, the following equations describe the motion of the electron stream in the electrostatic field $E_0(x)=-d\phi/dx$:

$$\frac{\partial n}{\partial t} + \frac{\partial}{\partial x}(nv)=0, \quad \frac{\partial v}{\partial t} + v\,\frac{\partial v}{\partial x} =-\frac{eE}{m_e}, \quad \frac{\partial E_1}{\partial x} =-\frac{en_1}{\varepsilon_0}$$

$n=n_o(x)+n_1(x,t)$ is the electron density, $v=v_o(x)+v_1(x,t)$ the velocity, and $E=E_o(x)+E_1(x,t)$ the electric field. In the stationary state it follows that the electron density and velocity are $n_B/G(x)$ and $v_BG(x)$, respectively, where $G(x)=[1+2e\phi(x)/m_ev_B^2]^{1/2}$, and n_B and v_B denote the density and velocity at points where $\phi(x)=0$. It is straightforward to show [5] that the linearized versions of the equations above give the following equation for $E_1(x,t)$:

$$v_B^2\frac{\partial}{\partial x}\left[G^3(x)\frac{\partial E_1}{\partial x}\right]+2v_BG^2(x)\frac{\partial^2 E_1}{\partial x\partial t}+2v_BG(x)G'(x)\frac{\partial E_1}{\partial t}+G(x)\frac{\partial^2 E_1}{\partial t^2}+\omega_B^2E_1=0 \qquad (1)$$

where $\omega_B=(e^2n_B/\varepsilon_om_e)^{1/2}$. Furthermore, the energy conservation equation $\partial W_1/\partial t+\partial F_1/\partial x=0$ associated with eq. (1) involves the following expressions for the energy density and flux, respectively:

$$W_1(x,t)=\frac{1}{2}\varepsilon_o\omega_B^{-2}G(x)\left(\frac{\partial E_1}{\partial t}\right)^2-\frac{1}{2}\varepsilon_ov_B^2\omega_B^{-2}G^3(x)\left(\frac{\partial E_1}{\partial x}\right)^2+\frac{1}{2}\varepsilon_oE_1^2 \qquad (2a)$$

$$F_1(x,t)=\varepsilon_ov_B^2\omega_B^{-2}G^3(x)\frac{\partial E_1}{\partial x}\frac{\partial E_1}{\partial t}+\varepsilon_ov_B\omega_B^{-2}G^2(x)\left(\frac{\partial E_1}{\partial t}\right)^2 \qquad (2b)$$

In the following we will use two different approaches to the equation.

2.1. Lagrangian description. We take a certain point, fixed in the stream (initially at $x=x_o$) and describe the field at this point as a function of x. This field we denote by $E_1(x;x_o)$. Using the expression for the stationary electron trajectories, it is then easily shown [5] that eq. (1) turns into

$$v_B^2\frac{d}{dx}\left[G^3(x)\frac{d}{dx}E_1(x;x_o)\right]+\omega_B^2E_1(x;x_o)=0 \qquad (3)$$

Moreover, the energy density, kinetic plus potential, of the electron oscillations, as observed in the moving coordinate system, is given by

$$W_1(x;x_o)=\frac{1}{2}\varepsilon_ov_B^2\omega_B^{-2}G^3(x)\left[\frac{d}{dx}E_1(x;x_o)\right]^2+\frac{1}{2}\varepsilon_oE_1^2(x;x_o) \qquad (4)$$

2.2. Coupled mode description. We let $E_1(x,t)$ be given by $a(x)\exp(-i\omega t)$ and define the normal modes $a_\pm(x)$ according to

$$a_\pm(x)=(ik_\mp a-da/dx)/(ik_\mp-ik_\pm)$$

where $k_{\pm}(x)=[\omega\mp\omega_B G^{-1/2}(x)]/v_B G(x)$ are the wave numbers for the fast (pos.energy) and slow (neg.energy) waves. By use of eq. (1) we then find that

$$da_{\pm}/dx-ik_{\pm}a_{\pm}=\frac{3}{4}G'(x)G^{-1}(x)(a_{\mp}-a_{\pm}) \tag{5}$$

It is convenient to introduce the normal mode 'amplitudes' $A_{\pm}(x)$ defined by

$$A_{\pm}(x)=a_{\pm}(x)G^{3/4}(x)\exp\left[-i\int_0^x k_{\pm}(u)du\right] \tag{6}$$

in which case the (averaged) energy flux (2b) for the fast and slow modes are proportional to $\pm|A_{\pm}(x)|^2$, respectively, and the total flux is

$$\langle F_1\rangle=\frac{1}{2}\varepsilon_0\omega v_B\omega_B^{-1}\left(|A_+(x)|^2-|A_-(x)|^2\right)$$

Moreover, we introduce the dimensionless parameter

$$\lambda_B(x)=4\pi eE_0(x)/m_e v_B(x)\omega_B(x) \tag{7}$$

where $v_B(x)$ and $\omega_B(x)$ denote local values of the streaming velocity and plasma frequency, respectively, and the variable

$$y=\omega_B(2\pi v_B)^{-1}\int_0^x G^{-3/2}(u)du$$

Then the coupled mode equations (5) can be written

$$dA_{\pm}/dy=-\frac{3}{8}\lambda_B(y)\exp\left[\pm 4\pi iy\right]A_{\mp} \tag{8}$$

The parameter λ_B, that essentially determines the strength of the coupling, can be given a simple physical interpretation, namely the relative change in kinetic energy, during one plasma period, for the streaming electrons.

3. <u>Weak field</u>. When $|\lambda_B(x)|\ll 1$ the mode coupling is clearly negligible, and the situation is rather similar to the field-free case. However, in order to conserve the energy flux $\pm|A_{\pm}(x)|^2$, the amplitude of the

oscillations must vary slowly according to $|a_{\pm}(x)| \sim G^{-3/4}(x)$, by virtue of (6). This result can be given another physical explanation by use of the Lagrangian description in sec. 2.1. One observes that the expression (4) for the energy density can be given the form of a Hamiltonian for a harmonic oscillator, i.e.

$$W_1(x;x_o)=p^2/2M(x)+\frac{1}{2}M(x)\Omega^2(x)q^2 \tag{9}$$

where

$$q=E_1(x;x_o),\quad p=\varepsilon_o v_B^2 \omega_B^{-2} G^3(x)\frac{d}{dx}E_1(x;x_o)$$

$$M(x)=\varepsilon_o v_B^2 \omega_B^{-2} G^3(x),\quad \Omega(x)=\omega_B v_B^{-1} G^{-3/2}(x)$$

Moreover, Hamilton's equations resulting from (9) are equivalent to eq. (3). Thus, for sufficiently slow variation of $M(x)$ and $\Omega(x)$, we can directly exploit the result from classical mechanics, which states that the quantity W_1/Ω, i.e.

$$Q=W_1(x;x_o)G^{3/2}(x) \tag{10}$$

is an adiabatic invariant. It is easy to show that the conditions for slow variation, $2\pi|d\Omega/dx|<<\Omega^2$ and $2\pi|dM/dx|<<M\Omega$, can equivalently be formulated $|\lambda_B(x)|<<1$, λ_B given by (7). If adiabatic conditions prevail, the energy density is locally proportional to the square of the amplitude $a(x;x_o)$, so that the invariance of Q requires $a(x;x_o)\sim G^{-3/4}(x)$, as we found before. This result explains some earlier calculations on electron oscillations on accelerated streams, e.g. [6].

4. Strong field. If the adiabatic condition $|\lambda_B(x)|<< 1$ is not fulfilled, the invariant (10) is in general violated, and at the same time mode coupling becomes significant. We now examine the details of the coupling in two special cases. (A similar treatment, with application to microwave tubes, has been given in [7].)

4.1. Constant λ_B potential. Such a potential can be realized by choosing

$$\phi(x)=m_e v_B^2\left[(1-3\omega_B\lambda_B x/8\pi v_B)^{4/3}-1\right]/2e$$

In this case eqs (8) are easily solved, and the solution, with $A_-(0)=0$, is:

$$\lambda_B<\lambda_B^C: |A_+(y)/A_+(0)|^2=1+(\beta^{-2}-1)\sin^2 2\pi\beta y$$

$$\lambda_B=\lambda_B^C: |A_+(y)/A_+(0)|^2=1+4\pi^2y^2$$

$$\lambda_B>\lambda_B^C: |A_+(y)/A_+(0)|^2=1+(\beta^{-2}+1)\sinh^2 2\pi\beta y$$

where $\lambda_B^C=16\pi/3\approx17$, $\beta=|1-(\lambda_B/\lambda_B^C)^2|^{1/2}$

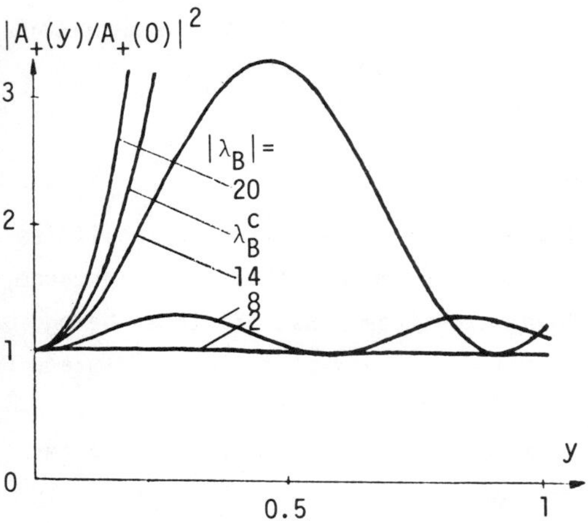

FIG.1. Energy flux for fast beam mode in a potential with λ_B = constant.

In fig. 1, $|A_+(y)/A_+(0)|^2$ is shown for a few values of λ_B. If $\lambda_B<\lambda_B^C$, the energy flux is oscillating between the fast and slow modes ($|A_-(y)|^2=|A_+(y)|^2-|A_+(0)|^2$), and the amplitude of the oscillations increases with λ_B. If $\lambda_B>\lambda_B^C$, the energy flux increases exponentially, and the situation is similar to an ordinary convective instability. The critical accelerating field above which this instability occurs is

$$E_0^C=4m_e v_B\omega_B/3e$$

4.2. Constant field. Choosing $\phi(x)=-E_0x$, we can write $\lambda_B(y)=\lambda_{B0}/(1-\frac{1}{4}\lambda_{B0}y)$. In this case the solution to the coupled mode eqs (8) can be given in terms of Bessel functions. With $z=1-\frac{1}{4}\lambda_{B0}y$, one finds that (again choosing $A_-(y=0)=0$)

$$|A_+(z)/A_+(1)|^2=\pi^2\sigma^2z\Big\{\Big[Y_1(\sigma)J_1(\sigma z)+$$
$$+Y_2(\sigma)J_2(\sigma z)-J_1(\sigma)Y_1(\sigma z)-J_2(\sigma)Y_2(\sigma z)\Big]^2+$$
$$+\Big[Y_1(\sigma)J_2(\sigma z)-Y_2(\sigma)J_1(\sigma z)-J_1(\sigma)Y_2(\sigma z)+$$
$$+J_2(\sigma)Y_1(\sigma z)\Big]^2\Big\}/16$$

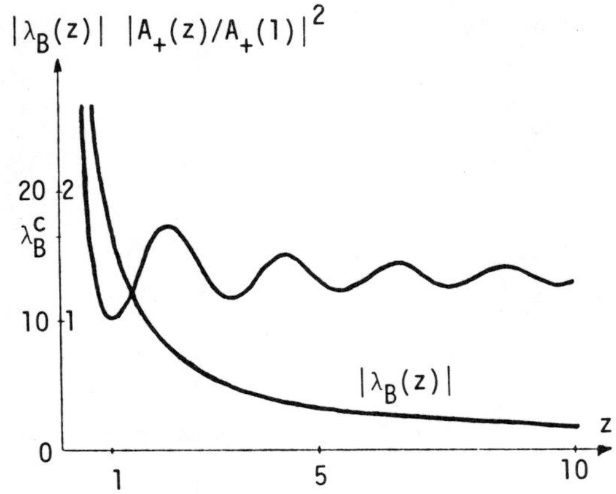

FIG.2. Energy flux for fast beam mode, and local λ_B in a constant electric field.

where $\sigma=8\pi/|\lambda_{BO}|$.
In fig. 2 this solution is shown together with $\lambda_B(z)=\lambda_{BO}/z$. σ is 3/2 in the figure, which means that $|\lambda_{BO}|=\lambda_B^C$. Thus $z\leq 1$ corresponds to the unstable range $\lambda_B\geq\lambda_B^C$ found in sec. 4.1., whereas $z>1$ corresponds to $\lambda_B<\lambda_B^C$, where the oscillating character of the energy flux is observed, and the amplitude of the oscillations decreases together with λ_B.

5. Stability discussion. It is well known that certain microinstabilities can be analysed within the framework of coupled mode theory. For instance, it is possible to interpret the ordinary two-stream instability as being caused by the coupling between the negative energy wave in the stream and the positive energy wave in the plasma [8]. However, if the plasma is penetrated by an electric field, maintained e.g. by one of the mechanisms mentioned in sec. 1, it would of course be necessary to include the additional mode coupling that we have dealt with here in a stability analysis. As a particular example let us take the double layer [3]. In fig. 3 a schematic picture of the phase space distribution of electrons for a typical layer is shown. The potential (not shown) is assumed to increase monotonically from a constant value for $x<0$ to another constant value for $x>\Delta x$. The upper shaded region shows a population of beam-like accelerated electrons, and the lower region a population of thermal electrons, of which some

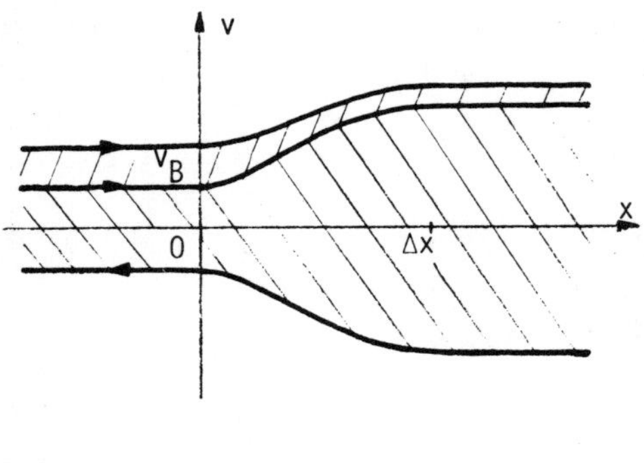

FIG.3. Schematic picture of phase-space distribution of electrons in a typical double layer.

are reflected by the potential. It is experimentally observed [3] that the critical drift velocity between the electrons and ions, above which the layer is formed, is of the order of the electron thermal velocity. Thus, in order to make an order of magnitude estimate of the parameter λ_B (7), let us use $v_B \sim (kT_e/m_e)^{1/2}$, which gives a minimum value for the drift velocity of the accelerated electron component. Moreover, the thickness of the layer is observed [3] to be of the order of a few Debye lengths, and we can, for convenience, use the estimate $E_0 \sim \Delta\phi/4\pi\lambda_D$, where $\Delta\phi$ is the total potential drop across the layer, and λ_D is the Debye length. Then, using $\omega_B = (kT_e/m_e)^{1/2}/\lambda_D$, we find that

$$\lambda_B \sim e\Delta\phi/kT_e$$

However, a characteristic feature of the double layer, inside which quasi-neutrality is not valid, is that $e\Delta\phi/kTe \sim O(1)$ or even $\gg 1$ for a strong shock [9] and we see that this in fact is an example of a situation where the electron oscillations are non-adiabatic, and mode coupling significant. In view of this result it seems doubtful whether local Penrose-stability [9] is sufficient for stability of this electron distribution. Indeed, it is quite possible that even if the Penrose-criterion predicts stability for all x, the following feedback mechanism can drive an absolute instability anyway. Let us consider a very simple case with a cold accelerated component and a flat ('water-bag') distribution of trapped electrons with $v_{th} \sim v_B$ (i.e. $fe(x,v)$ = constant in the lower shaded region in fig. 3). In fig. 4 the local dispersion relation is shown for: i) the accelerated electron component (assumed cold) and the thermal component ($v_{th} \sim v_B$) separately (solid lines), and ii) the entire system (dashed lines). The point C shows where the fast beam mode (F) couples to the plasma mode (P). This coupling is of evanescent type and the system is locally stable. However, it is interesting to evaluate the frequency at the coupling point C as a function of the potential ϕ in the plasma. It is very easy to

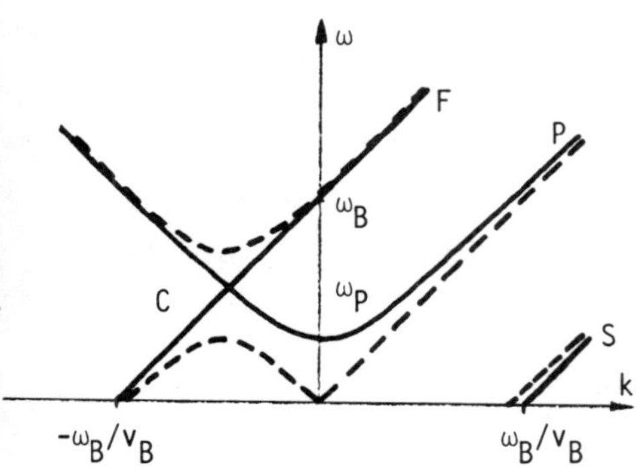

FIG.4. Local dispersion relation for fast (F) and slow (S) beam modes and plasma (P) modes. Dashed curves show dispersion relation for the coupled beam-plasma system.

show [5] that situations can exist such that the coupling frequency is the same in regions $x<0$ and $x>\Delta x$ of the plasma, but lower within the layer $(0<x<\Delta x)$. This result also applies for the cut-off frequency, i.e. the frequency at the minimum of the dashed curve above the point C in fig. 4. Then assume that such an equilibrium is perturbed so that e.g. the fast mode, at (or close to) the coupling frequency ω_0, carries the energy flux F_B into the layer. Here the coupling to the thermal component is weak (if $\omega_0 >$ local cut-off), but the coupling to the slow mode is strong, if λ_B is large. Since the slow mode has negative energy, the flux carried by the fast mode when it passes $x=\Delta x$ is always larger than F_B (e.g., gF_B, $g>1$). Due to the evanescent type of coupling for $x>\Delta x$, this flux will be entirely transferred to the plasma wave, which has negative group velocity and thus sends the flux gF_B back into the layer. In general, one must expect that some of this flux wi be reflected (i.e. coupled to the plasma wave with positive group velocity) and we can assume that the amount tgF_B, $t<1$, penetrates the layer entirely. However, this flux will also be entirely transferred to the fast-beam mode and carried into the layer again. Clearly, if $tg>1$, this feedback loop might increase the energy flux exponentially, and the configuration is absolutely unstable. On the other hand, if $tg<1$, the flux is likely to decay exponentia Preliminary calculations [5] indicate that one always has $tg<1$ in the limit $\Delta x \to 0$, whereas $tg>1$ indeed can be fulfilled for finite Δx and a properly chosen form of $\phi(x)$. However, a rigorous mathematical treatment of this phenomenon becomes very complicated, and our main effort at present is direct to finding a suitable approach that reduces the difficulties to a manageable level.

Acknowledgements. Part of this work was performed while I was a graduate student at the Royal Institute of Technology, Stockholm, and I wish to thank Professor E. T. Karlson for his guidance as thesis advisor, and Professor C.-G. Fälthammar for his kind interest and support.

The work was financially supported by the Swedish Atomic Research Board.

REFERENCES

[1] ALFVÉN, H., FÄLTHAMMAR, C.-G., Cosmic Electrodynamics, Clarendon Press, Oxford (1963) 169;
PERSSON, H., Phys. Fluids **6** (1963) 1756; **9** (1966) 1090.

[2] LINDBERG, L., KRISTOFERSON, L., Cosmic Electrodyn. 2 (1971) 305; Royal Inst. Technol. Stockholm Res. Rep. TRITA-EPP-72-32 (1972).

[3] BABIC, M., TORVÉN, S., Royal Inst. Technol. Stockholm Res. Rep. TRITA-EPP-74-02 (1974);
BLOCK, L.P., Cosmic Electrodyn. **3** (1972) 349; Proc. 30th Nobel Symp. on Physics of Hot Plasma in the Magnetosphere, Stockholm, 1975, Plenum Press, New York (1976);
QUON, B.H., WONG, A.Y., Phys. Rev. Lett. **37** (1976) 1393.
[4] SAGDEEV, R.Z., GALEEV, A.A., Nonlinear Plasma Theory, Benjamin, New York (1969) 94.
[5] WAHLBERG, C., Inst. of Technol. Uppsala Res. Rep. UPTEC 7678 R (1976).
[6] TIEN, P.K., FIELD, L.M., Proc. Inst. Radio Eng. **40** (1952) 688.
[7] HADDAD, G.I., ADAIR, J.E., IEEE Trans. Electron Devices, ED-12 (1965) 536.
[8] HASEGAWA, A., Plasma Instabilities and Nonlinear Effects, Springer-Verlag, Berlin (1975) 18.
[9] KNORR, G., GOERTZ, C.K., Astrophys. Space Sci. **31** (1974) 209.

Non-linearities

IAEA-SMR-32/8

STATISTICAL THEORY OF STRONG LANGMUIR TURBULENCE

V.N. TSYTOVICH
Lebedev Physics Institute,
Academy of Sciences of the USSR,
Moscow

K. KOMILOV, F.Kh. KHAKIMOV
V.I. Lenin Tajik State University,
Dushanbe,
Union of Soviet Socialist Republics

Abstract

STATISTICAL THEORY OF STRONG LANGMUIR TURBULENCE.
1. Basic principles. 2. Cubic non-linearity. 3. Linear theory. 4. Non-linear theory. 5. The energy-containing region. 6. The inertial region. 7. Spectra of inertial Langmuir oscillations. 8. Discussion.

1. BASIC PRINCIPLES

At the Second Kiev Conference, one of the authors [1] described the basic principles of the statistical theory of strong Langmuir turbulence and the first results obtained [2]. The present authors have greatly developed this theory during the period between the second and third Kiev conferences: the groundwork of the basic equation has been completed; the formulation of the equations has been simplified and their analytical solutions have been given. There are several arguments showing that the construction of the theory of strong Langmuir turbulence on a statistical level from the very beginning is desirable and even necessary:

(1) The irreproducibility of complex interactions appearing at the non-linear stage of modulational interactions is a necessary factor without which a plasma cannot be called turbulent. If, for example, statistics is due only to a random initial distribution of caverns, which contract dynamically and independently of each other, the picture is reproducible and hence non-turbulent.

(2) The different existing dynamic models seem to be inconsistent, making a number of additional assumptions without sufficient grounds (e.g. statistical independence of caverns).

(3) A complete kinetic description is necessary owing to the inseparability of the kinetic and hydrodynamic effects and their interlacing (e.g. the influence of scattering on modulational effects, and so on). A complete dynamic description of all such processes is very difficult, but it can be made easier by the statistical approach in which from the very beginning one is not interested in details when describing the process but only in the average characteristics of motions.

(4) The ability of the statistical theory to give the answers for a real three-dimensional case when computers are not effective.

In the formulation of the statistical theory [2, 3] we start with the assumption of spiky-type turbulence; in other words, we consider turbulence as never stationary but continuously undergoing temporal and spatial changes on the scale of modulation interactions. The Langmuir field E^{ℓ} is defined as a field whose frequencies are close to $\pm\omega_{pe}$ and the deviation of its frequencies $\Delta\omega$ from $\pm\omega_{pe}$ is small compared with ω_{pe}.

The spiky type of turbulence should be described by the correlator

$$\langle E_k^{\ell} E_{k'-k}^{\ell} \rangle = I_{k,k'}, \quad k = \{\vec{k}, \omega\}, \quad k' = \{\vec{k}', \omega'\} \tag{1}$$

The low-frequency density variations δn due to modulation perturbation are considered together with the Langmuir field. It is assumed that only after averaging on the spatial scale and on the time interval essentially larger than the scale and time of modulational interactions will the correlation function be

$$\overline{I_{k_1,k'}} = I_{k_1}\,\delta(k') \tag{2}$$

The spectrum of strong Langmuir turbulence is defined as an integral over the correlation curve

$$W_{\vec{k}} = \frac{1}{4\pi}\int I_{\vec{k},\omega}\,d\omega \tag{3}$$

The correlation function of the low-frequency density variation is given:

$$\overline{\delta n_{k'}\delta n_{k''}} = |\delta n|_{k'}^2\,\delta(k'+k'') \tag{4}$$

and the correlations of density variation with the Langmuir field are also introduced:

$$\overline{\delta n_{k'} I_{kk''}} = V_{k,k'}\,\delta(k'+k'') \tag{5}$$

$$\overline{I_{k,k'}I_{k_1,k''}} - \overline{I_{k,k'}}\,\overline{I_{k_1,k''}} = G_{k,k_1,k'}\,\delta(k'+k'') \tag{6}$$

To construct the theory it is necessary:

(1) To find the equations for all the introduced values;
(2) To prove the possibility of a break in the chain of correlation functions; and
(3) To find effective ways of solving the equations.

2. CUBIC NON-LINEARITY

The first variant of the statistical theory [1, 2] was based on the collisionless kinetic equations for the particles and Poisson's equation for the fields. The theory deals with the modulation perturbations $k' \ll k$ (k' is the wavenumber of the disturbances and k the wavenumber of Langmuir waves). It was found that the theory could be simplified by taking into account the cubic non-linearities only [4, 5]:

$$\frac{ik}{4\pi} \epsilon_k E = \int S_{12} E_1 E_2 d_{12} + \int \Sigma_{123} E_1 E_2 E_3 d_{123} \tag{7}$$

$$E = E_k, \quad E_1 = E_{k_1}, \ldots, \quad k = \{\vec{k}, \omega\}, \quad dk = \{d\vec{k}, d\omega\}$$

$$d_{12} = dk_1 dk_2 \delta(k - k_1 - k_2), \quad d_{123} = \delta(k - k_1 - k_2 - k_3) dk_1 dk_2 dk_3$$

The coefficients S_{12} and Σ_{123} are not yet fixed but later on they will be calculated by using kinetic equations. Equation (7) is more general in a definite sense than that used in Ref.[1] since the matrix elements can be calculated, for example, taking into account the binary particle collision or using any other equations. The possibility of using an expansion in field strength up to cubic non-linearity only follows from the result known in the kinetic theory that the matrix elements S and Σ are analytical functions of $\Delta\omega = \omega_1 - \omega_2$, and even they decrease with the growth of $\Delta\omega$. The case

$$\frac{\Delta\omega}{\omega_{pe}} \lesssim \frac{W}{nT} \ll 1 \tag{8}$$

can thus be considered in the frame of Eq. (7).

The first inequality of (8) is a necessary condition for strong Langmuir turbulence, where W is the energy density of turbulence. The second part of the inequality (8) is the condition when the expansion in the fields can be used. It should be emphasized that the $\Delta\omega$ in the first inequality (8) should be determined from the non-linear equation (7) as a function of W, i.e. an expansion in W is not used. This is the main point in which the theory of strong turbulence differs from the theory of weak turbulence (in which $\Delta\omega$ is assumed to be the difference between two linear Langmuir frequencies).

Since the matrix elements of S and Σ in general are complicated functions of $\Delta\omega$, the theory of strong turbulence finally contains any degrees W contrary to the weak turbulence theory in which only the W^2 effects are considered.

The effective method of constructing an equation containing the Langmuir field only has been developed by excluding from (7) the low-frequency virtual field and virtual fields with the frequencies close to $\pm 2\omega_{pe}$, not using any methods of averaging on phases or frequencies. In a non-isothermal plasma the low-frequency field could be the sound field, but in an isothermal plasma the low-frequency field becomes virtual too. As a result the equation [5]:

$$\epsilon_k E_k^+ = \int \widetilde{\Sigma}_{123} E_1^+ E_2^+ E_3^- d_{123} \tag{9}$$

has been obtained for the Langmuir field, where the indices + and – correspond to positive-frequency and negative-frequency parts of the Langmuir field. In the first approximation on the parameters

$$\alpha = \frac{\Delta\omega}{\omega_{pe}} \ll 1 \quad \text{and} \quad \beta = \frac{k^2 v_{Te}^2}{\omega_{pe}^2} \ll 1$$

only the processes through the virtual low-frequency wave are essential and from a collisionless kinetic equation we obtain

$$\widetilde{\Sigma}_{123} = - \frac{1}{4\pi n_0 T_{eff}} \frac{\epsilon_{2+3}^i}{\epsilon_{2+3}} \frac{(\vec{k}\vec{k}_1)(\vec{k}_2 \vec{k}_3)}{k k_1 k_2 k_3} \equiv \widetilde{\Sigma}_0 \tag{10}$$

From this expression it is easy to obtain the equations which were used in the analysis of strong Langmuir turbulence by Rudakov [6], Zakharov [7], Galeev et al. [8], Nishikawa et al. [9] and also by Kruer et al. and Thomson et al. [10], and Weinstock and Bezzerides [11], etc. These equations are obtained if ϵ_{2+3}^i and ϵ_{2+3}^ℓ are approximated by the expressions $-\omega_{pi}^2/(\omega-\omega_1)^2$ and $\omega_{pe}^2/((k-k_1)^2 v_{Te}^2)$, i.e. one can neglect the imaginary parts and consider them to be much greater than unity. The equations obtained from (10) are more exact in the sense that under the non-equilibrium distributions the velocity of sound in non-dimensional units is not equal to unity, but equals the ratio of two moments in the distribution function

$$v_s = \sqrt{\frac{T_{eff}}{T_e}} \tag{11}$$

where, in the one-dimensional case, for example,

$$T_{eff}^{-1} = -\frac{1}{m}\int \frac{1}{v}\frac{\partial \phi^{(e)}}{\partial v}\,dv; \quad T_e = \int m_e v^2 \phi^{(e)} dv \tag{12}$$

$\phi^{(e)}$ is the electron distribution function.

Our interest in the result (10) does not lie in this type of correction in the non-equilibrium plasma which, however, in a number of cases (especially for comparison with the experimental data on solitons) is essential, but lies in the possibility of regular calculation of the correction terms. Such additional terms are not small corrections, by the way. In fact,in the $\tilde{\Sigma}$ compensation ϵ^i/ϵ appears, which reflects the effect of local quasineutrality. Due to this compensation, the separate terms in $\tilde{\Sigma}$ are much larger than their sum, and small corrections to each of them may be of the order or greater than $\tilde{\Sigma}$. Such corrections are sometimes wrongly called electron non-linearities.

The calculations show that the effects of deviation from quasineutrality in the first order on β are of the same order as the so-called electron non-linearities. The contribution of the processes due to the virtual wave with double plasma frequency is also of the same order. The sum of these three corrections is strictly equal to zero in the one-dimensional case and it is not equal to zero in the 3D case. It has been shown that these corrections stop the growth of the three-dimensional caverns, in which $x \sim t^{2/3}$ on the level [12]

$$\frac{\delta n}{n} \sim \left(\frac{m_e}{m_i}\right)^{1/3} \tag{13}$$

i.e. a supersonic collapse [7] ($\delta n/n \sim 1$) is impossible. It has been found that the selfsimilar supersonic contraction [7] needs regularity of initial phases and phasing in the process of compression $\varphi \sim \tau^{1/3}$. The relative role of these self-phasing processes in a really turbulent plasma is not quite understood, but it is certain that many processes not included in Ref. [7] such as are described above are essential. Moreover, the other selfsimilar solutions found in Ref. [13] do not conserve the number of quanta as the energy in them tends to zero. The possibility exists that Langmuir waves, lost by one of the caverns, start then to form a new cavern. The description of such processes and the successive development of such schemes necessitates the statistical approach from the very beginning. It has also been made clear that any radiation of sound under compression can carry away the energy, comparable with the initial energy [14]. It was shown that spherical caverns are converted into spherical layers [15], the further autocompression of which can be stopped by powerful radiation of sound [16]. It has also been shown in Ref. [17] that the density perturbation remaining after cavern contraction dissipates into sound waves which stabilize the processes of compression afterwards. Thus the problem of complicated interactions with sound waves appears to be due to both processes.

The elucidated complexity of the dynamic processes can lead to irreproducibility of motions. We start with the assumption that the fields are random, and first of all consider a linear theory of modulation instability of a random pump. This description can be based on an equation with cubic non-linearity.

3. LINEAR THEORY

Analysis of the theory of modulation instability of a random field is of great importance for a number of reasons.

It seems that from the dynamic picture one can conclude that phase correlations may play an essential role in the modulation interactions. There is a regular correlation between the phases in Langmuir solitons, and a phasing of the oscillations exists in the processes of auto-compression of 3D caverns. The modulational instability was analysed mainly for the case of a regular pump. The first analysis of the modulation instability was, however, given for gas of cool plasmons (i.e. random pump) by Vedenov and Rudakov [18] and Gailitis [19] (also for random fields) from the principle of minimum energy. The further analysis of the modulation instability of random fields [20] has shown that the result of Ref. [18] corresponds mainly to a regular pump, but for the random pump one should take into account the non-linear plasma responses and renormalization. The analysis of instability of random fields based on Ref. [20] can give information on the importance of phase correlations in the modulation processes.

Originally in Ref. [20] the linear theory was constructed on the basis of a kinetic equation. It has been made clear that the equations for the modulation instabilities are modified in two respects:

(1) Instead of $1/\epsilon_{k-k_0}$ (in the dispersion equation for the regular pump)

$$1 = \frac{k^2 v_s^2}{\omega^2 - k^2 v_s^2} \frac{E_0^2}{8\pi n_0 T_e} \left\{ \frac{(\vec{k}, \vec{k} - \vec{k}_0)^2}{k^2 |\vec{k} - \vec{k}_0|^2 \epsilon_{k-k_0}} + \frac{(\vec{k}, \vec{k} + \vec{k}_0)^2}{k^2 |\vec{k} + \vec{k}_0|^2 \epsilon_{k+k_0}} \right\} \tag{14}$$

the total dielectric permeability appears:

$$\tilde{\epsilon}_k = \epsilon_k + \epsilon_k^N \tag{15}$$

where

$$\epsilon_k^N = -\frac{8\pi}{ik} \int \Sigma_{kk_1} I_{k_1} dk_1$$

is the non-linear dielectric permeability.

(2) Instead of E_0^2 (in the equation for the regular pump), $(1/(1+d_k))\int I_{k_0}\, dk_0$ appears, which is the natural generalization in the case of the presence of many modes of the random pump field; the coefficient $1/(1+d_k)$ corresponds to the renormalization of charge and takes into account the fact that the 'fur coat' surrounding the probe charge reacts differently in modulation perturbation of different frequencies and wave numbers. For $\omega_0 - \omega_0' \ll |\vec{k}_0 - \vec{k}_0'| v_{Ti}$,

$$d_k = \frac{1}{4\pi n_0 (T_e + T_i)} \int \frac{(\vec{k}_0, \vec{k}+\vec{k}_0)^2}{k_0^2 |\vec{k}+\vec{k}_0|^2} \frac{I_{\vec{k}_0, \omega_0}}{\tilde{\epsilon}_{k-k_0}} d\vec{k}_0\, d\omega_0 \qquad (16)$$

The results of the analysis of the dispersion equation obtained in the limit $k \ll k_0$ have been given in Ref.[21]. Recently, linear theory has been developed along the following lines:

(1) It has been shown [22] that dispersion equations can be obtained from the equation (7) for the cubic non-linearities;

(2) A method of deriving and solving the integral equations for three-dimensional modulation instabilities has been developed [23].

(3) The role of non-linear permeability (Eq.(15)) in the general case of weak anisotropic distribution of Langmuir waves for

$$k \ll k_* = k_i = \frac{k_d}{3} \sqrt{\frac{m_e}{m_i}}; \quad k_d = \frac{\omega_{pe}}{v_{Te}}$$

was analysed by Komilov [23b]. It is known that the non-linear frequency shift has the form

$$\delta\omega^N = -\frac{\omega_{pe}}{2} \int \frac{(\vec{k}\vec{k}_1)^2}{k^2 k_1^2} \frac{I_{k_1}\, dk_1}{4\pi n_0 T_e} \qquad (17)$$

The point is that the case of one-dimensional and strictly isotropic turbulence appears to be degenerate since the non-linear frequency shifts are constant, cancelled in $\Delta\omega$, and do not appear in the dispersion equation for the modulation perturbation. In the real case of noticeable or weak anisotropy, such cancellation is absent and ϵ_k, as has been shown [23], contributes essentially and changes the modulation instabilities (Fig.1).

(4) The role of ion-sound perturbations, leading to an additional contribution in ϵ_k^N, has been analysed and their stabilizing role has been found (see also Ref.[17]), owing to the imaginary part of ϵ_k^N.

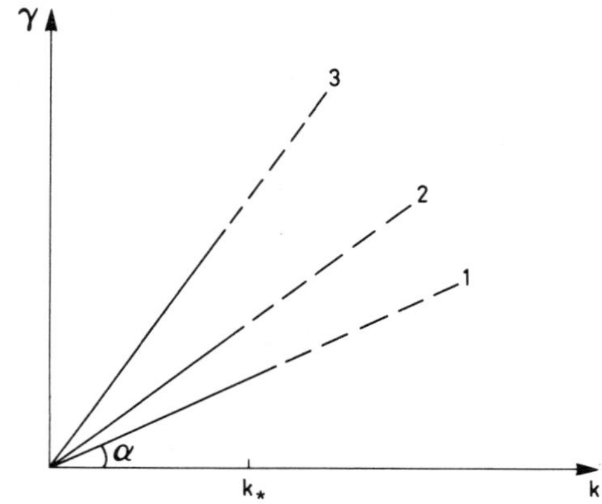

FIG.1. Modulation instabilities: $\alpha_1 \sim W^{1/4}$ [18]
$\alpha_2 \sim W^{1/2}$ [20]
$\alpha_3 \sim W$ [23]

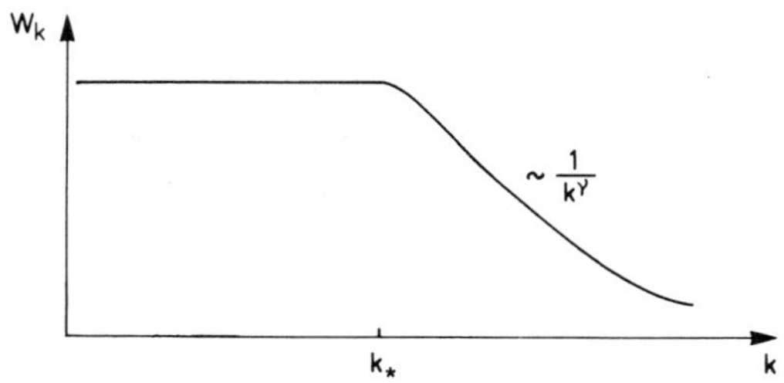

FIG.2. Modulation instabilities of spectra.

(5) The effect of resonant particles on the development of modulation instability, for example in the case of relativistic electronic beams, has been analysed by Tsytovich [24] and by Komilov and Tsytovich [25]. A considerable modification of the modulation instabilities has been found when the imaginary part of ϵ_k takes the main role.

(6) The present authors have analysed [26] the modulation instabilities of spectra having the maximum at $k \ll k_i$ and falling down rapidly enough at $k \gg k_i$, e.g. falling down according to a power law $1/k^{\nu}$ (Fig.2).

The modulation instability of long-wave oscillations $k \ll k_*$ (energy-containing region) and the short-wave oscillations $k \gg k_*$ (inertial region) has been analysed separately. Owing to the behaviour of the non-linear permeability, it is possible to divide the analysis into two separate analyses. It has been shown that the discrepancy in frequency shifts and in the ϵ_k^N exists for the case when the pump waves belong to the energy-containing region ($k_1 \ll k_i$) and the modulation perturbation belongs to the inertial region ($k \gg k_i$). The growth rate of instability

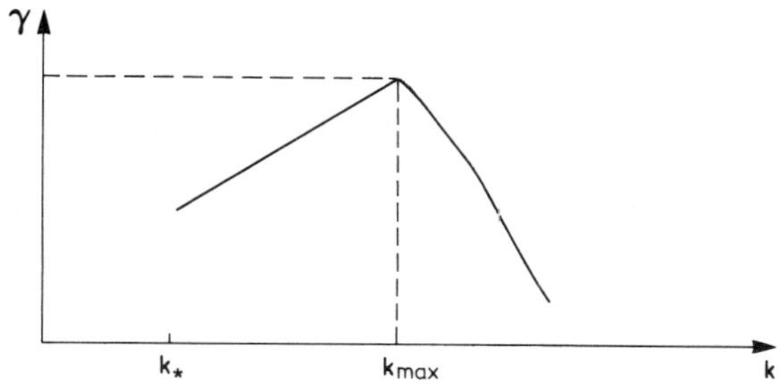

FIG.3. Growth rate of modulation instability. $W/nT \gg m_e/m_i$; $\gamma \sim k^{5/6}$ *at* $k \ll k_{max}$.

for the disturbances belonging to the energy-containing region grows with k more rapidly than in the case of regular fields of the same total energy and reaches the maximum value already at $k \sim k_i$, $\gamma_{max} \sim \omega_{pe}\sqrt{(m_e/m_i)(W/nT)}$, but at $k \gg k_i$ the growth rate tends to zero due to the above discrepancy between the non-linear shifts.

In the inertia region the modulation instability is determined by the inertial field ($k_1 > k_i$). The wavenumber at which the maximum growth rate is obtained:

$$k_{max} = k^{RP}\left(\frac{m_e/m_i}{W/nT}\right)^{1/6}, \quad \frac{W}{nT} \gg \frac{m_e}{m_i} \tag{18}$$

is somehow less than the

$$k^{RP} = k_i\left(\frac{m_e}{m_i}\frac{W}{nT}\right)^{1/4}$$

of the corresponding regular pump. The maximum growth rate also becomes less (Fig.3):

$$\gamma_{max} = \gamma_{max}^{RP}\left(\frac{m_e/m_i}{W/nT}\right)^{1/6}; \quad \gamma_{max}^{RP} = \omega_{pi}\sqrt{\frac{W}{nT}} \tag{19}$$

Thus the linear analysis shows that the random fields influence the values of maximum growth rates, though the instability is not essentially changed, i.e. phasing is not fundamental to the development of modulation perturbations. The linear analysis also shows that there are physical grounds for the separation of turbulent fields in the energy-containing $k \ll k_i$ and inertial $k \gg k_i$ regions, since their instabilities are independent. The analysis shows that this separation is also useful at the non-linear stage.

4. NON-LINEAR THEORY

Using the basic principles given above, the non-linear theory has been re-formulated on the basis of the equations with cubic non-linearity [3, 27]. Integral equations have been obtained for $I_{k,k'}$ containing $G_{k,k_1,k'}$ and $V_{k,k'}$; the integral equations for $G_{k,k_1,k'}$ containing $V_{k,k'}$ and $|\delta n|^2_{k'}$. These integral equations take into account:

(a) The non-linear permeability ϵ^N_k;
(b) The effect of the charge renormalization; and
(c) The fact that they do not fix the values $\Delta\omega$ which should be found as a result of the solution of the equations.

The solution of these integral equations through $V_{k,k'}$,and $I_{k,k'}$ through $G_{k,k_1,k'}$, permits us to express $V_{k,k'}$ through $G_{k,k_1,k'}|\delta n|^2_{k'}$. Finally, it is possible to find the equation for $I_{k,k'}$ containing some additional terms, linear in $|\delta n|^2_{k'}$. After averaging (b), we obtained the equation for I_k, containing the terms linear in $|\delta n|^2_{k'}$:

$$(\epsilon^\ell_k + \epsilon^N_k) I_k = \hat{I}_k \int \hat{\Lambda}_{k,k'} \frac{|\delta n|^2_{k'}}{n_0^2} dk' + I_k \int \widetilde{\Sigma}_{k,k_1,k'} I_{k_1} \frac{|\delta n|^2_{k'}}{n_0^2} dk' dk_1 \qquad (20)$$

where $\hat{I}_k \to \{I_k, I_{k-k'}\}$.

Since only the terms linear in $|\delta n|^2_{k'}$ are taken into account, an additional examination should be made by evaluating the following terms of higher order in $|\delta n|^2_{k'}$. The non-linear equation for $|\delta n|^2_{k'}$ has been obtained, containing linear and quadratic terms in $|\delta n|^2_{k'}$, the coefficients of which depend in a complex way on $I_{k,k'}$ and I_k:

$$\epsilon^M_{k'} \frac{|\delta n|^2_{k'}}{n_0^2} = \int S^M_{k',k'',k'''} \frac{|\delta n|^2_{k''}}{n_0^2} \frac{|\delta n|^2_{k'''}}{n_0^2} \delta(k' - k'' - k''') dk'' dk''' \qquad (21)$$

Here the operators $\hat{\Lambda}$, $\widetilde{\Sigma}$, S^M are functionally dependent on $I_{k,k'}$ and

$$\epsilon^M_{k'} = \epsilon^i_{k'} + (\epsilon^e_{k'} - 1) \frac{1 + d_k}{1 + d_k\left(1 + \dfrac{T_e}{T_e + T_i}\right)} \qquad (22)$$

The possibility of limitation by the terms quadratic in $|\delta n|^2_{k'}$ needed a more detailed basis. We emphasize here that the values $\Delta\omega$ are not fixed and they must be defined by the solutions of non-linear equations for $|\delta n|^2_{k'}$ and $I_{k,k'}$, and the value $\Delta\omega\ \omega_{pe}/(k^2 v^2_{Te})$ is arbitrary.

In this connection, the general method was developed in which the correlation effects of higher order in $|\delta n|^2_{k'}$ in the equation for $I_{k,k'}$ as well as in the equation for $|\delta n|^2_{k'}$ were taken into account. The solutions found by considering the lower orders in $|\delta n|^2_{k'}$ were used to estimate the contribution of terms of higher order in $|\delta n|^2_{k'}$. It was found that the criterion for neglecting them is

$$\alpha = \frac{W}{nT} \ll 1 \tag{23}$$

In the frame of the system of equations obtained, taking into account lower-order terms in $|\delta n|^2_{k'}$, the problem of their modification due to kinetic processes (e.g. the scattering processes) has been considered. It was found that the essential modification takes place due to additional terms, proportional to $|\delta n|^2_{k'}$ in the equation for $I_{kk'}$. They can lead to a change of direction of energy transfer if $|\delta n|^2_{k'}$ corresponds to the value reached at the stage of non-linear saturation. It has been shown that the scattering on electrons transfers the waves in one step to a large k-s. The scattering on ions remains differential and it does not differ essentially from the scattering that takes place in the absence of condensation. The balance of processes of scattering on electrons as well as on ions leads to the spectrum $1/k^2$ in both three- and one-dimensional cases [3, 4]. The modification of this spectrum due to Landau damping on the particle tails has been calculated and the necessary appearance of some groups of fast particles was shown [28] in accordance with observations from a number of existing experiments.

It was shown, however, that for not very large values of k the role of kinetic effects is rather small compared with hydrodynamic effects.

5. THE ENERGY-CONTAINING REGION

The equations for I_k in the energy-containing region were analysed in the case when the perturbations are of the modulation character $k \ll k_1$. It was shown [27, 29] that:

(1) The modulation disturbances lead to some additional non-linear frequency shift of Langmuir oscillations, depending integrally on the spectrum of the modulation perturbations,

$$\Delta\omega \cong \omega_{pe} \sqrt{\int \frac{|\delta n|^2_{k'}}{n_0^2}\, dk'} \tag{24}$$

(2) The non-linear frequency shift provides a decompensation in non-linear responses which leads to the stabilization of the modulation perturbations.

It was shown that the energy-containing vibrations become modulationally stable at

$$\int \frac{|\delta n|^2_{k'}}{n_0^2}\, dk' \approx \left(\frac{W}{nT}\right)^2 \tag{25}$$

(3) The analysis of the non-linear equation for $|\delta n|^2_{k'}$ shows that saturation of the modulational perturbations is obtained on the level corresponding to the relation (25).

(4) The results for the three- and one-dimensional cases differ only by a numerical factor of the order of unity [30].

(5) At the relatively low level of the modulation perturbations necessary to stabilize the modulational instability, the limitation by the lowest effects in $|\delta n|^2_{k'}$ is sufficient.

Thus the non-linear theory of the energy-containing field leads to a complete and fully self-consistent non-linear stage of modulation instability. On the basis of these concepts the theory of the radiation of a strong Langmuir turbulence at frequencies close to the double plasma frequency has been constructed [31].

6. THE INERTIAL REGION

The general scheme for the solution of the system for the integral equations is the following:

(1) It was shown that, with the necessary accuracy, the coefficient of the equation for $(\delta n/n)^2_{\vec{k},\omega}$ depends on $W_0 = \int I_{\vec{k},\omega}\, d\vec{k}\, d\omega/4\pi$ only. Thus the first step is the solution of the integral equations for $(\delta n/n)^2_{\vec{k},\omega}$ as a function of W_0.

(2) The second step is to find $V_{k,k'}$ as a function of $(\delta n/n)_{\vec{k},\omega}$.

(3) The third step is to find $G_{k,k_1,k'}$ as a function of $V_{k,k'}$.

(4) The fourth step is to find $I_{k,k'}$ as a function of $G_{k,k_1,k'}$. After averaging $I_{k,k'}$ it gives I_k and thus W_0 too. In the inertial field it is necessary to solve the non-linear equations for

$$\left(\frac{\delta n}{n_0}\right)^2_{\vec{k},\omega} = \frac{v_{Te}}{\omega_{pe}}\, v_{\lambda,k} \tag{26}$$

which one may easily consider as a function of $\lambda = \omega/kv_{Te}$ and k.

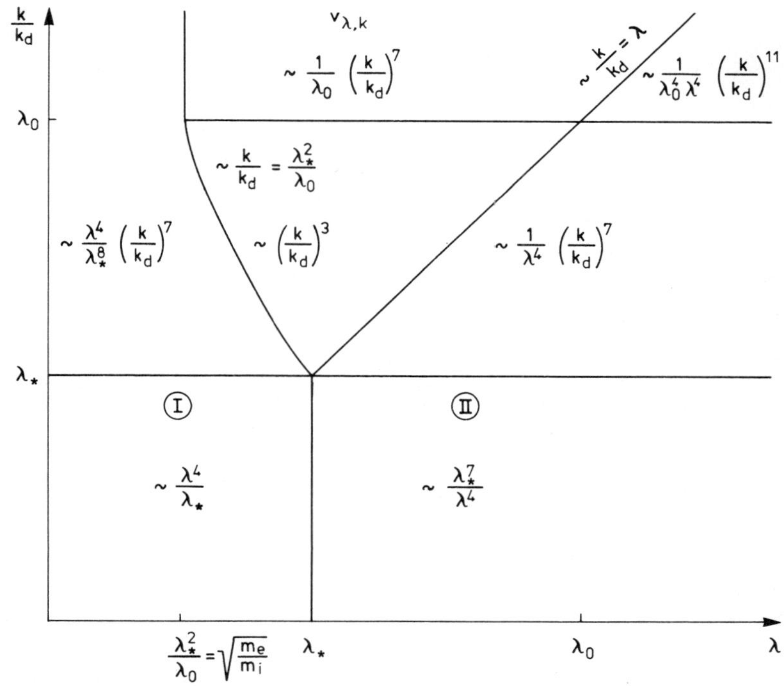

FIG.4. Phase plane $\kappa = k'/k_d$ *and* λ *breaking into various regions.*

Introducing

$$\lambda_0 = \sqrt{\frac{W_0}{n_0 T_e}}, \qquad \lambda_* = \lambda_0^{1/2}\left(\frac{m_e}{m_i}\right)^{1/4} \tag{27}$$

we find that the phase plane $\kappa = k'/k_d$ and λ breaks into a number of regions, according to Fig.4.

It must be noted that at $k/k_d \ll \lambda_*$ the introduced function $v_{\lambda,k}$ does not depend on k, i.e. $v_{\lambda,k} = v_\lambda$. The authors have obtained and solved the integral equations asymptotically in all the regions discussed (far from the lines dividing the regions in Fig.4) and have then joined them together. The result is shown in Fig.4. As an example we give here the following equation which is valid in regions I and II of the figure:

$$v_\lambda = \frac{81}{64}\,\frac{\lambda_0^4}{\left[\frac{m_e}{m_i}\left(1+\frac{3}{4}\left(\frac{\lambda_0}{\lambda}\right)^2\right)-\lambda^2\right]^2}\int \phi(\lambda,\lambda_1,\lambda_2)\,v_{\lambda_1}\,v_{\lambda_2}\,d\lambda_1\,d\lambda_2 \tag{28}$$

where

$$\phi(\lambda, \lambda_1, \lambda_2) = \frac{\lambda^2 + \lambda_1^2 + \lambda_2^2}{\lambda_1^4 \lambda_2^2} \left\{ \frac{\lambda_2^2}{\left|1 + \frac{3}{4}\left(\frac{\lambda_0}{\lambda_2}\right)^2\right|^2} + \frac{\lambda_1^2}{\left|1 + \frac{3}{4}\left(\frac{\lambda_0}{\lambda_1}\right)^2\right|^2} \right\}$$

$$\times \frac{|\lambda_1 - \lambda_2|}{|\lambda - \lambda_2||\lambda_1 - \lambda|} \tag{29}$$

The solutions obtained by joining together the asymptotic solution in regions I and II for $\lambda_* \ll \lambda_0$, i.e. $W/nT \gg m_e/m_i$, have the form

$$v_\lambda = \frac{56}{9 \ln \Lambda} \frac{\lambda_*^3 \lambda^4}{\lambda_*^4 + \frac{\lambda^8}{\lambda_*^4}}; \quad \Lambda = \frac{1}{2}\left(\frac{W/nT}{m_e/m_i}\right)^{1/4} \tag{30}$$

This result allows us to evaluate the level of modulation perturbations in the inertial region:

$$\int \left(\frac{\delta n}{n}\right)^2_{k'} dk' = \left(\frac{W}{nT}\frac{m_e}{m_i}\right)^{3/2} \tag{31}$$

It appeared to be lower than the modulation perturbation level in the energy-containing region. In the same manner, the solutions of corresponding equations in all the remaining regions have been found.

These results show that the modulation perturbations are in fact strong, i.e. they have no real determined frequency. The stabilization by the modulation perturbations in the example given is similar to the stabilization by the sound waves discussed previously. But the result (30) is valid in the case of isothermic plasma. The maximum of $I_{k,\omega}$ is close to the frequency corresponding to the linear growth rate, since this is the only value with the frequency dimension that enters into the problem.

7. SPECTRA OF INERTIAL LANGMUIR OSCILLATIONS

It has been shown that the energy-containing oscillations $I_k^{(0)}$ create an effective source Q_k, generating the inertial oscillations I_k^i. This source depends integrally on the modulation inertial perturbations:

$$I^i_{k,\omega} = \int \frac{\omega_{pe}^2}{k^2 v_{Te}^2} \frac{w^i_{k,\omega-\omega''} I^{(0)}_{k'',\omega''} dk'' d\omega''}{\tilde{\epsilon}_{k,\omega} 4\pi n_0 T_e |1 + d^i_{k,\omega-\omega''}|^2}$$

$$\frac{\omega_{pe}^2}{k^2 v_{Te}^2} w^i_{k,\omega} = \left(\frac{\delta n^i}{n}\right)^2_{k,\omega} \tag{32}$$

Since ω'' is rather close to ω_{pe}, and ω is close to ω_1, we have

$$W_k = W_0 \int \left(\frac{\delta n}{n}\right)^2_{k', \omega'} d\omega' \psi(k, k') \tag{33}$$

where

$$\psi(k, k') = \frac{1}{\left|1 + \frac{k^2 v_{Te}^2}{\omega^2} \lambda_0^2\right|^2 \left|\epsilon^{\ell}_{k, \omega_{pe} + \omega'} + \epsilon^{N}_{k, \omega_{pe} + \omega'}\right|^2} \tag{34}$$

$$\tilde{\epsilon}_{k, \omega' + \omega_{pe}} \approx \frac{2\omega' \omega_{pe} - 3k^2 v_{Te}^2}{\omega_{pe}^2} \tag{35}$$

Thus the correlation effects and the structure of the correlation curve is fully defined by the frequency dependence of the modulation perturbations, i.e. by the order of magnitude of the growth rate of the modulation instabilities. The integration (33) gives the spectrum $k < k_d \lambda_*$:

$$W_k = \frac{\alpha}{k^3}, \quad \alpha \approx 5.2\pi \frac{W_0}{\ln \Lambda} \left(\frac{m_e}{m_i}\right)^{3/2} \left(\frac{W_0}{n_0 T_e}\right)^{-1/2} k_d \tag{36}$$

By taking into account the renormalization (not taken into account in (36)), one finds that the spectrum is of the type:

$$W = \frac{\alpha'}{k^{\nu}}, \quad 2 \leqslant \nu < 3 \tag{37}$$

In the region of $k > k_d \lambda_*$ the spectrum (37) also appears. The spectrum indices are essentially less than those that appear in the models of non-interacting caverns.

Let us analyse the three-dimensional isotropic spectra. The analysis has shown that with accuracy up to numerical factors of the order of unity, which enter only in the level of the perturbations (i.e. in the coefficients α, α'), the one- and three-dimensional spectra have the same spectrum indices.

Thus a theory of the spectra of strong Langmuir turbulence with physical and mathematical bases has been created which is valid for both one- and three-dimensional cases. The latter is important for practical purposes and the first one-dimensional case is important for comparison with the numerical calculations and the 1D models. In the 1D case the theory leads to the conclusions which are qualitatively in agreement with the 1D soliton model if the kinetic effects of a soliton breaking are taken properly into account.

8. DISCUSSION

The proposed theory gives concrete conclusions concerning the correlation effects, spectra of modulation perturbations, spectra of turbulence, and spectra of fast particles, which are easy enough for an experimental test. The developed theory is logically and physically complete and mathematically well grounded. Further development of the theory is necessary, especially in order to investigate the acceleration of fast tail particles.

The theory has many astrophysical applications: shock waves, annihilation of magnetic fields, sporadic radio-emission, pulsar emission, and acceleration of cosmic rays.

REFERENCES

[1] TSYTOVICH, V.N., in Proc. 2nd Int. Conf. on Plasma Theory, Kiev, 1974, Naukova Dumka (1975).

[2] KHAKIMOV, F.Kh., TSYTOVICH, V.N., Zh. Ehksp. Teor. Fiz. **64** (1973) 1261.

[3] KHAKIMOV, F.Kh., TSYTOVICH, V.N., P.N. Lebedev Phys. Inst. Rep. 103 (1975).

[4] KHAKIMOV, F.Kh., TSYTOVICH, V.N., Zh. Ehksp. Teor. Fiz. **67** (1976) 312.

[5] TSYTOVICH, V.N., Proc. Nobel Symp. on Nonlinear Effects in Plasma, Göteborg 1976; P.N. Lebedev Phys. Inst. Rep. 181 (1976).

[6] RUDAKOV, L.D., Dokl. Akad. Nauk USSR **207** (1972) 821.

[7] ZAKHAROV, V.E., Zh. Ehksp. Teor. Fiz. **62** (1972) 1745.

[8] GALEEV, A.A., et al., Fiz. Plazmy **1** (1975) 10.

[9] NISHIKAWA, K., LEE, I.C., LUI, H.C., Comments on Plasma Phys. and Contr. Fusion **2** (1975) 63.

[10] (a) KRUER, W.L., ESTABROOK, K.G., THOMSON, J.J., Univ. Calif. Rep. UCRL 74947 (1974);
(b) THOMSON, J.J., FANEL, R.C., KRUER, W.L., Phys. Rev. Lett. **31** (1973) 918.

[11] WEINSTOCK, I., BEZZERIDES, B., Phys. Fluids **18** (1975) 251.

[12] KHAKIMOV, F.Kh., TSYTOVICH, V.N., Zh. Ehksp. Teor. Fiz. **70** (1976) 1785.

[13] LITWAK, A.G., FRAIMAN, G.M., YUNAKOVSKII, A.D., Pis'ma Zh. Ehksp. Teor. Fiz. **19** (1974) 23.

[14] VERJAEV, A.A., TSYTOVICH, V.N., Izv. Vyssh. Uchebn. Zaved., Radiofiz. **20** (1977) 163

[15] DEGTJAREV, L.M., ZAKHAROV, V.E., RUDAKOV, L.I., Fiz. Plazmy **2** (1976) 3.

[16] LITWAK, A.G., FRAIMAN, G.M., Zh. Ehksp. Teor. Fiz. **69** (1975) 4.

[17] GALEEV, A.A., SAGDEEV, R.Z., SHAPIRO, V.D., SHEVCHENKO, V.I., Pis'ma Zh. Ehksp. Teor. Fiz. **24** (1976) 125.

[18] VEDENOV, A.A., RUDAKOV, L.I., Dokl. Akad. Nauk USSR **159** (1964) 767.

[19] GAILITIS, A., Izv. Akad. Nauk Latv. SSSR **4** (1965) 13.

[20] TSYTOVICH, V.N., Zh. Ehksp. Teor. Fiz. **57** (1969) 141.

[21] TSYTOVICH, V.N., Theory of Turbulent Plasma, Atomizdat, Moscow (1971).

[22] KHAKIMOV, F.Kh., TSYTOVICH, V.N., Dokl. Akad. Nauk Tajik SSR **19** (1976) 3.

[23] (a) KOMILOV, K., KHAKIMOV, F.Kh., TSYTOVICH, V.N., Zh. Tekhn. Fiz. **44** (1974) 971;
(b) KOMILOV, K., Dokl. Akad. Nauk Tajik SSR **19** (1976) 17.

[24] TSYTOVICH, V.N., Proc. 12th Int. Conf. Phenomena in Ionized Gases, Eindhoven, 1975, Part 2, North-Holland, Amsterdam (1975) 141.
[25] KOMILOV, K., TSYTOVICH, V.N., Fiz. Plazmy **4** (1978), in press.
[26] KOMILOV, K., KHAKIMOV, F.Kh., TSYTOVICH, V.N., Fiz. Plazmy **4** (1978), in press.
[27] KHAKIMOV, F.Kh., TSYTOVICH, V.N., P.N. Lebedev Phys. Inst. Rep. 43 (1976).
[28] KHAKIMOV, F.Kh., TSYTOVICH, V.N., Zh. Tekhn. Fiz. (1976) 1400.
[29] TSYTOVICH, V.N., KHAKIMOV, F.Kh., Proc. 12th Int. Conf. on Phenomena in Ionized Gases, Eindhoven, 1975, North-Holland, Amsterdam (1975) 191.
[30] KOMILOV, K., KHAKIMOV, F.Kh., TSYTOVICH, V.N., P.N. Lebedev Phys. Inst. Rep. 47 (1977)
[31] KOMILOV, K., et al., Zh. Ehksp. Teor. Fiz. **46** (1976) 680.

THE PROBLEM OF COLLAPSE IN PLASMAS

B. BUTI
Physical Research Laboratory,
Ahmedabad, India

Abstract

THE PROBLEM OF COLLAPSE IN PLASMAS.

The phenomena of the collapse of plasma waves are discussed from fluid-model and kinetic-model points of view.

1. INTRODUCTION

It is well known that many of the instabilities that arise in plasmas lead to turbulence in the system. The theory of weak turbulence is quite well understood [1, 2] but, unfortunately, there has not yet been any proper theory for strong plasma turbulence. Since the strongly turbulent regime can give rise to plasma heating, there has recently been a great deal of interest in developing an adequate theory for strong plasma turbulence. Sufficiently strong excitation of plasma waves can give rise to such strong turbulence. Moreover, the phenomenon of non-linear modulational instability of plasma waves leads from a weakly turbulent regime to a strongly turbulent regime. So one way to understand the strong turbulence would be to study the non-linear evolution of these modulational instabilities. Some attempts have been made in this direction [3–7] and it has been shown that the phenomenon of collapse, which is quite a well-known feature in gravitational systems, can also occur in strongly turbulent plasmas. In the latter case, by collapse one means that the size of the three-dimensional density depletion (caviton) reduces to Debye length. When this happens, the trajectories of the particles start to interact, and the wave energy is consequently dissipated. So collapse offers a non-linear dissipation mechanism with the difference that, unlike the linear dissipation mechanisms, it has an amplitude threshold which is determined by the spectral distribution of the waves under consideration.

It has been shown by many authors that the non-linear Schrödinger (NS) equation characterizes a variety of weakly non-linear plasma waves. In one dimension, the NS equation admits localized stationary solutions: the envelope solitons or envelope holes for modulationally unstable or stable situations respectively [8, 9]. The cavitons, which are infinitely increasing local density depletions, however, cannot exist in one dimension. The existence of these cavitons has been demonstrated both by experiments [10, 11] and by numerical simulation [12], but so far we have not been able to show the identity of these cavitons with the exact solutions of the 3D NS equation. In fact, because of the mathematical complexity involved, work dealing with the 3D NS equations has been very scanty.

Zakharov [3] studied the time evolution of spherical 3D Langmuir waves in strongly turbulent plasmas and showed that in finite time the Langmuir waves undergo a collapse. His model will be discussed in Section 2. Tsytovich [5, 13], on the other hand, followed the statistical approach for the strongly turbulent systems. His paper is published in these Proceedings, so we shall not elaborate on his theory. By using the multiple space time scale method of Krylov, Bogolyubov and Mitropolsky (KBM) [14], we self-consistently derived the 3D NS equations for non-linear ion-acoustic and Langmuir waves [6, 7] and derived the sufficient and necessary conditions for the collapse to occur. In that analysis of Langmuir waves we had neglected the ion dynamics. Recently, we [15] have extended our investigations to include the effects of electron inertia and ion dynamics. In contrast to our previous work, where we had used fluid equations, in the present analysis the Vlasov equation has been used. A general discussion on the problem of collapse of plasma waves is given in Section 5.

2. ZAKHAROV'S MODEL

By averaging over the fast time ω_{pe}^{-1} and by assuming quasineutrality in the slow motions, Zakharov arrived at a simplified dynamic model for 3D Langmuir waves. His basic equations are

$$\nabla \cdot \left(i \frac{\partial \vec{E}}{\partial t} + \frac{3}{2} \omega_{pe} \lambda_D^2 \nabla \nabla \cdot \vec{E} \right) = \frac{1}{2} \omega_{pe} \nabla \cdot \left(\frac{\delta n}{n_0} \vec{E} \right) \tag{1}$$

and

$$\left(\frac{\partial^2}{\partial t^2} - c_s^2 \nabla^2 \right) \delta n = \frac{1}{8\pi M} \nabla^2 |\vec{E}|^2 \tag{2}$$

where $\vec{E}$ is the complex amplitude of the high-frequency electric field; δn is the ion density perturbation; n_0 is the equilibrium plasma density; $c_s = [K(T_e + \frac{3}{2} T_i)/M]^{1/2}$ is the sound speed; and $\omega_{pe} = (4\pi n_0 e^2/m)^{1/2}$ and $\lambda_D = (K T_e / 4\pi n_0 e^2)^{1/2}$ are the electron plasma frequency and the Debye length respectively. In writing Eq.(1), electronic non-linearities have been neglected; this is valid only for fast processes with characteristic time $\gtrsim (\omega_{pe} k^2 \lambda_D^2 W)^{-1}$ (see [16]), with W as the wave energy normalized to the thermal energy nKT. Equation (2), however, is valid if $W \gg (m T_i / M T_e)$ for $k^2 \lambda_D^2 \ll m/M$.

For $k^2 \lambda_D^2 \ll m/M$, i.e. for $v_g^2 \ll c_s^2$ ($\vec{v}_g$ being the group velocity of Langmuir waves), Eqs (1) and (2) give the following NS equation:

$$\nabla \cdot \left(i \frac{\partial \vec{E}}{\partial t} + \frac{3}{2} \omega_{pe} \lambda_D^2 \nabla \nabla \cdot \vec{E} \right) = - \frac{1}{16\pi M} \frac{\omega_{pe}}{n_0 c_s^2} \nabla \cdot (\vec{E} | \vec{E} |^2) \tag{3}$$

For a spherically symmetric case, $\vec{E} = \partial\phi/\partial\vec{r}$, Eq.(3) reduces to

$$i \frac{\partial \phi}{\partial \tau} + \frac{\partial}{\partial \rho} \left(\frac{1}{\rho^2} \frac{\partial}{\partial \rho} \rho^2 \phi \right) + |\phi|^2 \phi = 0 \tag{4}$$

where the dimensionless quantities ρ, τ, ϕ are defined as

$$r = \sqrt{\frac{3}{2}} \lambda_D \rho \;;\; t = \tau \omega_{pe}^{-1} \;;\; E = 4 c_s (\pi n_0 M)^{1/2} \phi$$

One can easily show that Eq.(4) has the integrals of motion given by

$$I_1 = \int_0^\infty d\rho \, \rho^2 |\phi|^2$$

and

$$I_2 = \int_0^\infty d\rho \left[|\nabla (\rho\phi)|^2 + 2 |\phi|^2 - \frac{1}{2} \rho^2 |\phi|^4 \right] \tag{5}$$

Moreover, Eq.(4) satisfies the relation

$$\frac{d^2 A}{d\tau^2} = 6 I_2 - 2 \int_0^\infty d\rho |\nabla (\rho\phi)|^2 - 4 \int_0^\infty d\rho \, \rho^2 |\phi|^4 < 6 I_2 \tag{6}$$

with A defined by

$$A = \int_0^\infty d\rho \, \rho^4 |\phi|^2$$

On integrating (6) we find that the positive definite quantity A is given by the relation

$$A < 3I_2\tau^2 + C_1\tau + C_2 \tag{7}$$

where C_1 and C_2 are constants of integration. From (7) it is clear that, for $I_2 < 0$, this relation can be satisfied only for a finite time; at this time ($\tau = \tau_0$, say) the solution of the initial problem goes through a singularity. In other words, Langmuir waves go through a collapse if $I_2 < 0$, i.e. when $W > k^2\lambda_D^2$, which is the condition for strong turbulence as well as for modulational instability. It is interesting to note that as $\tau \to \tau_0$, the fluctuations in the ion density become too large ($\delta n/n_0 \sim m/M$) and consequently our Eqs (3) and (4) break down. For the same reason, the self-similar solution discussed in Ref. [3] also breaks down when the collapse occurs.

This analysis is restricted to very long wavelength turbulence, i.e. for $k^2\lambda_D^2 \ll m/M$. The question one would like to ask is what happens when $k^2\lambda_D^2 \gtrsim m/M$? Does the collapse still occur? We shall try to answer this in the next Sections 3 and 4 by self-consistently deriving the 3D NS equation with the help of fluid equations and the Vlasov equation.

3. FLUID MODEL FOR THE NS EQUATION

In deriving Eqs (1) and (2), which are the basic equations for Zakharov's model, we had neglected the electronic non-linearities. These have to be taken into account when the intensity of the oscillations becomes large, especially near the region of collapse. Moreover, this model cannot be used for low-frequency phenomena because of the fast time-averaging used. By employing the multiple space time scale method [14], which has been successfully used for 1D problems [17–22], we remove these restrictions in our model.

3.1. Ion-acoustic waves

To illustrate the procedure, we restrict ourselves in this section to the wave-number range $k^2\lambda_D^2 \gg m/M$. In this case, we can use the 3D fluid equations and neglect the electron inertia. On assuming that the slow variations of the finite amplitude of the waves in a weakly non-linear plasma are given by

$$\frac{\partial \vec{a}}{\partial t} = \epsilon \vec{A}_1 + \epsilon^2 \vec{A}_2 + \ldots \tag{8a}$$

and

$$\frac{\partial a_i}{\partial x_j} = \epsilon B_{ij}^{(1)} + \epsilon^2 B_{ij}^{(2)} + \ldots \tag{8b}$$

we find that the resonant secularities to order ϵ^2 and ϵ^3 can be removed by imposing the conditions [6, 7]

$$\vec{A}_1 + \vec{v}_g \cdot \overset{\Rightarrow}{B}^{(1)} = 0 \tag{9}$$

and

$$i \frac{\partial \Phi}{\partial \tau} + \overset{\Rightarrow}{P} : \frac{\partial^2 \Phi}{\partial \vec{\xi}\, \partial \vec{\xi}} = Q |\Phi|^2 \Phi + R\Phi \tag{10}$$

respectively. Equation (10) is the desired 3D NS equation. For cold ions and isothermal electrons, the various quantities appearing in Eqs (9) and (10) are

$$\vec{v}_g = (\omega^3/k^4)\vec{k} \;=\; \text{group velocity}$$

$$\Phi = \frac{(\vec{k} \cdot \vec{a})}{k}$$

$$\tau = \epsilon^2 t, \quad \vec{\xi} = \epsilon(\vec{x} - \vec{v}_g t)$$

$$P_{\alpha\beta} = \frac{1}{2} \frac{\partial v_{g\alpha}}{\partial k_\beta} = \frac{\omega^3}{2k^4} \left[\delta_{\alpha\beta} - \frac{1}{k^2} k_\alpha k_\beta (1 + 3\omega^2)\right] \tag{11}$$

$$Q = \frac{\omega^5}{12k^6} (3 + 3k^2 + k^4)^{-1} (3k^{10} + 6k^8 - 6k^6 - 29k^4 - 30k^2 - 12) \tag{12}$$

and

$$R = \frac{\omega}{2k^2} \left[\eta_1 (\omega^2 - k^2) + \frac{2k^2}{\omega} (\vec{\gamma}_1 \cdot \vec{k})\right] \tag{13}$$

In Eq.(13), η_1 and $\vec{\gamma}_1$ are absolute constants to be determined from the initial conditions. In writing Eqs (11) and (12), use has been made of the linear dispersion relation

$$\omega^2 - k^2 + k^2 \omega^2 = 0 \tag{14}$$

ω and k are normalized to ion plasma frequency and electron Debye length. The R-term in Eq.(10) can be eliminated by the simple transformation

$\Phi \rightarrow \Phi \exp(-iR\tau)$. So from now on we shall drop the R-term from Eq.(10). We can easily show that Eq.(10) is modulationally unstable only if $(\vec{\vec{P}} : \vec{\kappa}\vec{\kappa})Q < 0$, $\vec{\kappa}$ being the characteristic wavenumber of the modulating wave [7]. For $\vec{\vec{P}}$ and Q defined by Eqs (11) and (12), we find that for oblique modulation, namely $(\vec{\kappa} \cdot \vec{\kappa}) = \kappa k \cos\theta$, ion-acoustic waves are unstable for $k > 1.47$ if $\theta \leqslant 55°$ and for $0 < k < 1.47$ if $\theta \geqslant 59.5°$.

Now to investigate the time evolution of this modulational instability, let us invoke spherical symmetry, in which case Eq.(10) reduces to

$$i \frac{\partial \Phi}{\partial \tau} + \frac{\omega^3}{k^4} \left(\frac{1}{\xi} \frac{\partial \Phi}{\partial \xi} - \frac{3}{2} \omega^2 \frac{\partial^2 \Phi}{\partial \xi^2} \right) = Q |\Phi|^2 \Phi \tag{15}$$

On making the simple transformations

$$\Phi = \rho^{\mu} |Q|^{-1/2} \Psi \quad \text{and} \quad \xi = \frac{3\omega^5}{2k^4} \rho$$

Eq.(15) for $\omega > 0$ takes the form

$$i \frac{\partial \Psi}{\partial \tau} - \nabla_{\rho}^2 \Psi + \frac{\mu}{\rho^2} (\mu - 1) \Psi = \eta \rho^{2\mu} |\Psi|^2 \Psi \tag{16}$$

where $\eta = 1$ for $Q > 0$ and -1 for $Q < 0$; $\mu = 1 + (3\omega^2)^{-1}$; and $\nabla_{\rho}^2 = (\partial^2/\partial\rho^2) + (2/\rho$
As in the previous section, it can easily be shown that Eq.(16) has the following integrals of motion:

$$I_1 = \int_0^{\infty} d\rho\, \rho^2 |\Psi|^2 \tag{17a}$$

and

$$I_2 = \int_0^{\infty} d\rho\, \rho^2 \left[|\nabla \Psi|^2 + \frac{\mu}{\rho^2} (\mu - 1) |\Psi|^2 - \frac{1}{2} \eta \rho^{2\mu} |\Psi|^4 \right] \tag{17b}$$

Once again, we observe that a positive definite quantity

$$A = \int_0^{\infty} d\rho\, \rho^4 |\Psi|^2$$

satisfies the relation

$$\frac{1}{2}\frac{d^2A}{d\tau^2} = 12 I_2 - 8\mu(\mu-1)\int_0^\infty d\rho\,|\Psi|^2 + (3+2\mu)\eta\int_0^\infty d\rho\,\rho^{2\mu+2}|\Psi|^4 \tag{18}$$

By definition $\mu > 1$, so the second term on the right-hand side of Eq.(18) is negative. For $\eta = -1$ $(Q < 0)$, the third term is also negative. In this case Eq.(18) gives

$$A < 12 I_2 \tau^2 + C_1' \tau + C_2' \tag{19}$$

From the arguments given in Section 2 we see that the ion-acoustic spherical clump will collapse if $I_2 < 0$. However, for $Q < 0$, according to Eq.(17b) I_2 is always positive and hence, for $k^2\lambda_D^2 \gg m/M$, there is no collapse of ion-acoustic waves. For $k^2\lambda_D^2 \lesssim m/M$, we cannot neglect the electron inertia; what happens in this case we shall investigate in the next section. It is worth pointing out that I_2 can be negative provided $Q > 0$; so the necessary (but not sufficient) condition for collapse is that $Q > 0$, which is the condition (for $\omega > 0$) for the linear modulational instability of plasma waves in one dimension. The reason $Q > 0$ is not the sufficient condition for collapse is because the relation (19) is no longer valid.

3.2. Langmuir waves

Following the procedure outlined for ion-acoustic waves, one finds that for adiabatic electrons, on retaining the electronic non-linearities, the finite-amplitude Langmuir waves are also described by the NS equation (10). For $m/M \ll k^2\lambda_D^2 < 1$, i.e. when the ion dynamics and the kinetic effects are neglected, the coefficients of the dispersive and non-linear terms are given by

$$P_{\alpha\beta} = \frac{1}{2\omega^3}\left[\delta_{\alpha\beta}(1+k^2) - k_\alpha k_\beta\right] \tag{20}$$

and

$$Q = \frac{k^4}{3\omega}(16k^2 + 15) > 0 \tag{21}$$

where ω and k are related by the dispersion relation $\omega^2 = (1 + k^2)$. Here ω is normalized to ω_{pe} and k to $(4\pi n_0^2 e^2/\gamma p_0)^{1/2}$, γ being the ratio of specific heats and p_0 the equilibrium electron pressure. For $\vec{\vec{P}}$ and Q given by (20) and (21), $(\vec{\vec{P}} : \vec{\kappa}\vec{\kappa})Q > 0$; so Langmuir waves are modulationally stable. For a spherically symmetric case, Eq.(10), along with Eqs (20) and (21), gives

$$i \frac{\partial \Phi}{\partial \tau} + \frac{1}{\omega}\left(\frac{1}{\xi}\frac{\partial \Phi}{\partial \xi} + \frac{1}{2\omega^2}\frac{\partial^2 \Phi}{\partial \xi^2}\right) = Q|\Phi|^2 \Phi \tag{22}$$

which,on using the transformations $\Phi = \rho^{\mu'} |Q|^{-1/2} \Psi$ and $\xi = \rho/(2\omega^3)$, yields the two integrals of motion, namely I_1 as defined by Eq.(17a) and

$$I_2' = \int_0^\infty d\rho\, \rho^2 \left[|\nabla \Psi|^2 + \frac{\mu'}{\rho^2} (\mu' - 1) |\Psi|^2 + \frac{1}{2} \eta \rho^{2\mu'} |\Psi|^4 \right] \tag{23}$$

with $\mu' = (1 - \omega^2) = -k^2$. The μ'-term is absent in Zakharov's model, which is valid in the limit $k \to 0$. We may remark that for $k^2 \leqslant 1$, I_2' is a well-behaved integral. Once again we can show that collapse is possible only if $I_2' < 0$, i.e. if $\eta = -1$, which corresponds to $Q < 0$. However, for the weakly turbulent case $(m/M \sim W \ll k^2 \lambda_D^2)$ that we have considered here, $Q > 0$. Hence the possibility of collapse exists only for the case $k^2 \lambda_D^2 \lesssim m/M$, in which case the ion dynamics plays an equally important role. In fact, as shown in Section 4, the two non-linearities, namely the non-inertial (finite-k contribution) and the inertial (ion-dynamical effects of order m/M), compete with each other and the collapse occurs when the inertial non-linearity dominates the non-inertial one.

4. KINETIC MODEL FOR THE NS EQUATION

Here we give the unified analysis which would include, over and above the wavenumber regimes considered in the fluid model and in Zakharov's model, the wave-particle interactions also. However, we consider only the electrostatic waves in collisionless plasmas; in this case the governing equations are the Vlasov-Poisson equations:

$$\frac{\partial f_\alpha}{\partial t} + \vec{v} \cdot \frac{\partial f_\alpha}{\partial \vec{x}} + \frac{e_\alpha}{m_\alpha} \vec{E} \cdot \frac{\partial f_\alpha}{\partial \vec{v}} = 0 \tag{24}$$

and

$$\nabla \cdot \vec{E} = 4\pi \sum_\alpha e_\alpha \int d^3 v f_\alpha \tag{25}$$

Once again we use the KBM [14] method. For a weakly non-linear system, we can use the expansions

$$f = f_0 + \epsilon f_1 + \epsilon^2 f_2 + \dots \tag{26}$$

and

$$\vec{E} = \epsilon \vec{E}_1 (a, \bar{a}, \psi) + \epsilon^2 \vec{E}_2 (a, \bar{a}, \psi) + \dots \tag{27}$$

where a is the complex amplitude of the wave and $\psi = (\vec{k} \cdot \vec{x} - \omega t)$. The slow space-time variations of a are given by

$$\frac{\partial a}{\partial \vec{x}} = \epsilon \vec{B}^{(1)} + \epsilon^2 \vec{B}^{(2)} + \dots$$

$$\frac{\partial a}{\partial t} = \epsilon A_1 (a, \bar{a}) + \epsilon^2 A_2 (a, \bar{a}) + \dots$$

where $A_n (a, \bar{a})$ and $\vec{B}^{(n)}(a, \bar{a})$ are to be determined from secularity removal conditions to various orders in ϵ. A similar procedure was used by Einaudi and Axford [23] to determine the time evolution of the particle distributions. To lowest order, Eqs (24) and (25) have the solutions

$$\vec{E}_1 = \vec{k}\, a e^{i\psi} + \text{c.c.}$$

and

$$f_{1\alpha} = ia \frac{m_e}{m_\alpha} \frac{e^{i\psi} \nu}{(\vec{k} \cdot \vec{v} - \omega)} \left(\vec{k} \cdot \frac{\partial f_{0\alpha}}{\partial \vec{v}} \right) + \text{c.c.} \tag{28}$$

$\nu = +1$ for ions and -1 for electrons. The subscript α stands for the charged species – electrons or ions. Moreover, the various quantities k, v, ω and E are normalized to λ_D^{-1}, $v_e = (KT_e/m)^{1/2}$, ω_{pe} and $(m v_e \omega_{pe}/e)$. From Eq.(28) it is obvious that $f_{1i} \sim (m/M) f_{1e}$. To find out higher order solutions, we have to order the ion distribution function properly depending on the order of the non-linearity parameter ϵ which could be taken as the strength of the non-inertial non-linearity for the Langmuir waves. To illustrate this point, let us take the case $\epsilon \sim (m/M)^{1/3}$; in this case the proper expansion for f_i would be:

$$f_i = f_{0i} + \epsilon^4 f_{1i} + \epsilon^5 f_{2i} + \dots \tag{29}$$

The f_{ni} defined by Eq.(29) are to be considered on the same footing as f_{ne}. The dispersion relation for the Langmuir waves is then given by

$$k^2 = \vec{k} \cdot \int \frac{d^3 v}{(\vec{k} \cdot \vec{v} - \omega)} \frac{\partial f_{0e}}{\partial \vec{v}} \tag{30}$$

and, on following the procedure outlined in the previous section, we can show that the NS equation, which arises from the secularity-removal condition to order ϵ^3, does not become affected by the ion dynamics at all. However, for the case $\epsilon \sim (m/M)^{1/2}$ we should take

$$f_i = f_{0i} + \epsilon^3 f_{1i} + \epsilon^4 f_{2i} + \ldots$$

We can now show [24] that the resonant secularity-removal condition to order ϵ^2 demands that

$$A_1 + \vec{v}_g \cdot \vec{B}^{(1)} = 0 \tag{31}$$

where $\vec{v}_g = (\vec{k} + \vec{\Lambda}_e)/\chi_{2e}$, with

$$\chi_{n\alpha} = \vec{k} \cdot \int \frac{d^3 v}{(\vec{k} \cdot \vec{v} - \omega)^n} \frac{\partial f_{0\alpha}}{\partial \vec{v}} \quad ; \quad n \geqslant 1 \tag{32}$$

$$\vec{\Lambda}_\alpha = \int \frac{d^3 v \, \vec{v}}{(\vec{k} \cdot \vec{v} - \omega)^2} \left(k \cdot \frac{\partial f_{0\alpha}}{\partial \vec{v}} \right) \tag{33}$$

Similarly, the resonant secularity, to order ϵ^3, can be removed by satisfying the relation

$$i \frac{\partial a}{\partial \tau} + \vec{\vec{P}}' : \frac{\partial^2 a}{\partial \vec{\xi} \partial \vec{\xi}} = Q' |a|^2 a + R' a \tag{34}$$

with

$$\vec{\vec{P}}' = - \frac{1}{\chi_{2e}} \left[\vec{v}_g \vec{v}_g \chi_{3e} - \vec{v}_g \frac{d\vec{\Lambda}_e}{d\omega} + \vec{\vec{S}}_e \right] \tag{35}$$

$$Q' = + \frac{k^4}{2\chi_{2e}} \left(3\,\chi_{5e} - \frac{1}{3k^2}\,\chi_{3e}^2 \right) \tag{36}$$

$$R' = -\chi_{1i}/\chi_{2e}$$

and

$$\overset{\Rightarrow}{S}_e = \int \frac{d^3v\ \vec{v}\vec{v}}{(\vec{k}\cdot\vec{v}-\omega)^3} \left(\vec{k} \cdot \frac{\partial f_{0e}}{\partial \vec{v}} \right)$$

We may note that the ion contribution appears only in the linear term which can be removed from the NS equation by the transformation $a \to a \exp(-iR'\tau)$. As long as R' is real, the stability of Eq.(34) is assured provided $(\overset{\Rightarrow}{P'} : \vec{\kappa}\vec{\kappa})\,Q' > 0$. In deriving Eq.(34) we have made use of relations (30) and (31) and of the fact that $\mathrm{Im}\,\omega \ll \mathrm{Re}\,\omega$.

For Maxwellian distribution for both the species and for $\vec{k} = k\,\hat{e}_z$, we can show that $P'_{\alpha\beta} = 0$ for $\alpha \neq \beta$:

$$P'_{xx} = P'_{yy} = -\frac{1}{I_2}\frac{\mu^2}{\omega^2}\left[Z'(\mu) + \frac{\mu}{2} Z''(\mu) \right]$$

$$P'_{zz} = -\frac{1}{k^2 I_2}\left[1 + 3\mu Z(\mu) + 3\mu^2 Z'(\mu) + \frac{\mu^3}{2} Z''(\mu) \right.$$

$$- \frac{2k}{\omega} v_g\,\mu \left\{ Z(\mu) + 2\mu Z'(\mu) + \frac{\mu^2}{2} Z''(\mu) \right\}$$

$$\left. + \frac{\mu^2}{\omega^2} k^2 v_g^2 \left\{ Z'(\mu) + \frac{\mu}{2} Z''(\mu) \right\} \right]$$

$$Q' = + \frac{k^4}{2I_2} \left(3\,I_5 + \frac{1}{3k^2}\,I_3^2 \right)$$

$$R' = -\frac{1}{I_2}\frac{v_e^2}{v_i^2}\left[1 + \frac{\mu v_e}{v_i}\, Z\left(\frac{\mu v_e}{v_i}\right) \right]$$

with $\mu = (\omega/\sqrt{2}k)$, $Z \equiv Z(\mu)$ as the plasma dispersion function, Z^n as the nth derivative of Z, and

$$I_n = \left(\frac{\mu}{\omega}\right)^{n-1} \left[\frac{1}{(n-2)!} Z^{n-2}(\mu) + \frac{\mu}{(n-1)!} Z^{n-1}(\mu) \right] \tag{37}$$

For $\omega^2 \gg k^2$, these relations can be further simplified and we finally obtain $P'Q' \sim (-k^4/\omega^4)$ and

$$R' = \frac{1}{2\omega}\left(1 - \frac{3}{\mu^2} + \frac{3}{2\mu^2}\frac{v_i^2}{v_e^2}\right)$$

Here P' represents both P'_{xx} and P'_{zz}. Since $P'Q' < 0$, we immediately conclude that Langmuir waves in strongly turbulent plasmas are modulationally unstable. Without elaborating on the integrals of motion for the spherically symmetric case of Eq.(34) (because the arguments are similar to those given in the earlier sections), we can assert that for $k^2\lambda_D^2 \lesssim (m/M)$ Langmuir waves do undergo a collapse.

Let us check the case $\epsilon \sim (m/M)$. There,

$$f_i = f_{0i} + \epsilon^2 f_{1i} + \epsilon^3 f_{2i} + \ldots$$

Our preliminary calculations show that the system in this case is governed by a modified NS equation; these details will be reported elsewhere.

For low-frequency waves, on following a similar procedure, we arrive at NS equation (34) but with the coefficients P', Q', R' defined as

$$\overset{\Rightarrow}{P}{}' = \frac{1}{S_2}\sum_\alpha \left[\vec{v}_g\vec{v}_g\,\chi_{3\alpha} - \vec{v}_g\,\frac{d\vec{\Lambda}_\alpha}{d\omega} + \overset{\Rightarrow}{S}_\alpha\right]$$

$$Q' = +\frac{k^2}{2S_2}\left(3k^2 S_5 + \frac{1}{3}S_3^2\right)$$

$$R' = 0$$

with $S_n = \sum_\alpha I_{n\alpha}$ and $\vec{v}_g = -(\vec{k} + \sum_\alpha \vec{\Lambda}_\alpha)/S_2$. $I_{ne} = I_n$, as defined in Eq.(37), and $I_{ni} = (v_e^2/v_i^2)\, I_n\ (\mu \to \mu')$; $\mu' = (\mu\, v_e/v_i)$.

A detailed discussion of the non-linear stability of these waves, including electron Landau damping, is published separately. A somewhat similar problem has been investigated by Nishikawa et al. [25].

We may remark that the NS equation given by Eq.(34) has Landau-damping effects embedded in it and we do not have to introduce these effects in the equation from outside as done by Nicholson and Goldman [26]. Similar inclusion was made by Ott and Sudan [27] for the KDV equation.[1]

[1] Landau damping is outside the scope of this paper.

5. DISCUSSION AND CONCLUSIONS

The modulational stability of a turbulent plasma is governed by the appropriate NS equation. From the analysis given in this paper we find that in a spherically symmetric system the collapse occurs only if the system is linearly modulationally unstable. So, physically, one can view the formation of the cavitons as the time evolution of the modulational instability. In the case of Langmuir turbulence, as the intensity of the oscillation grows, more and more plasma is expelled from the caviton owing to the high-frequency pressure. This continues until the size of the caviton is reduced to that of the Debye length. At this stage, the wave intensity becomes so large that the trajectories of the electrons start interacting and consequently leave the caviton, carrying with them the wave energy. So the collapse offers a very efficient non-linear dissipation mechanism.

In the initial stages of the development of modulational instability, one can ignore the wave-particle effects but, later, when the collapse takes place, energy is transferred to waves with larger wavenumbers and consequently the region $k^2\lambda_D^2 \gtrsim 1$ becomes important because the further dissipation takes place through Landau damping. The turbulence spectrum in this range is also expected to fall off much faster than k^{-2} [4]. Since our NS equation includes the wave-particle interactions, this can give us the turbulence spectrum in the range $k^2\lambda_D^2 \gtrsim 1$.

So far, the occurrence of collapse is known only in spherical systems. It would be interesting to find out what happens when spherical symmetry is relaxed. In the case of oscillating two-stream instability, even in one dimension, one finds the spatial collapse of the electric field [28].

Our analysis shows that for spherical collapse, there is a threshold that depends on the spectral distribution of the waves. In the case of Langmuir turbulence, collapse occurs if $k^2 < k_c^2 \sim m/M$. This is true for the subsonic case; for the supersonic case, however, one has to consider the ion-wave radiation. The coefficient of the cubic non-linear term in the NS equation, Q, in this case becomes time-dependent and therefore we cannot expect to have the integrals of motion which do exist in the subsonic case. Moreover, the kinetic model presented here has to be generalized to include the non-local effects. The weak collisions can be included in this theory very easily, but for purely growing modes the KBM method does not work; for that we must look for a different technique.

REFERENCES

[1] DAVIDSON, R.C., Methods in Nonlinear Plasma Theory, Academic Press, New York (1972).
[2] TSYTOVICH, V.N., Nonlinear Effects in Plasma Turbulence, Plenum Press (1970); Introduction to the Theory of Plasma Turbulence, Pergamon, Oxford (1972).

[3] ZAKHAROV, V.E., Sov. Phys. – JETP **35** (1972) 908.
[4] KINGSEP, A.S., RUDAKOV, L.I., SUDAN, R.N., Phys. Rev. Lett. **31** (1973) 1482.
[5] TSYTOVICH, V.N., in Proc. 37th Nobel Symp. on Plasma Physics, Stockholm, 1976, Plenum Press, New York (1977).
[6] BUTI, B., Phys. Rev. Lett. **38** (1977) 498.
[7] BUTI, B., in Recent Advances in Plasma Physics (BUTI, B., Ed.), Indian Academy of Sciences (1977).
[8] HASEGAWA, A., Plasma Instabilities and Nonlinear Effects, Springer-Verlag, New York (1975).
[9] KARPMAN, V.I., Nonlinear Waves in Dispersive Media, Pergamon, Oxford (1975).
[10] KIM, H.C., STENZEL, R.L., WONG, A.Y., Phys. Rev. Lett. **15** (1974) 886.
[11] WONG, Q.Y., QUON, B.H., Phys. Rev. Lett. **34** (1975) 1499.
[12] MORALES, G.J., LEE, Y.C., Phys. Rev. Lett. **38** (1974) 1016.
[13] TSYTOVICH, V.N., these Proceedings, paper IAEA-CN-32/8.
[14] BOGOLYUBOV, N.N., MITROPOLSKY, Y.A., Asymptotic Methods in the Theory of Nonlinear Oscillations, Hindustan Publishing Co., Delhi (1961).
[15] BUTI, B., Astrophys. Space Sci. (May 1978).
[16] ZAKHAROV, V.E., Sov. Phys. – JETP **24** (1967) 455.
[17] SHARMA, A.S., BUTI, B., J. Phys. A **9** (1976) 1823.
[18] KAKUTANI, T., SUGIMOTO, N., Phys. Fluids **17** (1974) 1617.
[19] BUTI, B., IEEE Trans. Plasma Sci. **PS4** (1976) 291.
[20] SHARMA, A.S., BUTI, B., IEEE Trans. Plasma Sci. **PS4** (1976) 283.
[21] SHARMA, A.S., BUTI, B., IEEE Trans. Plasma Sci. **PS5** (1977) 293.
[22] SHARMA, A.S., BUTI, B., Pramana, in press.
[23] EINAUDI, F., AXFORD, W.I., J. Plasma Phys. **1** (1967) 483.
[24] BUTI, B., in preparation.
[25] NISHIKAWA, K., et al., Phys. Rev. Lett. **33** (1974) 148.
[26] NICHOLSON, D.R., GOLDMAN, M.V., Phys. Fluids **19** (1976) 1621.
[27] OTT, E., SUDAN, R.N., Phys. Fluids **12** (1969) 2388; **13** (1970) 1432.
[28] MORALES, G.J., LEE, Y.C., WHITE, R.B., Phys. Rev. Lett. **32** (1974) 457.

IAEA-SMR-32/10

TURBULENCE, CLUMPS AND THE BETHE-SALPETER EQUATION*

J.A. KROMMES
Institute for Advanced Study,
Princeton, New Jersey,
United States of America

Abstract

TURBULENCE, CLUMPS AND THE BETHE-SALPETER EQUATION.

The theory of two-body correlations (Bethe-Salpeter equation) is used to unify renormalized plasma weak turbulence theory, particle clumping, and the direct interaction approximation.

There exists to date no satisfactory statistical turbulence theory which is both "first principles" and practically computable (analytically or numerically). Well-known approximations are, for plasmas, the weak turbulence theory (WTT)[1] and, for fluids, the direct interaction approximation (DIA)[2]. However, the WTT ignores clumping,[3] and the DIA is not invariant to random Galilean transformations.[4] Recently, Dupree proposed theories which he claimed improved upon both of these defects.[3,5] Characteristic of Dupree's approach was the use of exact two-body, rather than one-body, propagators. However, Dupree's closure approximations were intuitive, in part, and the relation of his work to the WTT, DIA, Kraichnan's higher order Eulerian closure,[4] and the exact evolution of the physical observables (statistical averages) was unclear. In this paper, we discuss that relation by using the powerful variational formalism of Martin, Siggia and Rose (MSR).[6]

* Work supported by USERDA, Grant No. E(11-1-3237). A preliminary version was published in Bull. Am. Phys. Soc. **21** (1976) 1023.

MSR presented an important formulation of the exact theory of quadratically nonlinear stochastic equations. They also gave a systematic sequence of approximations for the self-energy term in the equation for the one-body propagator. Those equations are of disparate appearance from Dupree's, and somewhat obscure two-body dynamics. However, the MSR (vertex-renormalized) approximations can, in fact, be cast in a natural two-body form. By developing and interpreting a "thermodynamic" relation presented but not stressed by MSR, we will show that two-body Eulerian closures are closely related to the MSR "first vertex renormalization" and to a theory discussed much earlier by Kraichnan.[4] For fluids, that theory is recognizably not Galilean invariant and had been rejected. Statements that closures based on the two-body formulation should be naturally Galilean invariant are apparently in error, without further qualification.

The two-body "scattering amplitude"--a moment of which gives the self-energy--is determined by the formally exact Bethe-Salpeter equation.* Low-order approximations to that equation have unifying statistical interpretations. Thus, the two-body theories are Gaussian, in a precise technical sense , in a space whose state vector describes two-body ($\psi\psi$) stochastic variables. The less detailed (generic[6]) DIA is Gaussian in a one-body (ψ) space. As examples of DIA of interest for plasmas, we have the Orszag-Kraichnan equations[8] (Vlasov DIA), the Krommes-Oberman (KO equation for equilibrium fluctuations,[9] and Dubois' space-time formulation of WTT.[10] The MSR techniques allow a more succinct ar we believe, more understandable derivation of these approximations

* For the quantum version see, for example, Ref.7.

A generic stochastic equation of sufficient generality** is

$$\partial_{t_1}\psi(1) = U(12)\psi(2) + \tfrac{1}{2}\bar{U}(123)\psi(2)\psi(3)$$

(summation convention). The essence of the MSR procedure is to interrelate the observables by finding the simultaneous changes induced in them by external forces. Interactions with the system of a one-body force η_1 and a two-body force η_2 are described by the generating functional[12]

$$S\{\eta_1, \eta_2\} \equiv \exp_+[\eta_1(1)\Phi(1) + \tfrac{1}{2}\eta_2(12)\Phi(1)\Phi(2)]$$

Here "+" denotes time ordering; Φ is the operator-doubled vector[6] $\Phi \equiv (\psi, \hat{\psi} \equiv -\delta/\delta\psi) \equiv (\Phi_+, \Phi_-)$. The cumulants

$$F(1) \equiv \frac{\delta}{\delta\eta_1(1)} \ln \langle S \rangle \quad , \quad G(12) \equiv \frac{\delta}{\delta\eta_1(2)} F(1)$$

become in the limit $\eta_i = 0$ the observables $F = \langle\Phi\rangle$, $G = \langle\delta\Phi\delta\Phi\rangle_+$ (e.g., $G_{++} \equiv C = \langle\delta\psi\delta\psi\rangle$, $G_{+-} \equiv R = \langle\psi\hat{\psi}\rangle_+ = \langle\delta\psi/\delta\eta_1^{(-)}\rangle$). We have exactly[6]

$$(G_o^{-1} - \gamma F - \Sigma)G = 1 \tag{1a}$$

$$\Sigma = \tfrac{1}{2}\gamma\, GG\,\Gamma = \tfrac{1}{2}\gamma\, R_4\, \gamma \tag{1b}$$

$$\Gamma = \gamma + (\delta\Sigma/\delta F)_{\eta_2} = (1_s - I_4\, GG)^{-1}\gamma \tag{1c}$$

$$I_4 \equiv (\delta\Sigma/\delta G)_F \;, \quad R_4 \equiv (\delta G/\delta\eta_2)_F \tag{1d,e}$$

$$(1_s - GG\, I_4)R_4 = (GG)_s \tag{1f}$$

where "s" means symmetrized. The free propagator is defined as $G_o \equiv (i\sigma\partial_t - \gamma_2 - \eta_2)^{-1}$; see MSR for the matrix $i\sigma$ and construction of the bare coupling matrices γ_2 and $\gamma_3 \equiv \gamma$ from the scalar U's. In (1a), F is the self-consistent field (linearized Vlasov)

** Generalization to the case of random coefficients and driving forces is straightforward (see Ref.11).

term, Σ the self-energy term (generalized collision matrix). $R_4(12;34)$ is the two-body "scattering" amplitude, described by the Bethe-Salpeter Eq. (1f) (not pointed out by MSR). I_4 describes irreducible two-body interactions. $G^{3/2}\Gamma \sim \langle\delta\psi^3\rangle\langle\delta\psi^2\rangle^{-3/2}$ is the intrinsic (measurable) skewness; it would vanish if the system were exactly Gaussian.

MSR stated that expansion of $\delta\Sigma/\delta F$ in powers of Γ, the effective nonlinear interaction, was preferable to one based on the bare coupling γ and wrote (1c) as

$$\Gamma \simeq \gamma + \Gamma G \Gamma G G \Gamma \tag{2}$$

This finds a clearer interpretation in terms of the Bethe-Salpeter equation. Define the Legendre transforms

$$\mathscr{L}_1 \equiv \ln\langle S\rangle - \eta_1 F \ ; \quad \mathscr{L}_2 \equiv \mathscr{L}_1 - \eta_2 G$$

We have[6,11] $\Gamma \equiv \delta^3\mathscr{L}_1/\delta F^3$; define also the "two-body" extended vertex

$$\Gamma^{(2)} \equiv \delta^3\mathscr{L}_2/\delta G^3 = -(\delta R_4^{-1}/\delta G)_F$$

From (1d), (1b), and $R_4\gamma = G^2\Gamma$, we get

$$I_4 = \tfrac{1}{2}\Gamma G G \Gamma^{(2)} G G \Gamma$$

while from (1f), we find

$$\Gamma^{(2)} = \tfrac{1}{4}(G^{-1}G^{-1}G^{-1})_s + \delta I_4/\delta G \tag{3}$$

Neglect in (3) of $\delta I_4/\delta G$ -- i.e., neglect of two-body vertex renormalization--leads to $I_4 \simeq I_4^{(o)} \equiv \Gamma G \Gamma$ and, from (1c), to (2). This procedure allows irreducible two-body scattering (e.g. clumps, hard sphere collisions) to be important and treated nonperturbatively*** by approximating (neglecting non-Gaussian corrections to) the form of the two-body interaction

*** Thus, ordering of the bare two-body interaction for discrete plasma $\underset{\sim}{E}_{12}\cdot\underset{\sim}{\partial} + (1\longleftrightarrow 2)$ is non-uniform in spatial gradients. Though of nominal order $1/n\lambda_D^3 \ll 1$, the term is not small for large-angle (Boltzmann) scattering ($|\underset{\sim}{x}_1 - \underset{\sim}{x}_2| \ll \lambda_D$). See Ref. 13.

rather than its effect--that is, by approximating the inverse R_4^{-1} instead of the resonant function R_4. It defines a DIA in a two-body space--DIA(2). The usual DIA≡DIA(1) approximates one-body scattering G^{-1} by ignoring one-body vertex renormalization: $\Gamma \approx \gamma$; this neglects non-Gaussian corrections to the one-body self-energy. Equivalently, it ignores irreducible two-body scattering entirely: $R_4 \approx (G\,G)_s$.

DIA(1) is a "renormalized weak coupling" approximation. For equilibrium plasma, DIA(1) reduces to the KO equation.[9] This neglects Boltzmann collisions, but has been quite success-ful for treating collective hydrodynamic effects important at large magnetic fields.

The approximation (2), or its higher order versions obtained by iterating (3), assumes small skewness (near Gaussian). It fails when intrinsically non-Gaussian processes are important. Thus, the detailed structure of a divided phase space including both adiabatic and stochastic regions, in the technical sense of nonoverlapping or overlapping resonances,[14] is not accessible from low order truncations of the vertex expansion. For the plasma, these would predict incorrectly that nonresonant and trapped particles diffuse for times longer than a trapping time. (Study of a model problem shows that DIA(2) reduces the spurious diffusion from DIA(1), but does not eliminate it.[15]) This problem is closely related to the absence in DIA(1,2) of Galilean invariance, essential for proper description of high Reynolds number fluid flow. The difficulties arise from the Eulerian nature of the correlation matrix G . We shall return to this point later.

We illustrate DIA(1) by outlining a new space-time formulation of WTT which shortens and elucidates the work of Dubois.[10]

[Dubois' functional derivatives are taken w.r.t. $(\underset{\sim}{x}, t)$; the MSR variations are in the full phase space $(\underset{\sim}{x}, \underset{\sim}{v}, t)$.] The Klimontovich function obeys

$$[\partial_t + \underset{\sim}{v} \cdot \underset{\sim}{\nabla} + (c^{-1}\underset{\sim}{v} \times \underset{\sim}{B}_o + \underset{\sim}{E}_o + \underset{\sim}{E}) \cdot \underset{\sim}{\partial}\,] N = \hat{\eta}$$

where $\underset{\sim}{\partial} \equiv (q/m)\partial/\partial\underset{\sim}{v}$, $\underset{\sim}{E}(1) = \underset{\sim}{\mathcal{E}}(12)N(2)$, $\underset{\sim}{\mathcal{E}}(12) \equiv -\underset{\sim}{\nabla}|\underset{\sim}{x}_1 - \underset{\sim}{x}_2|^{-1}(nq)_2$, and $\hat{\eta}$ is eventually set to 0. U_2, $\bar{U}_3$, are obvious.[6] From (1a), we find for $C(11') \equiv \langle \delta N(1)\delta N(1')\rangle$ and $R(11') \equiv \langle \delta N(1)/\delta\hat{\eta}(1')\rangle$ the Dyson equations

$$g^{-1} R + \underset{\sim}{\partial} f \cdot \underset{\sim}{\mathcal{E}} R = 1 \,, \quad C = R\,\tilde{\Sigma} R^T \qquad (4a,b)$$

where $f = \langle N \rangle$, "T" means transpose, the averaged particle "propagator" obeys

$$g^{-1} = g_o^{-1} - \Sigma \quad ; \quad g_o^{-1} = \partial_t 1 - U_2$$

and the scalar collision operators are

$$\Sigma = \bar{U}\, R\, C\, \bar{U} + O(\Gamma^3) \qquad (5a)$$

$$\tilde{\Sigma} = \tfrac{1}{2}\, \bar{U}\, C\, C\, \bar{U}^T + O(\Gamma^3) \qquad (5b)$$

R defines the plasma dielectric properties. By rearranging (4a), we find

$$R = g(1 - \underset{\sim}{\partial} f \cdot \underset{\approx}{\epsilon}^{-1} \cdot \underset{\sim}{\mathcal{E}}\, g) \qquad (6)$$

where the renormalized "dielectric function"

$$\underset{\approx}{\epsilon} = \underset{\approx}{1} + \underset{\sim}{\mathcal{E}}\, g \underset{\sim}{\partial} f$$

has appeared. WTT orders $C = O(\delta E^2)$ small. For nonresonant (NR) particles, then, $g \simeq g_o + g\,\Sigma g$ which gives in obvious notation $\varepsilon \simeq \varepsilon_\ell + \varepsilon_{n\ell}$. Using the explicit term in (5a), it is then easy to show that $\varepsilon_{n\ell}$ describes just the induced scattering.[1,10] It is clear that the present procedure effects the renormalization of the plasmon propagator trivially, in contrast to the complicated manipulations which have been proposed previously.[16] It is interesting to observe that ε is not identical, except to lowest order, to the coefficient ε^e in the statement of the averaged first order response $\langle \Delta \underset{\sim}{E} \rangle$ to an external field $\underset{\sim}{E}_e$: $\langle \Delta \underset{\sim}{E} \rangle = (\underset{\approx}{\varepsilon}^e)^{-1} \cdot \underset{\sim}{E}_e$. ε^e includes the effects of certain fluctuation-fluctuation beat terms absent in the definition of ε . The difference appears to be unimportant through $O(\langle \delta E^2 \rangle)$ in WTT. In general, however, in theories of fluctuations at finite (i.e., nonlinear) level, there seems to be no particular reason why ε^e should ever appear.

For resonant (R) particles, we cannot expand g . The induced scattering matrix elements are thus properly integrals over only NR $(\underset{\sim}{k}, \omega)$.[17] For R, it is common[18] to retain from Σ only the renormalized quasilinear term

$$\Sigma_{RQL}(1\,\bar{1}) = \underset{\sim}{\partial}_1 \cdot g(12) \langle \delta \underset{\sim}{E}(1) \delta \underset{\sim}{E}(2) \rangle \cdot \underset{\sim}{\partial}_2\, \delta(2-\bar{1})$$

This cannot be justified in general. Near equilibrium, KO have shown that the full $\Sigma[\mathrm{DIA}(1)]$ must be used.[9]

Equation (4b) is a balance equation which states that the fluctuations C arise from "incoherent noise"

$$\langle \delta\tilde{N}\delta\tilde{N} \rangle \equiv g\,\tilde{\Sigma}\,g^T \tag{7}$$

in a dielectric medium ϵ. Thus defining symbolically

$$\delta\tilde{\underset{\sim}{E}} \equiv \underset{\sim}{\mathcal{E}}\,\delta\tilde{N}\ ,\quad \delta N_c \equiv -g\underset{\sim}{\partial} f \cdot \underset{\approx}{\epsilon}^{-1} \cdot \delta\tilde{\underset{\sim}{E}}$$

we can write (4b) as

$$C = \langle \delta\tilde{N}\delta\tilde{N} \rangle + \langle \delta N_c \delta\tilde{N} \rangle + \langle \delta\tilde{N}\delta N_c \rangle + \langle \delta N_c \delta N_c \rangle \tag{8}$$

$$\langle \delta N\, \delta \underset{\sim}{E} \rangle = \langle \delta\tilde{\underset{\sim}{N}}\delta\tilde{E} \rangle \cdot (\underset{\approx}{\epsilon}^{-1})^T - g\,\underset{\sim}{\partial} f \cdot \langle \delta \underset{\sim}{E}\, \delta \underset{\sim}{E} \rangle \tag{9}$$

$$\langle \delta \underset{\sim}{E}\, \delta \underset{\sim}{E} \rangle = \underset{\approx}{\epsilon}^{-1} \cdot \langle \delta\tilde{\underset{\sim}{E}}\delta\tilde{\underset{\sim}{E}} \rangle \cdot (\underset{\approx}{\epsilon}^{-1})^T \tag{10}$$

These results are the first nonperturbative expression of Dupree's intuitive ideas.[3] Equation (9) at equal times describes the collisional (turbulent or discrete) evolution of the background:

$$[\partial_t + \underset{\sim}{v}\cdot\underset{\sim}{\nabla} + (\underset{\sim}{\mathcal{E}} f)\cdot\underset{\sim}{\partial}]f = -\underset{\sim}{\partial}\cdot\langle\delta\tilde{\underset{\sim}{E}}\,\delta\tilde{N}\rangle\cdot(\underset{\approx}{\epsilon}^{-1})^{T} + \underset{\sim}{\partial}\cdot g\langle\delta\underset{\sim}{E}\;\delta\underset{\sim}{E}\rangle\cdot\underset{\sim}{\partial}f \qquad (11)$$

which includes a friction term due to the noise as well as the usual renormalized quasilinear term. For a time-stationary system, Eq. (10) reduces to three-wave coupling if we assume DIA(1) and compute the C moments needed for $\tilde{\Sigma}$ in (7) by retaining, after velocity integration, only terms explicitly involving the coherent fluctuation δN_c. That is, $C \simeq \langle\delta N_c \delta N_c\rangle$,

$$\langle\delta N\delta\underset{\sim}{E}\rangle \simeq \langle\delta N_c \delta\tilde{\underset{\sim}{E}}\rangle + \langle\delta N_c \delta\underset{\sim}{E}_c\rangle = -g\underset{\sim}{\partial}f\cdot\langle\delta\underset{\sim}{E}\delta\underset{\sim}{E}\rangle$$

Dubois' form for the three-wave matrix element [his Eq.(70)] follows immediately, here with renormalized g and ϵ. Four-wave coupling follows from an obvious iteration of Eq. (9).

The iteration procedure in terms of the coherent fluctuation fails for particles sufficiently localized in phase space (including time).[3] This is particularly likely for the equal time C needed for (11). Consider the exact alternative form of (8) or (4b):

$$(g_o^{-1} - \Sigma)C - \tilde{\Sigma}g^{T} = \underset{\sim}{\partial}f\cdot\langle\delta\underset{\sim}{E}\delta\underset{\sim}{E}\rangle\cdot(g\underset{\sim}{\partial}f)^{T} + g^{-1}\,[\langle\delta N_c\,\delta\tilde{N}\rangle + \langle\delta\tilde{N}\delta N_c\rangle] \qquad (12)$$

Dupree's clump theory[3,19] follows from (12) with the following approximations (which appear reasonable, but which should be studied further). Construct from (12) the equation for C(t,t), assume DIA(1), retain only the diffusion terms of Σ and $\tilde{\Sigma}$ in the Markovian approximation (MA) (appropriate for a continuous spectrum and scales of C greater than the autocorrelation scales), and neglect the clump self-fields [bracketed terms in (12)]. Then ($\underset{\sim}{B}_o = 0$ for simplicity)

$$\{\partial_t + [\underset{\sim}{v}_1 \cdot \underset{\sim}{\nabla} + \underset{\sim}{\partial}_1 \cdot (\underset{\approx}{D}_{11} \cdot \underset{\sim}{\partial}_1 + \underset{\approx}{D}_{12} \cdot \underset{\sim}{\partial}_2) + (1 \leftrightarrow 2)]\} C(\underset{\sim}{1}, \underset{\sim}{2}, t) = (\underset{\approx}{D}_{12} + \underset{\approx}{D}_{21}) : (\underset{\sim}{\partial}_1 f)(\underset{\sim}{\partial}_2 f)$$

$$\underset{\approx}{D}_{ij}(1,2) = (2\pi)^{-4} \int d\underset{\sim}{k} d\omega \left\langle \delta \underset{\sim}{E} \delta \underset{\sim}{E} \right\rangle_{\underset{\sim}{k}\omega} \exp[-i\underset{\sim}{k} \cdot (\underset{\sim}{x}_i - \underset{\sim}{x}_j)] g_k(\underset{\sim}{v}_j) \tag{13}$$

which is Dupree's Eq. (49).[3] The cross-diffusion terms D_{12}, D_{21} on the left come from $\tilde{\Sigma}$. Solution of (13) gives the clump contribution to $\langle \delta\tilde{N}\delta\tilde{N} \rangle$, a process additional to the n-wave processes. Integrals in the n-wave terms should exclude the phase space volume over which $\langle \delta\tilde{N}\delta\tilde{N} \rangle_{cl}$ is significant. Notice that since we assumed DIA(1), clumping is contained in the equations of Orszag and Kraichnan.[8] We believe that this observation is new.

Solution of (13) is effected by a Green's function technique.[3] For consistency, one verifies that this function is just the two-body propagator predicted by the (++--) component of (1f) with the weak coupling approximation $I_4^{(o)} \simeq \gamma G \gamma$. It is not necessary to invoke this fact in solving (13), though it aids understanding. Replacing Γ by γ in $I_4^{(o)}$ means that we ignore "clumping of clumps", presumably a weak effect.

Nonlinearity dominates for the Navier-Stokes fluid at high Reynolds number, or the 2-D guiding center plasma; weak coupling is inappropriate. Furthermore, DIA(1) is not invariant to random Galilean transformations,[4] hence fails to properly describe the inertial and dissipative subranges. Dupree argued[5] that "the use of an average two-particle propagator ... appears to incorporate [Galilean invariance] in a natural way." In essence, he argued that such a theory would contain two variables naturally descriptive of an eddy: a center-of-mass coordinate $\underset{\sim}{R}$ and a relative coordinate $\underset{\sim}{\rho}$. Convection of an inertial range eddy would be described by $\underset{\sim}{R}$, motion of which would effectively define a Lagrangian frame in which internal eddy distortions, described by $\underset{\sim}{\rho}$, could be computed. The eddy lifetime would be the time for two fluid elements to relatively diffuse, in $\underset{\sim}{\rho}$, an eddy size apart. Aspects of this relative motion were discussed in an unrenormalized context by Krommes.[20] A systematically renormalized theory which treats on equal footing both centrix and relative motions is DIA(2). In fact, straightforward calculation which we omit shows that Dupree's equations follow, with various practical approximations like MA, from (1f) with $I_4^{(o)} \approx \gamma G \Gamma$, a simplified DIA(2). The result is a relative diffusion equation, quite similar to (13), for the two-body propagator $R_4(++\,;--)$, except that the diffusion coefficients are defined in terms of R_4 (the Γ in $\gamma G \Gamma$) instead of the weak coupling factorization $R_4 \approx (G\,G)_s$. This does introduce a plausible eddy lifetime, and is Galilean invariant. This is interesting, since we find that DIA(2) has been previously discussed and rejected.

DIA(2) in the vertex form (2) was first derived by Kraichnan, who used opaque graphical techniques to discuss his model stochastic oscillator.[21] He showed that, for the stationary two-time $R(t-t')$, DIA(2) was numerically more precise than DIA(1). However, he later generalized that model to an idealized convection problem which modeled correctly the Kolmogoroff ansatz that "the large-scale structures unquestionably convect the small scales without distorting them."[4] He concluded that for the equal time moments pertinent to energy transfer, the Eulerian DIA(2) was still seriously inadequate; an infinite number of Eulerian are needed to describe pure convection. In other words, pure convection is intrinsically non-Gaussian. Kolmogoroff spectra appear to no order in Eulerian vertex-renormalized perturbation theory.[22] As Kraichnan notes,[4] it is true that DIA(2) should be very accurate if the wavenumber integrals are modified ad hoc to remove the spurious convection effects. However, this asystematic procedure assumes, rather than justifies, the Kolmogoroff hypotheses.[23] We echo Kraichnan's call for a Lagrangian theory.[22]

In DIA(2), as described by Eq. (2), the equal time moments depend intrinsically on the two-time correlations. This feature was retained by Kraichnan in his study,[4] and is, in fact, the basis for his pessimistic conclusions concerning Eulerian theory. Dupree, on the other hand, makes the MA and derives closed equations for the equal time quantities. In contrast to the strict DIA(2), his results are Galilean invariant. However, the MA is not readily justifiable here. It is not clear that Dupree's (complicated) equations will be quantitatively more accurate than simpler Galilean invariant approximations which have been proposed; for example, truncation of wavenumber integrals,[4] or the Test Field Method.[23,24] Clearly, more research is called for. The recent work of Weinstock may be of considerable help here.[25]

There are situations for which the vertex expansion is apparently well suited. Such are problems of stochastic instability[14] at high stochasticity. These can be treated by identifying ψ as the stochastic Liouville function $L(J,\theta,t)$ of the relevant actions J and angles θ . The relevant averaging functional $\langle\cdot\rangle$ forces the rotational invariance appropriate to the stochastic limit:

$$\langle A(J_1..J_n,\theta_1..\theta_n,t)\rangle \equiv \int d\theta' \bar{A}(J_1..J_n,\theta_1-\theta_n+\theta',..\theta_{n-1}-\theta_n+\theta',\theta',t)$$

for arbitrary function A of n angles (e.g. $A\equiv\delta L\delta L$); superbar denotes average over initial conditions. Using this prescription, a successful DIA(1) renormalization of a stochastic instability problem of relevance to plasma heating has already been performed.[15] Similar notions permit systematic treatment of ergodic magnetic field line motion.

The functional thermodynamics of MSR is of great utility. However, it does not resolve the current interest problems of Galilean invariance and adiabaticity. The equations are a supplement to, rather than a substitute for, physical intuition.

ACKNOWLEDGEMENTS

We are grateful to P. Channell, D. Chernin, C. Oberman and M.N. Rosenbluth for informative discussions and critical comments on the manuscript. We are especially indebted to T. Dupree for a clarifying discussion concerning Ref. [5].

REFERENCES

[1] See TSYTOVICH, V.N., An Introduction to the Theory of Plasma Turbulence, Pergamon, Oxford (1972) and references therein.
[2] KRAICHNAN, R.H., J. Fluid Mech. **5** (1959) 497.
[3] DUPREE, T.H., Phys. Fluids **15** (1972) 334.
[4] KRAICHNAN, R.H., Phys. Fluids **7** (1964) 1723.
[5] DUPREE, T.H., Phys. Fluids **17** (1974) 100.
[6] MARTIN, P.C., SIGGIA, E.D., ROSE, H.A., Phys. Rev. **A8** (1973) 423.
[7] NISHIJIMA, K., Fields and Particles, Benjamin, New York (1969) Ch. 7.
[8] ORSZAG, S.A., KRAICHNAN, R.H., Phys. Fluids **10** (1967) 1720.
[9] KROMMES, J.A., OBERMAN, C., J. Plasma Phys. **16** (1976) 193.

[10] DUBOIS, D.F., Phys. Fluids **19** (1976) 1764.
[11] DEKER, U., HAAKE, F., Phys. Rev. **A11** (1975) 2043.
[12] DE DOMINICIS, C., MARTIN, P.C., J. Math. Phys. **5** (1964) 14.
[13] FRIEMAN, E.A., BOOK, D.L., Phys. Fluids **6** (1968) 1700.
[14] For a review, see ZASLAVSKIJ, G.M., CHIRIKOV, B.V., Sov. Phys.-Usp. **14** (1972) 549.
[15] KROMMES, J.A., SMITH, G.R., Bull. Am. Phys. Soc. **21** (1976) 1023, Abstract 7T4.
[16] MAKHANKOV, V.G., TSYTOVICH, V.N., Nucl. Fusion **10** (1970) 405.
[17] CHOI, D., HORTON, W., Phys. Fluids **17** (1974) 2048.
[18] RUDAKOV, L.I., TSYTOVICH, V.N., Plasma Phys. **13** (1971) 213.
[19] DUPREE, T.H., WAGNER, C.E., MANHEIMER, W.M., Phys. Fluids **18** (1975) 1167.
[20] KROMMES, J.A., Proc. Ann. Mtg. Theoretical Aspects of CTR (Berkeley, 1974), Lawrence Livermore Lab. Rep. Conf.-740403, Abstract F2.
[21] KRAICHNAN, R.H., J. Math. Phys. **2** (1961) 124.
[22] KRAICHNAN, R.H., Adv. Math. **16** (1975) 305.
[23] KRAICHNAN, R.H., J. Fluid Mech. **47** (1971) 513.
[24] KROMMES, J.A., Test-Field Method for 2-D Weakly Ionized Turbulent Plasma, Inst. for Advanced Study, Rep. C00-3237-71 (1977).
[25] WEINSTOCK, J., Phys. Fluids **20** (1977) 1631.

CTR theory

IAEA-SMR-32/11

SPHERICAL IMPLOSION DUE TO COALESCED WEAK SHOCKS IN PLASMA

Method of approximate self-similar solution

S.S. JHA
Tata Institute of Fundamental Research,
Bombay

L.K. CHAVDA
Physics Department,
S. Gujarat University, Surat,
India

Abstract

SPHERICAL IMPLOSION DUE TO COALESCED WEAK SHOCKS IN PLASMA: METHOD OF APPROXIMATE SELF-SIMILAR SOLUTION.

A method to determine an approximate solution for the problem of spherical implosion in an ideal gas is described in detail. This solution may be used to investigate the nature of spherical implosion in a plasma, considered to be an ideal gas with specific heat ratio $C_p/C_v = 5/3$, due to coalesced weak shocks. Before coalescing, the process is almost isentropic, and the linear acoustic approximation can give a fairly accurate result. However, when the coalesced front reaches the centre of the sphere, the non-linearity becomes dominant. The solution near the central singular point is then determined by the method of self-similar solutions of the fluid equations. Such shocks may be produced in a plasma because of surface ablation generated by pulsed laser beams or pulsed ion beams or the soft X-ray generated by the absorption of relativistic electron beams. For a given deposited shock energy, the resulting thermonuclear fusion gain for a DT plasma has also been calculated.

1. INTRODUCTION

Before describing our analytical method of finding an approximate solution to the problem of spherical implosion due to coalesced weak shocks in a plasma, we begin by presenting the exact mathematical statement of the problem. At time $t = t_i$, let an ideal gas occupy a spherical region of radius R, with pressure p_0, density ρ_0 and gas velocity $u_0 \cong 0$. Let a uniform radial pressure p(t) be applied at that time at its surface and switched off later at time t_f. In particular, we are interested in considering the special pressure profile p(t) which can generate a large number (m) of weak non-overtaking compression shocks, moving with increasing speed towards the centre, such that all these shocks merge at a given radius r_m from the centre, precisely at the same time t_m. The problem is to find the distribution of pressure p(r,t), mass density ρ(r,t) and the gas velocity u(r,t) as functions of the radial distance r and time t, for $t > t_m$. On physical grounds, the front of the coalesced shocks is expected to move towards the centre, with pressure and density increasing behind the front. After reaching the centre, the front will of course get reflected, but at any fixed radial distance r in the gas, density and pressure will continue to rise until the reflected shock front passes through the point. If t = 0 is defined to be the time of reflection of the front from the centre r = 0, it is obvious that for positive times the solution behind the reflected front is of the explosive type.

Recent interest in solving the kind of model problem described above has arisen in connection with the investigation of an inertial confinement scheme [1] for very high density thermonuclear plasmas due to ablative absorption of energy from pulsed lasers or pulsed relativistic electron beams or from suitable ion beams. While considering the feasibility of such a heating and confinement scheme, it became clear quite early in the game that in order to minimize the level of the input energy required for obtaining a viable fusion gain factor, one has to compress a solid D T pellet of size, for example, 10^{-1} cm, by a factor 10^3 or 10^4 before it is heated to thermonuclear temperatures (for details see Ref. [2]). Using the method of self-similar solutions near the centre of the sphere, the problem of spherical implosion and reflection of a single strong shock has been solved numerically by Guderley [3] and by Landau and Stanyukovich (see Ref. [4]) more than thirty years ago. Guderley's solution was for the gas with specific heat ratio $\gamma = 7/5$; since then Goldman [5], among many other workers, has obtained the numerical solution for $\gamma = 5/3$. However, the maximum compression for $\gamma = 5/3$, after the reflection of a single strong shock, is found to be only about 32. Based on Goldman's result, and assuming as a first approximation the fully ionized D T plasma (equal densities of D and T) to behave as an ideal gas of $\gamma = 5/3$, Brueckner and Jorna [2] have calculated the resulting fusion gain factor as a function of the input shock energy to find

$$E_{SHOCK} \cong 40.9 \left[\frac{E_{FUSION}}{E_{SHOCK}}\right]^3 \left[\frac{\rho_s}{\rho_0}\right]^2 \text{ megajoules} \qquad (1)$$

where the expression has been normalized to the solid DT density $\rho_s = 0.213$ gram/cm³ (number density $n_s = 4.7 \times 10^{22}$ cm^{-3}). For obtaining a viable gain factor $G_F = E_{FUSION}/E_{SHOCK}$, the input shock energy thus turns out to be too large.

To circumvent the need for handling and depositing extremely high input energy in a very short time at the pellet, it has been suggested [1] that one should arrange to produce ablation of the type that gives rise to weak coalescing shocks, described earlier, instead of a single strong shock. In such a case, the compression process is almost adiabatic until all the shocks coalesce at the radius r_m at time t_m, thus minimizing the shock energy. As already discussed, there will be further compression and heating of the plasma as the merged front moves towards the centre, and there will be a sudden jump in its density and temperature to still higher values when the reflected front reaches the observation point. Since, immediately after the reflection of the front, the pressure and temperature become extremely large near the centre, it can initiate nuclear fusion reactions there if the initial parameters are chosen properly. The fusion gain can be maximized further if the radius of the resulting central hot region is comparable to the range of 3.5-MeV α-particles generated in the nuclear fusion reactions, because in such a situation further heating of the compressed pellet will be due to these absorbed α-particles. Consequently the reflected front will be accompanied by a powerful nuclear burn wave moving outward at a supersonic speed so that most of the fuel will be burnt before the disassembly time of the exploding plasma, with a minimum of input energy to be supplied by external means for compression and heating. This is the 'implosion-cum-self-heating' scheme [1]. For density $\rho = 10^3 \rho_s$, and the electron temperature T_e corresponding to 10 keV, the range [6] of 3.5-MeV α-particles is indeed small ($\lesssim 10^{-3}$ cm) if the radius of the compressed pellet is about 10^{-2} cm, so the scheme seems to be viable.

It is clear from the statement of our mathematical problem that we are not including any energy absorption by the plasma due to the absorption of α-particles generated in the nuclear reactions. The solution of our problem can therefore only give the state of the plasma, i.e. the distribution of the density, temperature, etc., just before the nuclear burn wave is ignited. However, for systems in which fusion energy deposition (self-heating) in the plasma itself is negligible, we can also calculate the fusion gain factor as a function of the input shock energy for the case of weak coalescing shocks, similar to Eq.(1) for the case of a single strong shock. In the next few sections we introduce the non-dissipative fluid equations relevant to our problem, discuss the procedure of solving them analytically by using the method of approximate self-similar solutions [7], and finally apply the resulting solution to calculate the fusion gain factor in a DT plasma, in the absence of self-heating.

2. SELF-SIMILAR EQUATIONS AND APPROXIMATE INTEGRAL CURVES

2.1. Fluid equations

In the absence of any dissipation and external energy source, the fluid equations for one-dimensional spherical motion are

$$\frac{\partial \rho}{\partial t} = -\frac{1}{r^2}\frac{\partial}{\partial r}(r^2 u \rho) \tag{2}$$

$$\frac{\partial(\rho u)}{\partial t} = -\frac{1}{r^2}\frac{\partial}{\partial r}(r^2 \rho u^2) - \frac{\partial p}{\partial r} \tag{3}$$

$$\frac{\partial}{\partial t}\left(\epsilon\rho + \frac{\rho u^2}{2}\right) = -\frac{1}{r^2}\frac{\partial}{\partial r}\left[r^2 u\left(\rho\epsilon + \frac{\rho u^2}{2} + p\right)\right] \tag{4}$$

where u is the radial velocity and ϵ is the energy per unit mass of the gas. Because of the assumed spherical symmetry, the velocity u, the energy ϵ, pressure p and the density ρ are functions of r and t only, where r = 0 is chosen to be the centre of the sphere and t = 0 is chosen to be the time when the coalesced shocks reach the centre of the sphere and become reflected. The set of equations (2) – (4) has to be supplemented by the initial boundary conditions and the equation of state. For an ideal gas obeying the Clapeyron equation, we of course have

$$\epsilon = C_V T = \frac{C_V}{R}\frac{p}{\rho} = \frac{C_V}{(C_p - C_V)}\frac{p}{\rho} = \frac{1}{(\gamma-1)}\frac{p}{\rho} \tag{5}$$

In general, using the continuity equation (2) and the force equation (3), the energy balance equation (4) can be rewritten as

$$\frac{d\epsilon}{dt} = -p\frac{d}{dt}\left(\frac{1}{\rho}\right) + \dot{Q}_{ext} \quad , \quad \frac{d}{dt} \equiv \frac{\partial}{\partial t} + u\frac{\partial}{\partial r} \tag{6}$$

where, in the absence of external source term, $\dot{Q}_{ext} = 0$. Using the thermodynamic relation for entropy S,

$$T\,ds = d\epsilon + pd\left(\frac{1}{\rho}\right) \tag{7}$$

the energy equation (4) is therefore equivalent to

$$\frac{dS}{dt} = \frac{\dot{Q}_{ext}}{T} = 0 \tag{8}$$

implying an isentropic motion. Thus for an ideal gas the energy balance equation reduces to

$$\frac{\partial}{\partial t}(\ell n\, p\rho^{-\gamma}) + u\frac{\partial}{\partial r}(\ell n\, p\rho^{-\gamma}) = 0 \tag{4'}$$

since from (5) and (6) the ideal gas entropy is

$$S = C_V\,\ell n\, p\rho^{-\gamma} + \text{const} \tag{9}$$

Before we consider the self-similar method of tackling spherical shock reflection from the centre, let us see what is the difficulty in using the linear acoustic approximation for Eqs (2) – (4) to investigate this problem. In terms of the sound velocity c defined by

$$c^2 = \left(\frac{\partial p}{\partial \rho}\right)_S = \frac{\gamma p}{\rho} \tag{10}$$

if we consider only small variations of pressure,etc., such that

$$\begin{aligned} & p = p_0 + \delta p, \quad \rho = \rho_0 + \delta\rho, \quad u = 0 + u \\ & \delta p \ll p_0, \quad \delta\rho \ll \rho_0, \quad u \ll c_0 \end{aligned} \tag{11}$$

in the linear approximation, the velocity potential satisfies the spherical wave equation

$$\frac{1}{c_0^2}\,\frac{\partial^2 \phi}{\partial t^2} = \frac{1}{r^2}\,\frac{\partial}{\partial r}\left(r^2\,\frac{\partial \phi}{\partial r}\right) \tag{12}$$

where

$$u \equiv \frac{\partial \phi}{\partial r}; \quad \delta p = -\rho_0\,\frac{\partial \phi}{\partial t}; \quad \delta\rho = -\frac{\rho_0}{c_0^2}\,\frac{\partial \phi}{\partial t} \tag{13}$$

The general solution of (12) gives

$$\phi = \frac{1}{r}\,[F_1(r - c_0 t) + F_2(r + c_0 t)] \tag{14a}$$

$$\delta p = \frac{\rho_0 c_0}{r}\,[F_1'(r - c_0 t) - F_2'(r + c_0 t)] \tag{14b}$$

$$\delta\rho = \delta p / c_0^2 \tag{14c}$$

where the nature of arbitrary functions F_1 and F_2 is determined from the initial boundary conditions. According to the above solutions, the velocity, pressure and density grow indefinitely at the centre of the sphere, but that invalidates the linearization of the fluid equations. However, these solutions may still be applicable at large r while the disturbance is weak.

2.2. Initial boundary conditions

It is of course well known that in the absence of dissipation any compression wave always develops into a shock wave, which is a surface of discontinuity in pressure, velocity and density of the fluid. If a shock wave is moving with a radial velocity D_1 towards the region specified by p_0, ρ_0 and u_0, from the region specified by p_1, ρ_1 and u_1, a simple integration of Eqs (2)–(4) in the frame moving with velocity D_1 leads to the relations:

$$(u_1 - D_1)\rho_1 = (u_0 - D_1)\rho_0 \tag{15a}$$

$$p_1 + \rho_1(u_1 - D_1)^2 = p_0 + \rho_0(u_0 - D_1)^2 \tag{15b}$$

$$\epsilon_1 + \tfrac{1}{2}(u_1 - D_1)^2 + \frac{p_1}{\rho_1} = \epsilon_0 + \tfrac{1}{2}(u_0 - D_1)^2 + \frac{p_0}{\rho_0} \tag{15c}$$

Let us now consider a large number of non-overtaking shocks moving radially towards the centre of the spherical gas, so that the first shock moves between the region labelled 0 (the undisturbed region, with pressure p_0, density ρ_0, and $u_0 = 0$) and the region labelled 1 (pressure p_1, density ρ_1, and gas velocity u_1, depending very weakly on r and t), with shock speed D_1 (negative), etc. In general, the nth shock wave moves with velocity D_n (negative) across regions labelled $n-1$ and n. Equations (5) and (15) then give the well-known jump conditions (see e.g. [8]) across each shock wave, in terms of the pressure increment across the front [7]. For a single strong shock wave with $p_1/p_0 \gg 1$, this reduces to

$$\frac{\rho_1}{\rho_0} \cong \frac{\gamma + 1}{\gamma - 1} = 4 \quad \text{for} \quad \gamma = 5/3 \tag{16a}$$

$$u_1 \cong -c_0 \left[\frac{2}{\gamma(\gamma + 1)}\right]^{1/2} \left(\frac{p_1}{p_0}\right)^{1/2} \tag{16b}$$

$$D_1 \cong -c_0 \left[\frac{\gamma + 1}{2\gamma}\right]^{1/2} \left(\frac{p_1}{p_0}\right)^{1/2} \tag{16c}$$

On the other hand, if we have a large number of weak shocks (total number m) such that

$$\frac{p_n}{p_{n-1}} - 1 = |\delta| \ll 1, \quad 1 \leqslant n \leqslant m \tag{17}$$

successive iterations of discontinuities across these shocks lead to [7]

$$\frac{\rho_m}{\rho_0} \cong \left(\frac{p_m}{p_0}\right)^{1/\gamma} \tag{18a}$$

$$u_m \cong -\frac{2c_0}{(\gamma-1)}\left(\frac{p_m}{p_0}\right)^{\frac{\gamma-1}{2\gamma}} \tag{18b}$$

$$D_m \cong -c_0\left(\frac{\gamma+1}{\gamma-1}\right)\left(\frac{p_m}{p_0}\right)^{\frac{\gamma-1}{2\gamma}} \tag{18c}$$

where we have assumed that

$$\left(\frac{p_m}{p_0}\right)^{\frac{\gamma-1}{2\gamma}} \gg 1 \tag{18d}$$

Thus the process is almost adiabatic, until these shocks coalesce at the time t_m at a distance r_m from the centre. Equations (18) determine the boundary conditions at the coalesced shock front at time t_m for its subsequent motion.

2.3. Self-similar equations

Although the exact integration of the non-linear fluid equations (2)–(4) for the spherical case is not possible, the solution near the singular point r = 0, where the linear approximation fails, can be investigated quite accurately. In fact, if t = 0 is the time for the coalesced weak shock,or the single shock, as the case may be, to reach the centre, it is physically consistent to assume that the position of the incident shock front for small negative t can be expanded as

$$R_I(t) = \sum_q a_q(-t)^q \cong \xi(-t)^\alpha, \quad t \to 0 \tag{19}$$

where q's are arbitrary positive numbers and where α is the smallest value of q that appears in the expansion. Thus, for a sufficiently small neighbourhood of the singular point t = 0, we can keep only one term in the expansion, i.e. represent it by a parabola of order α. In such a case we have two basic scales, ρ_0 and $R_I(t)$, for density and length; $\dot{R}_I(t)$ can be taken to be the velocity scale, and $\rho_0\,\dot{R}_I^2$ as the pressure scale. Because of the scale invariance of the fluid equations [8], we can then define reduced dimensionless fluid variables in such a way that these are functions of $r/R_I(t)$ only. In other words, if $R_I(t) = \xi(-t)^\alpha$, the motion is self-similar. If $|t_0|$ is the time taken for the incident front to reach the centre from a point r in the sphere ($r < r_m$), i.e. if

$$R_I(t_0) = r = \xi\,|t_0|^\alpha, \quad |t_0| = \left(\frac{r}{\xi}\right)^{1/\alpha} \tag{20}$$

we can introduce the similarity variable s defined by

$$s = \frac{t}{|t_o|} \tag{21a}$$

$$(-s)^{-\alpha} = \frac{r}{R_I(t)} \quad \text{(for } s < 0 \text{ only)} \tag{21b}$$

so that the position of the incident front is represented by s = −1. Note that

$$\begin{aligned} \dot{R}_I(t) &= D_I(r)(-s)^{\alpha-1} \\ D_I(r) &= -\alpha\xi^{1/\alpha}\, r^{-(1/\alpha-1)} \end{aligned} \tag{22}$$

$$\frac{r}{t} = \frac{1}{\alpha s}\,|D_I(r)| \tag{23}$$

where $|D_I(r)|$ is the incident front speed. Since at the origin it is expected to go to infinity (all shock energy concentrated at a point), in our case α is expected to be less than 1. If we extend the similarity argument to the distance r_m, where weak shocks coalesce, $|D_I(r)|$ is related to D_m of (18c) by

$$\frac{|D_I(r)|}{|D_m|} = \left(\frac{r_m}{r}\right)^{(1/\alpha)-1} \tag{24}$$

In our problem, for time $t > 0$, one can also show for the reflected wave that

$$R_R(t) = \frac{\xi}{s^{*\alpha}}\, t^{\alpha} \tag{25a}$$

$$\dot{R}_R(t) = D_R(r)s^{\alpha-1}, \quad D_R(r) = \frac{|D_I(r)|}{s^*} \tag{25b}$$

$$\left(\frac{s}{s^*}\right)^{-\alpha} = \frac{r}{R_R(t)} \quad \text{(for } s > 0) \tag{25c}$$

where s^* is a constant greater than 1. The position of the reflected shock front is now given by $s = s^*$, which is greater than 1, because the absolute speed of the reflected front, which has to move outwards in the gas moving inwards, is slower than the incident front.

In terms of the reduced dimensionless variables G(s), V(s) and Z(s), defined by

$$\rho = \rho_0 G(s) \tag{26a}$$

$$u = \frac{r}{t}\ V(s) = \frac{|D_I(r)|}{\alpha s}\ V(s) \tag{26b}$$

$$c^2 = \frac{r^2}{t^2}\ Z(s) = \frac{|D_I(r)|^2}{\alpha^2 s^2}\ Z(s) \tag{26c}$$

the fluid equations (2), (3) and (4′) reduce [7, 8] to

$$\frac{dV}{d\,\ell n\,|s|} + (V-\alpha)\frac{d\,\ell n\,G}{d\,\ell n\,|s|} = 3\alpha v \tag{27a}$$

$$(V-\alpha)\frac{dV}{d\,\ell n\,|s|} + \frac{Z}{\gamma}\frac{d\,\ell n\,G}{d\,\ell n\,|s|} + \frac{1}{\gamma}\frac{dZ}{d\,\ell n\,|s|} = \frac{2Z\alpha}{\gamma} + V(V-1)\,\alpha \tag{27b}$$

$$(\gamma-1)\,Z\frac{d\,\ell n\,G}{d\,\ell n\,|s|} - \frac{dZ}{d\,\ell n\,|s|} = -\alpha\frac{2Z(V-1)}{(V-\alpha)} \tag{27c}$$

Note that, in terms of the reduced variables,

$$p = \rho_0\ |D_I(r)|^2\ \frac{Z(s)\ G(s)}{\alpha^2 s^2 \gamma} \tag{28}$$

$$k_B T \equiv \theta = \tfrac{1}{2}\ M\ |D_I(r)|^2\ \frac{Z(s)}{\alpha^2 s^2 \gamma} \tag{29}$$

where M/2 is the average mass of the particles in the gas. At the incident shock front (s = −1) the boundary conditions (18) reduce to

$$G(-1) \cong \left(\frac{p_m}{p_0}\right)^{1/\gamma} \quad ; \quad V(-1) \cong \frac{2\alpha}{\gamma+1} \quad ; \quad Z(-1) \cong \frac{(\gamma-1)^2\alpha^2}{(\gamma+1)^2} \tag{30}$$

With the rearrangement of terms, the self-similar gas equations (27) can be reduced to [7]

$$-\alpha\frac{d\,\ell n\,|s|}{dV} = \frac{(\alpha-V)^2 - Z}{Z\left[3V - \dfrac{2(1-\alpha)}{\gamma}\right] - V(1-V)\,(\alpha-V)} \equiv \frac{\Delta(Z,V)}{\Delta_1(Z,V)} \tag{31a}$$

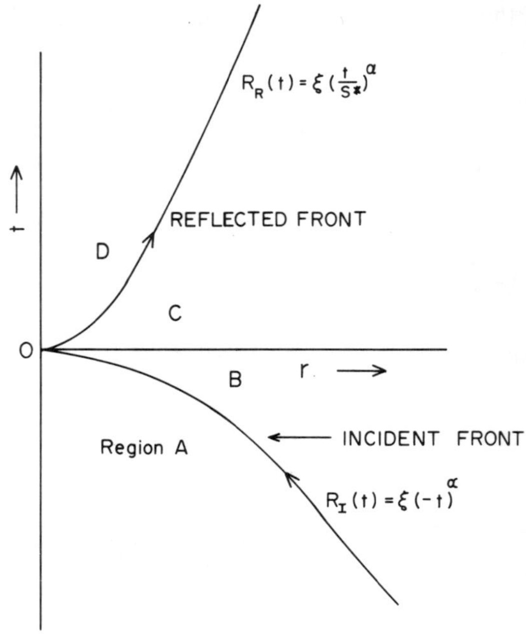

FIG.1. r-t diagram showing the incident and reflected shock fronts, for $t > t_m$, $r < r_m$. *Here* $t = 0$ *is the time of reflection of the front from the centre, where the shock speed goes to infinity. The reflected front, however, moves slower than the incident front.*

$$\frac{d\,\ell n\,Z}{dV} = \frac{(\gamma-1)}{(\alpha-V)} + \frac{(3\gamma V - V - 2)}{(\alpha - V)} \frac{\Delta(Z,V)}{\Delta_1(Z,V)} \tag{31b}$$

$$G = \text{const}\left[(\alpha-V)^{2(1-\alpha)}\, s^{-6\alpha}\, Z^{3\alpha}\right]^{\frac{1}{(3\alpha\gamma-2-\alpha)}} \tag{31c}$$

We have to consider solutions of Eqs (31a) and (31b) in regions A, B, C, D of Fig.1. Of course in the region A, which is ahead of the incident shock front, the gas is undisturbed and the solution is already known ($G_A = 1$, $u = 0$, $p = p_0$). In the region B, which is behind the incident shock front, one has to determine the integrals of (31) in such a way that they satisfy the boundary conditions (30) at the shock front. Note that Z(s) is always positive in all regions, since from (26c) it has the same sign as the square of the sound speed. However, V(s) can have both signs. In the region B, since u is negative and $t < 0$ (or $s < 0$), because of Eq. (26b) V(s) has to be positive here. The solution in the region C which is ahead of the reflected shock front ($t > 0$) can be obtained from causality argument (analytically continuing from $-t$ to $+t$), once the solution in the region B is known, because it is still unaffected by the reflected front. Note that now $s > 0$, i.e. $t > 0$, so that in this region $V(s)$ becomes negative. In fact $V(s)$ changes sign smoothly from +ve to −ve, going through zero at $s = 0$, the boundary between the regions B and C. However, u does not go to zero there. The solution in the region D which is behind the reflected shock front has to be determined by solving (31) again, but now at the reflected shock front (at $s = s^*$), one has to satisfy the regular shock jump conditions across the region C. In terms of the reduced variables these jump conditions (obtained fromEq.(15)), at $s = s^*$, are:

$$V_D = V_C + \frac{Z_C - (V_C - \alpha)^2}{\left(\frac{2}{\gamma+1}\right)^{-1} (V_C - \alpha)} \tag{32a}$$

$$G_D = G_C \frac{(V_C - \alpha)}{(V_D - \alpha)} \tag{32b}$$

$$Z_D = \frac{(\gamma-1)^2}{(\gamma+1)^2} \left[1 + \frac{2 Z_C}{(\gamma-1)(V_C-\alpha)^2}\right] \left[\frac{2\gamma}{(\gamma-1)} (V_C-\alpha)^2 - Z_C\right] \tag{32c}$$

where

$$\frac{1}{s^*} = \lim_{r\to 0} \left[\frac{D_R(r)}{|D_I(r)|}\right] = \left[1 + \frac{u(r,t=0)}{|D_I(r)|}\right]_{r\to 0} = \left[1 + \frac{V(s)}{\alpha s}\right]_{s\to 0} \tag{33}$$

Also, in the region D, the solution must be of explosive type, i.e. $d\,\ell n\, |s|^{-\alpha}/dV > 0$, and a particle located at the origin must remain at rest. In this region, u is positive, and hence $V(s)$ is also positive.

2.4. Integral curves

To obtain integral ($Z - V$) curves from (32a) and (32b) in different regions, let us examine these more carefully. We know that physical variables u, c and ρ must be bounded at a finite r, so that the solution must satisfy the condition

$$V \to 0, \quad Z \to 0, \quad \text{as } s \to 0 \tag{34a}$$

In fact, for small V, as $s \to 0$, it can be shown by examining the self-similar equations that

$$Z \sim V^2 \quad (s \to 0) \tag{34b}$$

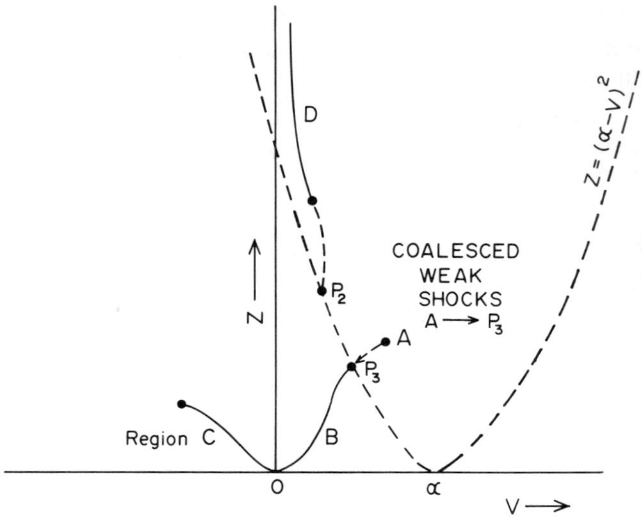

FIG.2. Possible physical integral curves in different regions of the Z-V plane, and the location of singular and initial boundary points. Only solid curves represent possible solutions.

Further, since the solutions of the gas equations must be single valued, ℓn |s| as a function of V should not have any extrema, i.e. d ℓn |s| dV should not become zero. However, the determinant $\Delta(Z,V)$ is equal to zero on the parabola (see Fig.2),

$$Z = (\alpha - V)^2 \tag{35a}$$

The boundary point A in the Z – V plane, at s = −1, is usually above this parabola (for the coalesced weak shocks it is almost on the parabola itself), and since the integral curve has to pass through the origin, according to (34a), it implies that it must cross the parabola. To ensure single-valuedness (see (31a)), it must therefore cross the parabola on the point where

$$\Delta_1(Z,V) = Z\left[3V - \frac{2(1-\alpha)}{\gamma}\right] - V(1-V)(\alpha - V) = 0 \tag{35b}$$

also. Of course, $V = \alpha$, $Z = 0$ is the trivial solution of the self-similar gas equations, but this is useless since it cannot satisfy the boundary conditions (30) at the shock front. However, Eqs (35a) and (35b) determine two more singular points P_2 and P_3 (see Fig.2) as solutions of the equations

$$(\alpha - V_0)\left[3V_0 - \frac{2(1-\alpha)}{\gamma}\right] = V_0(1-V_0), \qquad Z_0 = (\alpha - V_0)^2 \tag{36}$$

In fact, it is the larger of the two roots V_0 which determines the singular point P_3 where the integral curve in the region B has to cross the parabola (35a) in going to the origin [3]. The smaller positive root V_0 determines the singular point P_2 which lies on the integral curve in the region D if extended towards the parabola. In actual practice, the physical integral curve in the region D starts beyond this point to satisfy the shock conditions across the regions C and D at $s = s^*$.

2.5. Solution in region B

While finding the integral curve in the region B, the value of α gets determined uniquely, since for only a particular value of α can one go smoothly from the boundary point A to the singular point P_3 and then to the origin 0, as drawn in Fig.2. But for coalesced weak shocks, the boundary point A at $s = -1$ (determined by Eq.(30)) is approximately on the parabola itself; it must therefore coincide with the singular point P_3 of Eq. (36). This immediately determines α uniquely. We find

$$\alpha = \frac{(\gamma + 1)(2\gamma - 1)}{(4\gamma^2 - \gamma - 1)} = 0.736842 \text{ for } \gamma = \tfrac{5}{3} \tag{37}$$

For more general initial boundary conditions, e.g. for the case of a single strong shock, A is not on the parabola, and α can be determined only after integrating Eqs (31a) and (31b).

It has to be noted that one can immediately integrate (31b) if the right-hand side of (31a) can be written as a function of V alone. In fact, at the point P_3, a self-consistent evaluation of the limiting value shows

$$F(Z,V) \equiv \frac{\Delta(Z,V)\, V(V-1)}{\Delta_1(Z,V)\,(\alpha - V)} \longrightarrow 1 \text{ (approx.)} \tag{38}$$

and the same function at the origin behaves as

$$F(Z,V) \xrightarrow[\substack{V \to 0 \\ Z \to \alpha V^2}]{} 1 \tag{39}$$

Thus, it is not a bad approximation to make an expansion of F(Z,V) in V, which retains the forms (38) and (39) at P_3 and 0, respectively. In the general case, since F(Z,V) differs from 1 at the boundary point A, a few terms in the expansion are always necessary to fit that point also. However, for the case of coalesced shock waves, in which the point A coincides with the point P_3, it is enough to assume

$$\frac{\Delta(Z,V)}{\Delta_1(Z,V)} \cong \frac{(\alpha-V)}{V(V-1)} \tag{40}$$

everywhere. Integrations of Eqs (31) and the boundary conditions (30) then immediately lead to

$$Z(s) = |K_1| \frac{V^2(1-V)^{3\gamma-3}}{(\alpha-V)^{\gamma-1}} \tag{41a}$$

$$G(s) = |K_2| \frac{(1-V)^3}{(\alpha-V)} \tag{41b}$$

$$s = -|K_3| V(1-V)^{(1-\alpha)/\alpha} \tag{41c}$$

$$|K_1| = \frac{(\gamma-1)^2}{4} \left[\frac{\alpha(\gamma-1)}{(\gamma+1)} \quad \frac{(\gamma+1)^3}{(\gamma+1-2\alpha)^3}\right]^{\gamma-1} \tag{41d}$$

$$|K_2| = \left(\frac{p_m}{p_0}\right)^{1/\gamma} \alpha\left[\frac{\gamma+1}{\gamma+1-2\alpha}\right]^3 \frac{(\gamma-1)}{(\gamma+1)} \tag{41e}$$

$$|K_3| = \left[\frac{\gamma+1-2\alpha}{2\alpha}\right] \left[\frac{\gamma+1}{\gamma+1-2\alpha}\right]^{1/\gamma} \tag{41f}$$

where α is given by Eq.(37). It can be verified that the solution (41) when substituted in the function F(Z,V) of (38) shows that in the entire range, $-1 < s < 0$, F differs from 1 by at most 6%. Hence it should be an acceptable approximate solution in this region. Note that as $V \to 0$, $s \to 0$, as required.

2.6. Solution in region C

As already discussed, if the solution is available for the region B, causality demands that the same solution holds in the region C, if we change $-t$ to $+t$, i.e. $-|s|$ to $+|s|$. It of course implies that V changes sign now from positive to negative values. From Eq. (33), the solution in the region C can be continued only till $s = s^*$, where

$$s^* = \lim_{s \to 0} \left[1 + \frac{V(s)}{\alpha s}\right]^{-1} = \frac{\alpha|K_3|}{\alpha|K_3| - 1} \tag{42}$$

if we use the approximate solution (41). For $\gamma = 5/3$, Eqs (37) and (41f) lead to $s^* \cong 2.277$ for the case of coalesced weak shocks. We have found that the extended solution (41) in the region C is also quite good, in the sense that F(Z,V) of Eq.(38) still does not differ too much from 1 in this range, but the value of s^* is a sensitive quantity and its exact value may well be slightly different from the value that we obtained.

Once the value of s^* is known, from the solution (41) for the region C (with s positive, V negative), we can immediately obtain V_C, G_C and Z_C at $s=s^*$. Equations (32a), (32b) and (32c) then give the boundary values at the shock front in the region D, i.e. V_D, G_D and Z_D. For $\gamma = 5/3$, we find

$$V_C \cong -0.80, \quad G_C \cong 31\left[\frac{\gamma-1}{\gamma+1}\right]\left(\frac{p_m}{p_0}\right)^{1/\gamma}, \quad Z_C \cong 0.25 \tag{43}$$

$$V_D \cong 0.19, \quad G_D \cong 90\left[\frac{\gamma-1}{\gamma+1}\right]\left(\frac{p_m}{p_0}\right)^{1/\gamma}, \quad Z_D \cong 0.90 \tag{44}$$

2.7. Solution in region D

The integral curve in the region D has to be the explosive type, i.e. here $\Delta(Z,V)/\Delta_1(Z,V)$ of Eq.(31a) has to be positive. Also, as $s \to \infty$, either $t \to \infty$ or $r \to 0$. From (26c) one has therefore $Z(s) \to \infty$, as $s \to \infty$, whereas from (26b), V(s) is finite or zero as $s \to \infty$. According to Fig.2, note that V cannot exceed the root V_0 corresponding to the singular point P_2. In fact, an examination of (31a) shows that as $s \to \infty$, Δ/Δ_1 will be positive only if

$$\lim_{s\to\infty} Z(s) \to \infty$$

$$\lim_{s\to\infty} V(s) \longrightarrow \frac{2(1-\alpha)}{3\gamma} \equiv V(\infty) \cong 0.105 \text{ for } \gamma = 5/3 \tag{45}$$

$$\lim_{s\to\infty} 3Z\,[V-V(\infty)] < V(\infty)\,[1-V(\infty)]\;\;[\alpha-V(\infty)] \tag{46}$$

Since the range of variation of V(s) from $V_D = 0.19$ at $s = s^*$ to $V(\infty) = 0.105$ as $s \to \infty$ is very narrow, one can assume without too much error that it has a constant average value

$$V = \frac{V_D}{2} + \frac{(1-\alpha)}{3} = V_a = \text{const} \tag{47}$$

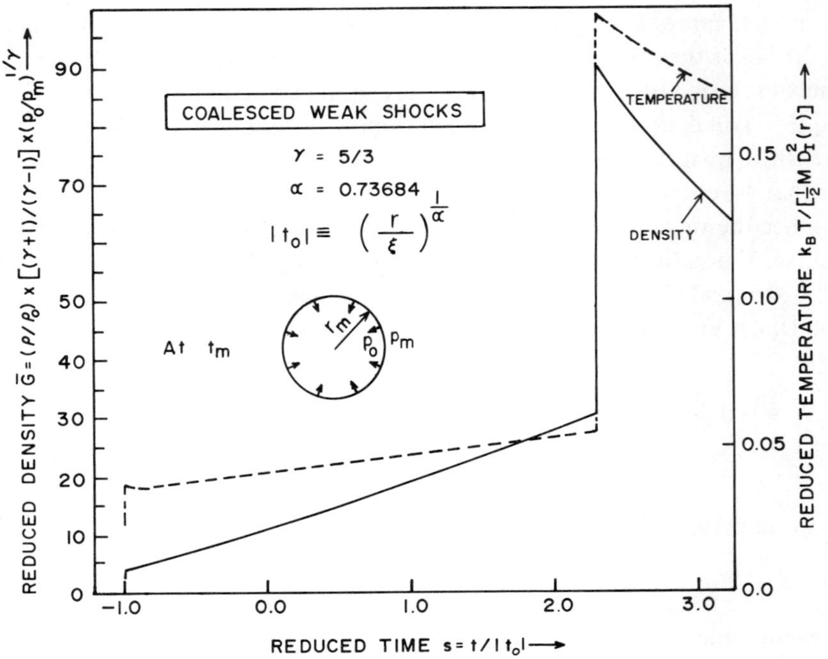

FIG.3. Plot of reduced mass density (left scale) and reduced temperature (right scale) as a function of the similarity variable $s = t/|t_0|$; $|t_0| \equiv (r/\xi)^{1/\alpha}$. *Here* ξ *is related to the shock-speed parameter* (p_m/p_o) *through* $D_I(r)$ *and* D_m *(see text).*

throughout the range. This immediately leads to the solution

$$G = G_D \left(\frac{s}{s^*}\right)^{-\frac{3V_a\alpha}{\alpha - V_a}}, \quad Z = Z_D \left(\frac{s}{s^*}\right)^{\frac{\alpha(2+V_a-3\gamma V_a)}{(\alpha - V_a)}} \tag{48}$$

in the region D. A more detailed solution, which takes care of the variation of V approximately, is given in Ref. [9]. For most purposes, however, the approximate solution given by Eqs (47) and (48) should be good enough.

The solutions for all the regions for density ρ and temperature T as functions of s are plotted in Fig.3, essentially in terms of the shock parameter (p_m/p_o) at the coalescing radius r_m. Here M/2 is the average mass of the ion-electron system in the gas, so that $p/\rho = k_B T/(M/2)$, and $D_I(r)$ is related to the shock parameter (p_m/p_o) at r_m by Eqs (18c) and (24). One finds that the density of the gas after shock reflection can be much higher than the case of a single strong shock, depending upon the value of the shock parameter (p_m/p_o) for the coalesced weak shocks at r_m. For initial compression $(p_m/p_o)^{3/5} \cong 5 \times 10^2$, the final compression

ρ/ρ_0 reaches the value 10^4. The heating is also maximum near r = 0, for $t \gtrsim 0$, i.e. just after the shock reflection from the centre, and the shock parameter can always be adjusted so that the temperature reaches 5 to 10 keV. Note that from our self-similar solutions as a function of s, all the gas variables can be determined easily for any r and t in the physical region.

3. FUSION GAIN CALCULATION FOR A DT PLASMA

We now apply our approximate solutions obtained earlier to the case of a DT plasma. In the first approximation [2], we consider the fully ionized DT plasma to be an ideal gas, with common electron and ion temperatures T, and the specific heat ratio $\gamma = 5/3$. With the number densities $n_D = n_T = n/2$, $n_e = n$, one therefore has

$$p = n_i k_B T + n_e k_B T = 2nk_B T \tag{49a}$$

$$\frac{p}{\rho} = \frac{2nk_B T}{(Mn+m_e n)} \cong \frac{k_B T}{M/2} \equiv \frac{\theta}{M/2} \tag{49b}$$

where M is the average mass of D and T nuclei and M/2 is approximately the average mass of the ion-electron system.

In the absence of self-heating and absorption of the α-particles in the pellet, in each nuclear reaction

$$D + T \longrightarrow n + \alpha + W\ (= 17.6\ \text{MeV}) \tag{50}$$

17.6 MeV of the fusion energy is released. If $n\langle\sigma v\rangle/2$ is the rate of this reaction, which depends upon temperature and ρ, and hence on r and t in the pellet, the fusion energy released during the whole process is

$$E_{FUSION} = 4\pi W \int_0^{r_{max}} dr\, r^2 \int dt\, \frac{n^2}{4} \langle\sigma v\rangle_\theta \tag{51}$$

Since the integrand is highly peaked near s = s*, when the temperature is maximum, one can approximately rewrite Eq. (51) in the form

$$E_{FUSION} = \frac{\pi \rho_0^2 W}{\xi^{1/\alpha}} \int_0^\infty dr\, r^{(2+1/\alpha)} \langle\sigma v\rangle_{s^*}(r) \int_{s^*}^\infty ds\, G^2(s) \tag{52}$$

Similarly, the input shock energy

$$E_{SHOCK} = 4\pi \int_0^{r_{max}} dr\, r^2 \left[\tfrac{1}{2}\rho u^2 + \frac{p}{\gamma-1}\right]$$

$$= 4\pi \rho_0 \alpha\, \xi^5\, t^{(5\alpha-2)} \int_{s^*}^{\infty} ds\, \frac{[\tfrac{1}{2}GV^2 + \tfrac{9}{10}ZG]}{s^{(5\alpha+1)}} \tag{53}$$

Since from Eqs (29) and (24)

$$k_B T(r,s^*) \equiv \theta(r,s^*) = \frac{M}{2}\, (r)^{2-2/\alpha}\, \xi^{2/\alpha} \left[\frac{Z(s^*)}{s^{*2}\gamma}\right]$$

one has

$$r = \left[\frac{2\, s^{*2}\gamma}{M\, Z(s^*)\, \xi^{2/\alpha}}\right]^{\frac{\alpha}{2\alpha-2}} \theta^{\frac{\alpha}{2\alpha-2}} \tag{54}$$

so that the integration over r in (52) can be converted [2] into the integration over θ. In the temperature range of interest, $\langle\sigma v\rangle$ has the form [2]

$$\langle\sigma v\rangle \cong \frac{\tau_0}{\theta^{2/3}} \exp(-B/\theta^{1/3}) \tag{55}$$

where $B \cong 19.33\ (\mathrm{keV})^{1/3}$ and $\sigma_0 \cong 4 \times 10^{-12}\ [\mathrm{cm^3/s}]\ [\mathrm{keV}]^{2/3}$. One can then use the saddle-point method [2] to integrate over θ in the fusion energy calculation. For our case, we find that the saddle-point temperature

$$\theta_{sd} = 0.56\ \mathrm{keV} \tag{56a}$$

with the corresponding value of the radius

$$r_{sd} \equiv \left[\frac{2\, s^{*2}\gamma}{M\, Z(s^*)\, \xi^{2/\alpha}}\right]^{\frac{\alpha}{2\alpha-2}} \theta_{sd}^{\frac{\alpha}{2\alpha-2}} \tag{56b}$$

This leads to

$$E_{FUSION} = \frac{4\pi}{3} r_{sd}^3 \frac{\rho_0^2 \langle\sigma v\rangle \text{ (at } \theta_{sd})W}{4} |t_0(r_{sd})|$$

$$\times \frac{9\alpha}{(1-\alpha)} \left[\frac{\pi(1-\alpha)}{13-\alpha}\right]^{1/2} \int_{s^*}^{\infty} ds\, G^2(s) \tag{57}$$

Using similar arguments, the value of time t appearing in the expression (53) can be taken to be equal to [2]

$$t \cong |t_0(r_{sd})| = \left(\frac{r_{sd}}{\xi}\right)^{1/\alpha} \tag{58}$$

After using our solutions in the region D, one can perform the integration over s for calculating both E_{FUSION} and E_{SHOCK}. We find that in our case of weak coalesced shocks,

$$E_{SHOCK} \cong 35 \left[\frac{E_{FUSION}}{E_{SHOCK}}\right]^3 \left(\frac{\rho_s}{\rho_0}\right)^2 \left(\frac{p_m}{p_0}\right)^{-2/\gamma} \text{ MJ} \tag{59}$$

as opposed to the expression (1) for a single shock. Thus, depending on the compression ratio $(p_m/p_0)^{1/\gamma}$, at the time of coalescing at $r = r_m$, coalesced weak shocks can be much more efficient.

We conclude this discussion by arguing that our approximate method of self-similar solutions can also be used for other complicated initial boundary conditions, to get an idea of the temperature and density profiles as a function of t and r, for spherical systems. Our solution in the region D has of course to be modified if there is absorption of α-particles after the reflection of the shock from the centre. For such a situation, our solution in the region D still gives the vital boundary values to be used in that calculation. Our method has recently been extended to combinations of weak shocks and a single strong shock, in spherical as well as cylindrical geometries [9].

REFERENCES

[1] NUCKOLLS, J., WOOD, T., THIESSEN, A., ZIMMERMAN, G., Nature (London) **236** (1972) 139.

[2] BRUECKNER, K.A., JORNA, S., Rev. Mod. Phys. **46** (1974) 325.

[3] GUDERLEY, G., Luftfahrtforsch. **19** (1942) 302.

[4] STANYUKOVICH, K.P., Unsteady Motion of Continuous Media, Pergamon, London (1960).

[5] GOLDMAN, E.B., Plasma Phys. **15** (1973) 289.

[6] FRALEY, G.S., LINNEBUR, E.J., MASON, R.J., MORSE, R.L., Phys. Fluids **17** (1974) 474.

[7] JHA, S.S., CHAVDA, L.K., Phys. Rev. **A15** (1977) 1289; **A16** (1977) 1743.

[8] ZELDOVICH, Y.B., RAIZER, Y.P., Physics of Shock Waves and High Temperature Hydrodynamic Phenomena, Academic Press, New York (1967), Vols 1 and 2.

[9] JHA, S.S., CHAVDA, L.K., Pramana **10** 4 (1978)

THEORY FOR STEADY STATE OF RF PLUGGING

T. WATANABE
Institute of Plasma Physics,
Nagoya University, Nagoya

H. HOJO, K. NISHIKAWA
Faculty of Science,
Hiroshima University, Hiroshima,
Japan

Abstract

THEORY FOR STEADY STATE OF RF PLUGGING.

General formulation of the steady state obtained by an r.f. confinement of plasma is presented within the framework of the collisionless two-fluid model. The effect of plasma flow is taken into account by introducing modified scalar and vector potentials, modification being due to the centrifugal force and the Coriolis force, respectively. A simple variation principle is derived to determine the steady state using the Clebsch representation for the modified vector potential. For the axisymmetric case, expressions or equations for quasilinear modification of the average quantities are derived in the weak-field approximation. Results of numerical calculation for the r.f. plugging applied to a line cusp are presented and discussed.

1. INTRODUCTION

There are two trends in the research on radio-frequency (r.f.) confinement of plasmas [1]. One is to confine the plasma by an r.f. field alone, i.e. to establish the entire confinement of the plasma by the ponderomotive force of the r.f. field of a cavity eigenmode. Experiments [2] and computer simulations [3, 4] were carried out in the region of radio-frequencies higher than the cut-off frequency, and the results indicate that a steady confinement is possible when the r.f. field pressure, $|\vec{E}|^2/8\pi$, becomes comparable to the plasma kinetic pressure, nT [4]. For a thermonuclear fusion plasma of density $10^{14}/cm^3$ and temperature 20 keV, the field amplitude required for the confinement becomes 3.8 mV/cm.

The second method is to confine the plasma by a combined use of an r.f. field and a static magnetic field in a linear machine. Here the r.f. field is used to stopper the particle outflux along the magnetic field. This method has the following two advantages over the first. First, in the presence of a magnetic field there exist several eigenmodes inside the plasma, and by adjusting the applied radio-frequency to one of the eigenmode frequencies one can produce a resonant enhancement of the r.f. field inside the plasma, thereby permitting an efficient confinement by a relatively weak applied r.f. field. Here, caution is required to avoid the cyclotron resonance which destroys the adiabaticity of the particle motion and hence the confinement efficiency. The second advantage is that because of the magnetic field we can strongly reduce the electron response to the r.f. field perpendicular to the line of force and thereby give it greater effect on the ions than on the electrons.

To make use of these advantages, a device has been invented at the Nagoya Institute of Plasma Physics to produce a plasma with an end thickness or radius of the order of the ion Larmor radius [5]. Such a narrow plasma sustains an eigenmode of frequency lying approximately in the middle of ω_c and $2\omega_c$, where ω_c is the ion-cyclotron frequency; the electric field is polarized to the direction nearly perpendicular to the magnetic field, so that the electron response to this field is more strongly reduced than the ion response. Although a proper description of this eigenmode requires a full kinetic treatment, the mode in a thin-sheet plasma will to a certain extent be simulated by the fundamental mode of the electrostatic ion-cyclotron wave in the absence of the finite ion Larmor radius effect [6, 7]. We take this view here and develop a theory for r.f. plugging of the plasma within the framework of the fluid description. We also neglect the dissipation effect since we are interested in the adiabatic r.f. confinement.

In Section 2, we first write out the basic equations in the collisionless two-fluid mode. In an open-ended system it is necessary to take into account the effect of plasma flow. To this end, we introduce modified scalar and vector potentials, modification being due to the centrifugal force and Coriolis force, respectively. Then in Section 3, we introduce the Clebsch representation for the modified vector potential and derive a variation principle to determine the steady state of the r.f. confinement self-consistently. In Section 4, we consider the situation where both the plasma and the r.f. field are axially symmetric and calculate the quasilinear modification of various average quantities by introducing a generalized oscillating displacement vector and by restricting ourselves to the case of a weak r.f. field. Results of a numerical analysis for the r.f. plugging applied to a line cusp are presented in Section 5, where we model the system by a stationary plasma of an infinite extension without flow. A brief discussion of the results is presented in the final section.

2. BASIC EQUATIONS

We consider a plasma that can be described by a collisionless two-fluid model. That is, we neglect all kinetic effects, such as the finite Larmor radius effect, the resonant wave-particle interaction effect, the collisional effect, etc. The basic equations are then given by the following set:

$$\dot{\vec{v}}_s + \vec{v}_s \cdot \vec{\nabla}\, \vec{v}_s = \frac{q_s}{m_s}[\vec{E} + \frac{1}{c}\vec{v}_s \times \vec{B}] - \frac{1}{m_s n_s}\vec{\nabla} P_s \qquad (1)$$

$$\dot{n}_s + \vec{\nabla} \cdot (n_s \vec{v}_s) = 0 \qquad (2)$$

$$\dot{\vec{E}} = c\, \vec{\nabla} \times \vec{B} - 4\pi \sum_s q_s\, n_s\, \vec{v}_s \qquad (3)$$

$$\vec{\nabla} \cdot \vec{E} = 4\pi \sum_s q_s\, n_s \qquad (4)$$

$$\dot{\vec{B}} = - c\, \vec{\nabla} \times \vec{E} \qquad (5)$$

$$\vec{\nabla} \cdot \vec{B} = 0 \qquad (6)$$

where the dot denotes the time derivative, the subscript s depicts the particle species (s = e for the electron and s = i for the ion), and the other notation is standard. We assume an equation of state which relates the plasma pressure P_s to the density n_s and other physical quantities, by which the above set of equations is closed.

The ponderomotive force of the r.f. field is given by

$$\vec{F}_s \equiv m_s\, \overline{\vec{v}_s \cdot \vec{\nabla}\vec{v}_s} - (q_s/c)\, \overline{\vec{v}_s \times \vec{B}} \qquad (7)$$

where the bar denotes the time average. The component of this force along the magnetic field (or more precisely the modified magnetic field to be defined later) acts to stopper the end outflux of the plasma. More generally, this force modifies the average plasma properties, such as the average density, the average flow velocity, the average magnetic field, the average electrostatic potential, etc., and the modification of these average quantities in turn changes the propagation characteristics of the r.f. field. Our task is then to solve the entire problem self-consistently.

As mentioned in Section 1, proper consideration of the effect of the plasma flow is necessary for theoretical analysis of the r.f. confinement in an open-ended system. Such an analysis can be substantially facilitated if one introduces the modified scalar and vector potentials, i.e. by noting that the inertia term of Eq. (1) can be divided into two parts as

$$\vec{v}_s \cdot \vec{\nabla}\, \vec{v}_s = \frac{1}{2}\vec{\nabla}|\vec{v}_s|^2 - \vec{v}_s \times (\vec{\nabla} \times \vec{v}_s) \qquad (8)$$

where the first term on the right-hand side corresponds to the centrifugal force and the second term to the Coriolis force. We define the modified scalar and vector potentials by the relations

$$\chi_s = \Psi + \frac{m_s}{2q_s} |\vec{v}_s|^2 \tag{9}$$

$$\vec{\alpha}_s = \vec{A} + \frac{m_s c}{q_s} \vec{v}_s \tag{10}$$

where Ψ and $\vec{A}$ are the usual electrostatic and vector potentials, respectively. Associated with the modified vector potential, we can introduce the modified magnetic field $\vec{\Omega}_s$ by the relation

$$\vec{\Omega}_s \equiv \vec{\nabla} \times \vec{\alpha}_s = \vec{B} + \frac{m_s c}{q_s} \vec{\nabla} \times \vec{v}_s \tag{11}$$

which obviously satisfies the condition

$$\vec{\nabla} \cdot \vec{\Omega}_s = 0 \tag{12}$$

Noting that $q_s \vec{\alpha}_s/c$ is the canonical momentum of the fluid, we can now rewrite the equation of motion (1) as

$$\dot{\vec{\alpha}}_s = -c \vec{\nabla} \chi_s + \vec{v}_s \times \vec{\Omega}_s - \frac{c}{q_s n_s} \vec{\nabla} P_s \tag{13}$$

Our basic equations are then given by Eqs (2) – (5), (12) and (13), supplemented by the relations (9) – (11).

3. VARIATION PRINCIPLE FOR STEADY STATE

Noting the relation (12), we first assume the existence of a modified flux function ψ_s defined by the relation

$$\vec{\Omega}_s \cdot \vec{\nabla} \psi_s = 0 \tag{14}$$

for each particle species. Then, assuming a periodicity in time, we show in this section that the steady-state solution of the above set of equations can be derived from a simple variation principle similar to that of Seligar and Whitham for the usual fluid [8].

To this end, we introduce the Clebsch representation for the modified vector potential, as is done for the velocity in Ref. [8]: instead of the three components of $\vec{\alpha}_s$, we use ψ_s, λ_s and μ_s, defined by the relation

$$\vec{\alpha}_s = \frac{1}{2\pi}\{\psi_s \vec{\nabla}\lambda_s + \vec{\nabla}\mu_s\} \tag{15}$$

as the independent variables. In this representation the modified magnetic field can be written as

$$\vec{\Omega}_s = \frac{1}{2\pi} \vec{\nabla}\psi_s \times \vec{\nabla}\lambda_s \tag{16}$$

from which we find that the equations

$$\psi_s = \text{const} \quad \text{and} \quad \lambda_s = \text{const} \tag{17}$$

determine the modified magnetic field line.

We now rewrite Eq. (13) using this representation. We first note the following relations:

$$\begin{aligned} \dot{\vec{\alpha}}_s &= \frac{1}{2\pi}\{\dot{\psi}_s\vec{\nabla}\lambda_s + \psi_s \vec{\nabla}\dot{\lambda}_s + \vec{\nabla}\dot{\mu}_s\} \\ &= \frac{1}{2\pi}\{\dot{\psi}_s\vec{\nabla}\lambda_s - \dot{\lambda}_s\vec{\nabla}\psi_s + \vec{\nabla}[\psi_s\dot{\lambda}_s + \dot{\mu}_s]\} \end{aligned} \tag{18}$$

$$\begin{aligned} \vec{v}_s \times \vec{\Omega}_s &= \frac{1}{2\pi}\vec{v}_s \times (\vec{\nabla}\psi_s \times \vec{\nabla}\lambda_s) \\ &= \frac{1}{2\pi}\{(\vec{v}_s\cdot\vec{\nabla}\lambda_s)\vec{\nabla}\psi_s - (\vec{v}_s\cdot\vec{\nabla}\psi_s)\vec{\nabla}\lambda_s\} \end{aligned} \tag{19}$$

$$n_s^{-1}\vec{\nabla}P_s = \vec{\nabla}\int^{P_s}\frac{dP_s}{n_s} - \left(\vec{\nabla}\int^{P_s}\frac{dP_s}{n_s}\right)_{P_s=\text{const}} \tag{20}$$

and assume an equation of state in the form

$$n_s = n_s(P_s, \psi_s, \lambda_s, t) \tag{21}$$

Substituting Eqs (18) – (21) into Eq.(13) then yields

$$\frac{1}{2\pi c}\{\dot{\psi}_s\vec{\nabla}\lambda_s - \dot{\lambda}_s\cdot\vec{\nabla}\psi_s + (\vec{v}_s\cdot\vec{\nabla}\psi_s)\vec{\nabla}\lambda_s - (\vec{v}_s\cdot\vec{\nabla}\lambda_s)\vec{\nabla}\psi_s\}$$

$$= -\vec{\nabla}\left\{\frac{\psi_s\dot{\lambda}_s + \dot{\mu}_s}{2\pi c} + \chi_s + \frac{1}{q_s}\int\frac{dP_s}{n_s}\right\}$$

$$+\left(\vec{\nabla}\psi_s\frac{\partial}{\partial\psi_s} + \vec{\nabla}\lambda_s\frac{\partial}{\partial\lambda_s}\right)\frac{1}{q_s}\int\frac{dP_s}{n_s(P_s, \psi_s, \lambda_s, t)} \tag{22}$$

In the first line on the right-hand side, we replace $\vec{\nabla}$ by

$$\vec{\nabla}\psi_s \quad \partial/\partial\psi_s + \vec{\nabla}\lambda_s \quad \partial/\partial\lambda_s + \vec{\nabla}\mu_s \quad \partial/\partial\mu_s$$

and then equate the coefficients of $\vec{\nabla}\psi_s$, $\vec{\nabla}\lambda_s$ and $\vec{\nabla}\mu_s$ separately. The coefficient of $\vec{\nabla}\mu_s$ can immediately be integrated to give

$$\frac{1}{2\pi c}(\psi_s\dot{\lambda}_s + \dot{\mu}_s) + \chi_s + \frac{1}{q_s}\int^{P_s}\frac{dP_s}{n_s} = 0 \tag{23}$$

where we absorbed the integration constant in the indefinite integral. Using Eq.(23), we can then write the coefficients of $\vec{\nabla}\lambda_s$ and $\vec{\nabla}\psi_s$ as follows:

$$\dot{\psi}_s + \vec{v}_s\cdot\vec{\nabla}\psi_s = \frac{2\pi c}{q_s}\frac{\partial}{\partial\lambda_s}\int\frac{dP_s}{n_s(P_s, \psi_s, \lambda_s, t)} \tag{24}$$

$$\dot{\lambda}_s + \vec{v}_s\cdot\vec{\nabla}\lambda_s = -\frac{2\pi c}{q_s}\frac{\partial}{\partial\psi_s}\int\frac{dP_s}{n_s(P_s, \psi_s, \lambda_s, t)} \tag{25}$$

Equations (23) – (25), combined with the continuity equation (2) and the Maxwell equations (3) and (4), determine the steady state in the present representation.

We now show that these sets of equations can be derivable from the variation principle:

$$\delta L[\psi_s, \lambda_s, \mu_s, \Psi, \vec{A}] = 0 \tag{26}$$

where

$$L = \int dt \int d^3\vec{x}\left[\frac{|\vec{E}|^2 - |\vec{B}|^2}{8\pi} + \sum_s P_s\right] \tag{27}$$

To show this, we use Eq.(23) to calculate δP_s, i.e. we first take the variation of Eq. (23) to obtain

$$\frac{1}{2\pi c}\left(\delta\psi_s \dot{\lambda}_s + \psi_s \delta \dot{\lambda}_s + \delta\dot{\mu}_s\right) + \delta\Psi + \frac{m_s}{q_s}\vec{v}_s\cdot\delta\vec{v}_s$$

$$+ \frac{1}{q_s n_s}\delta P_s + \frac{1}{q_s}\left(\delta\psi_s \frac{\partial}{\partial\psi_s} + \delta\lambda_s \frac{\partial}{\partial\lambda_s}\right)\int \frac{dP_s}{n_s} = 0 \tag{28}$$

where we used the definition (9). Using Eqs (10) and (15), we have

$$(m_s c/q_s)\,\delta\vec{v}_s = -\delta\vec{A} + \frac{1}{2\pi}(\delta\psi_s \vec{\nabla}\lambda_s + \psi_s \vec{\nabla}\delta\lambda_s + \vec{\nabla}\delta\mu_s) \tag{29}$$

which we substitute into Eq. (28). Using also the relations

$$n_s(\psi_s \delta\dot{\lambda}_s + \delta\dot{\mu}_s) = \frac{\partial}{\partial t}(n_s\psi_s\delta\lambda_s + n_s\delta\mu_s) - \delta\lambda_s\frac{\partial}{\partial t}(n_s\psi_s) - \delta\mu_s\dot{n}_s$$

$$n_s\vec{v}_s\cdot(\psi_s\vec{\nabla}\delta\lambda_s + \vec{\nabla}\delta\mu_s) = \vec{\nabla}\cdot(n_s\psi_s\vec{v}_s\delta\lambda_s + n_s\vec{v}_s\delta\mu_s) - \delta\lambda_s\vec{\nabla}\cdot(n_s\psi_s\vec{v}_s) - \delta\mu_s\vec{\nabla}\cdot(n_s\vec{v}_s)$$

we then obtain the following expression for δP_s:

$$\begin{aligned}
\delta P_s = &- \delta\psi_s n_s\left\{\frac{\partial}{\partial\psi_s}\int\frac{dP_s}{n_s} + \frac{1}{2\pi c}(\dot{\lambda}_s + \vec{v}_s\cdot\vec{\nabla}\lambda_s)\right\} \\
&+ \delta\lambda_s q_s\left\{-\frac{1}{q_s}\frac{\partial}{\partial\lambda_s}\int\frac{dP_s}{n_s} + \frac{1}{2\pi c}\left[\frac{\partial}{\partial t}(n_s\psi_s) + \vec{\nabla}(n_s\psi_s\vec{v}_s)\right]\right\} \\
&+ \delta\mu_s\frac{q_s}{2\pi c}[\dot{n}_s + \vec{\nabla}\cdot(n_s\vec{v}_s)] \\
&- \delta\Psi\, q_s n_s + \delta\vec{A}\cdot q_s n_s\vec{v}_s/c \\
&- \frac{\partial}{\partial t}\left\{\frac{n_s}{2\pi c}(\psi_s\delta\lambda_s + \delta\mu_s)\right\} - \left\{\vec{\nabla}\frac{n_s\vec{v}_s}{2\pi c}(\psi_s\delta\lambda_s + \delta\mu_s)\right\}
\end{aligned} \tag{30}$$

The last line in Eq.(30) vanishes when we integrate δP_s over the space and the time assuming that $\delta\lambda_s$ and $\delta\mu_s$ vanish on the surface and noting the periodicity in time.

The variation of the electromagnetic field pressure can be calculated as follows:

$$\delta\left(\frac{|\vec{E}|^2 - |\vec{B}|^2}{8\pi}\right) = \frac{\vec{E}}{4\pi}\cdot\left(-\vec{\nabla}\,\delta\Psi - \frac{1}{c}\,\delta\dot{\vec{A}}\right)$$
$$- \frac{\vec{B}}{4\pi}\cdot(\vec{\nabla}\times\delta\vec{A})$$
$$= \delta\Psi\,\frac{\vec{\nabla}\cdot\vec{E}}{4\pi} + \delta\vec{A}\cdot\left(\frac{1}{4\pi c}\,\dot{\vec{E}} - \frac{1}{4\pi}\,\vec{\nabla}\times\vec{B}\right)$$
$$- \frac{\partial}{\partial t}\left(\frac{1}{4\pi c}\,\vec{E}\cdot\delta\vec{A}\right) - \vec{\nabla}\cdot\left(\frac{\vec{E}\,\delta\Psi + \delta\vec{A}\times\vec{B}}{4\pi}\right) \tag{31}$$

Again the last line gives no contribution to the space-time integral under appropriate boundary conditions.

Substituting Eqs (30) and (31) into Eqs (26) and (27) then yields

$$\delta L = \int dt \int d^3\vec{x}\,\Bigg\{\sum_s\Bigg[\delta\psi_s\,\frac{q_s n_s}{2\pi c}\Bigg\{-\frac{2\pi c}{q_s}\,\frac{\partial}{\partial\psi_s}\int\frac{dP_s}{n_s}$$
$$- \dot{\lambda}_s - \vec{v}_s\cdot\vec{\nabla}\lambda_s\Bigg\}$$
$$+ \delta\lambda_s\,\frac{q_s}{2\pi c}\Bigg\{-\frac{2\pi c}{q_s}\,n_s\,\frac{\partial}{\partial\lambda_s}\int\frac{dP_s}{n_s}$$
$$+ \frac{\partial}{\partial t}(n_s\psi_s) + \vec{\nabla}(n_s\psi_s\vec{v}_s)\Bigg\}$$
$$+ \delta\mu_s\,\frac{q_s}{2\pi c}\,\{\dot{n}_s + \vec{\nabla}\cdot(n_s\vec{v}_s)\}\Bigg]$$
$$+ \delta\Psi\left(-\sum_s q_s n_s + \frac{1}{4\pi}\,\vec{\nabla}\cdot\vec{E}\right)$$
$$+ \delta\vec{A}\left(\frac{1}{c}\sum_s q_s n_s\vec{v}_s + \frac{1}{4\pi c}\,\dot{\vec{E}} - \frac{1}{4\pi}\,\vec{\nabla}\times\vec{B}\right)\Bigg\} \tag{32}$$

Obviously, $\delta L/\delta \mu_s$, $\delta L/\delta \Psi$ and $\delta L/\delta \vec{A}$ yield Eqs (2), (3) and (4), respectively, while Eq. (25) is derived from $\delta L/\delta \psi_s$. Using Eq. (2) and $\delta L/\delta \lambda_s$ then gives Eq. (24). This completes our proof.

4. QUASILINEAR THEORY FOR THE AXISYMMETRIC CASE

To obtain more explicit analytical results, we consider in this section the situation where both the plasma and the r.f. field are axially symmetric and the amplitude of the r.f. field is sufficiently weak. In particular, we derive explicit expressions or equations for the quasilinear modification of various time-averaged quantities.

Before considering the weak-field limit, we first derive some useful formulas valid for a general axisymmetric system. By axisymmetry, we mean that all quantities are independent of ϕ in the cylindrical co-ordinate representation, (r,ϕ,z). Then the divergence-free vector, such as $\vec{\Omega}$, can in general be written in the form $\vec{\nabla} L \times \vec{\nabla}\phi + M\vec{\nabla}\phi$, where L and M are certain functions of r, z and t. In particular, we have from Eq.(14):

$$\vec{\Omega} = \frac{1}{2\pi} \vec{\nabla}\psi \times \vec{\nabla}\phi + r\Omega_\phi \vec{\nabla}\phi \tag{33}$$

where we suppressed the suffix s designating the particle species. Comparing Eq. (33) with Eq. (11), we find

$$\psi = 2\pi r \left[A_\phi + \frac{mc}{q} v_\phi\right] \tag{34}$$

The Clebsch representation for the modified vector potential can be rewritten in a more physical way as

$$\vec{\alpha} = \frac{1}{2\pi} \psi\vec{\nabla}\phi + h(r,z)\vec{\nabla}\psi + \vec{\nabla}u(r,z) \tag{35}$$

where the first term denotes the ϕ-component, the second term the component across the modified magnetic surface and the last term the 'poloidal' component. Comparing Eq. (35) with Eq. (15), we have

$$\lambda = \phi - 2\pi h \tag{36}$$

$$\mu = 2\pi (u + h\psi) \tag{37}$$

Equation (36) implies that all quantities are independent of λ. Using these relations, we can immediately rewrite the equations of motion (23) – (25) as

$$\dot{u} + h\dot{\psi} = -c\chi - \frac{c}{q}\int^P \frac{dP}{n} \tag{38}$$

$$\dot{\psi} + \vec{v}\cdot\vec{\nabla}\psi = 0 \tag{39}$$

$$\dot{h} + \vec{v}\cdot\vec{\nabla}h = \frac{V_\phi}{2\pi r} + \frac{c}{q}\frac{\partial}{\partial\psi}\int^P \frac{dP}{n} \tag{40}$$

In many cases of interest, the equation of state is given by the form $P = P(n, \psi, t)$ rather than by the form of Eq. (21). One can then write the pressure contributions in the form

$$\int^P \frac{dP}{n} = \int^n \frac{dn}{n}\frac{\partial P}{\partial n}$$

$$\frac{\partial}{\partial\psi}\int^P \frac{dP}{n} = \left(\frac{\partial}{\partial\psi}\int^n \frac{dn}{n}\frac{\partial P}{\partial n}\right)_P$$

$$= \left(\frac{\partial}{\partial\psi}\int^n \frac{dn}{n}\frac{\partial P}{\partial n}\right)_n + \frac{1}{n}\frac{\partial P}{\partial n}\left(\frac{\partial n}{\partial\psi}\right)_P$$

$$= \left(\frac{\partial}{\partial\psi}\int^n \frac{dn}{n}\frac{\partial P}{\partial n}\right)_n - \frac{1}{n}\left(\frac{\partial P}{\partial\psi}\right)_n$$

We now consider the case of a weak field. As in Ref. [9], we divide the variables into the static parts denoted by a bar and the oscillating parts denoted by the suffix 1, and calculate the latter in the quasilinear approximation. To simplify the argument, we assume that the time dependence of the system arises solely from the oscillating field and that the pressure P is a function of the density alone, $P = P(n)$. We can then write

$$\int^P \frac{dP}{n} = H(n) + f(\psi) \tag{41}$$

where $f(\psi)$ is the integration constant which, by assumption, does not explicitly depend on time. The static and the oscillating components of the equations of motion and continuity are then written as

$$\bar{\vec{v}}\cdot\vec{\nabla}\bar{\psi} + \overline{\vec{v}_1\cdot\vec{\nabla}\psi_1} = 0 \tag{42}$$

$$\bar{\vec{v}}\cdot\vec{\nabla}\bar{\lambda} + \overline{\vec{v}_1\cdot\vec{\nabla}\lambda_1} = -\frac{2\pi c}{q}\,\overline{f'(\psi)} \tag{43}$$

$$\frac{1}{2\pi c}\,\overline{\psi_1\dot{\lambda}_1} + \bar{\chi} + \frac{1}{q}(\bar{H} + \bar{f}) = 0 \tag{44}$$

$$\vec{\nabla}\cdot(\bar{n}\bar{\vec{v}} + \overline{n_1\vec{v}_1}) = 0 \tag{45}$$

$$\dot{\psi}_1 + \bar{\vec{v}}\cdot\vec{\nabla}\psi_1 + \vec{v}_1\cdot\vec{\nabla}\bar{\psi} = 0 \tag{46}$$

$$\dot{\lambda}_1 + \bar{\vec{v}}\cdot\vec{\nabla}\lambda_1 + \vec{v}_1\cdot\vec{\nabla}\bar{\lambda} = -\frac{2\pi c}{q}\,f''(\bar{\psi})\,\psi_1 \tag{47}$$

$$\dot{n}_1 + \vec{\nabla}\cdot(\bar{n}\,\vec{v}_1 + n_1\,\bar{\vec{v}}) = 0 \tag{48}$$

The following calculations can be considerably simplified if one introduces a generalized oscillating-displacement vector $\vec{Q}$ whose three components are determined by the following three equations:

$$\psi_1 = -\vec{Q}\cdot\vec{\nabla}\bar{\psi}\ , \qquad \lambda_1 = -\vec{Q}\cdot\vec{\nabla}\bar{\lambda}\ , \qquad n_1 = -\vec{\nabla}\,(\bar{n}\vec{Q}) \tag{49}$$

or alternatively, using Eqs (46) – (48),

$$\bar{\vec{v}}\cdot\vec{\nabla}\psi_1 + \vec{v}_1\cdot\vec{\nabla}\bar{\psi} = \dot{\vec{Q}}\cdot\vec{\nabla}\bar{\psi} \tag{50}$$

$$\bar{\vec{v}}\cdot\vec{\nabla}\lambda_1 + \vec{v}_1\cdot\vec{\nabla}\bar{\lambda} + \frac{2\pi c}{q}\,f''(\bar{\psi})\,\psi_1 = \dot{\vec{Q}}\cdot\vec{\nabla}\bar{\lambda} \tag{51}$$

$$\vec{\nabla}\cdot(\bar{n}\,\vec{v}_1 + n_1\bar{\vec{v}}) = \vec{\nabla}\cdot(\bar{n}\,\dot{\vec{Q}}) \tag{52}$$

An explicit expression for $\dot{\vec{Q}}$ obtained by the first-order perturbational calculation is derived in Appendix I (see Eq. (A–8)). There we also show the following relations which are valid correct to the second order with respect to the oscillating field amplitude:

$$\bar{n}\,\vec{v}_1 + n_1\bar{\vec{v}} = \bar{n}\dot{\vec{Q}} + \vec{\nabla}\times(\vec{Q}\times\bar{n}\bar{\vec{v}}) \tag{53}$$

$$\bar{\vec{v}}\times\vec{\Omega}_1 + \vec{v}_1\times\bar{\vec{\Omega}} = \dot{\vec{Q}}\times\bar{\vec{\Omega}} - \frac{c}{q}\,\vec{\nabla}\,[f'(\bar{\psi})\,\psi_1] \tag{54}$$

Note that these relations are valid even for non-axisymmetric systems.

We now calculate the quasilinear modification of various average quantities. We first calculate the fluid variables in terms of the oscillating variables, the average flux function $\overline{\psi}$, and the electromagnetic variables; we then derive an equation to determine $\overline{\psi}$ from the self-consistency with Eq. (34); we finally derive the equations to determine the average electromagnetic variables.

We start from the static force balance along $\vec{\overline{\Omega}}$, i.e. Eq. (44), which determines the average density. Using the first two equations of (49), we have from Eq. (44)

$$\frac{1}{2c}\,\overline{(\vec{Q}\times\dot{\vec{Q}})}\cdot\vec{\overline{\Omega}} + \overline{\chi} + \frac{1}{q}\,(\overline{H} + \overline{f}) = 0 \tag{55}$$

We calculate $\overline{H}$ and $\overline{f}$ correct to the second order as

$$\overline{H} + \overline{f} = H(\overline{n}) + f(\overline{\psi}) + \frac{1}{2}\,[H''(\overline{n})\overline{n_1{}^2} + f''(\overline{\psi})\overline{\psi_1{}^2}\,] \tag{56}$$

while $\overline{\chi}$ can be written, using its definition (9), as

$$\overline{\chi} = \overline{\Psi} + \frac{m}{2q}\,(\,|\vec{\overline{v}}|^2 + \overline{|\vec{v}_1|^2}\,) \tag{57}$$

Substitution of Eqs (56) and (57) into Eq. (55) gives an equation to determine $\overline{n}$ in terms of the other quantities. In the special case of isothermal variation, i.e. $P = nT$ with $T = \text{const}$, we obtain

$$T\log\frac{\overline{n}}{N(\overline{\psi})} = -q\overline{\Psi} - \frac{1}{2}\,[\frac{q}{c}\,(\vec{Q}\times\dot{\vec{Q}})\cdot\vec{\overline{\Omega}} + m(|\vec{\overline{v}}|^2 + \overline{|\vec{v}_1|^2})\,]$$
$$- \overline{n_1{}^2}\,T/\overline{n}^2 + f''(\overline{\psi})\overline{\psi_1{}^2} \tag{58}$$

where $N(\overline{\psi})$ is an arbitrary function of $\overline{\psi}$, and $\vec{\overline{v}}$, $\overline{\Psi}$ and $\vec{\overline{\Omega}}$ are calculated below.

We next calculate the average flux density $\overline{n\vec{v}}$. Because of Eq. (45) and the axial symmetry, we can write

$$\overline{n\vec{v}} = \vec{\nabla}\Gamma\times\vec{\nabla}\phi + G\,\vec{\nabla}\phi \tag{59}$$

We first take its component across the modified magnetic surface:

$$\overline{n\vec{v}}\cdot\vec{\nabla}\overline{\psi} = (\vec{\nabla}\Gamma\times\vec{\nabla}\phi)\cdot\vec{\nabla}\,\overline{\psi} = \vec{\nabla}\Gamma\cdot(\vec{\nabla}\phi\times\vec{\nabla}\overline{\psi}) \tag{60}$$

Noting the relation

$$\vec{\nabla}\cdot\overline{(n\vec{v}\psi)} = -\overline{(\frac{\partial}{\partial t}n\psi)} = 0 \tag{61}$$

which follows from Eqs (2) and (39), and using Eqs (45) and (53), we can rewrite the left-hand side of Eq. (60) as

$$\overline{n\vec{v}}\cdot\vec{\nabla}\overline{\psi} = -\vec{\nabla}\cdot\overline{[(n\vec{v})_1\psi_1]}$$

$$= -\vec{\nabla}\cdot\overline{[\psi_1\{n\dot{\vec{Q}} + \vec{\nabla}\times(\vec{Q}\times\overline{n}\,\overline{\vec{v}})\}]} \tag{62}$$

As shown in Appendix II, the right-hand side can be rewritten correct to the second order as

$$-\vec{\nabla}\cdot\overline{[\psi_1\{\overline{n}\dot{\vec{Q}} + \vec{\nabla}\times(\vec{Q}\times\overline{n}\,\overline{\vec{v}})\}]}$$

$$= \frac{1}{2}(\vec{\nabla}\overline{\psi}\times\vec{\nabla}\phi)\cdot\vec{\nabla}[r^2\overline{n}\,\overline{(\dot{\vec{Q}}\times\vec{Q})}\cdot\vec{\nabla}\phi]$$

$$-(\vec{\nabla}\psi_1\times\vec{\nabla}\phi)\cdot\vec{\nabla}(\vec{Q}\cdot\vec{\nabla}\Gamma) \tag{63}$$

We calculate the last term in Eq. (63) by approximating Γ by its lowest-order expression, $\Gamma_0(\overline{\psi})$. Then we have, using the first of Eqs (49),

$$\vec{\nabla}(\vec{Q}\cdot\vec{\nabla}\Gamma) \doteq \vec{\nabla}[\vec{Q}\cdot\vec{\nabla}\overline{\psi}\,\Gamma_0'(\overline{\psi})]$$

$$= -\vec{\nabla}[\Gamma_0'(\overline{\psi})\psi_1]$$

$$= -\Gamma_0''(\overline{\psi})(\vec{\nabla}\overline{\psi})\psi_1 - \Gamma_0'(\overline{\psi})\vec{\nabla}\psi_1 \tag{64}$$

Substitution of Eq. (64) into the last term of Eq. (63) yields

$$-\overline{(\vec{\nabla}\psi_1\times\vec{\nabla}\phi)\cdot\vec{\nabla}(\vec{Q}\cdot\vec{\nabla}\Gamma)}$$

$$= -\Gamma_0''(\overline{\psi})\vec{\nabla}\overline{\psi}\cdot[\vec{\nabla}(\frac{1}{2}\overline{\psi_1^2})\times\vec{\nabla}\phi]$$

$$= (\vec{\nabla}\overline{\psi}\times\vec{\nabla}\phi)\cdot\vec{\nabla}(\frac{1}{2}\Gamma_0''(\overline{\psi})\overline{\psi_1^2}) \tag{65}$$

Using Eq. (65) in Eq. (63) and substituting the result into Eq. (62) and then into Eq. (60), we finally obtain the 'poloidal' flux Γ as

$$\Gamma = \Gamma_0(\overline{\psi}) + \frac{1}{2}\Gamma_0''(\overline{\psi})\overline{\psi_1{}^2} - \frac{1}{2}r^2\,\overline{n}\,(\overline{\dot{\vec{Q}}\times\vec{Q}})\cdot\vec{\nabla}\phi \tag{66}$$

This determines the r, z-components of the average flux. If we calculate $\overline{n\vec{v}\cdot\vec{\nabla}\lambda}$ in a similar way, we can obtain the 'toroidal' flux G in the form (see Appendix II):

$$G = 2\pi r^2\vec{\nabla}\phi\cdot[\{\Gamma_0'(\overline{\psi}) + \frac{1}{2}\Gamma_0'''(\overline{\psi})\overline{\psi_1{}^2}\}\vec{\Omega} + \Gamma_0''(\overline{\psi})\overline{\psi_1\vec{\Omega}_1}]$$
$$- \frac{2\pi r^2 c}{q}\overline{n}\{f'(\overline{\psi}) + \frac{\overline{n_1\psi_1}}{\overline{n}}f''(\overline{\psi}) + \frac{1}{2}f'''(\overline{\psi})\overline{\psi_1{}^2}\} \tag{67}$$

On the other hand, G can be written as

$$G\vec{\nabla}\phi = \overline{n}\;\overline{v}_\phi + \overline{n_1 v_{1\phi}} \tag{68}$$

where $\overline{v_\phi}$ is expressed in terms of $\overline{\psi}$ and $\overline{A_\phi}$ by Eq. (34). If we also note the relation

$$\overline{\Omega}_\phi = \overline{B}_\phi - \frac{mc}{q}\Big[\frac{\partial}{\partial z}\{\frac{1}{r}\frac{\partial\Gamma}{\partial z} + \overline{n_1 v_{1r}})\frac{1}{\overline{n}}\}$$
$$+ \frac{\partial}{\partial r}\{\frac{1}{r}\frac{\partial\Gamma}{\partial r} - \overline{n_1 v_{1z}})\frac{1}{\overline{n}}\}\Big] \tag{69}$$

which follows from Eqs (11) and (59), then Eq. (67) can be regarded as an equation which determines the average flux function $\overline{\psi}$ in terms of the oscillating variables and the electromagnetic variables.

Our final task is to determine the average electromagnetic variables by Maxwell's equations. Taking the time average of Maxwell's equations, we immediately get

$$\nabla^2\overline{\Psi} = -4\pi\sum_s q_s\,\overline{n}_s \tag{70}$$

$$\overline{B}_\phi = \frac{4\pi}{cr}\{\sum_s q_s\,\Gamma_s + \frac{I}{2\pi}\} \tag{71}$$

$$r\vec{\nabla}\cdot[\frac{1}{r^2}\vec{\nabla}(r\overline{A}_\phi)] = -\frac{4\pi}{cr}\sum_s q_s G_s \tag{72}$$

where I is the external current that produces the static magnetic field in the absence of the r.f. field. The right-hand sides of these equations are known functions of $\overline{\Psi}$, $\overline{B}_\phi$, $\overline{A}_\phi$ and the oscillating variables, so by solving these equations we can entirely determine the quasilinear modification in the axisymmetric system in terms of the oscillating variables.

5. NUMERICAL ANALYSIS FOR RF PLUGGING AT LINE CUSP

In this section, we show some preliminary results of numerical analysis for the steady state obtained by an r.f. field applied to a line cusp.

In general, the steady state in an open-ended system is obtained by a balance between the particle supply by a plasma source and the particle loss due to the end outflux. The object of the r.f. plugging is to reduce this end outflux by the ponderomotive force, and thereby to increase the plasma density in the interior region of the machine for the given plasma source. The reduction of the end outflux can be achieved either by the density depression at the end or by the slowing-down of the outward fluid velocity. Here, we are interested in the former effect, i.e. the effect of the r.f. field on the density depression at the end of the machine. We study this effect by neglecting the plasma flow velocity along the magnetic field. Note that our primary interest is in the possibility of exciting a strong r.f. field inside the plasma at the line cusp due to the resonance with an eigenmode. Using this resonance, we expect an efficient density depression at the end, and hence the resulting suppression of the end outflux, even by a relatively weak external r.f. field (the r.f. field pressure is less than the plasma kinetic pressure).

In the experiment of r.f. plugging at the line cusp [7], an r.f. field is excited by condenser plates. We therefore assume that the r.f. field $\vec{E}_1$ is electrostatic,

$$\vec{E}_1 = - \vec{\nabla} \Psi_1 \tag{73}$$

We consider the frequency in the region $\omega_c < \omega < 2\omega_c$. In this frequency range, the electron motion across the magnetic field is neglected, whereas along the magnetic field we assume that the effective phase velocity $\omega/k_{\parallel}$ is less than the electron thermal speed, so that the electrons obey the Boltzmann distribution. Neglecting the ponderomotive force along the magnetic field also, we then have

$$n_e = N_e(\psi_e) \exp [e\Psi/T_e] \tag{74}$$

To simplify the calculation, we entirely neglect the ion thermal effect on the wave propagation. For the motion along the magnetic field, this can be justified if

$$|(\omega - n\omega_c)/k_{\parallel} v_T| \gg 1 \qquad (n=0,1,2,\cdots)$$

where v_T is the ion thermal speed. For the motion across the magnetic field, we must, strictly speaking, take into account the finite ion Larmor radius effect, since the experimental scale length across the magnetic field at the line cusp is of the order of the ion Larmor radius. However, an eigenmode analysis for a sheet plasma including the ion Larmor motion shows that the frequency of the fundamental mode is close to that of the electrostatic ion-cyclotron wave with no finite Larmor radius effect [7].

Figure 1 shows the geometry of the model. The plasma thickness in the absence of the r.f. field is taken to be the ion Larmor radius ρ, and the condenser plates are placed along the magnetic field line at a distance 3ρ from the centre of the plasma. We choose the x-axis across the plasma sheet and the z-axis along the magnetic field, and assume uniformity in the y-direction. The temperature is assumed to be constant for both the electron and the ion.

Choosing the electrostatic potential in the form

$$\Psi = \overline{\Psi} + \Psi_1 \cos \omega t \tag{75}$$

we have from Eq.(74)

$$n_e = N_e(\overline{\psi}_e)\exp(e\overline{\Psi}/T_e)\{I_0(\Phi) + 2I_1(\Phi)\cos\omega t\} \tag{76}$$

where $\Phi = e\Psi_1/T_e$, I_0 and I_1 are the modified Bessel functions of zeroth and first order, respectively, and we have neglected the higher harmonics. Note that although we consider the situation where the electric field pressure is less than the plasma pressure, Φ can become greater than unity, whence we keep the higher-order terms with respect to Φ. We determine $\overline{\Psi}$ from the charge neutrality condition, $\overline{n_e} = \overline{n_i}$, where $\overline{n_i}$ is calculated by the static ion force balance along the magnetic field:

$$\overline{n_i} = N_i(\overline{\psi}_i)\exp\left[-\frac{e\overline{\Psi}}{T_i} - \frac{T_e^2 (d\Phi/dx)^2}{4\, T_i m_i (\omega^2 - \omega_c^2)}\right] \tag{77}$$

where we used the following expression for the ponderomotive force on the ions [9]:

$$\frac{m_i}{2}\overline{v_{i_1}^2} + \frac{1}{2c}(Q_{i_1} \times \dot{Q}_{i_1})\cdot B_0 = -\frac{e^2}{4m_i}\frac{(d\Psi_1/dx)^2}{\omega^2 - \omega_c^2} \tag{78}$$

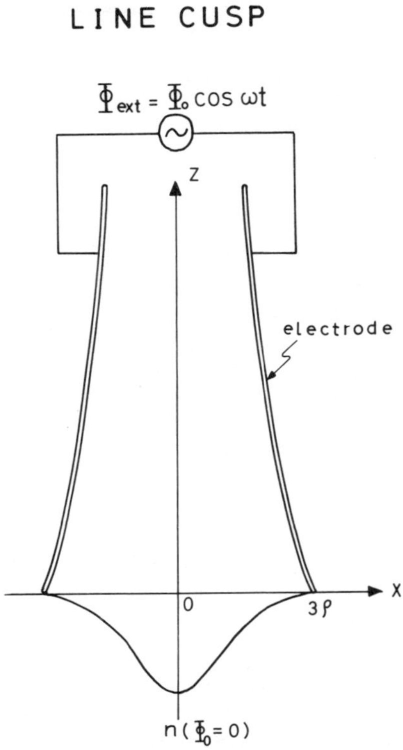

FIG.1. Geometry used in the numerical analysis of r.f. plugging at a line cusp.

Noting that $\overline{\psi_e} = \overline{\psi_i} \equiv \overline{\psi}$ because of the absence of the plasma flow, we then get

$$\overline{n_e} = \overline{n_i} = N(\overline{\psi})\,[I_0(\Phi)]^{\eta} \exp\left[-\frac{T_e\eta\,(d\Phi/dx)^2}{4\,m_i(\omega^2 - \omega_c^2)}\right] \quad (79)$$

where $\eta = T_e/(T_e + T_i)$ and $N = N_e^{\eta} N_i^{\eta-1}$. Our variation principle can then be written as

$$\delta L = \delta \int dx \left\{ \overline{n}_e (T_e + T_i) + \frac{1}{16\pi}\left(\frac{d\Phi}{dx}\right)^2 \frac{T_e^2}{e^2} \right\} = 0 \quad (80)$$

The Euler equation of this variation principle becomes a second-order non-linear differential equation for Φ:

$$-\frac{1}{8\pi}\frac{d^2\Phi}{dx^2} + \frac{e^2}{2m_i}\frac{d}{dx}\left[\frac{I_0(\Phi)^{\eta}}{\omega^2-\omega_c^2}\, N \exp\left\{-\frac{T_e\eta\,(d\Phi/dx)^2}{4m_i(\omega^2-\omega_c^2)}\right\}\frac{d\Phi}{dx}\right]$$

$$+\frac{Ne^2}{T_e}[I_0(\Phi)]^{\eta-1}\, I_1(\Phi) \exp\left\{-\frac{T_e\eta\,(d\Phi/dx)^2}{4m_i(\omega^2-\omega_c^2)}\right\} = 0 \quad (81)$$

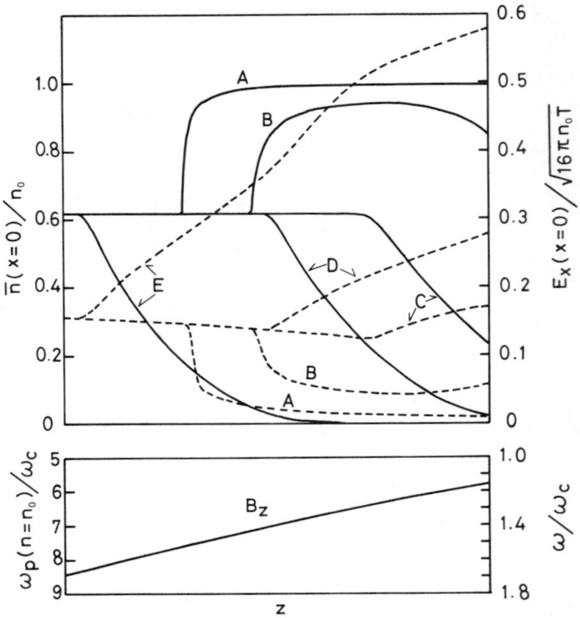

FIG.2. The upper trace shows the normalized density (solid line) and the normalized r.f. field (dotted line) at the centre (x = 0) of the plasma sheet plotted against z for various values of the applied potential Φ_0: A for $\Phi_0 = 1$, B for $\Phi_0 = 3$, C for $\Phi_0 = 5$, D for $\Phi_0 = 10$ and E for $\Phi_0 = 20$. The lower trace shows the magnetic field strength in terms of $\omega/\omega_c(z)$ and $\omega_p(n = n_0)/\omega_c$.

The coefficient of $d^2\Phi/dx^2$ of this equation is given by

$$C(x) \equiv \omega^2 - \omega_c^2 - \omega_p^2(x)\left[1 - \frac{T_e\eta\,(d\Phi/dx)^2}{2\,m_i\,(\omega^2-\omega_c^2)}\right] \quad (82)$$

where $\omega_p(x)$ is the local ion plasma frequency:

$$\omega_p^2(x) = \frac{4\pi N_e^2}{m_i}\exp\left[-\frac{T_e\eta\,(d\Phi/dx)^2}{4\,m_i\,(\omega^2-\omega_c^2)}\right] I_0(\Phi)^\eta \quad (83)$$

At the point where C(x) vanishes, the differential equation becomes singular and one has to take into account the thermal effects, such as the higher-order dispersion effect due to the finite ion Larmor radius, the dissipation effect due to the wave-particle and the collisional processes, etc. For the case of weak r.f. field, the vanishing of C(x) takes place at the lower hybrid resonance point,

$\omega^2 = \omega_c^2 + \omega_p^2(x)$. Here, instead of taking into account the ion thermal effect, we introduce a phenomenological dissipation effect by replacing C(x) by

$$\tilde{C}(x) = [C^2(x) + \nu^2]/C(x) \tag{84}$$

By varying the value of ν between 0.1 ω_c and ω_c, we found that the overall structure of the steady state is insensitive to the value of ν.

Choosing $N(\bar{\psi})$ in the form

$$N(\bar{\psi}) = n_0 \exp[-x^2/2\rho^2] \tag{85}$$

and taking $T_e = T_i$, we have numerically solved Eq.(81) by the shooting method. The results are shown in Figs 2 and 3. The upper trace of Fig.2 shows the normalized density, $\bar{n}/n_0$, and the normalized field amplitude, $|E_1|/\sqrt{16\pi n_0 T}$, at the centre (x = 0) of the plasma sheet plotted against z, where the frequency ω normalized by the local cyclotron frequency $\omega_c(z)$ is shown in the lower trace. Note that ω/ω_c increases as we go to the interior of the plasma. When the applied r.f. potential is small, i.e. $\Phi_0 \lesssim 1$, we can observe little effect of r.f. plugging: $\bar{n}(x=0)/n_0 \cong 1$ in the outer region where ω/ω_c is close to unity. In this region, the electric field is even smaller than the vacuum field because of the plasma shielding effect. Near the point at which $\omega = 1.4\ \omega_c$, where the applied field resonates with the fundamental mode of the electrostatic ion cyclotron wave, the electric field inside the plasma is enhanced over the vacuum field. In the region where $\omega/\omega_c \gtrsim 1.49$, we find a transition of the r.f. field from the fundamental mode to higher wavenumber modes. Although in this region we can observe a substantial density depression ($\bar{n}(x=0)/n_0 \sim 0.62$) due to the ponderomotive force of the higher wavenumber modes, the present treatment, which neglects the ion thermal effect, is not justifiable for these short-wavelength modes. Physically, we can expect a substantial reduction of the ponderomotive force effect due to the ion Larmor motion if the wavelength becomes less than the Larmor radius.

For a stronger field, i.e. for $\Phi_0 \gtrsim 5$, we can observe strong enhancement of the r.f. field and the resulting density depression in the outer region of the plasma. This result is physically expected since the density depression lowers the resonance frequency of the electrostatic ion cyclotron wave, as can be seen from the dispersion relation,

$$(\omega/\omega_c)^2 = 1 + k_\perp^2\rho^2/[1 + k_\perp^2\lambda_D^2]$$

where $k_\perp$ is the perpendicular wavenumber and λ_D is the electron Debye length. In the inner region, on the other hand, the density depression stays nearly constant ($\bar{n}/n_0 \sim 0.62$) independently of the applied potential. The constancy of the density depression in this region implies a balance between the ponderomotive force effect and the electron shielding effect, represented by $I_0(\Phi)^\eta$, in Eq. (79).

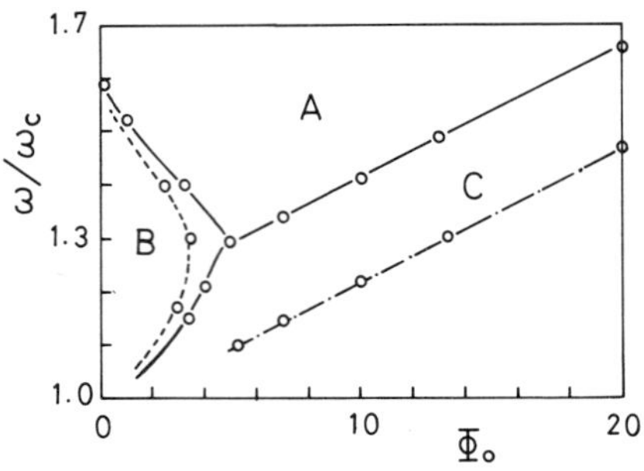

FIG.3. The normalized central density, $\bar{n}(x=0)/n_0$, for various values of Φ_0 and ω/ω_c. In region A, $\bar{n}/n_0 \sim 0.62$; in region B, $\bar{n}/n_0 > 0.62$, and in region C, $\bar{n}/n_0 < 0.62$.
- - - - - - - - : an equidensity line for $\bar{n}/n_0 = 0.9$
— · — · — : $\bar{n}/n_0 = 0.1$

In Fig.3, we plot the normalized central density, $\bar{n}(x=0)/n_0$, for various values of Φ_0 and ω/ω_c. Here in region A, the balance discussed above between the ponderomotive force effect and the electron shielding effect is obtained, and hence the density depression is kept nearly constant ($\bar{n}/n_0 \sim 0.62$). In region B, the field cannot penetrate efficiently inside the plasma because of the electron shielding. The most appropriate region for r.f. plugging is denoted by region C, where the density can be substantially reduced by a strong enhancement of the r.f. field inside the plasma. For frequencies close to ω_c, however, the present model cannot be applied since the ion thermal motion is neglected.

6. SUMMARY AND DISCUSSION

In this paper, we have investigated the steady r.f. confinement of a plasma in a magnetic field using the collisionless two-fluid mode. We showed in Section 3 that the steady state can be determined by solving a simple variation principle. This variation principle is valid for arbitrary amplitude of the r.f. field. In Section 4, we considered an axially symmetric system and calculated the quasi-linear modification of the average physical quantities, such as the density, flow flux and electromagnetic fields. The calculation was facilitated by the introduction of a generalized oscillating displacement vector. The equation determining the modified flux function (modification due to the Coriolis force associated with the plasma flow) is a direct generalization of the corresponding equation in the absence of the r.f. field. Numerical analysis for a simple model described in Section 5 indicates that, under certain conditions, a strong enhancement of the r.f. field inside the plasma can be produced by resonance with one of the plasma eigenmodes. Such an enhanced field causes a strong depression of the plasma density and thereby suppresses the plasma outflux along the magnetic field.

In the numerical analysis presented here, the plasma flow is neglected. A more appropriate model is to fix the plasma flow flux at the boundary and then demonstrate the density increase by the application of an r.f. field. It is also necessary to take into account the effect of the ion thermal motion. Within the framework of the collisionless fluid model, these effects are taken into account in the present formulation, and numerical analyses including these effects are now under way.

Stability of the steady state determined by the present formalism is our future problem. However, we can present some arguments concerning the stability of a sheet plasma with r.f. field. First, the eigenmode in a magnetized plasma, such as the electrostatic ion cyclotron wave, tends to be localized in a high-density region, whereas the ponderomotive force in the frequency region $\omega_c < \omega < 2\omega_c$ tends to expel the plasma from the high-field region. Therefore, such an eigenmode is likely to be stable against the modulational instability. Secondly, since the dispersion relation of the electrostatic ion cyclotron wave is insensitive to the density variation, its parametric coupling to other electrostatic waves will be relatively weak. Thirdly, for a cusp configuration with characteristic gradient length of the magnetic field less than $\omega^{-1}\sqrt{T_e/m_e}$, the electrons can always follow the ion motion by the Debye shielding, so that the r.f. field can hardly produce a charge separation. Then the plasma can practically be described by the one-fluid model and therefore the parametric coupling to other modes is strongly reduced unless the r.f. field has a very short wavelength. Finally, there is a possibility of microscopic instabilities due to resonant wave-particle interactions, since there is population inversion in the cusp region. Investigation of these microinstabilities will be left to the future.

ACKNOWLEDGEMENTS

We are grateful to Professor K. Takayama, Director of the Institute of Plasma Physics, Nagoya University, for his continuous encouragement throughout this work. We are also indebted to the members of the r.f. confinement group of the Institute, particularly Drs T. Sato, H. Obayashi and T. Hatori, for many valuable comments and fruitful discussions. Numerical calculations were carried out by FACOM M-190 at the Computer Centre of the Nagoya Institute of Plasma Physics and HITAC 8700 at the Computer Centre of Hiroshima University.

This work was partially supported by a Grant in Aid of Scientific Research from the Ministry of Education, Science and Culture of Japan.

APPENDIX I

In this Appendix, we show the relations (53) and (54) and derive an explicit expression for $\vec{Q}$ using first-order perturbational calculation.

We start from relation (54). Using Eq. (19), we have

$$(\vec{v}\times\vec{\Omega})_1 = \frac{1}{2\pi}\{(\overline{\vec{v}}\cdot\vec{\nabla}\lambda_1 + \vec{v}_1\cdot\vec{\nabla}\overline{\lambda})\vec{\nabla}\psi + \vec{v}\cdot\vec{\nabla}\overline{\lambda}\;\vec{\nabla}\psi_1$$

$$- (\overline{\vec{v}}\cdot\vec{\nabla}\psi_1 + \vec{v}_1\cdot\vec{\nabla}\overline{\psi})\vec{\nabla}\lambda - (\overline{\vec{v}}\cdot\vec{\nabla}\overline{\psi})\vec{\nabla}\lambda_1\}$$

$$\doteqdot \frac{1}{2\pi}\{[\dot{\vec{Q}}\cdot\vec{\nabla}\overline{\lambda} - \frac{2\pi c}{q} f''(\overline{\psi})\psi_1]\vec{\nabla}\overline{\psi} - \frac{2\pi c}{q} f'(\overline{\psi})\vec{\nabla}\psi_1$$

$$- \dot{\vec{Q}}\cdot\vec{\nabla}\overline{\psi}\;\vec{\nabla}\overline{\lambda}\}$$

where we used Eqs (51), (43), (50) and (42) and neglected the third-order terms. Using the relations

$$\dot{\vec{Q}}\cdot\vec{\nabla}\overline{\lambda}\;\vec{\nabla}\overline{\psi} - \dot{\vec{Q}}\cdot\vec{\nabla}\overline{\psi}\;\nabla\overline{\lambda} = \dot{\vec{Q}}\times(\vec{\nabla}\overline{\psi}\times\vec{\nabla}\overline{\lambda})$$

$$\doteqdot \dot{\vec{Q}}\times 2\pi\overline{\vec{\Omega}}$$

$$f''(\overline{\psi})\psi_1\;\vec{\nabla}\overline{\psi} + f'(\overline{\psi})\vec{\nabla}\psi_1 = \vec{\nabla}\,[f'(\overline{\psi})\psi_1]$$

we immediately derive Eq. (54).

To derive Eq. (53), we calculate $\overline{n}(\dot{\vec{Q}}-\vec{v}_1)\cdot\nabla\overline{\psi}$, $\overline{n}(\dot{\vec{Q}}-\vec{v}_1)\cdot\nabla\overline{\lambda}$ and $\vec{\nabla}\cdot[\overline{n}(\dot{\vec{Q}}-\vec{v}_1)]$ by using Eqs (50) – (52). Using Eqs (49), (42) and (45) and neglecting the third-order terms, we first have

$$\overline{n}(\dot{\vec{Q}} - \vec{v}_1)\cdot\vec{\nabla}\overline{\psi} = \overline{n}\vec{v}\cdot\vec{\nabla}\psi_1 = -\overline{n\vec{v}}\cdot\vec{\nabla}(\vec{Q}\cdot\vec{\nabla}\overline{\psi})$$

$$\doteqdot -\vec{\nabla}\cdot[\overline{n\vec{v}}\;(\vec{Q}\cdot\vec{\nabla}\overline{\psi})]$$

$$= +\vec{\nabla}\cdot[\vec{\nabla}\overline{\psi}\times(\vec{Q}\times\overline{n\vec{v}}) - \vec{Q}\cdot(\overline{n\vec{v}}\cdot\vec{\nabla}\overline{\psi})]$$

$$\doteqdot \vec{\nabla}\cdot[\vec{\nabla}\overline{\psi}\times(\vec{Q}\times\overline{n\vec{v}})]$$

$$= -[\vec{\nabla}\times(\vec{Q}\times\overline{n\vec{v}})]\cdot\vec{\nabla}\overline{\psi} \qquad \text{(A–1)}$$

from which we get, using $\overline{\vec{v}} \cdot \vec{\nabla}\overline{\psi} \doteq 0$,

$$[\bar{n}\dot{\vec{Q}} - (\bar{n}\vec{v}_1 + n_1\overline{\vec{v}}) + \vec{\nabla}\times(\vec{Q}\times\overline{n\vec{v}})]\cdot\vec{\nabla}\overline{\psi} = 0 \qquad (A\text{–}2)$$

Similarly, we have

$$\bar{n}(\dot{\vec{Q}} - \vec{v}_1)\cdot\vec{\nabla}\bar{\lambda} = \frac{2\pi c}{q}\bar{n}\, f''(\bar{\psi})\psi_1 + \overline{n\vec{v}}\cdot\vec{\nabla}\lambda_1$$

$$= \frac{2\pi c}{q}\bar{n}\, f''(\bar{\psi})\psi_1 + \vec{\nabla}\cdot[\vec{\nabla}\bar{\lambda}\times(\vec{Q}\times\overline{n\vec{v}}) - \vec{Q}\cdot(\overline{n\vec{v}}\cdot\vec{\nabla}\bar{\lambda})]$$

$$= \frac{2\pi c}{q}\bar{n}\, f''(\bar{\psi})\psi_1 - \vec{\nabla}\bar{\lambda}\cdot\vec{\nabla}\times(\vec{Q}\times\overline{n\vec{v}}) - \vec{\nabla}\cdot(\bar{n}\vec{Q})\,\overline{\vec{v}}\cdot\vec{\nabla}\bar{\lambda}$$

$$- \bar{n}\vec{Q}\cdot\vec{\nabla}(\overline{\vec{v}}\cdot\vec{\nabla}\bar{\lambda})$$

$$= [-\vec{\nabla}\times(\vec{Q}\times\overline{n\vec{v}}) + n_1\overline{\vec{v}}]\cdot\vec{\nabla}\bar{\lambda} \qquad (A\text{–}3)$$

where in the last line we used the relation (see Eq.(43))

$$\bar{n}\vec{Q}\cdot\vec{\nabla}(\overline{\vec{v}}\cdot\vec{\nabla}\bar{\lambda}) \doteq -\bar{n}\vec{Q}\cdot\frac{2\pi c}{q}f''(\bar{\psi})\vec{\nabla}\bar{\psi}$$

$$= \frac{2\pi c}{q}\bar{n}\, f''(\bar{\psi})\psi_1$$

From (A–2) and (A–3), we get

$$\bar{n}\vec{v}_1 + n_1\overline{\vec{v}} = \bar{n}\dot{\vec{Q}} + \vec{\nabla}\times(\vec{Q}\times\overline{n\vec{v}}) + p\,\vec{\nabla}\bar{\psi}\times\vec{\nabla}\bar{\lambda} \qquad (A\text{–}4)$$

where, from Eq.(52), p must be a function of $\bar{\psi}$ and $\bar{\lambda}$ alone, $p = p(\bar{\psi},\bar{\lambda})$. Note that we are assuming that the time dependence arises solely from the oscillating field. Taking the time average of Eq.(A–4), we then find that p=0, which proves relation (53). Note that these relations are derived without recourse to the axial symmetry.

To derive the first-order expression for $\dot{\vec{Q}}$, we use the axial symmetry of the system in which we can write the zeroth-order velocity as [10]

$$\vec{v}_0 = \left[-\frac{F(\psi_0)}{\bar{n}r}\frac{1}{r}\frac{\partial\psi_0}{\partial z},\quad v_{0\phi},\quad \frac{F(\psi_0)}{n_0 r}\frac{\partial\psi_0}{\partial r}\right] \qquad (A\text{–}5)$$

where the suffix 0 stands for the unperturbed quantity. Then using the first of Eqs (A–1) and (A–3) and the relations

$$\vec{\Omega}_0 \cdot \vec{\nabla}\psi_1 \doteqdot - \vec{\Omega}_1 \cdot \vec{\nabla}\psi_0 \,, \qquad \vec{\Omega}_0 \cdot \vec{\nabla}\lambda_1 \doteqdot - \vec{\Omega}_1 \cdot \vec{\nabla}\psi_0$$

we get

$$(\dot{\vec{Q}} - \vec{v}_1) \cdot \vec{\nabla}\overline{\psi} \doteqdot \frac{2\pi}{n_0} F \vec{\Omega}_0 \cdot \vec{\nabla}\psi_1 \doteqdot - \frac{2\pi}{n_0} F \vec{\Omega}_1 \cdot \vec{\nabla}\psi_0 \qquad \text{(A–6)}$$

$$(\dot{\vec{Q}} - \vec{v}_1) \cdot \vec{\nabla}\overline{\lambda} \doteqdot \frac{2\pi c}{q} f''(\psi_0)\psi_1 + \frac{2\pi}{n_0} F \vec{\Omega}_0 \cdot \nabla\lambda_1$$

$$\doteqdot \left[\frac{2\pi c}{q} r^2 f''(\psi_0)\psi_1 \vec{\nabla}\phi - \frac{2\pi}{n_0} F \vec{\Omega}_1 \right] \cdot \vec{\nabla}\lambda_0 \qquad \text{(A–7)}$$

where we used relation (36) which yields $\vec{\nabla}\lambda \cdot \vec{\nabla}\phi = r^{-2}$. From Eqs (A–6) and (A–7) we then get (noting $\vec{\nabla}\psi_0 \cdot \vec{\nabla}\phi = 0$)

$$\dot{\vec{Q}} = \vec{v}_1 - \frac{2\pi F}{n_0} \vec{\Omega}_1 + \frac{2\pi c}{q} \psi_1 r^2 f''(\psi_0) \vec{\nabla}\phi + p \vec{\nabla}\psi_0 \times \vec{\nabla}\lambda_0 \qquad \text{(A–8)}$$

where, for the same reason as in (A–4) and from Eq.(52), the arbitrary function p must be equal to

$$p = \frac{1}{n_0} \left[\frac{n_1}{n_0} F(\psi_0) - \psi_1 F'(\psi_0) \right]$$

APPENDIX II

In this Appendix, we derive relations (63) and (67). We start from the first term on the left-hand side of Eq. (63). Using the first of Eqs (49), we can write

$$- \vec{\nabla} \cdot (\psi_1 \overline{n} \dot{\vec{Q}}) = \vec{\nabla} \cdot [(\vec{Q} \cdot \vec{\nabla}\overline{\psi}) n \dot{\vec{Q}}]$$

$$= \frac{1}{2} \vec{\nabla} \cdot [\vec{\nabla}\overline{\psi} \times (\dot{\vec{Q}} \times \vec{Q}) \overline{n}] \qquad \text{(A–9)}$$

Noting that only the ϕ-component of the vector $(\dot{\vec{Q}} \times \vec{Q})$ contributes to Eq.(A–9), we get

$$- \vec{\nabla} \cdot (\psi_1 \overline{n} \dot{\vec{Q}}) = \frac{1}{2} \vec{\nabla} \cdot [\vec{\nabla}\overline{\psi} \times \{(\dot{\vec{Q}} \times \vec{Q}) \cdot (r\vec{\nabla}\phi)\overline{n} r \vec{\nabla}\phi\}]$$

$$= \frac{1}{2} (\vec{\nabla}\overline{\psi} \times \vec{\nabla}\phi) \cdot \vec{\nabla} \{r^2 \overline{n} (\dot{\vec{Q}} \times \vec{Q}) \cdot \vec{\nabla}\phi\} \qquad \text{(A–10)}$$

which gives the first term on the right-hand side of Eq.(63). The second term on the left-hand side of Eq.(63) can be calculated as follows:

$$- \vec{\nabla} \cdot [\psi_1 \vec{\nabla} \times (\vec{Q} \times \overline{n\vec{v}})] = - \vec{\nabla}\psi_1 \cdot [\vec{\nabla} \times (\vec{Q} \times \overline{n\vec{v}})]$$

$$= \vec{\nabla} \cdot [\vec{\nabla}\psi_1 \times (\vec{Q} \times \overline{n\vec{v}})]$$

$$= \vec{\nabla} \cdot [\vec{\nabla}\psi_1 \times \{(\vec{Q} \times \overline{n\vec{v}}) \cdot (r\vec{\nabla}\phi) r \vec{\nabla}\phi\}] \qquad \text{(A–11)}$$

where in the last line we again used the fact that only the ϕ-component of the vector $(\vec{Q} \times \overline{n\vec{v}})$ can contribute. In the last expression of Eq. (A–11), we approximate $\overline{n\vec{v}}$ by $\overline{n\vec{v}} = \vec{\nabla}\Gamma \times \vec{\nabla}\phi + G\vec{\nabla}\phi$, where the last term, $G\vec{\nabla}\phi$, gives no contribution to Eq. (A–11). Then using the relation

$$\vec{Q} \times (\vec{\nabla}\Gamma \times \vec{\nabla}\phi) = - (\vec{Q} \cdot \vec{\nabla}\Gamma) \vec{\nabla}\phi \qquad \text{(A–12)}$$

we can immediately get the second term on the right-hand side of Eq.(63).

Calculation of the 'toroidal' flux G can be carried out in a similar way. Using Eq.(36), we first have

$$\overline{(n\vec{v})} \cdot \vec{\nabla}\overline{\lambda} = (\vec{\nabla}\Gamma \times \vec{\nabla}\phi + G\vec{\nabla}\phi) \cdot \vec{\nabla}(\phi - 2\pi\overline{h})$$

$$= \frac{G}{r^2} - (\vec{\nabla}\overline{\lambda} \times \vec{\nabla}\phi) \cdot \nabla\Gamma \qquad \text{(A–13)}$$

The left-hand side can be calculated as

$$\overline{n\vec{v}} \cdot \vec{\nabla}\overline{\lambda} = \vec{\nabla} \cdot (\overline{n\vec{v}}\ \overline{\lambda}) = \vec{\nabla} \cdot \overline{(n\vec{v}\lambda)} - \vec{\nabla} \cdot [\overline{(n\vec{v})_1 \lambda_1}]$$

$$= -\frac{\partial}{\partial t}\overline{(n\lambda)} - \frac{2\pi c}{q} \overline{n\ f'(\psi)}$$

$$- \vec{\nabla} \cdot [\overline{\lambda_1 \{n\dot{\vec{Q}} + \vec{\nabla} \times (\vec{Q} \times \overline{n\vec{v}})\}}] \qquad \text{(A–14)}$$

where in the last equality we used Eqs (25), (41) and (53). The first term in the last expression vanishes while the last term can be calculated in the same way as for Eq. (63):

$$- \vec{\nabla} \cdot [\lambda_1 \{ \overline{n\dot{\vec{Q}}} + \vec{\nabla} \times (\vec{Q} \times \overline{n\vec{v}}) \}]$$

$$= \frac{1}{2} (\vec{\nabla}\bar{\lambda} \times \vec{\nabla}\phi) \cdot \vec{\nabla} \{ r^2 \bar{n} (\dot{\vec{Q}} \times \vec{Q}) \cdot \vec{\nabla}\phi \}$$

$$- (\vec{\nabla}\lambda_1 \times \vec{\nabla}\phi) \cdot \vec{\nabla} (\vec{Q} \cdot \vec{\nabla}\Gamma) \qquad \text{(A–15)}$$

Substituting Eq. (A–15) into Eq. (A–14) and then into Eq. (A–13), and using Eqs (66) and (64), we get

$$\frac{G}{r^2} = (\vec{\nabla}\bar{\lambda} \times \vec{\nabla}\phi) \cdot \vec{\nabla} [\Gamma_0(\bar{\psi}) + \Gamma_0''(\bar{\psi}) \overline{\frac{\psi_1^2}{2}}] - \frac{2\pi c}{q} \overline{n f'(\psi)}$$

$$+ \overline{(\vec{\nabla}\lambda_1 \times \nabla\phi) \cdot \{ \Gamma_0''(\bar{\psi})(\vec{\nabla}\bar{\psi})\psi_1 + \Gamma_0'(\bar{\psi})\vec{\nabla}\psi_1 \}}$$

$$= \vec{\nabla}\phi \cdot \{ - \Gamma_0'(\bar{\psi}) (\vec{\nabla}\bar{\lambda} \times \vec{\nabla}\bar{\psi} + \overline{\vec{\nabla}\lambda_1 \times \vec{\nabla}\psi_1})$$

$$- \frac{1}{2} \Gamma_0'''(\bar{\psi}) \overline{\psi_1^2} (\vec{\nabla}\bar{\lambda} \times \vec{\nabla}\bar{\psi})$$

$$- \Gamma_0''(\bar{\psi}) \overline{\psi_1 (\vec{\nabla}\bar{\lambda} \times \vec{\nabla}\psi_1 + \vec{\nabla}\lambda_1 \vec{\nabla}\bar{\psi})} \}$$

$$- \frac{2\pi c}{q} \overline{n f'(\psi)} \qquad \text{(A–16)}$$

Using Eq.(16) and neglecting the third-order terms, we finally get Eq. (67).

REFERENCES

[1] MOTZ, H., WATSON, C.J.H., "The radio-frequency confinement and acceleration", Advances in Electronics and Electron Physics (MARTON, L., Ed.), Academic Press, New York (1967).

[2] SHELBY, C.F., HATCH, A.J., Phys. Rev. Lett. **25** (1972) 834.

[3] HATCH, A.J., SHOHET, J.L., Phys. Fluids **17** (1974) 232.

[4] GITMER, S.J., SHOHET, J.L., Phys. Fluids **20** (1977) 1019.

[5] ADATI, K., et al., Ann. Rev. Inst. Plasma Phys., Nagoya Univ. (1975) 44.

[6] WATANABE, T., HATORI, T., Inst. Plasma Phys., Nagoya Univ. Res. Rep. IPPJ–132 (1972).
[7] HATORI, T., et al., "An adiabatic RF-plugging scheme for a controlled fusion reactor", Plasma Physics and Controlled Nuclear Fusion Research 1974 (Proc. 5th Int. Conf. Tokyo, 1974) **2**, IAEA, Vienna (1975) 663.
[8] SELIGAR, R.L., WHITHAM, G.B., Proc. R. Soc. (London) **A305** (1968) 1.
[9] WATANABE, T., LEE, Y.C., NISHIKAWA, K., Plasma Physics, Nonlinear Theory and Experiment (Proc. 30th Nobel Symp. 1976), Plenum Press, New York, London (1977) 142.
[10] WATANABE, T., NISHIKAWA, K., Inst. Plasma Phys., Nagoya Univ. Res. Rep. IPPJ–282 (1977), Phys. Fluids **21** (1978) 390.

ON THE EVOLUTION OF TOKAMAK PLASMA EQUILIBRIA

G.V. PEREVERSEV, V.D. SHAFRANOV,
L.E. ZAKHAROV
Kurchatov Institute of Atomic Energy,
Academy of Sciences of the USSR,
Moscow,
Union of Soviet Socialist Republics

Abstract

ON THE EVOLUTION OF TOKAMAK PLASMA EQUILIBRIA.

The problem of the evolution of the magnetic configuration of a toroidal tokamak-type system is shown to be soluble separately from the plasma diffusion process, in the case of a prescribed plasma pressure. In other words, to calculate both poloidal and toroidal field diffusion, a knowledge of the 'effective' transverse conductivity $\sigma_\perp$ (which is responsible for anomalous diffusion) is not necessary. The knowledge of $\sigma_\parallel$ is essentially sufficient. Since the latter is described by the classical Spitzer formula for a stable stage of the discharge, the problem of evolution acquires a reliable basis.

1. INTRODUCTION

According to widespread opinion, an optimized tokamak-type system should have a small-aspect ratio and non-circular cross-section. The shape of the magnetic surfaces of such a system is sensitive to the plasma pressure and current density distributions. For example, all magnetic surface cross-sections are similar in an elliptical plasma column if current density is uniform, while they are found to be more rounded near the axis in the case of a peaked current. Therefore, the temporal dependence of the plasma parameters leads to the evolution of the magnetic configuration. Knowledge of the evolution is necessary to control the equilibria and stability of the plasma.

A number of papers [1–10] have been concerned with the study of methods to solve the evolution problem by neglecting the inertia term. Among these, only those papers which do not require a knowledge of the transport processes have a reliable basis. These, first of all, are the papers by Grad et al. [1–4], showing that, in the framework of ideal MHD, such a 'dissipative' problem as bifurcation of the equilibria, of the type where the plasma is split into two columns in the

Doublet device, may be solved. The solutions on different sides of the separatrix are matched by the appropriate boundary conditions. The second important problem, which does not involve the transport processes, is that of the evolution due to very rapid growth of the plasma pressure. This is the concept of the so-called forced-flux conserving tokamak [8–10], which aims at achieving a high ratio of plasma pressure to the pressure of the magnetic fields (of the order of 10–20%). Considerations based on flux conservation are valid when the duration of the process $t \ll t_s = \sigma a^2/4$, where σ is the electrical conductivity and a is the transverse dimension of the plasma. If $t > t_s$, dissipative transport should be taken into account. Unfortunately, the available classical or neoclassical transport theories do not agree with the experimental data, and the approaches based on these theories are therefore only of methodological interest.

This paper formulates a simplified approach to the dissipative evolution of equilibria – a prescribed pressure method. It requires knowledge of only one component, parallel to the magnetic field, of the generalized Ohm's law, i.e. essentially only the longitudinal conductivity is necessary. As seen from experiment, the latter is classical Spitzer-Coulomb collisional conductivity taking place during the most interesting hydrodynamically stable phase of discharges.

2. EVOLUTION EQUATION

The method of prescribed pressure in the problem of the evolution of equilibria is as follows. Instead of obtaining doubtful data concerning the plasma pressure distribution, using an inadequate theory of diffusion and thermal conductivity, we shall take the pressure to be a given function of t,a where $a = a(\vec{r}, t)$ is the magnetic surface equation. For a given p(a,t), the toroidal and poloidal magneti fields are not independent; they are related by the equilibrium equation:

$$\text{rot}\,\vec{B} \times \vec{B} = \nabla p \tag{1}$$

It follows that we need only one equation to find the temporary dependence of the equilibrium parameters.

Let us consider a generalized Ohm's law:

$$\vec{E} + \vec{v} \times \vec{B} = \frac{\vec{R}}{en_e} \tag{2}$$

where $\vec{R}$ implies all forces of non-electrical origin–friction between different kinds of particles, thermo-electromotive force, etc. Taking the scalar product of the vector equation (2) with $\vec{B}$, $\vec{B} \times \nabla p$, ∇p, we obtain, taking into account condition (1),

$$\vec{E}\cdot\vec{B} = \frac{\vec{R}\cdot\vec{B}}{en_e} \tag{3}$$

$$\vec{E}\times\vec{B}\cdot\nabla p - |\vec{B}|^2\vec{v}\cdot\nabla p = \frac{1}{en_e}\vec{R}\times\vec{B}\cdot\nabla p \tag{4}$$

$$\vec{E}\cdot\nabla p + \vec{v}\cdot\vec{B}\times\nabla p = \frac{1}{en_e}\vec{R}\cdot\nabla p \tag{5}$$

Expressing the electric field in terms of vector and scalar potentials:

$$\vec{E} = -\frac{\partial\vec{A}}{\partial t} - \nabla\phi, \quad \vec{A}\cdot\nabla p = 0 \tag{6}$$

one finds from Eq.(3) a magneto-differential equation for ϕ:

$$\vec{B}\cdot\nabla\phi = -\vec{B}\cdot\frac{\partial\vec{A}}{\partial t} - \frac{1}{en_e}\vec{R}\cdot\vec{B} \tag{7}$$

The solubility condition leads to the equation:

$$\left\langle\vec{B}\cdot\frac{\partial\vec{A}}{\partial t}\right\rangle = -\frac{1}{en_e}\langle\vec{B}\cdot\vec{R}\rangle \tag{8}$$

where the angle brackets denote averaging over the layer volume between two narrow magnetic surfaces. After solution of Eq.(7) for ϕ, Eqs (4) and (5) serve to determine the normal and tangential components of the velocity relative to $\vec{B}$. Therefore only Eq.(8) describes the temporal evolution. Taking into account the dominating role of Eq.(8) in this approach, it will be referred to as the evolution equation. In the case of $\langle\vec{R}\cdot\vec{B}\rangle/en_e = \langle\vec{j}\cdot\vec{B}\rangle/\sigma_{\parallel}$, which is valid at least in the collision-dominated region, the evolution equation is written in terms of transverse and longitudinal magnetic fluxes and electric currents Ψ, Φ, I, J [11]:

$$\Phi'\dot{\Psi} - \Psi'\dot{\Phi} = \frac{1}{\sigma_{\parallel}}(JI' - IJ') \tag{9}$$

Here $\Phi' = \partial\Phi(a,t)/\partial a$, $\dot{\Phi} = \partial\Phi(a,t)/\partial t$, and so on. Currents J, I are related to the flux derivatives Ψ', Φ' through the metric tensor g_{ik} of the proper co-ordinate system with 'straight' field lines [12, 13, 3],

$$\begin{aligned} J &= \alpha_{22}\Psi' + \alpha_{23}\Phi' \\ -I &= \alpha_{23}\Psi' + \alpha_{33}\Phi' \\ \alpha_{ik} &= \langle g_{ik}/g\rangle, \quad g = \mathrm{Det}\, g_{ik} \end{aligned} \tag{10}$$

Therefore Eq.(9) includes two unknown functions Ψ and Φ. The integral equilibrium equation may be the second equation for them:

$$I'\Phi' - J'\Psi' = p'V' \tag{11}$$

The general method for solving the problem of evolution with time-dependent coefficient $\langle g_{ik}/g\rangle$ is the iterative method [2]. For each time step, with the help of equilibrium equation (1), one finds the co-ordinate dependence solution $V(\vec{r},t) = \text{const}$ and $\alpha_{ik}(V,t_n)$ if $\Psi(V,t_n)$, $\Phi(V,t_n)$ are known. After this the set of one-dimensional equations (9), (11) allows us to determine $\Psi(a, t_n + \Delta t)$, $\Phi(a, t_n + \Delta t)$ for the next time step.

3. DIFFERENT FORMS OF THE EVOLUTION EQUATION

Evolution equation (9) can be divided into two equations by artificially introducing the flux convective transfer velocity $v(a) = \langle\vec{v}\cdot\nabla a\rangle$ (in the general case, it has nothing in common with the plasma diffusion rate):

$$\begin{aligned} \frac{\partial\Psi}{\partial t} + v\Psi' &= -\frac{1}{\sigma_{\|}}\frac{I}{\Phi'}J' \\ \frac{\partial\Phi}{\partial t} + v\Phi' &= -\frac{1}{\sigma_{\|}}\frac{J}{\Psi'}I' \end{aligned} \tag{12}$$

In an axisymmetric case, when $\alpha_{23} = 0$, these equations look like two independent equations of diffusion with convection for Ψ and Φ. In reality, however, one of them determines v. The vector analogue of these equations has the form

$$\frac{\partial\vec{A}}{\partial t} - \vec{v}\times \text{rot}\,\vec{A} = -\frac{1}{\sigma_{\|}}\,\text{rot rot}\,\vec{A} \tag{13}$$

Evolution equation (9), taking account of the equilibrium equation (11), may be rewritten in a form resolved with respect to J', I':

$$J' = \frac{1}{J\Psi' - I\Phi'}\left[\sigma_{\|}(\dot{\Psi}\Phi' - \dot{\Phi}\Psi')\Phi' - Jp'V'\right] \tag{14}$$

$$I' = \frac{1}{J\Psi' - I\Phi'}\left[\sigma_{\|}(\dot{\Psi}\Phi' - \dot{\Phi}\Psi')\Psi' - Ip'V'\right] \tag{15}$$

The vector analogue of these equations is

$$\vec{j} = \frac{1}{|\vec{B}|^2} \{\sigma_{\|} \vec{B}(\vec{E} \cdot \vec{B}) + \vec{B} \times \nabla p\} \tag{16}$$

The role of the equation $\vec{j} = \text{rot}\,\vec{B}$ is played by Eq.(10).

Let us take $a \equiv \Phi$. Then $\Phi' = 1$, $\dot{\Phi} = 0$, and the evolution equation (9) takes the form

$$\frac{\partial \Psi(\Phi, t)}{\partial t} = -\frac{I^2}{\sigma_{\|}} \frac{\partial}{\partial \Phi} \left(\frac{J}{I}\right) \tag{17}$$

Differentiation with respect to Φ yields the diffusion equation in the Φ-space for the rotational transform $\mu = d\Psi/d\Phi$:

$$\frac{\partial \mu}{\partial t} = \frac{\partial}{\partial \Phi} \left\{ \frac{(\alpha_{33} + \alpha_{23}\mu)^2}{\sigma_{\|}} \frac{\partial}{\partial \Phi} \left(\frac{\alpha_{22}\mu + \alpha_{23}}{\alpha_{33} + \alpha_{23}\mu} \right) \right\} \tag{18}$$

Similarly, taking $a \equiv \Psi$, we obtain

$$\frac{\partial \Phi(\Psi, t)}{\partial t} = -\frac{J^2}{\sigma_{\|}} \frac{\partial}{\partial \Psi} \left(\frac{I}{J}\right) \tag{19}$$

Differentiating with respect to Ψ, we derive the diffusion equation with variables Ψ, t for rotation number $q = d\Phi/d\Psi$:

$$\frac{\partial q}{\partial t} = \frac{\partial}{\partial \Psi} \left\{ \frac{(\alpha_{22} + \alpha_{23} q)^2}{\sigma_{\|}} \frac{\partial}{\partial \Psi} \left(\frac{\alpha_{33} q + \alpha_{23}}{\alpha_{22} + \alpha_{23} q} \right) \right\} \tag{20}$$

It should be particularly emphasized that although the equations are formally similar to those that could be obtained from the simplest Ohm's law $\vec{j} = \sigma(\vec{E} + \vec{v} \times \vec{B})$, they are applicable to a more general form of the generalized Ohm's law: $(\vec{j}_{\|}/\sigma_{\|}) + (\vec{j}_{\perp}/\sigma_{\perp}) = \vec{E} + \vec{v} \times \vec{B} + \vec{R}_{\perp}/en_e$. In this case if, for example, v implies the plasma velocity, as it does in other papers devoted to the evolution problem, then both Eqs (12) would acquire some more terms (see e.g. Ref. [14]). If, however, v is excluded we shall again arrive at our evolution equation (9).

4. AXIAL SYMMETRY

In the case of axial symmetry it is easy to obtain the equations [3, 15]:

$$\alpha_{23} = 0; \quad \alpha_{22} = \frac{V'(a)}{4\pi^2}\left\langle \frac{|\nabla a|^2}{X^2} \right\rangle; \quad \alpha_{33} = \frac{4\pi^2}{V'(a)}\left\langle \frac{1}{X^2} \right\rangle^{-1} \tag{21}$$

$$J = \frac{V'}{4\pi^2}\left\langle \frac{|\nabla a|^2}{X^2} \right\rangle \frac{\partial \Psi}{\partial a}; \quad I = -\frac{4\pi^2}{V'\left\langle \frac{1}{X^2} \right\rangle} \frac{\partial \Phi}{\partial a} \tag{22}$$

(X is the distance from the symmetry axis.)

Evolution equation (9) and its variations (17)–(20) are of the form:

$$\Phi'\dot{\Psi} - \Psi'\dot{\Phi} = \frac{\Phi'/V'}{\sigma_{\|}\langle 1/X^2 \rangle} \frac{\partial}{\partial a}\left[V'\left\langle \frac{|\nabla a|^2}{X^2} \right\rangle \frac{\partial \Psi}{\partial a} \right] - \frac{\Psi' V'}{\sigma_{\|}}\left\langle \frac{|\nabla a|^2}{X^2} \right\rangle \frac{\partial}{\partial a}\left[\frac{\partial \Phi/\partial a}{V'\langle 1/X^2 \rangle} \right] \tag{23}$$

$$\frac{\partial \Psi(\Phi, t)}{\partial t} = \frac{1}{\sigma_{\|}\left(\frac{\partial V}{\partial \Phi}\right)^2 \left\langle \frac{1}{X^2} \right\rangle^2} \frac{\partial}{\partial \Phi}\left[\left(\frac{\partial V}{\partial \Phi}\right)^2 \left\langle \frac{1}{X^2} \right\rangle \left\langle \frac{|\nabla \Phi|^2}{X^2} \right\rangle \frac{\partial \Psi}{\partial \Phi} \right] \tag{24}$$

$$\frac{\partial \mu}{\partial t} = \frac{\partial}{\partial \Phi}\left\{ \frac{\frac{\partial}{\partial \Phi}\left[\left(\frac{\partial V}{\partial \Phi}\right)^2 \left\langle \frac{1}{X^2} \right\rangle \left\langle \frac{|\nabla \Phi|^2}{X^2} \right\rangle \mu \right]}{\sigma_{\|}\left(\frac{\partial V}{\partial \Phi}\right)^2 \left\langle \frac{1}{X^2} \right\rangle^2} \right\} \tag{25}$$

$$\frac{\partial \Phi(\Psi, t)}{\partial t} = \frac{1}{\sigma_{\|}}\left(\frac{\partial V}{\partial \Phi}\right)^2 \left\langle \frac{|\nabla \Psi|^2}{X^2} \right\rangle^2 \frac{\partial}{\partial \Psi} \frac{\partial \Phi/\partial \Psi}{\left(\frac{\partial V}{\partial \Psi}\right)^2 \left\langle \frac{1}{X^2} \right\rangle \left\langle \frac{|\nabla \Psi|^2}{X^2} \right\rangle} \tag{26}$$

$$\frac{\partial q}{\partial t} = \frac{\partial}{\partial \Psi}\left\{ \frac{V'^2(\Psi)}{\sigma_{\|}} \left\langle \frac{|\nabla \Psi|^2}{X^2} \right\rangle^2 \frac{\partial}{\partial \Psi} \frac{q}{V'^2 \left\langle \frac{1}{X^2} \right\rangle \left\langle \frac{|\nabla \Psi|^2}{X^2} \right\rangle} \right\} \tag{27}$$

The shape of the magnetic surfaces can be determined by using the equation

$$X^2 \,\mathrm{div} \frac{\nabla \Psi}{X^2} = 2\pi X j_\phi \tag{28}$$

where, according to the equilibrium conditions,

$$2\pi X j_{\phi} = -\frac{4\pi^2 X^2 p'}{\Psi'} - \frac{I I'}{\Psi'} \qquad (29)$$

Making use of the identity (the consequence of Eq.(16))

$$I' = \frac{1}{\langle |B| \rangle^2 V'} \{\Psi'(JI' - IJ') - Ip'V'\} \qquad (30)$$

the right-hand side of the latter relation can be written so that it will contain a combination of the quantities $(JI' - IJ')/\Phi' = C$, which enters into the evolution equation:

$$2\pi X j_{\phi} = -4\pi^2 \frac{p'}{\Psi'}\left[X^2 - \frac{\langle B_T^2 \rangle}{\langle B^2 \rangle \langle 1/X^2 \rangle}\right] + \frac{\langle B_T^2 \rangle}{\langle B^2 \rangle \langle 1/X^2 \rangle} C \qquad (31)$$

where B_T is the toroidal field and B the total magnetic field, so

$$\frac{\langle B_T^2 \rangle}{\langle |B|^2 \rangle} = \frac{1}{1 + \mu^2 \langle 1/X^2 \rangle \langle |\nabla V|^2/X^2 \rangle (2\pi)^{-4}} \qquad (32)$$

For rapid processes, when the relationship between the fluxes Ψ and Φ varies only slightly, or does not vary at all as is usually assumed for a flux-conserving tokamak [10], the right-hand side of the evolution equation should be used as

$$C = \frac{\partial \Phi}{\partial V} \left\langle \frac{1}{X^2} \right\rangle^{-2} \frac{\partial}{\partial V} \left[\left\langle \frac{1}{X^2} \right\rangle \left\langle \frac{|\nabla V|^2}{X^2} \right\rangle \mu\right] \qquad (33)$$

For slow processes it is better to use

$$C = \sigma_{\parallel} \frac{\partial \Psi}{\partial t}(\Phi, t) \qquad (34)$$

in Eq.(31).

In particular, in a steady-state case, when the power of the particle sources and energy and the inductive electromotive force are chosen so that the configuration does not vary with time (in this case $\partial\Psi/\partial t = \mathscr{E}_0$ = const) the equilibrium equation takes the form (which coincides with that obtained by Taylor [6]):

$$\frac{\partial^2 \Psi}{\partial X^2} - \frac{1}{X}\frac{\partial \Psi}{\partial X} + \frac{\partial^2 \Psi}{\partial Z^2} = -4\pi^2 p'(\Psi)\left[X^2 - \frac{\langle B_T^2 \rangle}{\langle B^2 \rangle \langle 1/X^2 \rangle}\right] + \frac{\langle B_T^2 \rangle}{\langle B^2 \rangle} \sigma_{\parallel} \mathscr{E}_0 \qquad (35)$$

When the external conditions are fixed, the steady-state solution depends on two functions, $p(\Psi)$ and $\sigma_{\|}(\Psi)$.

5. CYLINDRICAL SYMMETRY

To illustrate the features of the problem concerned with the evolution of equilibria, let us consider the simplest situation of a plasma column in cylindrical symmetry. In this case there is no problem in finding the system of magnetic surfaces (i.e. the main point considered in our analysis), but one of the points, the penetration of the magnetic field into the plasma, becomes well understood.

To describe the evolution in a cylinder let us take the diffusion equation (14) with the equilibrium condition (11), in which the cylindrical radius r will be taken as a co-ordinate:

$$\sigma_{\|}\left[B_z^2\frac{\partial\Psi}{\partial t}-\frac{B_\theta B_z}{2\pi r}\frac{\partial\Phi}{\partial t}\right]=\frac{B^2}{r}\frac{\partial}{\partial r}r\frac{\partial\Psi}{\partial r}+B_\theta\frac{\partial p}{\partial r} \tag{36}$$

$$\frac{B_\theta}{r}\frac{\partial}{\partial r}rB_\theta+B_z\frac{\partial B_z}{\partial r}=-\frac{dp}{dr} \tag{37}$$

In this case

$$B_\theta=\frac{\partial\Psi}{\partial r},\qquad B_z=\frac{1}{2\pi r}\frac{\partial\Phi}{\partial r},\qquad B^2=B_\theta^2+B_z^2$$

If $\partial\Phi/\partial t$ is assumed to be given in Eq.(36), both $\partial\Psi/\partial t$ and the poloidal field distribution may be determined for the next step of time. Equilibrium equation (37) determines a new distribution of the longitudinal magnetic field B_z, thus allowing us to obtain $\partial\Phi/\partial t$. This quantity may be substituted into Eq.(36) to calculate a corrected value $\partial\Psi/\partial t$. Such a procedure of sequential approximations is repeated until the necessary accuracy is achieved, after which the next step in time may be taken. This technique, which allows us to separate the one-dimensional diffusion from steady-state equilibrium equations, may be also applied in a toroidal case, in which the diffusion equation remains one-dimensional, while its coefficients and the derivative $\partial\Phi/\partial t$ should be calculated using a two-dimensional, stationary equilibrium equation. When there are no toroidal effects, the characteristic features of the problem appear only at $B_\theta \sim B_z$, i.e. under conditions typical of the Zeta and screw-pinch devices.

Let us first consider the process of current growth in a cylindrical plasma column surrounded by a conducting shell. For simplicity we shall take $p = 0$, $\sigma_{\|} = \text{const}$, the total flux to be conserved. As has been shown by Kadomtsev [16],

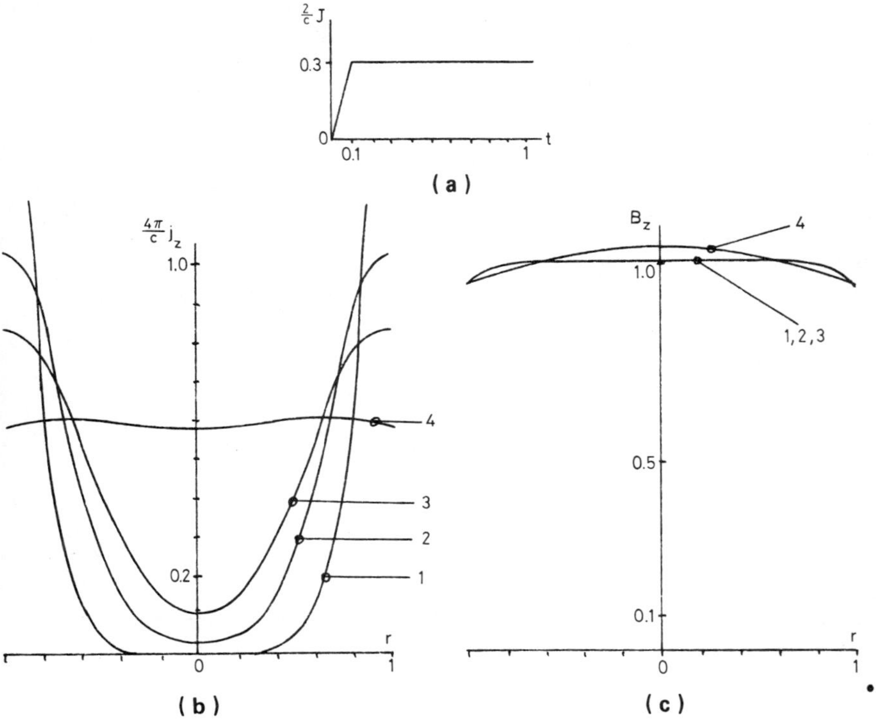

FIG.1. The process of current growth in a straight cylindrical plasma column. The case of weak current:

$$B_z|_{t=0} = 1; \quad \frac{J_a}{2\pi} = 0.3; \quad \sigma_{\|} = \text{const}; \quad \frac{\sigma_{\|}}{4} = 1$$

(a) Total current versus time.
(b) Current density and
(c) longitudinal magnetic field distributions. 1: t = 0.1; 2: t = 0.2; 3: t = 0.3; 4: t = 1.

under these conditions the steady-state solution of the evolution is characterized by the parameter $\eta = aJ/\Phi$, where a is the shell radius, and J_a, Φ_a are the total longitudinal current and flux.

Figures 1 and 2 show the evolution of the profiles j_z and B_z for two finite values of $\eta = 0.6$ and $\eta = 2.5$. In the case of a strong longitudinal field ($\eta = 0.6$) the equilibrium conditions do not influence the penetration of the current. In the case of a weak field ($\eta = 2.5$) the increase in the current is followed by the contraction of B_z, and the current penetrates the plasma more rapidly.

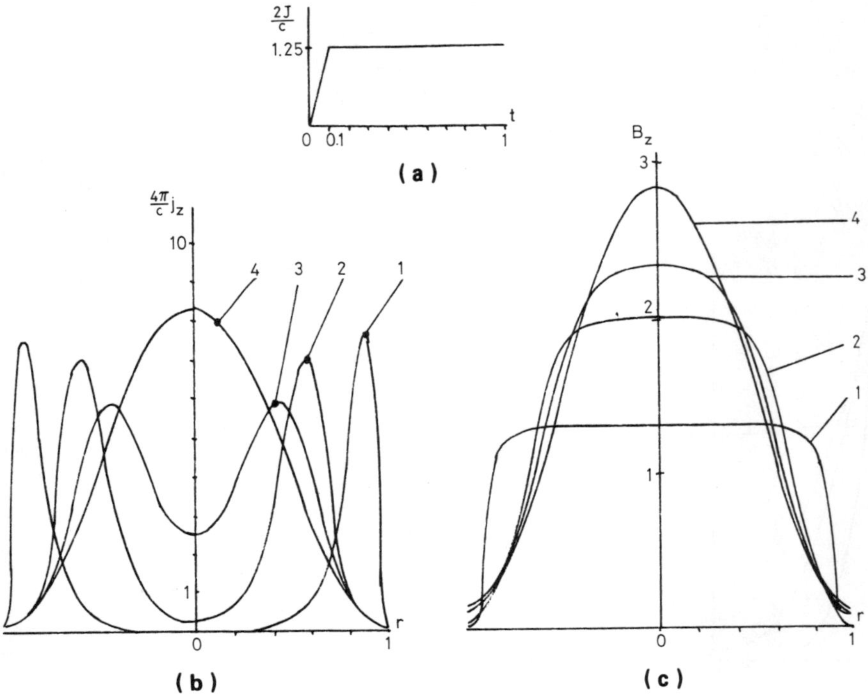

FIG.2. The case of sufficiently strong current: $B_z|_{t=0} = 1$; $\frac{J_a}{2\pi} = 1.25$.
(a) Current versus time.
(b) Current and
(c) longitudinal magnetic field distributions. 1: t = 0.1; 2: t = 0.2; 3: t = 0.3; 4: t = 1.

The physical picture of this behaviour is quite clear: as the current increases, a force appears which tends to compress the plasma (pinch effect). As a result, it should move with inertial velocity, i.e. instantly (under the assumption considered) inwards until the longitudinal field captured by it compensates this inward force. The field configurations resulting from such a motion are independent of the intermediate stages and are reflected in Eqs (36) and (37), in which the motion is eliminated; there is no inertial term. The effect of motion to increase the diffusion rate appears in changing the diffusion coefficient $1/\sigma_{\parallel}$ to $(B^2/B_z^2)\cdot(1/\sigma_{\parallel})$, in which the fields are determined by the equilibrium conditions. In particular, as the current increases, the longitudinal field at the plasma periphery decreases, resulting in an efficient and rapider penetration of the poloidal field.

As another example of the solution of Eqs (36) and (37) let us consider the evolution of equilibria with rapid plasma heating. Figure 3 presents the results of the calculation. When the pressure is sharply increased, the configuration re-arranges, the fluxes being conserved, i.e. the dependence $\Psi(\Phi)$ remains

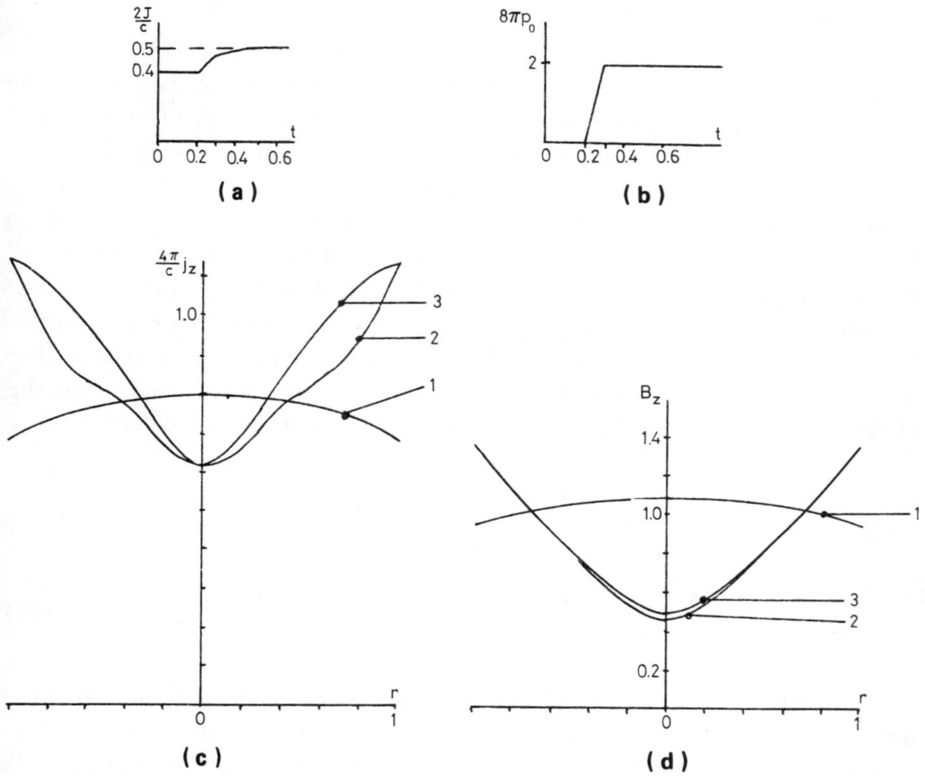

FIG.3. Evolution of cylindrical equilibria during fast growing of plasma pressure, the boundary electrical field being constant.

$$B_z|_{t=0} \approx 1; \quad \frac{J_a}{2\pi} = 0.4; \quad \sigma_\parallel = \text{const}; \quad \frac{\sigma_\parallel}{4} = 1; \quad p(r,t) = p_0(t)(1-r^2)$$

(a) Current versus time.
(b) Plasma pressure versus time.
(c) Current density and
(d) longitudinal magnetic field distributions. 1: t = 0.2; 2: t = 0.3; 3: t = 0.3.

unchanged. After this, a slower relaxation process takes place until the steady state is achieved.

Note that the calculations have shown that a term with $\partial\Phi/\partial t$ entering into the diffusion equation is found not to be essential (when the total flux Φ_a is conserved) even at large currents; the equilibrium conditions appreciably influence only the coefficients which contain magnetic fields.

6. CONCLUSION

We have shown that to solve the problems of evolution, assuming a prescribed plasma pressure p(V), it is sufficient to know the expression $\langle \vec{E}\cdot\vec{B}\rangle$ (Eq.(8)), i.e. essentially the longitudinal conductivity $\sigma_{\|}(\Psi)$. Realization of the solution is possible by both the Grad method [4] and the Taylor method [6]. In the first, as well as in Ref. [11], in which the evolution equations in the form of Eq.(9) were first written, the time derivatives relate to moving magnetic surfaces. Therefore the evolution equation (of the diffusion kind) is one-dimensional, but it includes the unknown 'surface' functions $\langle g_{ik}/g\rangle$. To find these functions it is necessary to know $\Psi(r, t)$. The dependence of Ψ on $\vec{r}$ is determined from an equation of the elliptic kind, which does not include time derivatives. In Taylor's method the problem is solved in a steady co-ordinate system and it contains the derivatives $\partial\Psi(\vec{r},t)/\partial t$ at a fixed point.

These derivatives are related in the following way. If a magnetic surface $\Psi(\vec{r}, t) = \hat{\Psi}(t)$ bounds a plasma volume (with a moving boundary!),

$$\left\langle \frac{\partial\Psi(\vec{r},t)}{\partial t}\right\rangle = \frac{d\hat{\Psi}}{dt}\,; \qquad \left\langle \frac{\partial V(\vec{r},t)}{\partial t}\right\rangle = 0 \tag{38}$$

Effective numerical procedures to solve the evolution problem need to be developed in order to solve subsequent problems of total simulation of the plasma processes, since the requirements for accurate location of the magnetic surfaces in the vacuum chamber increase with plasma temperature. The transport equations needed for determining the pressure are reduced to one-dimensional averaged continuity and thermal conductivity equations:

$$\frac{\partial n}{\partial t} + \frac{\partial}{\partial V}\left(n\langle \vec{v}\nabla V\rangle\right) = \langle \Gamma\rangle \tag{39}$$

$$\frac{3}{2}n_i\left(\frac{\partial T_i}{\partial t} + \langle \vec{v}\nabla V\rangle\frac{\partial T_i}{\partial V}\right) + n_i T_i\frac{\partial}{\partial V}\langle \vec{v}\nabla V\rangle + \frac{\partial}{\partial V}\langle \vec{q}_i\nabla V\rangle$$

$$= \frac{3m}{M}\frac{n_e}{\tau_e}(T_e - T_i) + \langle Q_\Gamma^i\rangle \tag{40}$$

$$\frac{3}{2}n_e\left(\frac{\partial T_e}{\partial t} + \langle \vec{v}\nabla V\rangle\frac{\partial T_e}{\partial V}\right) + n_e T_e\frac{\partial}{\partial V}\langle \vec{v}\nabla V\rangle + \frac{\partial}{\partial V}\langle \vec{q}_e\nabla V\rangle$$

$$= \frac{\langle j^2\rangle}{\sigma_{\|}} - \frac{3m}{M}\frac{n_e}{\tau_e}(T_e - T_i) + \langle Q_\Gamma^e\rangle \tag{41}$$

Densities n_i, n_e and temperatures T_e, T_i are the surface functions in this case, i.e. $n_i = n_i(V, t)$, etc. The averaged heat fluxes $\langle \vec{q}_i \nabla V \rangle$, $\langle \vec{q}_e \nabla V \rangle$ and the plasma expansion rate $\langle \vec{v} \nabla V \rangle$ for the collisional regime in general geometry are given in Ref. [14]; $\langle \Gamma \rangle$, Q^i_Γ, Q^e_Γ – the supplementary sources of the particles and heat – depend on a particular problem.

ACKNOWLEDGEMENTS

The authors would like to thank L.M. Degtyarev, A.P. Favorskij and P.N. Vabishchevich for useful discussions of the problems considered in the paper.

REFERENCES

[1] GRAD, H., HOGAN, J., Phys. Rev. Lett. **24** (1970)1337.
[2] GRAD, H., HU, P., STEVENS, D., Proc. Natl. Acad. Sci. USA **72** (1975) 3789.
[3] PAO, Y.-P., Phys. Fluids **19** (1976) 1177.
[4] GRAD, H., et al., "Classical plasma diffusion", Plasma Physics and Controlled Nuclear Fusion Research 1976 (Proc. 6th Int. Conf. Berchtesgaden, 1976) **2**, IAEA, Vienna (1977) 355.
[5] NÜHRENBERG, J., Nucl. Fusion **11** (1971) 655.
[6] TAYLOR, J.B., Culham Lab. Rep. PPN 10/75 (1975).
[7] CONNER, J.W., Culham Lab. Rep. PPN 29/75 (1975).
[8] CLARKE, J.F., SIGMAR, D.J., Oak Ridge Natl. Lab. Rep. ORNL/TM-5599 (1976).
[9] DORY, R.D., PENG, Y.K.M., Oak Ridge Natl. Lab. Rep. ORNL/TM-5555 (1976).
[10] CALLEN, J.D., et al., "Tokamak plasma magnetics", Plasma Physics and Controlled Fusion Research 1976 (Proc. 6th Int. Conf. Berchtesgaden, 1976) **2**, IAEA, Vienna (1977) 369.
[11] SHAFRANOV, V.D., Rev. Plasma Phys. **2** (1966) 103.
[12] SHAFRANOV, V.D., Nucl. Fusion **8** (1968) 253.
[13] BATEMAN, G., Nucl. Fusion **13** (1973) 227.
[14] SHAFRANOV, V.D., CONNOR, J.W., WATSON, K.J.H., Sov. J. Plasma Phys. **2** (1976).
[15] HINTON, F.L., HAZELTINE, R.D., Rev. Mod. Phys. **48** (1976) 239.
[16] KADOMTSEV, B.B., Nucl. Fusion, Suppl. **3** (1962) 969 (in Russian).

EQUILIBRIUM OF A TOROIDAL PLASMA

J. Pantuso SUDANO, L.C. Sandoval GOES
Instituto Technológico de Aeronáutica,
São José dos Campos,
Estado de São Paulo,
Brasil

Abstract

EQUILIBRIUM OF A TOROIDAL PLASMA.

A class of exact MHD toroidal equilibrium solutions is found which holds good for arbitrary aspect ratio. The main features of this solution are: (a) near the magnetic axis the flux function surfaces are confocal ellipses of known eccentricities; and (b) far from the magnetic axis there exists a separatrix with a form similar to a lemniscate.

The MHD equilibrium equation has already been solved for large-aspect ratio by several authors (see e.g. Shafranov [1]). Recently, Herrnegger [2] and Maschke [3] have presented an exact solution of the MHD equilibrium equation in terms of Coulomb wave functions. This paper is the continuation of the exact solution presented by Sudano [4].

We wish to solve the MHD equilibrium equations:

$$\mathrm{div}\vec{B} = 0 \tag{1}$$

$$\vec{J} = (c/4\pi)\ \mathrm{rot}\vec{B} \tag{2}$$

$$c^{-1}\vec{J} \times \vec{B} = \nabla p \tag{3}$$

inside an axisymmetric toroidal plasma. In cylindrical co-ordinates R, ϕ, Z, the magnetic field B and the current density J are written as usual in terms of the flux functions ψ and I by means of the expressions:

$$\vec{B} = (R_o/R)\vec{e}_\phi \times \nabla\psi + (R_o/R)\ I\ (R,Z)\ \vec{e}\ \phi \tag{4}$$

$$\vec{J} = J\phi\vec{e}_\phi - (c/4\pi)(R_o/R)\vec{e}_\phi \times \nabla I \tag{5}$$

In expressions (4) and (5), R_0 is the large radius of the plasma toroid. From the equilibrium equation (3), one shows that I and p are functions of ψ alone:

$$I = I(\psi), \qquad p = p(\psi) \tag{6}$$

The equilibrium equation (3) may be written in the well-known form:

$$\frac{\partial^2\psi}{\partial R^2} + \frac{\partial^2\psi}{\partial Z^2} - R^{-1}\frac{\partial\psi}{\partial R} = -4\pi\,\frac{R^2}{R_o^2}\,\frac{\partial p}{\partial\psi} - \frac{1}{2}\frac{\partial}{\partial\psi}\,I^2 \tag{7}$$

The toroidal current density may be put in the form:

$$J\phi = \frac{cR_o}{4\pi R}\left(-4\pi\frac{R^2}{R_o^2}\,\frac{\partial p}{\partial\psi} - \frac{1}{2}\,\frac{\partial}{\partial\psi}\,I^2\right) \tag{8}$$

If

$$I^2 = I_o^2 + \frac{1}{2}\beta\psi^2 \tag{9}$$

$$p = p_o + \alpha\psi \tag{10}$$

with constant I_0, β, p_0 and α, we show that a class of exact solutions can be found.

With this choice the equilibrium equation (7) becomes

$$\partial^2\psi/\partial R^2 + \partial^2\psi/\partial Z^2 - R^{-1}(\partial\psi/\partial R) = -(4\pi/R_o^2)\alpha R^2 - \frac{1}{2}\beta\psi \tag{11}$$

Equation (11) can be solved by writing

$$\psi(R,Z) = \psi_1(R) + \psi_2(Z) \tag{12}$$

Combining Eqs (11) and (12), we have

$$\psi_1^{11} - R^{-1}\psi_1^{1} + \frac{1}{2}\beta\psi_1 + (4\pi/R_o^2)\alpha R^2 = -\psi_2^{11} - \frac{1}{2}\beta\psi_2 = \lambda \tag{13}$$

Here,

$$\psi_1^{1} = (d/dR)\psi_1, \quad \psi_1^{11} = (d^2/dR^2)\psi_1, \quad \psi_2^{11} = d^2\psi_2/dZ^2$$

and λ is the separation constant. From Eq. (13) one obtains

$$\psi_2^{11} + \frac{1}{2}\beta\psi_2 + \lambda = 0 \tag{14}$$

$$\psi_1'' - R^{-1}\psi_1' + \frac{1}{2}\beta\psi_1 + (4\pi/R_o^2)\alpha R^2 = \lambda \tag{15}$$

We took solutions of (14) which are symmetric with respect to the plane $Z = 0$. We therefore put

$$\psi_2 = (2A_2/\beta)\cos(\beta/2)^{1/2} Z - 2\lambda/\beta \tag{16}$$

where A_2 is the integration constant. To solve Eq.(15) we write

$$\frac{1}{2}\beta\psi_1 - \lambda = RU(R) \tag{17}$$

and obtain from Eq.(15)

$$R^2\ddot{U} + R\dot{U} + (\frac{1}{2}\beta R^2 - 1)U = -(4\pi\alpha\beta/2R_o^2)R^3 \tag{18}$$

The solution of (18) may be written as

$$U = A_1 J_1\left[R(\beta/2)^{1/2}\right] + B_1 Y_1\left[R(\beta/2)^{1/2}\right] - (4\pi\alpha/R_o^2)R \tag{19}$$

Here J_1 and Y_1 are Bessel functions, and A_1, B_1 the constants of integration. Now from Eqs (17) and (19) one gets

$$\psi_1 = (2/\beta)R\left\{(A_1 J_1\left[(\beta/2)^{1/2} R\right] + B_1 Y_1\left[(\beta/2)^{1/2} R\right] - (4\pi\alpha/R_o^2)R\right\} + (2/\beta)\lambda \tag{20}$$

Thus, the exact solution we have given for the equilibrium equation (7) is

$$\psi(R,Z) = (2/\beta)R\{A_1 J_1\left[(\beta/2)^{1/2}R\right] + B_1 Y_1\left[(\beta/2)^{1/2}R\right] - (4\pi\alpha/R_o^2)R\} + (2A_2/\beta)\cos(\beta/2)^{1/2}Z \tag{21}$$

To determine the five parameters A_1, B_1, A_2, α and β, we assume that the curves $\psi(R,Z)$ = cte are confocal ellipses with eccentricity (e) near the magnetic axis and prescribe the value of $\psi(R_0, 0) = \psi_0$, with R_0 the radius of the magnetic axis.

The first condition may be put in the form

$$\psi(R_o + h, k) = \psi(R_o, 0) + \frac{1}{2} \left.\frac{\partial^2\psi}{\partial R^2}\right]_o h^2 + \frac{1}{2} \left.\frac{\partial^2\psi}{\partial z^2}\right]_o K^2 \tag{22}$$

where we impose that

$$\left.\frac{\partial\psi}{\partial R}\right]_o = \left.\frac{\partial\psi}{\partial Z}\right]_o = \left.\frac{\partial^2\psi}{\partial R\partial Z}\right]_o = 0 \tag{23}$$

$$\left.\frac{\partial^2\psi}{\partial R^2}\right]_o = (1 - e^2) \left.\frac{\partial^2\psi}{\partial Z^2}\right]_o \tag{24}$$

The conditions

$$\left.\frac{\partial\psi}{\partial Z}\right]_o = \left.\frac{\partial^2\psi}{\partial R\partial Z}\right]_o = 0$$

are satisfied identically. The conditions

$$\left.\frac{\partial\psi}{\partial R}\right]_o = 0 \text{ and } \left.\frac{\partial^2\psi}{\partial R^2}\right]_o = (1-e^2) \left.\frac{\partial^2\psi}{\partial Z^2}\right]_o$$

give, respectively:

$$A_1 J_o(\beta_o R_o) + B_1 Y_o(\beta_o R_o) = \frac{1}{\beta_o} \frac{8\pi\alpha}{R_o^2} \tag{25}$$

$$\frac{1}{\beta_o}\{A_1\left[J_o(\beta_o R_o) - \beta_o R_o J_1(\beta_o R_o)\right] + B_1\left[Y_o(\beta_o R_o)\right.$$

$$\left.- \beta_o R_o Y_1(\beta_o R_o)\right]\} - \frac{8\pi\alpha}{(\beta_o R_o)^2} = - A_2 (1 - e^2) \tag{26}$$

with $\beta_0 = (\beta/2)^{1/2}$. For the value $\psi(R, Z)$ at the magnetic axis we choose $\lambda = 0$ and $\psi_1(R_0) = 0$, so

$$\psi(R_o,0) = \psi_1(R_o) + \psi_2(0) = \psi_o$$

$$\psi(R_o,0) = \psi_2(0) = \frac{A_2}{\beta_o^2}$$

The condition $\psi_1(R_0) = 0$ gives

$$A_1 J_1(\beta_o R_o) + B_1 Y_1(\beta_o R_o) - \frac{4\pi\alpha}{R_o} = 0 \tag{27}$$

Finally we choose $\beta_0 R_0 = 1$ for convenience, and solve the subsequent system with A_2 as a parameter. A_2 and p_0 will be adjusted later to make the pressure null at the plasma edge and maximum at the magnetic axis:

$$p(\psi_b) = 0 \; ; \quad p(\psi_o) = \langle n_o k T_o \rangle$$

$$J_o(1) A_1 + J_1(1) B_1 - (8\pi/R_o)\alpha = 0$$

$$J_1(1) A_1 + Y_1(1) B_1 - (4\pi/R_o)\alpha = 0$$

$$\left[J_o(1) - J_1(1)\right] A_1 + \left[Y_o(1) - Y_1(1)\right] B_1 - (8\pi/R_o)\alpha$$

$$= - \frac{A_2}{R_o} (1-e^2)$$

The solution is:

$$A_1 = 2.59 \frac{A_2}{R_o} (1 - e^2) , \qquad \alpha = \frac{A_2}{4\pi} (1 - e^2)$$

$$B_1 = 0.179 \frac{A_2}{R_o} (1-e^2)$$

$$\psi(R,Z) = \frac{A_2}{\beta_o^2} \left\{ \frac{R}{R_o} (1-e^2) \left[2.59 \quad J_1(R/R_o) \right.\right.$$

$$\left.\left. + 0.179 \; Y_1(R/R_o) - \frac{R}{R_o} \right] + \cos\,(Z/R_o) \right\} \tag{28}$$

Putting

$$x = \frac{R}{R_o}, \; y = \frac{Z}{R_o}, \quad \psi_o = A_2/\beta_o^2 = A_2 R_o^2$$

$$J\phi^o = J\phi \;(R_o,0) = J\phi \;(\psi_o)$$

$$B\phi^o = B\phi \;(R_o,0) = B\phi(\psi_o)$$

$$X(x) = 2.59 \; J_1(x) + 0.179 \; Y_1(x)$$

we have

$$\psi(x,y) = \psi_o \; \{x \; (1 - e^2) \left[X\;(x) - x\right] + \cos\; y\} \tag{29}$$

$$J\phi(x,y) = \frac{c}{4\pi R_o^2} \frac{1}{x} \left[\psi - (1 - e^2)\; x^2 \; \psi_o\right]$$

$$J\phi(x,y) = J\phi^o \left[\frac{1-e^2}{2-e^2} \; X\;(x) + \cos\, y/(2-e^2)x^2\right] \tag{30}$$

$$B\phi(x,0) = B\phi^o \; \{(1-e^2) \left[X(x) - x\right] + x^{-1}\} \tag{31}$$

Proceeding in this way, we have fixed the magnetic axis and guaranteed its unicity within the variable plasma boundary. Any one of these surfaces inside the region limited by the separatrix can be chosen as the plasma boundary (as shown by Shafranov [1]), where the pressure is null [$p(\psi_b) = 0$]. The value of ψ_b is determined by the maximum dimension of the plasma major cross-section. It is also possible to find a solution in the case where the ellipses are elongated in the vertical direction (see Figs 1–4).

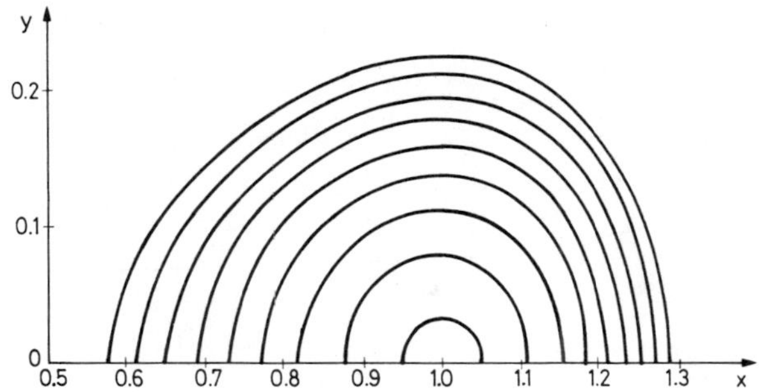

FIG. 1. Plot of magnetic flux surfaces near the magnetic axis for e = 0.5.

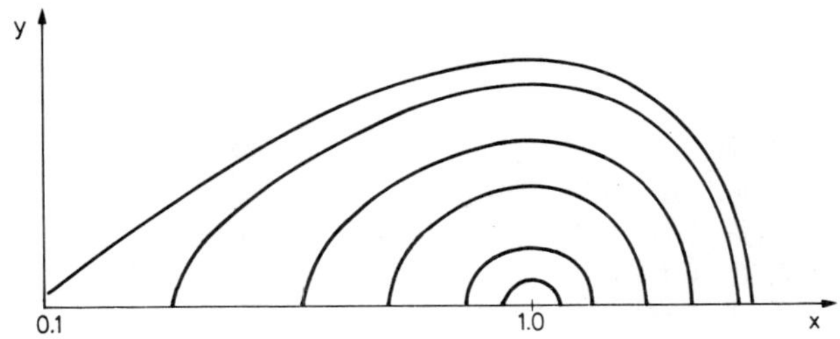

FIG. 2. Typical magnetic surfaces far from the magnetic axis for e = 0.5.

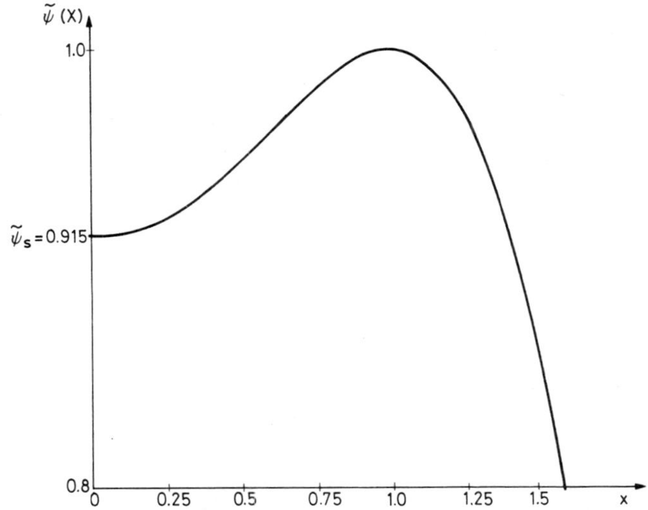

FIG. 3. Plot of $\tilde{\psi}(X) = \psi(R, 0)/\psi(R_0, 0)$.

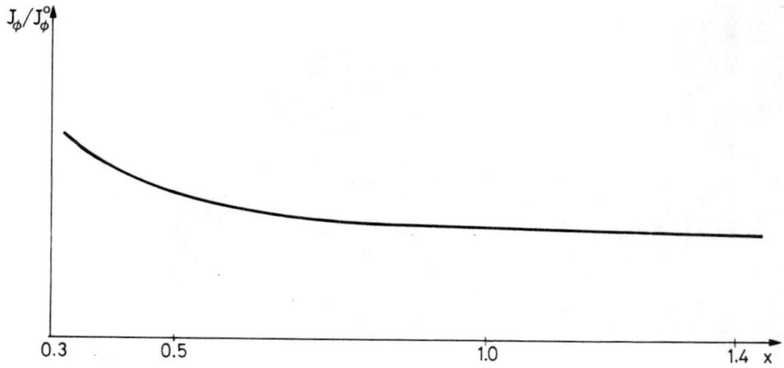

FIG.4. Plot of current density for Z = 0.

REFERENCES

[1] SHAFRANOV, V.D., Reviews of Plasma Physics **2** (LEONTOVICH, M.A., Ed.), Consultants Bureau, New York (1966) 103.

[2] HERRNEGGER, F., Proc. 5th Europ. Conf. on Controlled Fusion and Plasma Physics, Grenoble, 1972, Vol.1, p.26.

[3] MASCHKE, E.K., Plasma Phys. **15** (1973) 535.

[4] SUDANO, J.P., Phys. Fluids **17** (1975) 1915.

Appendix A

ICTP COLLEGE ON THEORETICAL AND COMPUTATIONAL PLASMA PHYSICS

PROGRAMME OF THE COLLEGE

Scope of the lectures

B. McNamara	Physics of single particles in magnetized plasmas
V. Tsytovich	Non-linear theory of waves
J. Killeen	Computational plasma physics
J. Brackbill	Magnetic confinement
K. Nishikawa	Kinetic theory (2 lectures)
D. Pfirsch	Transport theory
J. Byers	Electromagnetic simulation models
H. Okuda	Particle simulation (2 lectures)
J. Brackbill	Numerical methods in 3D MHD
K. Roberts	Construction of plasma physics codes
B. McNamara	The mirror machine
K. Roberts	Numerical methods in MHD
V. Tsytovich	Non-linear plasma kinetics
G. Laval	Quasi-linear theory (2 lectures)
D. Biskamp	MHD calculations
V. Shafranov	MHD (2 lectures)
V. Tsytovich	Non-linear theory
D. Pfirsh	Diffusion

Discussion on numerical methods

B. Kadomtsev	Non-linear phenomena (3 lectures)
V. Shafranov	Theory of equilibrium
R. Grimm	Tokamak equilibrium computations
K. Nishikawa	Parametric instabilities
Discussion on equilibria	
M.N. Rosenbluth	Resistive instabilities
R.B. White	2D ideal MHD in tokamaks
A. Langdon	Particle simulation (2 lectures)
D. Biskamp	MHD computations
Discussion on non-linear theory	
P. Rutherford	Plasma diffusion
M.N. Rosenbluth	Non-linear MHD in tokamaks
B. Waddell	Tearing mode calculations for tokamaks
J. Killeen	Fokker-Planck computations
R. Grimm	MHD stability computations
K. Roberts	MHD computations
C. Oberman	Advanced kinetic theory
A. Langdon	Electromagnetic particle code simulations
J. Brackbill	3D MHD simulations of Scyllac
K. Roberts	The reversed field pinch
J. Freeman	Electron beam fusion
A. Sitenko	Kinetic theory
B. Coppi	Tokamak instabilities (2 lectures)
D. Ryutov	Generation, focusing and accumulation of electron beams
D. Ryutov	Plasma heating by electron beams
A. Galeev	Magnetic reconnection in a space plasma

Appendix B

THIRD INTERNATIONAL ('KIEV') CONFERENCE ON PLASMA THEORY

PROGRAMME OF THE CONFERENCE

5 April

G. ECKER: Statistical mechanics of partially ionized gases
J.H. MISGUICH, R. BALESCU: A theory for correlations in strongly turbulent plasma

J. NÜHRENBERG: Stable MHD equilibria
G.O. SPIES, D.B. NELSON: Staircase model for global MHD modes
R.C. GRIMM: Numerical studies of MHD instabilities in tokamaks

M.B. TENDLER: Electron energy distribution in a high-pressure gas discharge sustained by a high-current electron beam
F.R. CROWNFIELD: Exact inhomogeneous 1D Vlasov equilibria
M.H.A. HASSAN: Anisotropic temperature relaxation in magnetized plasma

A. RAMANI, G. LAVAL: Electromagnetic instabilities in a plasma with a temperature gradient
A. BOURDIER, D. BABONNEAU, G. DI BONA, X. FORTIN: Relativistic effects on electromagnetic wave propagation in a cold inhomogeneous electron plasma
Y.-K.M. PENG: Poloidal field calculations for a D-shaped flux-conserving tokamak
N. SINGH: Resonance cones in a flowing magnetoplasma

A.G. SITENKO, I.P. YAKIMENKO, A.G. ZAGORODNY: Radiation by non-equilibrium plasma
R. CALINON, D. MERLINI: Particle correlations and gamma-instability in the charged-filaments model of plasma
M.P. RAVINDRA, G.C. DAS: Transport properties of multicomponent plasma using Baum's model
M.S. SAGOO: Electron recombination coefficient for dielectric droplets in a plasma

6 April

C. OBERMAN, J. KROMMES: Theory of transport in magnetized plasma
K.S. VISWANATHAN, G. RENUKA: Pitch angle diffusion by bounce resonance
R.W.B. BEST: Growing second-order wave in the Landau-Vlasov problem
R.K. VARMA: Interaction of moving magnetized plasma with a neutral gas

D.K. CALLEBAUT, A.H. KHATER: Variational principle including higher-order perturbations
C. UBEROI: Resonant absorption of Alfvén waves in non-uniform plasma
G. HASSELBERG, A. ROGISTER, A. EL NADI: Non-linear saturation of drift waves
G. HASSELBERG, A. ROGISTER: Collisional temperature drift – tearing instabilities in tokamak plasmas
T. OGINO, S. TAKEDA: Propagation of large-amplitude fast magnetosonic waves through cylindrical inhomogeneous plasmas

N.F. CRAMER: Parametric excitation of ion-cyclotron waves by a pump magnetic field in a high-temperature plasma
N.E. ANDREEV, K. SAUER: Jumps in the non-linear reflectivity of an overdense plasma layer
M.P. SRIVASTAVA: Absolute parametric instability in a turbulent inhomogeneous plasma
M. MOHAN, R. ACHARYA: Collisional absorption of two laser beams in plasma

B.V. WADDELL, G.L. JAHNS, J.D. CALLEN, H.R. HICKS: Interpretation of tokamak sawtooth oscillations
E.J. JORNADA, J.D. GAFFEY: Instabilities and plasma heating resulting from neutral beam injection in magnetized plasmas
R.S. SCHNEIDER, J.D. GAFFEY: Time-dependent alpha-particle distribution and plasma heating by alpha particles
Y.-K.M. PENG, D.J. SIGMAR, R.A. DORY, J.A. HOLMES: Properties of high-beta equilibria in flux-conserving tokamaks
A. RON: Fusion of superthermal ions in dense plasma

M. BRAMBILLA: Electron Landau damping of lower hybrid waves
M.-Y. YU, P. SHUKLA, K.H. SPATSCHEK: Modulation of propagating lower hybrid waves
H. ABE, H. KAJITANI, R. ITATANI: Particle simulation on the propagation and plasma heating of the lower hybrid waves in the non-uniform system
A. STERNLIEB: Non-linear aspects of the electromagnetic ion-cyclotron instability

7 April

H. TASSO: Dissipative MHD instabilities
D. SCHNACK, J. KILLEEN: Linear and non-linear calculations of the tearing mode
J.P. GOEDBLOED, J.P. FREIDBERG: Equilibrium and stability of diffuse tokamaks and screw pinches
K. NISHIKAWA: Strong turbulence effects in electron beam heating
T.P. KHAN: A functional approach to non-linear three-wave interaction in plasma
A.A. GALEEV, R.Z. SAGDEEV, V.D. SHAPIRO, V.I. SHEVCHENKO: Strong Langmuir turbulence and energy dissipation
M. ROSENBERG, G. KALMAN: Waves and instabilities in an electron-positron plasma in a strong magnetic field
C. WAHLBERG: Adiabatic and non-adiabatic electron oscillations in a static electric field
V. CADEZ, S. VUKOVIC: Electromagnetic surface wave instability due to inhomogeneous particle stream
S. VUKOVIC, V. CADEZ: Self-consistent absorption in overdense laser plasma

J.R. FREEMAN, L. BAKER, S.L. THOMPSON: MHD effects in electron beam fusion targets
A.A. GABR, A.M. EL NADI, T.A. EL KHALAFAWY, N.I. RUDNEV: Study of formation and distribution of plasma currents in hard-core theta pinch with azimuthal magnetic field
N.T. GLADD, R.C. DAVIDSON: Influence of strong inhomogeneities and magnetic shear on high-frequency mirror-drift-cone and convective-loss-cone instabilities
D.D. RYUTOV: Physics of plasma with $B \gg 1$.

K. KONNO, T. MITSUHASHI, Y.J. ICHIKAWA: Dynamical processes of the dressed ion acoustic solitons
B. BUTI: The problem of cóllapse of plasma waves
A.S. SHARMA, B. BUTI: Modulational instability of ion-acoustic waves in a two-electron-temperature plasma
M. MOHAN, A.S. SHARMA, B. BUTI: Modulational instability of electrostatic drift waves
N. TSAGAS: Backward-travelling waves in a co-axial accelerator

8 April

M. DOBROWOLNY, M. TESSAROTTO: Kinetic instabilities in the solar wind
V.J. PETVIASHVILI: Non-one-dimensional solitons and anomalous phenomena in plasma
A.B. KITSENKO, I.M. PANKRATOV: The non-linear theory of the excitation of monochromatic electromagnetic waves in a magnetoactive plasma by the relativistic beam of charged particles
H.D. ABOURDJANIA, A.B. KITSENKO, I.M. PANKRATOV: The non-linear theory of the interaction of the charged-particle beam with a plasma in the magnetic field
V.N. TSYTOVICH, K. KOMILOV, F.Kh. KHAKIMOV: Statistical theory of strong Langmuir turbulence
K. ELSASSER, H. SCHAMEL: Anomalous ion heating in the presence of strong Langmuir turbulence
R.N. SUDAN, M. KESKINEN: Theory of strongly turbulent convection of 2D low-pressure plasma

J. BYERS, B. COHEN, W. CONDIT, T. KAISER: Computer simulation of magnetic field reversal in mirror machines
S.S. JHA, L.K. CHAVDA: Solution for spherical implosion due to coalesced weak shocks in a plasma
T. WATANABE, K. NISHIKAWA: Self-consistent steady state of r.f. plugging of plasma
F.J. HELTON, T.S. WANG: MHD equilibria in non-circular tokamaks with field-shaping coils

M.N. ROSENBLUTH: Parametric instabilities in random media
J.A. KROMMES: Turbulence and clumps
P. DASGUPTA, B. DASGUPTA: Single-particle distribution in strongly turbulent plasmas

M. TESSAROTTO: Dissipative trapped particle modes in a toroidal reactor
G.V. PEREVERSEV, V.D. SHAFRANOV, L.E. ZAKHAROV: The evolution of tokamak plasma equilibria
M.E. RENSINK: Tandem mirror studies at Lawrence Livermore Laboratory
R.M. KULSRUD, A.B. HASSAN: Mass motions in toroidal plasmas

9 April

A. AIROLDI CRESCENTINI, A. OREFICE, R. POZZOLI: Quasi-linear evolution of the slide-away regime in toroidal plasmas

J.P. SUDANO: Toroidal plasma equilibrium

T. TUDA: Temperature gradient effects on heavy impurity-ion transport in a tokamak

M. COTSAFTIS: Non-linear dynamics of Joule-heated toroidal discharges

M.N. ROSENBLUTH, J.B. TAYLOR: Diffusion from large-amplitude drift cells

T. TANGE, K. NISHIKAWA: Theory of impurity diffusion in a turbulent plasma

D.C. BARNES, J.U. BRACKBILL, W. SCHNEIDER: Analytic and numerical studies of diffuse 3D equilibria in Scyllac

FACULTY

DIRECTORS OF THE ICTP COLLEGE ON THEORETICAL AND COMPUTATIONAL PLASMA PHYSICS

B.B. Kadomtsev	Kurchatov Institute for Atomic Energy, Academy of Sciences of the USSR, Moscow, USSR
B. McNamara	Lawrence Livermore Laboratory, University of California, Livermore, California, USA
M.N. Rosenbluth	Institute for Advanced Study, School of Natural Sciences, Princeton, New Jersey, USA

DIRECTORS OF THE THIRD INTERNATIONAL ('KIEV') CONFERENCE ON PLASMA THEORY

B.B. Kadomtsev	Kurchatov Institute for Atomic Energy, Academy of Sciences of the USSR, Moscow, USSR
V.N. Tsytovich	Lebedev Physics Institute, Academy of Sciences of the USSR, Moscow, USSR

EDITOR

Miriam Lewis	Division of Publications, IAEA, Vienna

LIST OF PARTICIPANTS

H. Abe	Department of Electronics, Kyoto University, Kyoto, Japan	Japan
I. Abonyi	Institute of Theoretical Physics, Roland Eötvös University, Puskin u.5–7, H-1088 Budapest, Hungary	Hungary
J.C. Adam	Centre de physique théorique, Ecole polytechnique, F-91128 Palaiseau Cedex, France	France
A. Ag	Central Research Institute for Physics, Hungarian Academy of Sciences, P.O. Box 49, H-1525 Budapest, Hungary	Hungary
S. Ahmed	Institute for Nuclear Research, Zaklad VII, Hoža 69, 00-681 Warsaw, Poland	Poland/ Bangladesh
A. Airoldi Crescentini	Laboratorio de Fisica del Plasma del CNR, Via Celoria 16, I-20133 Milan, Italy	Italy
R.N. Aiyer	Laser Section, Bhabha Atomic Research Centre, Trombay, Bombay 400 085, India	India
A.D. Alawneh	Department of Mathematics, Faculty of Science, University of Jordan, Amman, Jordan	Jordan
B. Alekseev	Moscow Aviation Institute, Volokolamskoje 4, A-80 Moscow, USSR	USSR

A.M. Al-Hassoun	Physics Department, College of Science, Baghdad University, Baghdad, Iraq	Iraq
A.K. Arora	Department of Physics, University of Rajasthan, Jaipur, India	Sweden/India
M. Aslam Baig	Department of Physics, University of Karachi, Karachi, Pakistan	Pakistan
S. Bardwell	Fusion Energy Foundation, PO Box 1943, New York, NY 10001, USA	USA
S. Beg	Department of Physics, University of Islamabad, Islamabad, Pakistan	Pakistan
R.W.B. Best	FOM-Instituut voor Plasmafysica, Postbus 7, Jutphaas, The Netherlands	FOM/Netherlands
D. Biskamp	Max-Planck-Institut für Plasmaphysik, D-8046 Garching-bei-München, Fed. Rep. Germany	Fed. Rep. Germany
J.L. Bobin	Commissariat à l'énergie atomique, Centre d'études de Limeil, BP.N° 27, F-94 Villeneuve-St-Georges, France	France
A. Bondeson	Institute of Physics, Göteborg, Sweden	Sweden
A. Bourdier	Commissariat à l'énergie atomique, Centre d'études de Limeil, BP N° 27, F-94 Villeneuve-St-Georges, France	France
J. Brackbill	Los Alamos Scientific Laboratory, MFE Division, Mail Stop 640, PO Box 1663, Los Alamos, NM 87545, USA	USA
M. Brambilla	Association Euratom-CEA, Centre d'études nucléaires, F-38 Grenoble, Cedex 85, France	Euratom/Italy

S. Bujarbarua	Department of Physics, Cotton College, Gauhati 781 001, Assam, India	India
B. Buti	Physical Research Laboratory, Navrangpura, Ahmedabad 380 009, India	India
J.A. Byers	Lawrence Livermore Laboratory, University of California, PO Box 808, Livermore, CA 94550, USA	USA
N. Byrne	Science Applications Inc., 1200 Prospect St., PO Box 2351, La Jolla, CA 92037, USA	USA
V.M. Cadez	Institute of Physics, Studentski Trg 16/V, PO Box 57, 11001 Belgrade, Yugoslavia	Yugoslavia
D. Callebaut	Physics Department, Universitair Instelling Antwerpen, Universiteitsplein 1, B-2610 Wilrijk, Belgium	Belgium
V. Canivell	Departamento de Fisica Teórica, Facultad de Ciencias, Universidad de Barcelona, Av. Generalísimo, Barcelona-14, Spain	Spain
E.A. Caponi	TRW, 1 Space Park, Redondo Beach, CA 90266, USA	USA/Argentina
P.J. Catto	Department of Mechanical and Aerospace Sciences, University of Rochester, Rochester, NY 14627, USA	USA
M.S.Z. Chaghtai	Department of Physics, Aligarh Muslim University, Aligarh, UP, India	India
B. Chakraborty	Department of Mathematics, Jadavpur University, Calcutta-32, India	India

B. Coppi	Department of Physics, Massachusetts Institute of Technology, Cambridge, MA 02139, USA	USA/Italy
N.F. Cramer	School of Physics, University of Sydney, New South Wales 2006, Australia	Australia
F.R. Crownfield	Department of Physics, College of William and Mary, Williamsburg, VA 23185, USA	USA
E.H. Da Jornada	Instituto de Fisica, Universidade Federal do Rio Grande do Sul, Av. Professor Luiz Englert, 90 000 Porto Alegre, Brazil	Brazil
P. Dasgupta	Department of Physics, University of Kalyani, Kalyani, West Bengal 741235, India	India
R. De Angelis	Centro di Studio sui Gas Ionizzati, Istituto de Elettrotecnica e di Elettronica, Via Gradenigo 6/a, I-35100 Padua, Italy	Italy
C.J. Diaz	Department of Physics, Universidad del Valle, AA 2188, Cali, Colombia	Colombia
D. Dillenburg	Instituto de Fisica, Universidade Federal do Rio Grande do Sul, Av. Professor Luiz Englert, 90 000 Porto Alegre, Brazil	Brazil
B. Duborgel	Commissariat à l'énergie atomique, Centre d'études de Limeil, BP N° 27, F-94 Villeneuve-St.-Georges, France	France
I.U.-R. Durrani	Nuclear Physics Division, PINSTECH, PO Nilore, Rawalpindi, Pakistan	Pakistan
G. Ecker	Theoretische Physik I, Ruhr-Universität Bochum, Universitätsstr. 150, D-4630 Bochum, Fed. Rep. Germany	Fed. Rep. Germany

A. El Nadi	Electrical Engineering Department, Cairo University, Giza, Egypt	Egypt
G. Ferrante	Istituto di Fisica, Università di Palermo, Via Archirafi 36, I-90123 Palermo, Italy	Italy
X. Fortin	Commissariat à l'énergie atomique, Centre d'études de Limeil, BP N° 27, F-94 Villeneuve-St.-Georges, France	France
V. Fouad Hanna	Electronic and Electrical Engineering Laboratory, National Research Centre, El-Tahrir Street, Dokki, Cairo, Egypt	Egypt
J.R. Freeman	Plasma Theory Division – 5241, Sandia Laboratories, Albuquerque, NM 87115, USA	USA
L. Friedland	Racah Institute of Physics, The Hebrew University of Jerusalem, Jerusalem, Israel	Israel
A.A. Gabr	Department of Physics, Faculty of Science, University of Cairo, Cairo, Egypt	Egypt
L.G.E.M. Gad	Atomic Energy Establishment, Cairo, Egypt	Egypt
J.D. Gaffey	Instituto de Fisica, Universidade Federal do Rio Grande do Sul, Av. Professor Luiz Englert, 90 000 Porto Alegre, Brazil	Brazil/USA
A.A. Galeev	Space Research Institute, Academy of Sciences of the USSR, Profsoyuznaya 88, 117810 Moscow, USSR	USSR
H. Gerhauser	Institut für Plasmaphysik der Kernforschungsanlage Jülich GmbH, Postfach 1913, D-5170 Jülich 1, Fed. Rep. Germany	Fed. Rep. Germany

C. Gillet	Institut für Plasmaphysik der Kernforschungsanlage Jülich GmbH, Postfach 1913, D-5170 Jülich 1, Fed. Rep. Germany	Belgium/ Fed. Rep. Germany
N.T. Gladd	Department of Physics and Astronomy, University of Maryland, College Park, MD 20742, USA	USA
J.P. Goedbloed	Association Euratom-FOM, FOM-Instituut voor Plasmafysica, Rijnhuizen Jutphaas, The Netherlands	Netherlands
H. Goldstein	Institut für Theoretische Physik IV, Ruhr-Universität Bochum, Postfach 102148, D-4630 Bochum, Fed. Rep. Germany	Fed. Rep. Germany
A. Gourdin Servenière	Centre de mathématiques appliquées, Ecole polytechnique, F-91128 Palaiseau Cedex, France	France
R. Grimm	Plasma Physics Laboratory, Princeton University, PO Box 451, Princeton, NJ 08540, USA	USA
K.-D. Harms	Universität Essen – GHS, Fachbereich 7 (Physik), Universitätsstr.2, D-4300 Essen 1, Fed. Rep. Germany	Fed. Rep. Germany
M.H.A. Hassan	Department of Physics, Faculty of Science, University of Khartoum, Khartoum, Sudan	Sudan
R.L. Hatfield	Department of Physics, University of Cincinnati, Cincinnati, OH 45221, USA	USA
G. Hazak	Department of Physics, Nuclear Research Centre – Negev, Atomic Energy Commission, PO Box 9001, Beer-Sheva, Israel	Israel
F.J. Helton	General Atomic, PO Box 81608, San Diego, CA 92138, USA	USA

R. Hoonsawat	Department of Physics, Mahidol University, Rama VI Road, Bangkok 4, Thailand	Thailand
A. Jayendran	Department of Physics, Faculty of Science, University of Khartoum, Khartoum, Sudan	Sudan
S.S. Jha	Tata Institute of Fundamental Research, Homi Bhabha Road, Colaba, Bombay 400 005, India	India
A. Kaleck	Institut für Plasmaphysik der Kernforschungsanlage Jülich GmbH, Postfach 1913, D-5170 Jülich 1, Fed. Rep. Germany	Fed. Rep. Germany
F. Kh. Khakimov	Department of Theoretical Physics, V.I. Lenin Tajik State University, Dushanbe, USSR	USSR
T.P. Khan	Department of Physics, Dinabandhu Andrews College, PO Garia, DT, 24-Parganas, West Bengal, India	India
A. Khater	Department of Mathematics, University of Assiut, Assiut, Egypt	Egypt
M.A.-S. Khidr	Department of Mathematics, University of Assiut, Assiut, Egypt	Egypt
J. Killeen	Lawrence Livermore Laboratory, University of California, PO Box 808, Livermore, CA 94550, USA	USA
A.B. Kitsenko	Kharkov Physical-Technical Institute, Kharkov 310108, USSR	USSR
K. Konno	Department of Physics and Atomic Energy Research Institute, College of Science and Technology, Nihon University, Kanda-Surugadai, Chiyoda-ku, Tokyo, Japan	Japan

J.A. Krommes	Plasma Physics Laboratory, PO Box 451, Princeton, NJ 08540, USA	USA
R.M. Kulsrud	Department of Physics, Princeton University, PO Box 451, Princeton, NJ 08540, USA	USA
J. Kyncl	Nuclear Research Institute, Řež, Czechoslovakia	Czechoslovakia
A.B. Langdon	Lawrence Livermore Laboratory, University of California, PO Box 808, Livermore, CA 94550, USA	USA/Canada
J. Larsson	Department of Plasma Physics, Umea University, Umea, Sweden	Sweden
G. Laval	Centre de physique théorique, Ecole polytechnique, F-91120 Palaiseau Cedex, France	France
K.J. Le Couteur	Department of Theoretical Physics, Research School of Physical Sciences, Australian National University, PO Box 4, Canberra, ACT 2600, Australia	Australia/UK
L. Lengyel	Max-Planck-Institut für Plasmaphysik, D-8046 Garching-bei-München, Fed. Rep. Germany	Fed. Rep. Germany
C.-S. Lim	Department of Mathematics, University of Singapore, Bukit Timah Road, Singapore 10	Singapore
L. Lo Cascio	Istituto di Fisica, Via Archirafi 36, I-90123 Palermo, Italy	Italy
J.J. Lodder	FOM – Instituut voor Plasmafysica "Rijnhuizen", Overeindseweg 2, Postbus 2, Nieuwegein, The Netherlands	FOM/Netherlands

L. Lofo	Department of Physics, Faculty of Sciences, Université Nationale du Zaïre, BP N° 190, Kinshasa XI, Zaïre	Zaire
A. Lopez Fraguas	Divisiốn de Fusión, Junta de Energía Nuclear, Madrid-3, Spain	Spain
W. Maajost	Theoretische Physik I, NB7/134, Ruhr-Universität Bochum, Universitätsstr. 150, D-4630 Bochum, Fed. Rep. Germany	Fed. Rep. Germany
M. Maheswaran	Department of Mathematics, University of Sri Lanka, Peradeniya, Sri Lanka	Sri Lanka
J.A. Markvoort	FOM – Instituut voor Atom en Molecuulfysica, Kruislaan 407, Amsterdam, The Netherlands	FOM/Netherlands
S.N. Masmanian	56 Templemere, Oatlands Drive, Weybridge, Surrey KT13 9PB, England	UK/Greece
M. Mastihuba	Department of Experimental Physics, Komensky University, Mlynska Dolina, 81631 Bratislava, Czechoslovakia	Czechoslovakia
L. Melendez-Lugo	Instituto Nacional de Energía Nuclear, Insurgentes Sur 1079, Mexico 18 DF, Mexico	Mexico
I. Mellado	Junta de Energía Nuclear, Avda. Complutense, Madrid-3, Spain	Spain
S. Migliuolo	Department of Physics and Astronomy, University of Rochester, Rochester, NY 14627, USA	USA/Italy
R.L. Miller	General Atomic, PO Box 81608, San Diego, CA 92138, USA	USA

J. Misguich	Association Euratom-CEA, Département de la physique du plasma et de la fusion contrôlée, Centre d'études nucléaires de Fontenay-aux-Roses, BP N° 6, F-92260 Fontenay-aux-Roses, France	France
M. Mohan	K.M. College, University of Delhi, Delhi-7, India	India
D. Moreau	Centre d'études nucléaires Grenoble, SPh – PFC – SIG, 85 X, F-38041 Grenoble Cedex, France	France
F. Moser	Institut für Plasmaforschung, Universität Stuttgart, Pfaffenwaldring 31, D-7000 Stuttgart 80, Fed. Rep. Germany	Fed. Rep. Germany
G. Murtaza	Department of Physics, Quaid-i-Azam University, PO Box 1090, Islamabad, Pakistan	Pakistan
G.F. Nalesso	Centro di Studio sui Gas Ionizzati, Istituto di Elettrotecnica e di Elettronica, Via Gradenigo 6/a, I-35100 Padua, Italy	Italy
U. Narain	Department of Physics, Meerut College, Meerut 250001, India	India
J.D. Neethling	Atomic Energy Board, Pelindaba, Private Bag X256, Pretoria 0001, South Africa	South Africa
K. Nishikawa	Faculty of Science, Hiroshima University, Higashi-Sendamachi, Hiroshima, Japan	Japan
E. Novelo	Instituto Politécnico Nacional, UPIICSA, THE 950, Col. Granjas Mexico, Mexico 8 DF, Mexico	Mexico

J. Nührenberg	Max-Planck- Institut für Plasmaphysik, D-8046 Garching-bei-München, Fed. Rep. Germany	Fed. Rep. Germany
C. Oberman	Plasma Physics Laboratory, Princeton University, James Forrestal Campus, PO Box 451, Princeton, NJ 08540, USA	USA
T. Ogino	Department of Electrical Engineering, Nagoya University, Furocho, Chikusa-ku, Nagoya, Japan	Japan
H. Okuda	Plasma Physics Laboratory, Princeton University, James Forrestal Campus, PO Box 451, Princeton, NJ 08540, USA	USA/Japan
A. Orefice	Laboratorio de Fisica del Plasma del CNR, Istituto di Fisica, Via Celoria 16, I-20133 Milan, Italy	Italy
S. Ortolani	Centro di Studio sui Gas Ionizzati, Istituto di Elettrotecnica e di Elettronica, Via Gradenigo 6/a, I-35100 Padua, Italy	Italy
F. Pegoraro	Scuola Normale Superiore, Piazza dei Cavalieri 7, I-56100 Pisa, Italy	Italy
Y.-K.M. Peng	Theory Section, Fusion Energy Division, Oak Ridge National Laboratory, PO Box Y, Oak Ridge, TN 37830, USA	USA
V.J. Petviashvili	I.V. Kurchatov Institute of Atomic Energy, Moscow 123182, USSR	USSR
D. Pfirsch	Max-Planck-Institut für Plasmaphysik, D-8046 Garching-bei-München, Fed. Rep. Germany	Fed.Rep.Germany
R. Pozzoli	Laboratorio de Fisica del Plasma del CNR, Istituto di Fisica, Via Celoria 16, I-20133 Milan, Italy	Italy

D.H. Priester	Division of Magnetic Fusion Energy, USERDA, Mail Station G-234, Washington, DC 20545, USA	USA
M. Pusterla	Istituto di Fisica, Università di Padova, Via Marzolo 8, I-35100 Padua, Italy	Italy
A. Ramani	Centre de physique théorique, Ecole polytechnique, F-91128 Palaiseau Cedex, France	France
R. Ramberger	Institut für Theoretische Physik, Universität Innsbruck, Innrain 52, A-6020 Innsbruck, Austria	Austria
K. Rashid	PINSTECH, Post Office Nilore, Rawalpindi, Pakistan	Pakistan
M.P.R. Ravindra	Department of Applied Mathematics, Indian Institute of Science, Bangalore 560012, India	India
G. Realini	Centro Comune di Ricerca, Euratom, Ispra, I-21020 Varese, Italy	Euratom/Italy
M.E. Rensink	Lawrence Livermore Laboratory, University of California, PO Box 808, Livermore, CA 94550, USA	USA
K.V. Roberts	Culham Laboratory, UKAEA Research Group, Abingdon, Oxfordshire OX14 3DB, England	UK
A. Rogister	Institut für Plasmaphysik der Kernforschungsanlage Jülich GmbH, D-5170 Jülich, Fed. Rep. Germany	Fed.Rep.Germany/ Belgium
A. Ron	Department of Physics, Israel Institute of Technology, Haifa, Israel	Israel

M. Rosenberg	Department of Theoretical Physics, University of Oxford, 12 Parks Road, Oxford, Oxfordshire OX1 3PQ, England	UK/USA
P.H. Rutherford	Plasma Physics Laboratory, Princeton University, James Forrestal Campus, PO Box 451, Princeton, NJ 08540, USA	USA
D.D. Ryutov	Pulsed Plasma Laboratory, Institute of Nuclear Physics, Novosibirsk 90, USSR	USSR
M.S. Sagoo	Department of Physics, Indian Institute of Technology, Haus Khaz, New Delhi-110029, India	India
M. San Miguel	Departamento de Física Teórica, Departamento de Ciencias, Universidad de Barcelona, Av. Generalísimo, Barcelona-14, Spain	Spain
K. Sauer	Zentralinstitut für Elektronenphysik, Akademie der Wissenschaften der DDR, 1199 Berlin – Adlershof, German Democratic Rep.	German Democratic Rep.
H. Schamel	Abteilung für Physik und Astronomie, Ruhr-Universität Bochum, Universitätsstrasse 150, Gebäude NB 7/135, Postfach 2148, D-4630 Bochum, Fed. Rep. Germany	Fed. Rep. Germany
T. Schep	FOM – Instituut voor Plasmafysica, Postbus 7, Nieuwegein, The Netherlands	FOM/Netherlands
R. Schneider	Instituto de Fisica, Universidade Federal do Rio Grande do Sul, Av. Professor Luiz Englert, 90 000 Porto Alegre, Brazil	Brazil
K. Schwörer	Institut für Plasmaforschung, Universität Stuttgart, Pfaffenwaldring 31, D-7000 Stuttgart 80, Fed.Rep.Germany	Fed. Rep. Germany

S.A. Selak	Geomagnetic Institute, 11306 Crocka – Belgrade, Yugoslavia	Yugoslavia
A. Shafee	Department of Physics, University of Dacca, Dacca 2, Bangladesh	Bangladesh
V.D. Shafranov	Khurchatov Institute for Atomic Energy, Academy of Sciences of the USSR, Moscow, USSR	USSR
M.Y. Shams	Optisches Institut, Technische Universität Berlin, Strasse des 17 Juni 135, West Berlin	Fed.Rep.Germany/ Pakistan
A.S. Sharma	Physical Research Laboratory, Navrangpura, Ahmedabad 380009, India	India
V.M. Shrestha	Physics Department, Tribhuvan University, Kirtipur Campus, Kathmandu, Nepal	Nepal
S. Shushurin	Moscow State University, Moscow V-234, USSR	USSR
N. Singh	Indian Institute of Technology, Kanpur 208016, India	India
A.G. Sitenko	Institute for Theoretical Physics, Ukrainian Academy of Sciences, 252 130 Kiev, USSR	USSR
A. Skorupski	Plasma Theory Group, Institute of Nuclear Research, Hoža 69, 00-681 Warsaw, Poland	Poland
G.O. Spies	Max-Planck-Institut für Plasmaphysik, 8046 Garching-bei-München, Fed. Rep. Germany	Fed.Rep.Germany
M.P. Srivastava	Department of Physics and Astrophysics, University of Delhi, Delhi-110007, India	India
A. Sternlieb	Centre d'études nucléaires Grenoble, DPh – PFC – SIG, 85 X, 38071 Grenoble Cedex, France	

M. Strauss	Physics Department, Nuclear Research Centre – Negev, Atomic Energy Commission, PO Box 9001, Beer-Sheva, Israel	Israel
R.N. Sudan	Laboratory of Plasma Studies, Cornell University, Upson Hall, Ithaca, NY 14853, USA	USA
J.P. Sudano	Depto. de Fisica do ITA/CTA, 1220 – São José dos Campos, São Paulo, Brazil	Brazil
A.S. Sulastri	Gama Research Centre, National Atomic Energy Agency of Indonesia, Jl. Babarsari, PO Box 8, Jogjakarta, Indonesia	Indonesia
W.N.-C. Sy	School of Physical Sciences, The Flinders University of South Australia, Bedford Park, SA 5042, Australia	Australia
M. Tagger	Centre d'études nucléaires de Fontenay-aux-Roses, DPh PFC/STGI – CEA, BP N° 6, F-92260 Fontenay-aux-Roses, France	France
N.A. Tahir	Department of Natural Philosophy, The University, Glasgow G12 8QQ, Scotland.	UK/Pakistan
K.L. Tan	Department of Physics, University of Singapore, Bukit Timah Road, Singapore 10	Singapore
T. Tange	Faculty of Science, Hiroshima University, Higashisenda-machi, Hiroshima, Japan	Japan
A. Taroni	Centro di Calcolo, Comitato Nazionale per l'Energia Nucleare, Via Mazzini 2, I-40138 Bologna, Italy	Italy

H. Tasso	Max-Planck-Institut für Plasmaphysik, D-8046 Garching-bei-München, Fed. Rep. Germany	Fed. Rep Germany/ Lebanon
M.B. Tayel	Department of Electrical Engineering, Faculty of Engineering, University of Alexandria, El-Hadara, Alexandria, Egypt	Egypt
J.B. Taylor	UKAEA, Culham Laboratory, Abingdon, Oxfordshire OX14 3DB, England	UK
M.B. Tendler	Institute of Technology, University of Uppsala, PO Box 534, S-751 21 Uppsala, Sweden	Sweden
M. Tessarotto	Istituto di Meccanica, Università di Trieste, Piazzale Europa 1, Trieste, Italy	Italy
J. Trnovec	Department of Experimental Physics, Komensky University, Mlynska Dolina, 81631 Bratislava, Czechoslovakia	Czechoslovakia
M. Trunk	Institut für Plasmaforschung, Universität Stuttgart, Pfaffenwaldring 31, D-7000 Stuttgart 80, Fed. Rep. Germany	Fed. Rep. Germany
N.F. Tsagas	(c/o G. Tsagas), Department of Mathematics, Polytechnic University of Saloniki, Saloniki, Greece	UK/Greece
T. Tuda	Tokai Research Establishment, Japan Atomic Energy Research Institute, Tokai-Mura, Naka-gun, Ibaraki-ken, Japan	Japan
C. Uberoi	Department of Applied Mathematics, Indian Institute of Science, Bangalore 560012, India	India
Q.N. Usmani	Department of Physics, Aligarh Muslim University, Aligarh 202001, India	India

L. Vadagnini	Via Armando Diaz 19, I-39100 Bolzano, Italy	Italy
P.M. Vandenplas	Laboratoire de physique des plasmas, Association Euratom-état belge, Ecole royale militaire, 30 avenue de la Renaissance, B-1040 Brussels, Belgium	Euratom/Belgium
A.E.P.M. van Maanen-Abels	Associatie Euratom-FOM, FOM-Instituut voor Plasmafysica "Rijnhuizen", Overeindseweg 2, Postbus 7, Nieuwegein, The Netherlands	FOM/Netherlands
R.K. Varma	Physical Research Laboratory, Navrangpura, Ahmedabad 380 009, India	India
M. Vecchione	Istituto di Astrofisica Teorica, Via M. Hermada 39, I-33100 Udine, Italy	Italy
F. Verheest	Instituut voor Theoretische Mechanika, Krijgslaan 271 – S9, B-9000 Ghent, Belgium	Belgium
K.S. Viswanathan	Department of Physics, University of Kerala, Kariavatom, Trivandrum 17, India	India
B.V. Waddell	Oak Ridge National Laboratory, PO Box Y, Oak Ridge, TN 37830, USA	USA
C. Wahlberg	Institute of Technology, University of Uppsala, PO Box 534, S-75121 Uppsala, Sweden	Sweden
M. Wardrop	Department of Theoretical Physics, University of Oxford, 12 Parks Road, Oxford, Oxfordshire OX1 3PQ, England	UK/Australia
T. Watanabe	Institute of Plasma Physics, Nagoya University, 464 Chikusaku Nagoya, Japan	Japan

M.L. Watkins	Culham Laboratory, Abingdon, Oxfordshire OX14 3DB, England	UK
M. Weenink	Technische Hogeschool, Eindhoven, The Netherlands	Netherlands
R.B. White	Plasma Physics Laboratory, Princeton University, PO Box 451, Princeton, NJ 08540, USA	USA
M.-Y. Yu	Theoretische Physik I, Ruhr-Universität Bochum, Universitätstr. 150, D-4630 Bochum, Fed. Rep. Germany	Fed. Rep. Germany/ UK
M.S. Yusaf	Department of Physics, Imperial College, Prince Consort Road, London SW7, England	UK/Pakistan
M. Zacone	Istituto di Fisica, Università di Palermo, Via Archirafi 36, I-90123 Palermo, Italy	Italy
M. Zales Caponi	TRW, 1 Space Park, Redondo Beach, CA 90266, USA	USA/Argentina

CONVERSION TABLE:
FACTORS FOR CONVERTING SOME OF THE MORE COMMON UNITS
TO INTERNATIONAL SYSTEM OF UNITS (SI) EQUIVALENTS

NOTES:

(1) SI base units are the metre (m), kilogram (kg), second (s), ampere (A), kelvin (K), candela (cd) and mole (mol).

(2) ▶ indicates SI derived units and those accepted for use with SI;
▷ indicates additional units accepted for use with SI for a limited time.
[*For further information see The International System of Units (SI), 1977 ed., published in English by HMSO, London, and National Bureau of Standards, Washington, DC, and International Standards ISO-1000 and the several parts of ISO-31 published by ISO, Geneva.*]

(3) The correct abbreviation for the unit in column 1 is given in column 2.

(4) ✱ indicates conversion factors given exactly; other factors are given rounded, mostly to 4 significant figures.
≡ indicates a definition of an SI derived unit: [ ] in column 3+4 enclose factors given for the sake of completeness.

Column 1 *Multiply data given in:*	Column 2	Column 3 *by:*	Column 4 *to obtain data in:*	
Radiation units				
▶ becquerel	1 Bq	(has dimensions of s^{-1})		
disintegrations per second (= dis/s)	1 s^{-1}	≡ 1.00 × 10^0	Bq	✱
▷ curie	1 Ci	= 3.70 × 10^{10}	Bq	✱
▷ roentgen	1 R	[= 2.58 × 10^{-4}	C/kg]	✱
▶ gray	1 Gy	[≡ 1.00 × 10^0	J/kg]	✱
▷ rad	1 rad	= 1.00 × 10^{-2}	Gy	✱
sievert *(radiation protection only)*	1 Sv	[= 1.00 × 10^0	J/kg]	✱
rem *(radiation protection only)*	1 rem	[= 1.00 × 10^{-2}	J/kg]	✱
Mass				
▶ unified atomic mass unit ($\frac{1}{12}$ of the mass of ^{12}C)	1 u	[= 1.660 57 × 10^{-27}	kg, approx.]	
▶ tonne (= metric ton)	1 t	[= 1.00 × 10^3	kg]	✱
pound mass (avoirdupois)	1 lbm	= 4.536 × 10^{-1}	kg	
ounce mass (avoirdupois)	1 ozm	= 2.835 × 10^1	g	
ton (long) (= 2240 lbm)	1 ton	= 1.016 × 10^3	kg	
ton (short) (= 2000 lbm)	1 short ton	= 9.072 × 10^2	kg	
Length				
statute mile	1 mile	= 1.609 × 10^0	km	
nautical mile (international)	1 n mile	= 1.852 × 10^0	km	✱
yard	1 yd	= 9.144 × 10^{-1}	m	✱
foot	1 ft	= 3.048 × 10^{-1}	m	✱
inch	1 in	= 2.54 × 10^1	mm	✱
mil (= 10^{-3} in)	1 mil	= 2.54 × 10^{-2}	mm	✱
Area				
▷ hectare	1 ha	[= 1.00 × 10^4	m^2]	✱
▷ barn *(effective cross-section, nuclear physics)*	1 b	[= 1.00 × 10^{-28}	m^2]	✱
square mile, (statute mile)^2	1 $mile^2$	= 2.590 × 10^0	km^2	
acre	1 acre	= 4.047 × 10^3	m^2	
square yard	1 yd^2	= 8.361 × 10^{-1}	m^2	
square foot	1 ft^2	= 9.290 × 10^{-2}	m^2	
square inch	1 in^2	= 6.452 × 10^2	mm^2	
Volume				
▶ litre	1 l *or* 1 ltr	[= 1.00 × 10^{-3}	m^3]	✱
cubic yard	1 yd^3	= 7.646 × 10^{-1}	m^3	
cubic foot	1 ft^3	= 2.832 × 10^{-2}	m^3	
cubic inch	1 in^3	= 1.639 × 10^4	mm^3	
gallon (imperial)	1 gal (UK)	= 4.546 × 10^{-3}	m^3	
gallon (US liquid)	1 gal (US)	= 3.785 × 10^{-3}	m^3	
Velocity, acceleration				
foot per second (= fps)	1 ft/s	= 3.048 × 10^{-1}	m/s	✱
foot per minute	1 ft/min	= 5.08 × 10^{-3}	m/s	✱
mile per hour (= mph)	1 mile/h	= 4.470 × 10^{-1}	m/s	
		= 1.609 × 10^0	km/h	
▷ knot (international)	1 knot	= 1.852 × 10^0	km/h	✱
free fall, standard, g		= 9.807 × 10^0	m/s^2	
foot per second squared	1 ft/s^2	= 3.048 × 10^{-1}	m/s^2	✱

This table has been prepared by E.R.A. Beck for use by the Division of Publications of the IAEA. While every effort has been made to ensure accuracy, the Agency cannot be held responsible for errors arising from the use of this table.

Column 1 *Multiply data given in:*	Column 2	Column 3 *by:*	Column 4 *to obtain data in:*
Density, volumetric rate			
pound mass per cubic inch	1 lbm/in^3	= 2.768×10^{4}	kg/m^3
pound mass per cubic foot	1 lbm/ft^3	= 1.602×10^{1}	kg/m^3
cubic feet per second	1 ft^3/s	= 2.832×10^{-2}	m^3/s
cubic feet per minute	1 ft^3/min	= 4.719×10^{-4}	m^3/s
Force			
▶ newton	1 N	[≡ 1.00×10^{0}	$m \cdot kg \cdot s^{-2}$] *
dyne	1 dyn	= 1.00×10^{-5}	N *
kilogram force (= kilopond (kp))	1 kgf	= 9.807×10^{0}	N
poundal	1 pdl	= 1.383×10^{-1}	N
pound force (avoirdupois)	1 lbf	= 4.448×10^{0}	N
ounce force (avoirdupois)	1 ozf	= 2.780×10^{-1}	N
Pressure, stress			
▶ pascal	1 Pa	[≡ 1.00×10^{0}	N/m^2] *
▷ atmosphere[a], standard	1 atm	= $1.013\,25 \times 10^{5}$	Pa *
▷ bar	1 bar	= 1.00×10^{5}	Pa *
centimetres of mercury (0°C)	1 cmHg	= 1.333×10^{3}	Pa
dyne per square centimetre	1 dyn/cm^2	= 1.00×10^{-1}	Pa *
feet of water (4°C)	1 ftH_2O	= 2.989×10^{3}	Pa
inches of mercury (0°C)	1 inHg	= 3.386×10^{3}	Pa
inches of water (4°C)	1 inH_2O	= 2.491×10^{2}	Pa
kilogram force per square centimetre	1 kgf/cm^2	= 9.807×10^{4}	Pa
pound force per square foot	1 lbf/ft^2	= 4.788×10^{1}	Pa
pound force per square inch (= psi)[b]	1 lbf/in^2	= 6.895×10^{3}	Pa
torr (0°C) (= mmHg)	1 torr	= 1.333×10^{2}	Pa
Energy, work, quantity of heat			
▶ joule (≡ W·s)	1 J	[≡ 1.00×10^{0}	N·m] *
▶ electronvolt	1 eV	[= $1.602\,19 \times 10^{-19}$	J, approx.]
British thermal unit (International Table)	1 Btu	= 1.055×10^{3}	J
calorie (thermochemical)	1 cal	= 4.184×10^{0}	J *
calorie (International Table)	1 cal_{IT}	= 4.187×10^{0}	J
erg	1 erg	= 1.00×10^{-7}	J *
foot-pound force	1 ft·lbf	= 1.356×10^{0}	J
kilowatt-hour	1 kW·h	= 3.60×10^{6}	J *
kiloton explosive yield (PNE) (≡ 10^{12} g-cal)	1 kt yield	≃ 4.2×10^{12}	J
Power, radiant flux			
▶ watt	1 W	[≡ 1.00×10^{0}	J/s] *
British thermal unit (International Table) per second	1 Btu/s	= 1.055×10^{3}	W
calorie (International Table) per second	1 cal_{IT}/s	= 4.187×10^{0}	W
foot-pound force/second	1 ft·lbf/s	= 1.356×10^{0}	W
horsepower (electric)	1 hp	= 7.46×10^{2}	W *
horsepower (metric) (= ps)	1 ps	= 7.355×10^{2}	W
horsepower (550 ft·lbf/s)	1 hp	= 7.457×10^{2}	W
Temperature			
▶ temperature in degrees Celsius, t where T is the thermodynamic temperature in kelvin and T_0 is defined as 273.15 K	$t = T - T_0$		
degree Fahrenheit	$t_{°F} - 32$	× $\left(\frac{5}{9}\right)$ gives	t *(in degrees Celsius)* *
degree Rankine	$T_{°R}$	× $\left(\frac{5}{9}\right)$ gives	T *(in kelvin)* *
degrees of temperature difference[c]	$\Delta T_{°R}$ (= $\Delta t_{°F}$)	× $\left(\frac{5}{9}\right)$ gives	ΔT (= Δt) *
Thermal conductivity[c]			
1 $Btu \cdot in/(ft^2 \cdot s \cdot °F)$ *(International Table Btu)*		= 5.192×10^{2}	$W \cdot m^{-1} \cdot K^{-1}$
1 $Btu/(ft \cdot s \cdot °F)$ *(International Table Btu)*		= 6.231×10^{3}	$W \cdot m^{-1} \cdot K^{-1}$
1 $cal_{IT}/(cm \cdot s \cdot °C)$		= 4.187×10^{2}	$W \cdot m^{-1} \cdot K^{-1}$

[a] atm abs, ata: atmospheres absolute;
atm (g), atü: atmospheres gauge.

[b] lbf/in^2 (g) (= psig): gauge pressure;
lbf/in^2 abs (= psia): absolute pressure.

[c] The abbreviation for temperature difference, deg (= degK = degC), is no longer acceptable as an SI unit.

HOW TO ORDER IAEA PUBLICATIONS

An exclusive sales agent for IAEA publications, to whom all orders and inquiries should be addressed, has been appointed in the following country:

UNITED STATES OF AMERICA	UNIPUB, P.O. Box 433, Murray Hill Station, New York, N.Y. 10016

In the following countries IAEA publications may be purchased from the sales agents or booksellers listed or through your major local booksellers. Payment can be made in local currency or with UNESCO coupons.

ARGENTINA	Comisión Nacional de Energía Atómica, Avenida del Libertador 8250, Buenos Aires
AUSTRALIA	Hunter Publications, 58 A Gipps Street, Collingwood, Victoria 3066
BELGIUM	Service du Courrier de l'UNESCO, 112, Rue du Trône, B-1050 Brussels
C.S.S.R.	S.N.T.L., Spálená 51, CS-113 02 Prague 1
	Alfa, Publishers, Hurbanovo námestie 6, CS-893 31 Bratislava
FRANCE	Office International de Documentation et Librairie, 48, rue Gay-Lussac, F-75240 Paris Cedex 05
HUNGARY	Kultura, Bookimport, P.O. Box 149, H-1389 Budapest
INDIA	Oxford Book and Stationery Co., 17, Park Street, Calcutta, 700016
	Oxford Book and Stationery Co., Scindia House, New Delhi-110001
ISRAEL	Heiliger and Co., 3, Nathan Strauss Str., Jerusalem
ITALY	Libreria Scientifica, Dott. Lucio de Biasio "aeiou". Via Meravigli 16, I-20123 Milan
JAPAN	Maruzen Company, Ltd., P.O. Box 5050, 100-31 Tokyo International
NETHERLANDS	Martinus Nijhoff B.V., Lange Voorhout 9-11, P.O. Box 269, The Hague
PAKISTAN	Mirza Book Agency, 65, Shahrah Quaid-e-Azam, P.O. Box 729, Lahore-3
POLAND	Ars Polona-Ruch, Centrala Handlu Zagranicznego, Krakowskie Przedmiescie 7, PL-00-068 Warsaw
ROMANIA	Ilexim, P.O. Box 136-137, Bucarest
SOUTH AFRICA	Van Schaik's Bookstore (Pty) Ltd., P.O. Box 724, Pretoria 0001
	Universitas Books (Pty) Ltd., P.O. Box 1557, Pretoria 0001
SPAIN	Diaz de Santos, Lagasca 95, Madrid-6
	Diaz de Santos, Balmes 417, Barcelona-6
SWEDEN	AB C.E. Fritzes Kungl. Hovbokhandel, Fredsgatan 2, P.O. Box 16358 S-103 27 Stockholm
UNITED KINGDOM	Her Majesty's Stationery Office, P.O. Box 569, London SE1 9NH
U.S.S.R.	Mezhdunarodnaya Kniga, Smolenskaya-Sennaya 32-34, Moscow G-200
YUGOSLAVIA	Jugoslovenska Knjiga, Terazije 27, POB 36, YU-11001 Belgrade

---

Orders from countries where sales agents have not yet been appointed and requests for information should be addressed directly to:

Division of Publications
International Atomic Energy Agency
Kärntner Ring 11, P.O.Box 590, A-1011 Vienna, Austria